Color Atlas and Textbook of
Diagnostic Microbiology

Elmer W. Koneman, M.D.
Executive Director
Colorado Association for Continuing
Medical Laboratory Education, Inc.
Denver, Colorado

Stephen D. Allen, M.D.
Assistant Professor of Clinical Pathology
Indiana University School of Medicine
Associate Director of Clinical Microbiology
Indiana University Medical Center
Indianapolis, Indiana

V. R. Dowell, Jr., Ph.D.
Chief Enterobacteriology Branch
Bacteriology Division
Bureau of Laboratories
Center for Disease Control
Atlanta, Georgia

Herbert M. Sommers, M.D.
Professor of Pathology
Northwestern University Medical School
Director of Clinical Microbiology
Northwestern Memorial Hospital
Chicago, Illinois

Color Atlas and Textbook of
Diagnostic Microbiology

Elmer W. Koneman, M.D.

Stephen D. Allen, M.D.

V. R. Dowell, Jr., Ph.D.

Herbert M. Sommers, M.D.

J. B. LIPPINCOTT COMPANY
Philadelphia • Toronto

ISBN 0-397-50405-5

Library of Congress Catalog Card Number 78-25980

Printed in the United States of America

1 3 5 4 2

Library of Congress Cataloging in Publication Data

Main entry under title:

Color atlas and textbook of diagnostic microbiology.

Includes index.
1. Micro-organisms—Identification. 2. Microbiology—Atlases. I. Koneman, Elmer W.
QR67.C64   616.01   78-25980
ISBN 0-397-50405-5

Preface

This Color Atlas and Textbook of Microbiology has been written to provide microbiology students, medical technologists, residents in pathology, and others interested in clinical microbiology with a practical introduction to the laboratory identification of microbial agents associated with infectious disease. This book is not intended to supplant currently available texts and manuals; rather, it is designed to show the student several different approaches used in microbiology laboratories to provide timely and relevant results to physicians.

The theme of this book is to emphasize the need to understand the basic biochemical and metabolic pathways which serve as the basis for the identification procedures used in the laboratory. Toward this end, charts outlining the theory, procedures, and interpretations for commonly used microbial identification characteristics are interspersed throughout the text instead of combined in an appendix at the end of the book.

Thirty-nine color plates, each having between six and twelve color prints, are used to illustrate the colonial and microscopic features and many of the biochemical characteristics of the more frequently encountered microorganisms. The color prints provide a cross reference for the text and are interrelated in such a way that individual plates often serve as a self-contained study guide. In addition to the color plates, black and white photographs and line drawings are included, particularly in Chapters 13 and 14,

where a number of illustrated tables portray the important diagnostic characteristics of the fungi and animal parasites of importance to humans.

The approach to the identification of certain groups of microorganisms is difficult because there are several effective methods in use. In those instances where no one identification scheme can be advocated for all purposes, several methods are described, including comments on the advantages and disadvantages of each so that microbiologists can determine the approach that is most suitable for their laboratory.

Because of limitations of space, not all areas in microbiology could be included in this text. Clinical virology is a subject too broad to introduce here. The recovery and identification of viruses is still limited to a few specialized laboratories and the details are better left for specific texts. Mycoplasma and chlamydia are mentioned only in passing and the clinical and laboratory identification of Legionnaire's disease is not covered. The spirochetes, including syphilitic and nonsyphilitic treponemes, are not discussed because most laboratories are not involved in the recovery and identification of these organisms. Similarly, the serology of infectious disease has not been treated as a separate subject; rather, a brief discussion of serologic techniques has been included in sections of those chapters where it is appropriate.

References at the conclusion of each

v

chapter have been kept few in number. We regret that it has not been possible to acknowledge the contributions of many clinical microbiologists. Much of the material presented in this text has been used in a variety of teaching programs, workshops, and seminars presented by the authors over the past decade. We would like to express our appreciation to the various faculty members and to the hundreds of participants and students in these programs who, through comments, suggestions and post-course evaluations, made significant contributions to the preparation of this text. Just as these evaluations were helpful in developing this text, so would comments, suggestions, and constructive criticisms further aid us in improving it.

Acknowledgments

We wish to acknowledge the following individuals who have made direct contributions to this text: Albert A. Balows, June Brown, William B. Cherry, William A. Clark, Betty R. Davis, James C. Feeley, Peter C. Fuchs, Lucille Georg, Sarah Gunderson, T. M. Hawkins, James D. Howard, Bilda L. Jones, George L. Lombard, Douglas Kellogg, Ernesta Tallot-Kelpsa, Edward J. Kennedy, Benny F. Mertens, J. Kenneth McClatchy, Glenn D. Roberts, James W. Smith, Thomas F. Smith, Benjamin F. Summers, Bobby E. Strong, Clyde Thornsberry, Beverly Waxler, Robert E. Weaver, and Edward R. Wilson. We also wish to acknowledge the contributions of Marion Scientific Corporation, Pasco Labs, Inc., and Micro-Media Systems, Inc.

ELMER W. KONEMAN, M.D.
STEPHEN D. ALLEN, M.D.
V. R. DOWELL, PH.D.
HERBERT M. SOMMERS, M.D.

Contents

Color Atlas and Textbook of
Diagnostic Microbiology

1 Introduction to Medical Microbiology

The purpose of this introductory chapter is to provide an overview of many of the specific functions and procedures that are carried out in the clinical and laboratory assessment of infectious diseases and the identification of microorganisms recovered in culture. From this overview the reader who is not familiar with the various functions of the clinical microbiology laboratory can gain some perspective into the many tasks that must be carried out. For individuals not directly involved in laboratory work, this orientation should prove helpful in demonstrating how their activities can best be applied to comply with the needs of the laboratory. For those just entering into clinical microbiology as a vocation, the orientation will provide a broader perspective of how their work in the laboratory best fits in the overall care of patients with infectious disease.

The remaining chapters will be concerned with the minute details involved in the classification, identification, and clinical significance of most of the microorganisms of importance to man.

The chief function of the clinical medical microbiology laboratory is to assist physicians in the diagnosis and treatment of patients with infectious disease. Excellence of patient care must remain the prime focus, and the work performed by the staff microbiologists should be directed toward the production of clinically useful results in as short a time as possible. The delay of microbiology reports beyond a time when the results can be of use in directing patient care is one of the major criticisms voiced by physicians on the performance of clinical laboratories.

The delivery of diagnostic laboratory service has become quite complex and requires the constant attention of the laboratory director, supervisors, and qualified personnel. Figure 1-1 is a schematic representation of the sequence of steps necessary in deriving a clinical and laboratory diagnosis of infectious disease. Note that the cycle begins with the patient who presents with signs or symptoms of infectious disease.

After the physician examines the patient, a tentative diagnosis is made and orders are written for laboratory tests to confirm or reject this diagnosis. The physician's orders most commonly are transcribed to a laboratory request slip and an appropriate specimen is collected from the patient for culture. Both the request slip and the specimen are promptly delivered to the laboratory.

The information on the request slip is entered into a laboratory log book and the specimen is processed; processing includes a direct visual examination, a microscopic examination if indicated, and inoculation of a small portion of the specimen into a carefully selected battery of primary isolation culture media. All inoculated media are placed in an incubator at an appropriate temperature and after a given time of incubation the bacteria or other microorganisms recovered are identified. The final results are entered on a laboratory report form

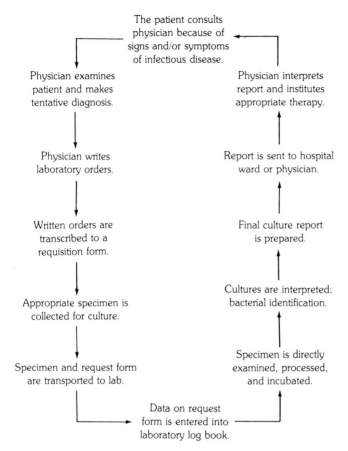

The patient consults physician because of signs and/or symptoms of infectious disease.

Physician examines patient and makes tentative diagnosis.

Physician interprets report and institutes appropriate therapy.

Physician writes laboratory orders.

Report is sent to hospital ward or physician.

Written orders are transcribed to a requisition form.

Final culture report is prepared.

Appropriate specimen is collected for culture.

Cultures are interpreted; bacterial identification.

Specimen and request form are transported to lab.

Specimen is directly examined, processed, and incubated.

Data on request form is entered into laboratory log book.

Fig. 1-1. The clinical and laboratory diagnosis of infectious disease. A schematic overview of the diagnostic cycle.

which is returned to the physician or to the patient's record, as may be appropriate. The physician in turn uses the information on the report form to further manage the patient and institute appropriate therapy.

Each step in the above cycle must be completed with accuracy and precision in as short a time as possible. Note that the laboratory is directly involved in only a portion of the cycle, and it is the obligation of the laboratory director and laboratory personnel to also be involved in decisions that will improve the efficiency of those functions external to the laboratory. Thus, although transcription of orders, the proper collection of specimens, specimen transport to the laboratory, and the posting and interpretation of final results are not under the direct control of laboratory personnel, they must assume some responsibility for seeing that these functions are properly carried out. Each step is of equal importance if optimal patient care is to be provided.[15]

THE DIAGNOSIS OF INFECTIOUS DISEASE

It is the physician's responsibility to suspect infectious disease in patients and to initiate studies to confirm or reject this suspicion. Patients with infectious disease may present with a variety of signs and symptoms, some overt and easy to recognize, others obscure and possibly misleading. Early Greek and Roman physicians recognized four cardinal signs of inflammation:

Dolor	=	pain
Calor	=	heat
Rugor	=	redness
Tumor	=	swelling

The diffuse redness and swelling of the throat or tonsils, serous or purulent discharges from wounds or mucous membranes, and the accumulation of pus in abscesses or body cavities, often resulting in pain, swelling, and increased heat to the area, are direct signs of infection calling for an immediate culture to establish the causative organism so that appropriate therapy may be started.

Cough, increased sputum production, burning on urination, and dysentery are indirect signs that infection may involve deep organ systems.

Fever, chills, flushing (vasodilation), and an increase in pulse rate may be general or systemic manifestations of infection, indicating that the infectious process may be extending beyond local confines.

Patients with subacute or chronic infections may present with minimal systemic symptoms and/or subtle signs such as intermittent low-grade fever, weight loss, easy fatigability, and lassitude. Toxic reactions to bacterial products that may affect the neuromuscular, cardiorespiratory, or gastrointestinal systems may be the initial symptoms of an underlying infectious disease.

X-ray manifestations of infectious disease include pulmonary infiltrates, fibrous thickening of cavity linings, the presence of gas and swelling in soft tissues, and the presentation of radiopaque masses or the accumulation of fluid within body cavities.

Laboratory values suggesting the presence of infectious disease in patients with minimal or early symptoms include an elevation in the erythrocyte sedimentation rate, peripheral blood leukocytosis or monocytosis, and alterations such as elevations in gamma globulin or the presence of type-specific antibodies (e.g., febrile agglutinins).

The physician must be aware of the various ways in which infectious disease may present and must utilize x-ray and laboratory techniques and tests to confirm his clinical impression. Appropriate cultures must be ordered, or, in most laboratories, techniques directed toward the serological detection of bacterial antigens or antibodies are available. Pathologists, microbiologists, and technologists are available in most institutions and communities to assist physicians by providing suggestions for the proper collection of specimens to achieve maximum recovery of microorganisms.

Table 1-1 summarizes the common sites of infections, clinical signs and symptoms, types of specimens to culture, and a list of organisms associated with infections at these sites. General symptoms, such as fever, chills, pain, and increase in cardiac rate, are not necessarily included since any or all of these may occur in patients with any infection. (The body sites of involvement and clinical manifestations of anaerobic infections are discussed in more detail in Chap. 10.)

The laboratory should be forewarned by the physician that certain microorganisms are suspected, particularly if a culture medium other than that commonly used is required for their recovery. For example, bacterial species belonging to the various genera of the Brucellaceae, Pasteurella, Moraxella, and Leptospira are among those requiring special culture techniques. The physician should always indicate on the laboratory request slip if an infection with mycobacteria or fungi is suspected because special culture media are required for their isolation. Also, since some laboratories do not routinely culture for members of the Hemophilus and Neisseria groups, physicians should indicate on the requisition form when these microorganisms are suspected.

SPECIMEN COLLECTION

The proper collection of a specimen for culture is possibly the most important step in the recovery of microorganisms responsible

(Text continues on p. 6)

Table 1-1. *The Diagnosis of Bacterial Infections at Different Body Sites*

Site of Infection	Presenting Signs and Symptoms	Specimens to Culture	Common Bacterial Species Associated With Infections
Urinary tract	Urinary bladder infection: Pyuria Dysuria Hematuria Pain and tenderness: suprapubic or lower abdomen Kidney infection: Back pain Tenderness: costovertebral angle (CVA)	Clean-catch midstream urine Catheterized urine Catheter bags: newborns and infants only Suprapubic aspiration of urine	Enterobacteriaceae: *Escherichia coli* Klebsiella species Proteus species Group D streptococci (enterococci) *Pseudomonas aeruginosa*
Respiratory tract	Upper tract—nose and sinuses: Headache Pain and redness: malar area Rhinitis X-ray: sinus consolidation, fluid levels, or membrane thickening	Acute: Nasopharyngeal swab Sinus washings Chronic: Sinus washings Surgical biopsy specimen	*Streptococcus pneumoniae* Streptococcus, beta Group A *Staphylococcus aureus* *Hemophilus influenzae* Klebsiella species and other Enterobacteriaceae
	Upper tract—throat and pharynx: Redness and edema of mucosa Exudation of tonsils Pseudomembrane formation Edema of uvula Gray coating of tongue: "strawberry tongue" Enlargement of cervical nodes	Swab of posterior pharynx Swab of tonsils (abscess) Nasopharyngeal swab	Streptococcus, beta Group A *Hemophilus influenzae* *Corynebacterium diphtheriae* *Neisseria meningitidis* *Bordetella pertussis*
	Lower tract—lungs and bronchi: Cough: bloody or profuse Chest pain Dyspnea Consolidation of lungs: Rales and rhonchi Diminished breath sounds Dullness to percussion X-ray infiltrates	Sputum (poor return) Blood Bronchoscopy secretions Transtracheal aspirate Lung aspirate or biopsy	*Streptococcus pneumoniae* *Hemophilus influenzae* *Staphylococcus aureus* *Klebsiella pneumoniae* and other members of Enterobacteriaceae Mycobacterium species Miscellaneous anaerobic bacteria
Gastrointestinal tract	Diarrhea Dysentery: Purulent Mucous Bloody Cramping abdominal pain	Stool specimen Rectal swab or rectal mucous Blood culture (typhoid fever)	Salmonella species Shigella species *Escherichia coli* (enterotoxigenic) *Vibrio comma* and *Vibrio parahaemolyticus* Yersinia species

Table 1-1 *The Diagnosis of Bacterial Infections at Different Body Sites* (Continued)

Site of Infection	Presenting Signs and Symptoms	Specimens to Culture	Common Bacterial Species Associated With Infections
Wounds	Discharge: serous or purulent Abscess: subcutaneous or submucous Redness and edema Crepitation (gas formation) Pain Ulceration or sinus formation	Aspirate of drainage Deep swab of purulent drainage Swab from wound margins or depths of ulcer Tissue biopsy material	*Staphylococcus aureus* *Streptococcus pyogenes* Clostridium species and other anaerobic bacteria Enterobacteriaceae species *Pseudomonas aeruginosa* Enterococcus species
Meningitis	Headache Pain in neck and back Stiff neck Straight leg raising: (positive Kernig's sign) Nausea and vomiting Stupor to coma Petechial skin rash	Spinal fluid Subdural aspirate Blood culture Throat or sputum culture	*Neisseria meningitidis* *Hemophilus influenzae* *Streptococcus pneumoniae* Streptococcus, beta Groups A and B (Group B in infants) Enterobacteriaceae: debilitated patients, infants, and postcraniotomy
Genital tract	Males: Urethral discharge: serous or purulent Burning on urination Terminal hematuria Females: Purulent vaginal discharge Burning on urination Lower abdominal pain, spasm, and tenderness Mucous membrane chancre or chancroid	Urethral discharge Prostatic secretions Uterine cervix Rectum (anal sphincter swab) Urethral swab Dark-field examination	*Neisseria gonorrhoeae* *Hemophilus ducreyi* *Treponema pallidum* (syphilis) *Hemophilus vaginalis* Nonbacterial: *Trichomonas vaginalis* *Candida albicans* T mycoplasma species Chlamydia
Bacteremia	Spiking fever Chills Cardiac murmur (endocarditis) Petechiae: skin and mucous membranes "Splinter hemorrhages" of nails Malaise	Blood: 3 or 4 cultures per day at 1-hour intervals or greater Urine Wounds Any suspected primary site of infection: Spinal fluid Respiratory tract Skin-umbilicus Skin-ear	Streptococcus species: Group A—all ages Alpha-hemolytic (endocarditis) Groups A, B, D—newborn *Staphylococcus aureus* *Streptococcus pneumoniae* *Escherichia coli* *Salmonella typhosa* (typhoid fever) Bacteroides species, and other anaerobic bacteria *Pseudomonas aeruginosa* *Listeria monocytogenes* *Hemophilus influenzae*
Eye	Conjunctival discharge: serous or purulent Conjunctival redness (hyperemia): "pinkeye" Ocular pain and tenderness	Purulent discharge Lower cul-de-sac Inner canthus	Hemophilus species Moraxella species *Neisseria gonorrhoeae* (newborns) *Staphylococcus aureus*

Table 1-1.   *The Diagnosis of Bacterial Infections at Different Body Sites (Continued)*

Site of Infection	Presenting Signs and Symptoms	Specimens to Culture	Common Bacterial Species Associated With Infections
Eye (Continued)			Streptococcus pneumoniae Streptococcus pyogenes Pseudomonas aeruginosa (report stat)

Middle ear	Serous or purulent drainage
	Deep pain in ear and jaw
	Throbbing headache
	Red or bulging tympanic membrane
		Acute:
		  No culture
		  Nasopharyngeal swab
		  Tympanic membrane aspirate
		Chronic: drainage of external meatus
			Acute:
			  Streptococcus pneumoniae
			  Hemophilus influenzae
			Chronic:
			  Pseudomonas aeruginosa
			  Proteus species

Bones and joints	Joint swelling Redness and heat Pain on motion Tenderness on palpation X-ray: synovitis or osteomyelitis	Joint aspirate Synovial biopsy Bone spicules or bone marrow aspirate	Staphylococcus aureus Hemophilus influenzae Streptococcus pyogenes Neisseria gonorrhoeae Streptococcus pneumoniae Enterobacteriaceae species

for infectious disease. A poorly collected specimen may be responsible for the failure to isolate the causative microorganism, and the recovery of contaminants can lead to an incorrect or even harmful course of therapy. For example, assume that *Klebsiella pneumoniae* has been recovered from the sputum of a patient with clinical pneumonia. If the sputum has been improperly collected and actually consists only of saliva, the *K. pneumoniae* may represent nothing more than a commensal inhabitant of the nasal sinuses, nares, or posterior pharynx, and may not reflect the true cause of the pneumonia. Treatment against *K. pneumoniae* may be improper since another organism with a different antibiotic susceptibility pattern may be responsible for the lower respiratory infection.

Basic Concepts of Proper Specimen Collection

1. *The culture specimen must be material from the actual infection site* and must be collected with a minimum of contamination from adjacent tissues, organs, or secretions.

The problem with salivary contamination of sputum samples or lower respiratory secretions was mentioned above. Other examples of problems encountered in specimen collection include the failure to culture the depths of a wound or draining sinus without touching the adjacent skin; inadequate cleansing of the paraurethral tissue and perineum prior to collecting a clean-catch urine sample from a female; contamination of an endometrial sample with vaginal secretions; and failure to reach deep abscesses with aspirating needles or cannulas.

2. *Optimal times for specimen collection must be established* for the best chance of recovery of causative microorganisms.

Knowledge of the natural history and pathophysiology of many of the infectious diseases is important in determining the optimal time for specimen collection. For example, as illustrated in Figure 1-2, in patients with typhoid fever the causative microorganism can be recovered optimally from the blood during the first week of illness. Culture of the feces or urine is usually positive during the second and third weeks of illness, although, in general, *Salmonella typhosa* is recovered best from the

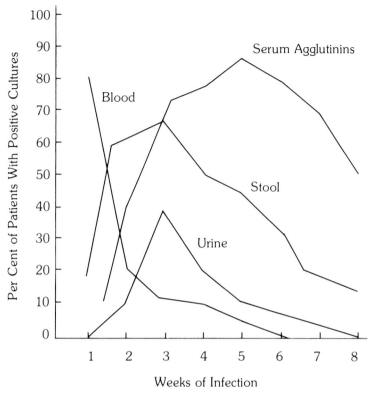

Fig. 1-2. Culture and serologic diagnosis of typhoid fever.

feces during the acute diarrheal stage of the disease. Serum agglutinins begin to rise during the second week of illness, reaching a peak during the fifth week and remaining detectable for many weeks following clinical remission of the disease.

Because of the high risk of contamination or overgrowth with more rapidly growing commensal bacteria, 24-hour collections of clinical materials for culture, particularly of sputum and urine, should be discouraged. On the other hand, Kaye[12] has shown that urine from normal individuals may be inhibitory or bactericidal for some microorganisms, particularly if the urine *p*H is 5.5 or less (acidic), if the osmolality is high, or if the urea concentration is increased. The ability of bacteria to grow in urine may represent a failure in the host defense mechanism.

The first early morning sputum and urine samples usually contain the highest concentration of infecting microorganisms because of the inactivity of the patient during the night. Historically, 24-hour collections of sputum or urine were necessary at a time when laboratory methods were inadequate to allow for a high recovery rate of microorganisms, particularly mycobacteria; however, current improvements in laboratory processing techniques of sputum and urine have made it possible to recover a high percentage of mycobacteria from samples of small volume. Random sampling is now recommended.

3. *A sufficient quantity of specimen must be obtained* to perform the culture techniques requested.

Guidelines should be established outlining what constitutes a sufficient volume of material for culture. In most cases of active bacterial infection, sufficient quantities of

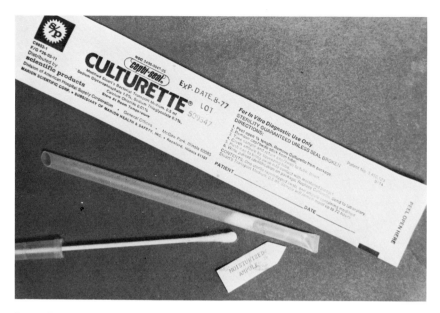

Fig. 1-3. Photograph of the Culturette culture collection system. After the swab is inoculated, it is replaced into the plastic tube. The ampule at the bottom of the tube is broken by squeezing between the thumb and forefinger, releasing Stuart transport medium, providing moisture for the cotton tip of the swab. The handle of the swab is attached to the cap, which makes a tight seal when the swab is fully extended into the tube.

pus or purulent secretions are produced so that volume is not a problem. In chronic or milder forms of infection; it may be difficult to procure sufficient material; the submission of a dry swab or scant secretions to the laboratory with the hope that something will grow is frequently an exercise in futility, and at considerable cost to the patient. However, such samples should not be categorically discarded; rather, a request should be made for a second sample since there are situations in which even a poor sample is better than no sample at all.

All too frequently 0.5 ml. or less of material labeled sputum is delivered to the laboratory with a request for routine, acid-fast, and fungus cultures. Such specimens may not represent pulmonary secretions from the site of infection, and the low volume is insufficient to enable the carrying out of all the procedures requested.

4. *Appropriate collection devices, specimen containers, and culture media must be used* to ensure optimal recovery of microorganisms.

Sterile containers should be used for the collection of all specimens. It is also important that containers be constructed for ease of collection, particularly if the patients are required to obtain their own specimens. Narrow-mouthed bottles are poorly designed for the collection of sputum or urine samples. The containers should also be provided with tightly fitting caps or lids to prevent leakage or contamination during transport.

Swabs are commonly used for the collection of specimens for culture. These are acceptable in most instances if certain precautions are taken. Because of residual fatty acids on the cotton fibers that may be inhibitory to some strains of fastidious bacteria, it is recommended that calcium alginate, Dacron, or polyester-tipped swabs be used. Specimens should not be allowed to remain in contact with the swab any longer than is necessary. Except for throat swabs where drying does not seem to affect the recovery of streptococci, it is recommended that swabs be placed into a transport medium or moist container to prevent the drying and death of bacteria. One commonly used tube,

Table 1-2. *Transport Containers for Anaerobic Specimens*

Container	Rationale or Description	Reference
Rubber-stoppered tubes or bottles for aspirates	Container filled with oxygen-free CO_2: 1. Dry bottles for fluid specimens 2. Container containing non-nutritive agar or liquid medium with a reducing agent and resazurin indicator*	Holdeman and Moore[10] Sutter, Vargo, and Finegold[21]
Syringe and needle for aspirates	Fresh exudate or liquid specimens can be transported to laboratory after bubbles are carefully expelled and the needle is inserted into a sterile rubber stopper. Should be used only if specimen can be taken to laboratory without delay	Sutter, Vargo, and Finegold[21]
Tissue transport containers	Tissue can be placed in sterile screwcapped vial and transported within: 1. The "mini-jar", a 35 mm. film container; steel wool and acidified $CuSO_4$ used as an oxygen absorbant, or 2. The Anaerobic Culture/Set†, a plastic bag with anaerobic gas generator, catlyst, and indicator.	Atteberry and Finegold[1]
Two-tube system for swabs	One tube contains sterile swab in oxygen-free CO_2 or N_2. Second tube contains either: 1. A few drops of a prereduced salts solution and oxygen-free gas* or 2. Semisolid agar containing a reducing agent and redox indicator. A deep tube of Stuart, Amies, or modified Cary-Blair medium can be used because the oxidation-reduction potential in the deeper portions is sufficiently low to preserve the viability of most clinically encountered anaerobes The BBL Port-A-Cul tube‡ (Fig. 1-5) is an example of this type. Even if the superficial portion of the medium is oxygenated as indicated by the indicator dye (Fig. 1-5), this reverts back to an anaerobic condition soon after the cap is replaced because of the action of the reducing agent in the medium.	Holdeman and Moore[10] Sutter, Vargo, and Finegold[21]

*Scott Laboratories, Inc., Fiskeville, R.I. 02823
†Marion Scientific Corp., Rockford, Ill. 61101
‡BBL. Division of Becton-Dickinson and Co., Cockeysville, Md. 21030

Culturette,* illustrated in Figure 1-3, has a sealed, self-contained, liquid-containing vial that can be broken when the swab is reinserted after taking a culture. This helps to keep the sample moist, and good recovery of most bacterial species from these tubes has been demonstrated for up to 24 hours. The use of culture tubes containing semisolid Stuart or Amies transport medium also serves as an adequate means for holding swab cultures during transport.

The use of swabs for collection of specimens for the recovery of anaerobic bacteria is discouraged; rather, aspiration with a needle and syringe is recommended. In either event, specimens once collected must be protected from exposure to ambient oxygen and kept from drying until they can be processed in the laboratory.

A number of transport containers suitable for anaerobic specimens are listed in Table 1-2, some of which are commercially available. For example, the Scott† two-tube system is illustrated in Figure 1-4, the BBL‡ Port-A-Cul tube in Figure 1-5, and the Marion Scientific Anaerobic culturette system in Figure 1-6.* These containers can also be used for transport of obligate aerobic and facultatively anaerobic microorganisms.

Regardless of the transport system used,

*Marion Scientific Corp. Rockford, Ill. 61101.
†Scott Laboratories Inc., Fiskeville, R.I. 02823.
‡BBL, Division of Becton-Dickinson and Co., Cockeysville, Md. 21030.

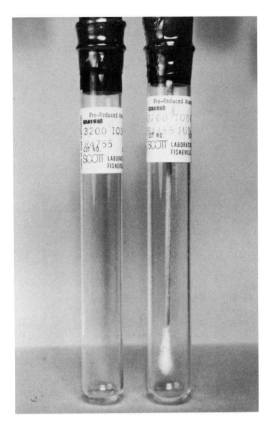

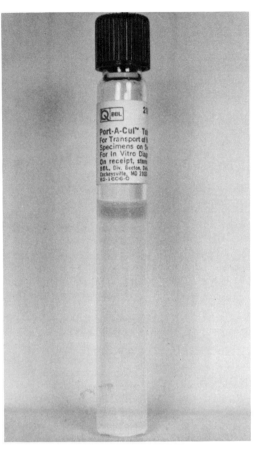

Fig. 1-4. Photograph of the Scott two-tube system. Both tubes have been evacuated of oxygen and the air within the tubes has been replaced with oxygen-free CO_2. The tube on the right includes a swab attached to the rubber stopper. When obtaining a culture, the swab is removed from the one tube, inoculated, and quickly placed into the second tube, with the stopper attached to the swab handle becoming the final seal for the second tube.

Fig. 1-5. Photograph of BBL's Port-A-Cul anaerobic transport tube. The tube contains semisolid holding medium, including a chemical reducing agent and a redox indicator. The darker band of shading at the top of the medium represents the red discoloration of the redox indicator after the cap has been removed and the tube exposed to the air for a few seconds.

the major principle is to keep the time delay between collection of specimens and inoculation of media to a minimum. This may be particularly important for the recovery of Shigella from patients with bacillary dysentery where rectal swabs have been used. These swabs should be inoculated directly to the surface of MacConkey medium or into GN enrichment broth. Urethral or cervical secretions obtained for the recovery of *Neisseria gonorrhoeae* should also be inoculated directly to the surface of chocolate agar or another appropriate medium.

5. *Whenever possible, obtain cultures prior to the administration of antibiotics.*

It is recommended that cultures be obtained prior to the administration of antibiotics. This is particularly true for isolation of organisms such as beta-hemolytic streptococci from throat specimens, *Neisseria gonorrhoeae* in genitourinary cultures, or *Hemophilus influenzae* and *Neisseria meningitidis* in meningitis. However, the administration of antibiotics does not necessarily preclude the recovery of other species of microorganisms from clinical specimens.

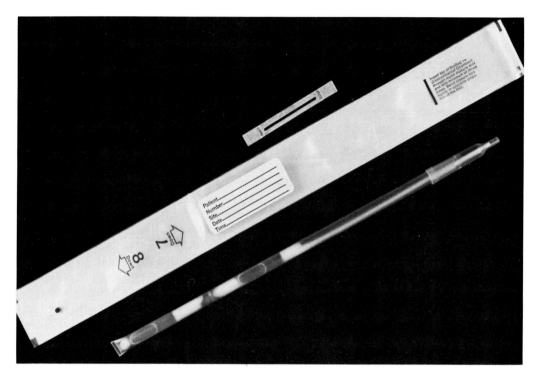

Fig. 1-6. Photograph of the Anaerobic Culturette System. The plastic tube contains a rayon-tipped swab and an inner sealed ampule that contains reduced cary-Blair medium and a hydrogen-CO_2 gas generating tablet. After the culture is obtained, the swab is reinserted into the plastic tube, which in turn is sealed in the outer cellophane envelope by means of the metal closure. The inner ampule is crushed, releasing the transport medium and hydrogen-CO_2 gas. Also enclosed are catalysts and desiccants that reduce residual oxygen, ensuring the viability of anaerobic bacteria for up to 48 hours.

The action of many antibiotics may be bacteriostatic, not bactericidal, and often microorganisms can be recovered when they are transferred to an environment devoid of the antibiotic (fresh culture medium). Also, the concentration of antibiotic may be below the minimal inhibitory concentration for the organism in question at the site of infection and recovery in culture is no problem. Thus, one should always make an attempt to culture these sites, although the results must be interpreted accordingly or qualified in the written report.

6. *The culture container must be properly labeled.*

In order for the laboratory microbiologist to perform the proper culture techniques and provide the physician with accurate and complete information, each culture container must have a legible label, with the following minimum information:

NAME _____

ID# _____

SOURCE _____

DOCTOR _____

DATE/HOUR _____

Figure 1-7 illustrates a culture tube with a label that has been properly filled out. The patient's full name should be used and initials should be avoided. The identification number may be the hospital number, clinic or office number, home address, or social security number, depending upon the circumstances. The physician's name or office title is necessary in the event that consultation or early reporting is required. The

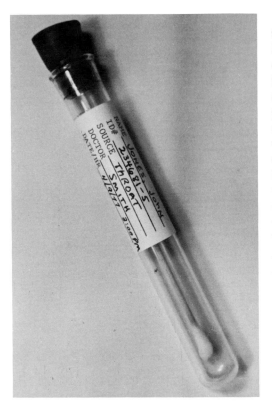

Fig. 1-7. Photograph of a culture transport tube illustrating a properly written identification label.

specimen source should be stated in the event that special culture media are required. The date and time of collection should appear on the label to ensure that the specimen is cultured within an acceptable length of time after it has been collected.

Collection of Specimens from Various Anatomical Sites

1. Throat and Nasal-Pharyngeal Swab Sample

The usual microbial flora of the throat and nasal pharynx consist of α-hemolytic streptococci, *Neisseria* species, *Staphylococcus epidermiditis*, *S. aureus*, *S. pneumoniae*, various Hemophilus species, "diphtheroids," and numerous species of anaerobic bacteria. In the majority of cases, throat swabs are obtained to recover Group A β-hemolytic streptococcus, which cause "pharyngitis."

The proper method for obtaining a throat sample is illustrated in Figure 1-8. A bright light from over the shoulder of the person obtaining the specimen should be focused into the opened oral cavity so that the swab can be usually guided into the posterior pharynx. The patient is instructed to breathe deeply and the tongue is gently depressed with a tongue blade. The swab is then extended between the tonsillar pillars and behind the uvula, taking care not to touch the lateral walls of the buccal cavity. Having the patient phonate an "ah" serves to lift the uvula and aids in reducing the gag reflex. The swab should be swept back and forth across the posterior pharynx to obtain an adequate sample. After the sample is collected, the swab should be placed immediately into a sterile tube or other suitable container for transport to the laboratory. Special techniques are required for recovery of *B. pertussis* (see Chap. 4), *Corynebacterium* (see Chap. 9), *N. gonorrhoeae*, and *N. meningitidis* (see Chap. 8).

2. Sputum and Lower Respiratory Cultures

It is difficult to prevent contamination of sputum samples with upper respiratory secretions. Having the patient gargle with water immediately prior to obtaining the specimen reduces the number of contaminating oropharyngeal bacteria. In cases where sputum production is scant, induction with nebulized saline (avoid saline for injection because it contains antibacterial substances)[20] through a positive-pressure respirator apparatus may be effective in producing a sample more representative of the lower respiratory tract.

Translaryngeal aspiration may be indicated in the following situations[9]:

1. The patient is debilitated and cannot spontaneously expectorate a sputum sample.
2. Routine sputum samples have failed to recover a causative organism in the face of clinical bacterial pneumonia.
3. An anaerobic pulmonary infection is suspected.

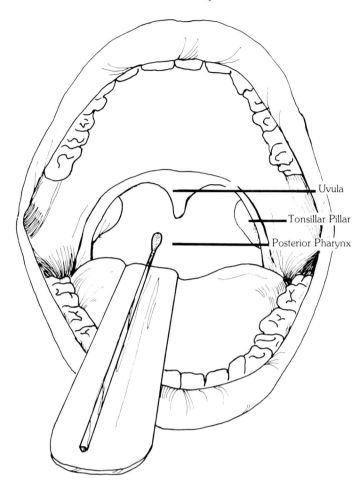

Uvula

Tonsillar Pillar

Posterior Pharynx

Fig. 1-8. Illustration of throat culture technique. The patient is asked to open his mouth widely and phonate an "ah." The tongue is gently depressed with a tongue blade and a swab guided over the tongue into the posterior pharynx. The mucosa behind the uvula and between the tonsillar pillars is swabbed with a gentle back-and-forth sweeping motion.

The translaryngeal aspiration technique is illustrated in Figure 1-9. After locally anesthetizing the skin, the cricothyroid membrane is pierced with a 14-gauge needle through which a 16-gauge polyethylene catheter is threaded into the lower trachea. Secretions are aspirated with a 20-cc. syringe. If secretions are scant, 2 to 4 ml. of sterile saline can be injected which will result in a paroxysm of coughing and usually produces an adequate specimen.

Fiberoptic bronchoscopy is another technique that may be employed for obtaining samples from infected abscesses or granulomas deep within the lung; however, the results may be difficult to interpret due to contamination of the instrument during transoral passage.[3] This technique has been found to be most useful in the recovery of mycobacteria in patients with closed lesions of tuberculosis and anaerobic bacteria in patients with lung abscesses. However, certain local anesthetics used for fiberoptic bronchoscopy have been shown to have distinct antibacterial activity requiring that these specimens be promptly cultured after collection.

3. Urine Specimens

For the optimal recovery of bacteria from the urinary tract, and to reduce potential contamination, it is imperative that careful attention be paid to the proper collection of urine samples. For best results, a nurse or a trained aide should personally supervise the collection of "clean-catch" samples from female patients.[14]

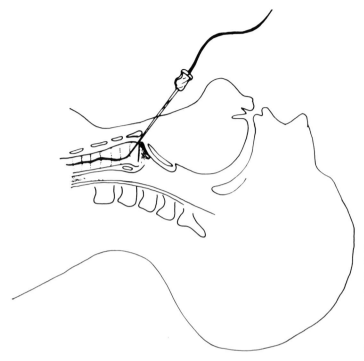

Fig. 1-9. Illustration of translaryngeal aspiration technique. A needle is passed percutaneously through the cricothyroid membrane. A polyethylene catheter is threaded through the needle and extended into the trachea.

For the proper collection of clean-catch urine samples from female patients, the periurethral area and perineum should be cleansed with soapy water and thoroughly rinsed with sterile saline or water. The labia should be held apart during voiding, and the initial few milliliters of urine passed into a bedpan or toilet bowl to flush out bacteria from the urethra. The midstream portion of urine is then collected into a sterile container (see Fig. 1-10).

Patients seen in a physician's office or in a clinic are frequently asked to obtain their own urine samples. This procedure is acceptable if the patients are given concise instructions as to how to properly collect the specimen. It is recommended that these instructions be printed on a card for the patient to read and follow rather than relying on verbal descriptions. It may be necessary for the nurse to read through the instructions with the patient, particularly if there is a language barrier. An example of such a card with directions as outlined by Kunin[14] is shown.

Instructions for Obtaining Clean-Catch Urine Specimens (Female)

1. Remove underclothing completely and sit comfortably on the seat, swinging one knee to the side as far as you can.
2. Spread yourself with one hand, and continue to hold yourself spread while you clean and collect the specimen.
3. Wash. Be sure to wash well and rinse well before you collect the urine sample. Using each of four separate 4″ × 4″ sterile sponges soaked in 10% green soap, wipe from the front of your body towards the back. Wash between the folds of the skin as carefully as you can.
4. Rinse. After you have washed with each soap pad, rinse with a moistened pad with the same front-to-back motion. Do not use any pad more than once.
5. Hold cup outside and pass your urine into the cup. If you touch the inside of the cup or drop it on the floor, ask the nurse to give you a new one.
6. Place the lid on the container or ask the nurse to do so for you.

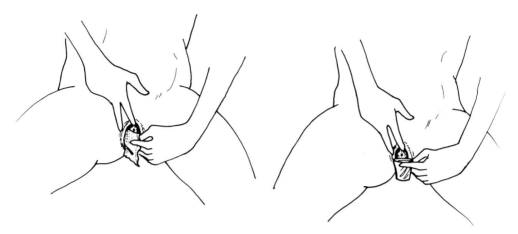

Fig. 1-10. Illustration of midstream clean-catch urine collection. In the illustration on the left, the labia are separated with the fingers and cleansed with a 4" × 4" gauze pad saturated with green soap. The midstream portion of the urine is collected into a sterile container as shown in the illustration on the right.

How well this procedure is being carried out can be monitored by noting the frequency with which urine colony counts in the intermediate range of 10,000 to 100,000 organisms/ml. of urine are reported. Patients without urinary tract infections should have no bacteria or very few colonies at most; those with infection most commonly have greater than 100,000 organisms/ml. Intermediate counts should be infrequent if the urine collection procedure is properly carried out by the patients. The recovery of three or more bacterial species also generally indicates that the specimen has been contaminated through faulty collection or delay in transport.

On occasion suprapubic aspiration of the urinary bladder may be required to obtain a valid urine specimen for culture, particularly from young children. The technique is illustrated in Figure 1-11. The suprapubic skin overlying the urinary bladder is disinfected as in preparation for surgery and an anesthetic is injected at the site of needle puncture. Once the urinary bladder is entered, urine is aspirated with a syringe and placed into an appropriate transport container.

Catheterized urine specimens may be used for culture. The "free flow" from the mouth of the catheter should be obtained. Urine from catheter bags is generally unsuitable for culture except from infants when special precautions have been taken. Urine specimens collected with Foley catheter tips are unsuitable for culture because catheter tips are invariably contaminated with urethral organisms.

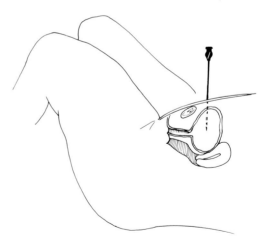

Fig. 1-11. Illustration of suprapubic urinary bladder aspiration. Through a percutaneous approach, a needle is directed into the urinary bladder just above the symphysis pubis. Urine can be removed with a syringe.

4. Wound Specimens

The surface of cutaneous wounds or decubitus ulcers frequently is colonized with environmental bacteria and swab samples often do not reflect the true cause of the infectious process. For that reason, the most desirable method of collecting cutaneous specimens is by aspirating loculated purulent material from the depths of the wound with sterile needle and syringe. The wound margins should be decontaminated as much as possible with surgical soap and application of 70 per cent ethyl or isopropyl alcohol. If material is obtained in the syringe, the needle cap can be replaced and the syringe sent to the laboratory for culture. If processing is delayed beyond 20 to 30 minutes, the specimen should be transferred to an anaerobic container.

If material cannnot be obtained with a needle and syringe and a swab must be used to collect the specimen, the wound margins should be gently separated with the thumb and forefinger of one hand (wearing a sterile glove) while extending the tip of the swab deep into the wound with the other hand, taking care not to touch the adjacent skin margins. The swab should be transported in an anaerobic container.

5. Stool Specimens

Laboratory confirmation of an intestinal infection caused by micoorganisms is usually made by detecting ova and parasites in direct saline or iodine mounts of fecal material or by recovering pathogenic bacteria from stool specimens. Samples obtained directly in sterile wide-mouthed containers should be covered with a tightly fitting lid.

Rectal swabs may be necessary in some instances for the recovery of Shigella species or *Neisseria gonorrhoeae.* For recovery of the latter organism, the swab should be inserted just beyond the anal sphincter, avoiding contamination with fecal material within the rectum.

Stool samples should be examined and cultured as soon as possible after collection. For the detection of motile trophozoites of protozoa, it is essential that the stool be examined while it is still warm. Stool specimens are not suitable for study of ova and parasites after a barium enema until 10 days have passed.

6. Cerebrospinal Fluid

Lumbar spinal puncture is the procedure used by physicians to obtain cerebrospinal fluid for culture and other laboratory studies. After properly disinfecting the skin of the lower back, patients are asked to lie on their side with the torso bent forward to separate the spinous processes of the lumbar vertebrae. Under local anesthesia, a long spinal needle is inserted into the spinal canal between the third and fourth lumbar vertebrae (Fig. 1-12). Cerebrospinal fluid (CSF) need not be aspirated in that it flows from the mouth of the needle under a pressure of approximately 90 to 150 mm. CSF in normal individuals.

Fluid is commonly collected into three tubes, the third of which is selected for culture. A total of 10 ml. is usually collected. If there is to be a delay in processing specimens, the fluid should be left at room temperature or placed in the incubator. Refrigeration is contraindicated because of the killing effect that chilling has on *Neisseria meningitidis* and *Hemophilus influenzae,* the two most common bacterial species causing meningitis.

7. Female Genital Tract

Vaginal cultures do not often produce meaningful results. In cases of suppurative vaginitis, direct wet mounts should be prepared at the bedside and examined microscopically shortly thereafter for the presence of *Trichomonas vaginalis* or the budding yeast forms of *Candida albicans. Hemophilus (Corynebacterium) vaginalis* is thought to cause primary vaginitis; other bacterial species, particularly members of the Enterobac-

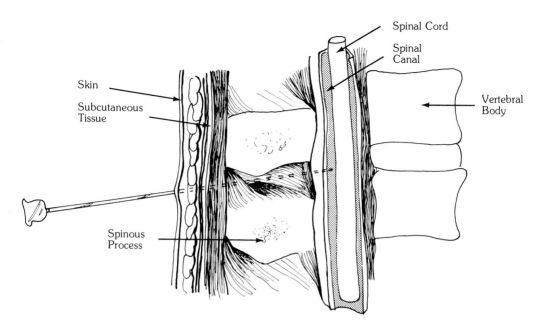

Fig. 1-12. Illustration of spinal tap technique. A long spinal needle is directed through the skin of the lumbar back, between the spinous processes of the vertebrae, through the intraspinous ligaments, and into the spinal canal.

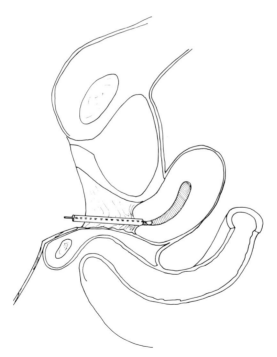

Fig. 1-13. Illustration of endometrial culture technique. Through a speculum, a catheter is introduced into the cervical os and a swab extended through the catheter into the endometrial cavity. This aids in preventing contamination of the swab by touching of the vaginal wall or the cervical os.

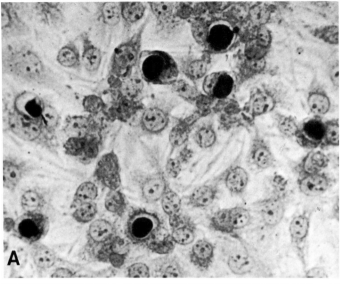

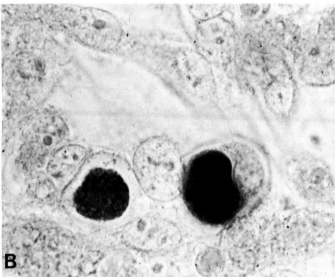

Fig. 1-14. Photomicrographs of intracellular inclusions of trachoma infection. A. Low-power photomicrograph of cell culture showing dense, semilunar to spherical intracytoplasmic inclusions seen in cases of trachoma-inclusion conjunctivitis (TRIC). Iodine stain. B. High-power photomicrograph of same preparation as A, showing the semilunar inclusions in greater detail. Note in the infected cell on the right that the inclusion is within the cytoplasm, and abuts against a faintly staining nucleus that lies within the "hoff" of the inclusion. Iodine stain.

teriaceae, are only rarely incriminated. The significance of anaerobic bacteria recovered from vaginal secretions is difficult to interpret because they are present as normal flora.

In cases of suspected endometritis, specimens should be obtained by direct vision through a vaginal speculum, inserting the tip of a culture swab through a narrow-lumen catheter that has been placed within the cervical os (Fig. 1-13). In this way, touching

of the cervical mucosa is prevented, reducing the chance for contamination.

8. Eye and Ear Cultures

Suppurative material from an infected eye should be collected from the lower cul-de-sac or from the inner canthus. Direct Gram's stains of the material obtained should always be prepared to determine the presence and type of bacteria in the collected material. If

Color Plates
1-1 *to* 1-3

Plate 1-1
Gram's Stain Evaluation of Sputum Smears

The quality of sputum samples can be evaluated by counting the relative numbers of squamous epithelial cells and segmented neutrophils per low-power field in a Gram's stained smear. The presence of squamous epithelial cells indicates a preponderance of oropharyngeal secretions. In contrast, bacterial pneumonia will produce large numbers of segmented neutrophils in the sputum.

A.
Squamous epithelial cells representative of oropharyngeal secretions.

B.
High-power view of mature squamous epithelial cell with adherent bacteria.

C.
Segmented neutrophils and mucin strands, highly suggestive of bacterial pneumonia.

D.
High-power view of segmented neutrophils seen in Frame C.

E.
Low-power view of squamous epithelial cells with large numbers of Gram-positive and Gram-negative bacteria. This may occur in sputum specimens that are not processed promptly, resulting in overgrowth with commensal bacteria.

F and G.
Low- and high-power views of ciliated columnar epithelial cells. These cells line the lower respiratory tract and their presence validates the lower respiratory origin of the specimen.

H and I.
Low- and high-power views of pulmonary alveolar macrophages, phagocytic cells found within the lung alveoli. Large numbers of alveolar macrophages containing yellow-staining granules of iron in sputum samples often indicate congestive heart failure.

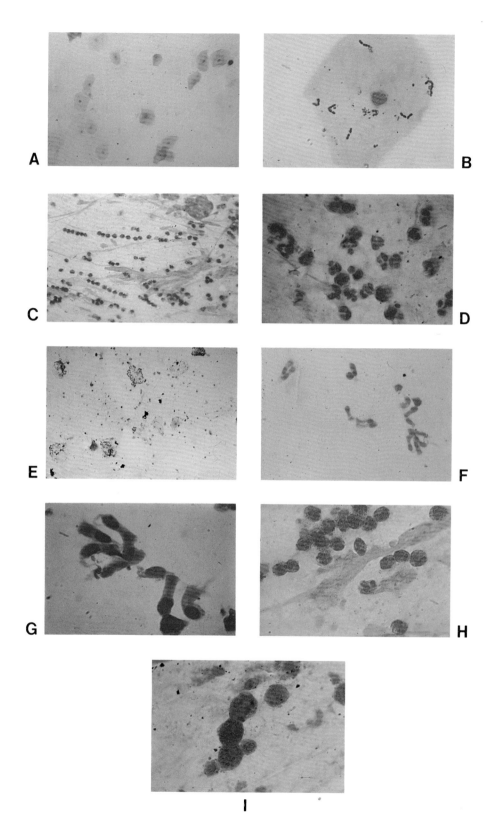

Plate 1-2
Presumptive Bacterial Identification
Based on Colonial Morphology

Microbiologists utilize various characteristics of bacterial colonies that grow on the surface of agar culture media for two purposes: to make a presumptive identification and as a guide in the selection of tests to determine differential characteristics for final identification. The size, shape, consistency, color, and pigment production by the colonies as well as the presence or absence of hemolytic reactions on blood agar are the criteria commonly used.

A.
Round, convex, yellow or white colonies on blood agar are suggestive of *Staphylococcus aureus* or *S. epidermidis.*

B.
Convex, opaque, smooth colonies suggestive of Micrococcus or Staphylococcus species.

C.
Colonies on blood agar showing distinct zones of hemolysis. The large size of the colonies compared to the zones of hemolysis suggests a hemolytic Staphylococcus species.

D.
Pinpoint, beta-nemolytic colonies suggestive of Streptococcus species.

E.
Higher-power view of Frame *D.* The small size of the bacterial colonies compared to the large size of the zone of beta hemolysis is highly suggestive of Streptococcus species. Compare this picture with the colonies of staphylococci shown in Frame *C.*

F.
Blood agar plate with flat alpha-hemolytic colonies. This culture is suggestive of *Streptococcus pneumoniae,* presumptively confirmed by the inhibition of bacterial growth around the optochin ("P") disk (diethylhydrocupereine).

G.
Opaque, convex, gray-yellow colonies suggestive of enteric bacilli belonging to the family Enterobacteriaceae. The colonies shown here are those of *Escherichia coli.*

H.
Spreading, smooth colonies with a distinct green pigmentation. This picture is suggestive of *Pseudomonas aeruginosa.*

I.
The swarming, wavelike colonial growth illustrated here is characteristic of either *Proteus vulgaris* or *Proteus mirabilis.*

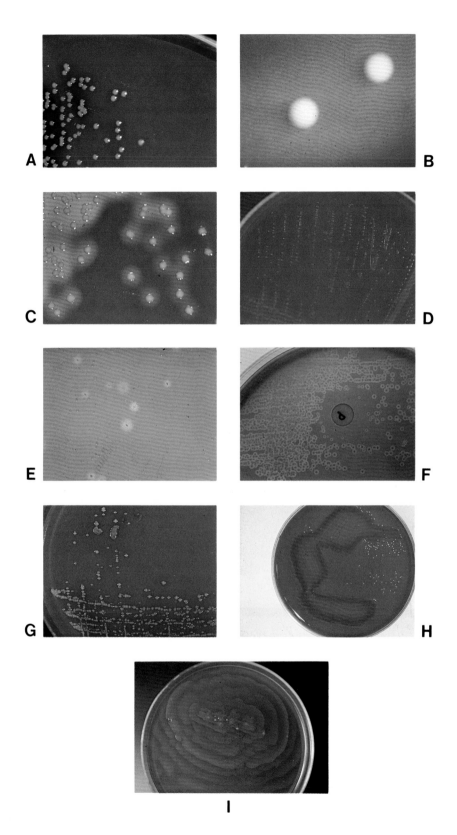

Plate 1-3
Bacterial Identification Based on Gram's Stain

The Gram's stain of bacteria is one of the most important characteristics determined by microbiologists in the presumptive identification of microorganisms. The morphology of the bacterial cells, their arrangement, and their staining characteristics are often distinctive enough to allow a presumptive identification in a Gram's stained smear. Several groups of bacteria are included in this plate.

A and B.
Gram-positive bacilli. In Frame *B,* the "Chinese letter" arrangement of the bacterial cells suggests the diphtheroidal bacilli belonging to the genus Corynebacterium (Chap. 9).

C.
Gram-positive cocci in chains, characteristic of Streptococcus species.

D.
Gram-positive cocci in clumps and tetrads, characteristic of Micrococcus species.

E.
Exudate containing gram-positive cocci in small clusters consistent with Staphylococcus species.

F.
A Gram's stained smear from a sputum specimen in which are seen several pairs of gram-positive diplococci characteristic of *Streptococcus pneumoniae.* The presence of a clear zone or capsule surrounding diplococci suggests that the organism is a virulent strain.

G.
Gram-negative bacilli. This picture is not distinctive of any one group of organisms but may be seen with members of the Enterobacteriaceae, the nonfermentative bacilli, and several species of "unusual bacilli."

H.
Smear of oropharyngeal secretions illustrating fusiform bacilli and spirochetes, often seen in Vincent's angina.

I.
Methylene-blue stain illustrating bacilli with metachromatic granules, characteristic of Corynebacterium species.

J.
Purulent urethral exudate illustrating intracellular, gram-negative diplococci, characteristic of *Neisseria gonorrhoeae.*

K.
Sputum smear stained by a modified acid-fast technique, illustrating branching, acid-fast filaments characteristic of Nocardia species.

L.
Fluorescent antibody stain illustrating the typical yellow glow of bacterial cells reacting specifically with the fluorescent antibody conjugate.

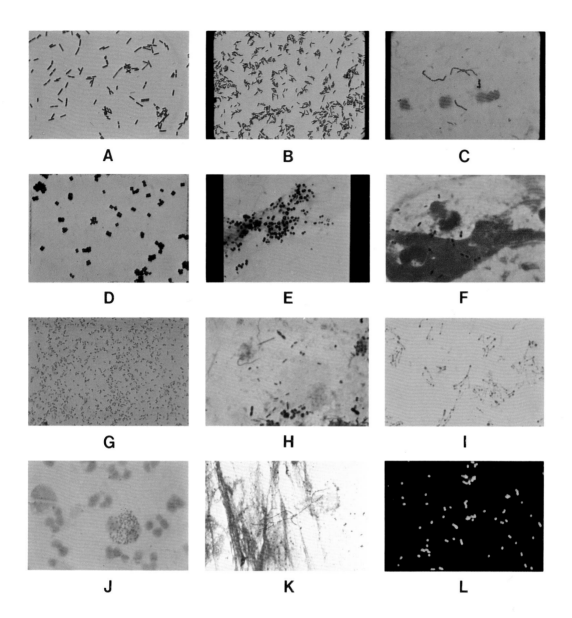

A

B

C

D

E

F

G

H

I

J

K

L

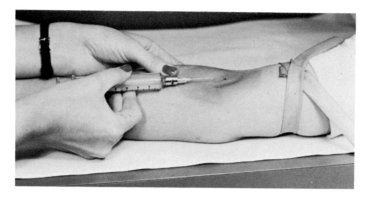

Fig. 1-15. Illustration of venipuncture technique for blood culture, using a sterile needle and syringe. A tourniquet is applied to the upper arm above the venipuncture site to distend the anticubital veins. The site had previously been prepared with tincture of iodine and alcohol. The blood is removed with the syringe and needle and injected into an appropriate blood culture bottle. To reduce the chance of skin contamination, it is recommended that a second syringe be used to draw the blood to be cultured, with the first syringe theoretically containing any organisms that were washed from the needle.

infection with *Chlamydia trachomatis* (trachoma) is suspected, corneal scrapings should be smeared on a glass microscope slide, air dried, stained with Wright's or Giemsa stains, and examined for the presence of the characteristic intracytoplasmic inclusions (Fig. 1-14).

Cultures of the external auditory canal generally do not reflect the bacterial cause of otitis media unless there has been recent rupture of the tympanic membrane. Tympanic membrane aspiration is rarely performed. In some cases of acute otitis media, the causative microorganism can be cultured from the posterior nasopharynx.

9. Blood Cultures

Blood cultures can be obtained either by using a needle and syringe (Fig. 1-15) or by the so-called "closed system" using a vacuum bottle and double-needle collecting tube (Fig. 1-16). In either instance the venipuncture site should be properly decontaminated to minimize skin contamination. Optimal skin preparation includes: (1) a 30-second prewash with green soap; (2) a rinse with sterile water; (3) an application of tincture of iodine which is allowed to dry; and (4) an alcohol wash to remove the iodine. The venipuncture site should not be palpated after this treatment unless a sterile glove is used.

On a routine basis samples for blood cultures should not be drawn at intervals less than one hour apart.[6] In cases of suspected bacterial endocarditis, three blood cultures with samples taken about 4 to 6 hours apart on three successive days are generally recommended.[6] It is also recommended that both aerobic and anaerobic bottles be obtained for optimal recovery of organisms.[6,23] Aerobic bottles should be vented in ambient air with a cotton-stoppered needle after the blood sample has been injected.

The culture medium should be nutritionally adequate to support the growth of fastidious organisms, particularly anaerobes, and several commercially available culture media have been found to be satisfactory. Examples of commercially available blood culture media include tryptic soy broth (Difco), thiol broth (Difco), Columbia broth (Difco), trypticase soy broth (Becton-

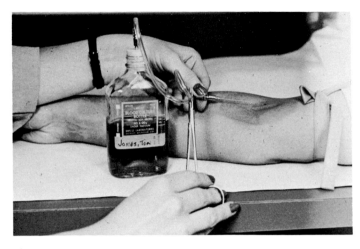

Fig. 1-16. Photograph illustrating the closed system of blood culture collection. The closed blood culture system consists of a vacuum blood culture bottle and a double-needle collection tube. The tube is first clamped with a hemostat and one needle placed into the stopper of the blood culture bottle. The opposite needle is used for the venipuncture. Again, note the tourniquet above the venipuncture site. When the needle enters the vein, the hemostat is released and blood is aspirated directly into the bottle. The vacuum is regulated so that exactly 10 ml. of blood is delivered into the bottle.

Dickinson), and thioglycollate medium (Becton-Dickinson). However, Reller[20] has determined that the volume of blood added to the bottle, and not the type of medium contained, is the most important factor in achieving maximal recovery of microorganisms. At least 10 ml. of blood should be collected (excepting in cases of infants and small children) and added to at least 90 ml. of culture medium.[20] Most blood culture bottles contain CO_2, and it has been established that the medium should include 0.03 to 0.05 per cent polyanetholsulfonate (SPS or Liquoid) to prevent clotting of the blood and reduce the effect of antibiotics and polymorphonuclear leukocyte activity in the medium.

10. Tissue and Biopsy Specimens

Tissue samples for culture should be delivered promptly to the laboratory in sterile gauze or in a suitably capped, sterile container. Formalized specimens are not suitable for culture unless the exposure time has been short and the culture is obtained from a portion of the tissue not exposed to formalin.

Bone marrow cultures may be helpful in the diagnosis of infectious granulomatous diseases such as brucellosis, histoplasmosis, and tuberculosis. Aspirated bone marrow samples can be placed directly into blood culture bottles, or they can be placed into a sterile vial containing heparin if centrifugation and harvest of the buffy coat is the culture technique desired.

SPECIMEN TRANSPORT

In a hospital setting a two-hour time limit between the time of collection and delivery of specimens to the laboratory is recommended.[5] This time limit poses a problem for specimens collected in physician's offices, and a transport medium is often required. Stuart, Amies, and Carey-Blair transport media are most frequently used.

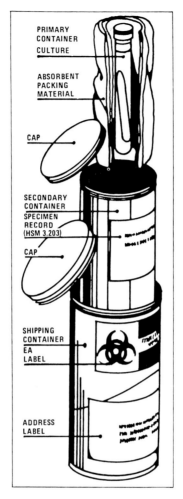

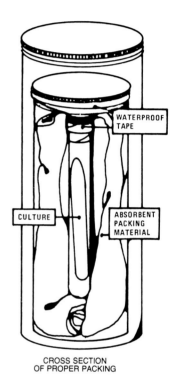

CROSS SECTION
OF PROPER PACKING

Fig. 1-17. Illustration of proper technique for packaging of biologically hazardous materials. (From laboratory methods in anaerobic bacteriology. CDC laboratory manual. DHEW publication No. (CDC) 74-8272, Center for Disease Control, Atlanta, 1974)

As seen from the formula for Stuart transport medium shown below, the medium is essentially a solution of buffers with carbohydrates, peptones, and other nutrients excluded. This medium is designed to preserve

Stuart Transport Medium

Sodium chloride	3.00 g.
Potassium chloride	0.20 g.
Disodium phosphate	1.15 g.
Monopotassium phosphate	0.20 g.
Sodium thioglycollate	1.00 g.
Calcium chloride, 1% aqueous	10.00 g.
Magnesium chloride, 1% aqueous	10.00 g.
Agar	4.00 g.
Distilled water	1.00 liter

$pH = 7.3$

the viability of bacteria during transport without significant multiplication of the microorganisms. Sodium thioglycollate is added as a reducing agent to improve recovery of anaerobic bacteria, and the small amount of agar provides a semisolid consistency to prevent oxygenation and spillage during transport.

All microbiology specimens to be transported through the United States mails must be packaged under strict regulations formulated by the Public Health Service. A complete list of etiologic agents that are included under these regulations is available upon request from the Center for Disease Control in Atlanta, Georgia.

Specimens must be packaged to withstand

leakage of contents, shocks, or pressure changes that may occur during handling. Figure 1-17 illustrates the proper technique for packaging and labeling of etiologic agents. The primary container (test tube, vial) must be fitted with a watertight cap and surrounded by sufficient packing material to absorb the fluid contents should a leak occur. This container in turn is placed in a secondary container, preferably constructed of metal with a tightly fitting screwcap lid. The primary and secondary containers in turn are enclosed in an outer shipping carton constructed of corrugated fiberboard, cardboard, or Styrofoam.

In addition to the address label, the outer container must also have the etiologic agents/biomedical material label affixed (Fig. 1-18). This label has a white background and the biohazard logo is in red or orange. The following notice must also be affixed to the outer container:

NOTICE TO CARRIER

This package contains *LESS THAN 50 ml. OF AN ETIOLOGIC AGENT, N.O.S.,* is packaged and labeled in accordance with the U. S. Public Health Service Interstate Quarantine Regulations (42 CFR, Section 72.25(c), (1) and (4), and *MEETS ALL REQUIREMENTS FOR SHIPMENT BY MAIL AND ON PASSENGER AIRCRAFT.*

This shipment is *EXEMPTED FROM ATA RESTRICTED ARTICLES TARIFF 6-D* (see General Requirements 386(d) (1) and from *DOT HAZARDOUS MATERIALS REGULATIONS* (see 49 CFR, Section 173, 386(d) (3)). *SHIPPER'S CERTIFICATES, SHIPPING PAPERS, AND OTHER DOCUMENTATION OR LABELING ARE NOT REQUIRED.*

Date   Signature of Shipper

CENTER FOR DISEASE CONTROL

ATLANTA, GEORGIA 30333

SPECIMEN PROCESSING

Each specimen received in the microbiology laboratory should be examined visually or microscopically to evaluate whether it is suitable for further processing. If there is evidence that the specimen has been improperly collected, if there is insufficient quantity of material, if the container is inappropriate, or if there was an excessive time delay in delivery, every attempt should be made to have a second sample collected.

Gross examination may provide valuable clues to the nature and quality of the specimens collected. Gross features to note include odor, purulent appearance of fluid specimens, and presence of gas or sulfur granules.[8] The finding of barium or other foreign materials may suggest that the specimen was contaminated with feces or bowel contents. Colored dyes, such as pyridium in urine specimens, or the presence of oily chemicals used for sputum induction or in bronchograms are other foreign materials that may hamper the recovery of certain microorganisms.

The importance of microscopic examinations of clinical materials has been emphasized by several authors.[8,10,22] Not only can the quality of the specimens be validated, but the observation of bacteria, mycelial elements, yeast forms, parasitic structures, or viral inclusions may provide sufficient information to render an immediate presumptive diagnosis leading to specific therapy. Direct microscopic examination may also give immediate presumptive evidence that anaerobic species of bacteria are present.

The examination of wet mounts of unstained materials by phase contrast or darkfield microscopy is useful for demonstrating motility, spirochetes, and endospores. Giemsa or Wright's stains may be helpful for observing bacterial forms that stain poorly or that have little contrast from background material.

Direct Gram's stains of clinical material can also be used to determine if a specimen

Fig. 1-18. Etiologic agents/ biomedical material label.

is representative of the site of infection. This technique has been applied to the evaluation of sputum samples. Based on the relative numbers of squamous epithelial cells and polymorphonuclear luekocytes in direct Gram's stains of sputum samples, Bartlett[5] has devised a grading system for evaluting sputum samples (see below). Negative numbers are assigned to a smear when one observes squamous epithelial cells, indicating contamination with oropharyngeal secretions (saliva). On the other hand, positive numbers are assigned for the presence of polymorphonuclear leukocytes, indicating the presence of active infection. The magnitude of these negative and positive numbers depends upon the relative numbers of epithelial cells and polymorphonuclear leukocytes as shown in the outline of Bartlett's grading system, below. A final score of 0 or less indicates either a lack of an inflammatory response or the presence of significant salivary contamination, invalidating the specimen. Representative photomicrographs of sputum Gram's stains illustrating this grading system are shown in Plate 1-1.

A similar grading system has been proposed by Murray and Washington,[16] as illustrated below. By this system, only specimens having less than 10 epithelial cells and more than 25 polymorphonuclear leukocytes per low-power field (Group 5) are considered

Bartlett's Grading System for Assessing the Quality of Sputum Samples

Number of Neutrophils/10× Field:	Grade
<10	0
10–25	+1
>25	+2
Presence of Mucus	+1

Number of Epithelial Cells/10× Field:	
10–25	−1
>25	−2
Total:	

Average the number of epithelial cells and neutrophils in about 20 or 30 separate 10/× microscopic fields. Calculate the total. A final score of 0 or less indicates absence of active inflammation or contamination with saliva. Repeat sputum specimens should be requested.

clinically relevant. In a recent clinical study Van Scoy[22] recommends that sputum samples containing more than 25 neutrophils be accepted for culture even if more than 10 epithelial cells are present (Group 4).

Murray and Washington's Grading System for Assessing the Quality of Sputum Samples

	Epithelial Cells/Low-Power Field	Leukocytes/ Low-Power Field
Group 1	>25	<10
Group 2	>25	10–25
Group 3	>25	>25
Group 4	10–25	>25
Group 5	<10	>25

The large number of epithelial cells in Groups 1 to 4 indicates contamination with oropharyngeal secretions and invalidates the samples. Only Group 5 specimens are considered clinically relevant.

Microscopic Techniques

A number of techniques may be used in the direct microscopic examination of clinical specimens, either to demonstrate the presence of microorganisms or to observe certain biochemical, physiologic, or serologic characteristics. The techniques more commonly used in clinical microbiology laboratories are outlined in Table 1-3.

Because of the refractive index of bacteria and other microorganismsmay be very similar to that of the mounting medium, the bacteria often may not be visible in bright-field illumination. Therefore, certain manipulations of the light source may be necessary. The simplest adjustment is to lower the condenser so that the focus of transmitted light is not directly on the object; rather the rays are scattered, producing a darkening of the background against which formerly invisible objects may be seen. Background illumination may also be reduced by closing the iris diaphragm on the condenser.

Special microscopic equipment, such as polarizing lenses, filters, or condensers of unique construction, is available to alter the nature of transmitted light rays.

Direct Stains

Because bacteria and other microorganisms are small and their protoplasm has a refractive index near that of water, biologic stains are generally required to adequately visualize them or to demonstrate the fine detail of their internal structures. The introduction of stains in the mid-nineteenth century was in large part responsible for the major advances that have occurred in clinical microbiology and in other fields of diagnostic microscopy during the past 100 years. Today we are so dependent on biologic stains that it is difficult to realize how progress in the study of bacteria could have been made before their introduction.

Stains consist of aqueous or organic preparations of dyes or groups of dyes that impart a variety of colors to microorganisms, plant and animal tissues, or other substances of biologic importance. Dyes not only serve as direct stains of biologic materials, but may also be used to demonstrate physiologic functions of microorganisms using so-called supravital techniques. Dyes also serve as indicators of pH shifts in culture media and as redox indicators to demonstrate the presence or absence of anaerobic conditions.

Almost all biologically useful dyes are derivatives of coal tar. The fundamental structure around which most dyes are chemically constructed is the benzene ring. The dyes differ from one another in the number and arrangement of these rings and in the substitution of hydrogen atoms with other molecules. For example, there are three key single substitutions for one hydrogen atom of benzene that comprise the basic structure of most dyes: (1) substitution of a methyl group to form toluene (methylbenzene); (2) substitution of a hydroxyl group to form phenol (carbolic acid); and (3) the substitution of an amine group to form aniline (phenylamine). The chemical formulas are:

Toluene (Methylbenzene)	Phenol (Carbolic acid)	Aniline (Phenylamine)

(Text continues on p. 27)

Table 1-3. *Techniques for Direct Examination of Unstained Specimens*

Methods and Materials	Purpose	Techniques
Saline Mount 1. Sodium chloride, aqueous, 0.85% 2. Glass microscope slides, 3″ × 1″ 3. Coverslips 4. Paraffin-Vaseline mixture (Vaspar)	To determine biologic activity of microorganisms, including motility or reactions to certain chemicals or serologic reactivity in specific antisera. The latter includes the quellung (capsular swelling) reaction used to identify different capsular types of *Streptococcus pneumoniae* and *Hemophilus influenzae*.	Disperse a small quantity of the specimen to be examined into a drop of saline on a microscope slide. Overlay a coverslip and examine directly with a 40× or 100× (oil immersion) objective of the microscope, lowering the condenser to reduce the amount of transmitted light. To prevent drying, ring the coverslip with a small amount of paraffin-Vaseline before overlaying the specimen drop on the slide.
Hanging-Drop Procedure 1. Hanging-drop glass slide (This is a thick glass slide with a central concave well on the under surface.) 2. Coverslip 3. Physiologic saline or water 4. Paraffin-Vaseline mixture	The hanging-drop mount serves the same purpose as the saline mount, except that there is less distortion from the weight of the coverslip and a deeper field of focus into the drop can be achieved. This technique is generally used for studying the motility of bacteria.	A small amount of paraffin-Vaseline mixture is placed around the lip of the well on the under surface of the hanging-drop slide. Cells from a bacterial colony to be examined are placed in the center of the coverslip, into a small drop of saline or water. The slide is inverted and pressed over the coverslip, guiding the drop of bacterial suspension into the center of the well. The slide is carefully brought to an upright position for direct examination under the microscope.
Iodine Mount 1. Lugol's iodine solution: Iodine crystals 5 g. Potassium iodide 10 g. Distilled water 100 ml. Dissolve KI in water and add iodine crystals slowly until dissolved. Filter and store in tightly stoppered bottle. Dilute 1:5 with water before use. 2. Microscope slides, 3″ × 1″ 3. Coverslips	Iodine mounts are usually used in parallel with saline mounts when examining feces or other materials for intestinal protozoa or helminth ova. The iodine stains the nuclei and intracytoplasmic organelles so that they are more easily seen. Iodine mounts cannot be used to the exclusion of saline mounts because iodine paralyzes the motility of bacteria and protozoan trophozoites.	A small amount of fecal matter or other material is mixed in a drop of the iodine solution on a microscope slide. This is mixed to form an even suspension and a coverslip is placed over the drop. The mount is then examined directly under a microscope. If this is to be delayed, or if a semipermanent preparation for future study is desired, the edges of the coverslip can be sealed with the paraffin-Vaseline mixture.
Potassium Hydroxide (KOH) Mount 1. Potassium hydroxide, 10% (aqueous) 2. Microscope slides, 3″ × 1″ 3. Coverslips	The KOH mount is used to aid in detecting fungus elements in thick mucoid material or in specimens containing keratinous material, such as skin scales, nails, or hair. The KOH dissolves the background keratin, unmasking the fungus elements to make them more apparent.	Suspend fragments of skin scales, nails, or hair in a drop of 10% KOH. Add coverslip over the drop and let sit at room temperature for about ½ hour. The mount may be gently heated in the flame of a Bunsen burner to accelerate the clearing process. Do not boil. The mount is placed under a microscope and examined for fungal hyphae or spores.

(Continued on p. 26)

Table 1-3. *Techniques for Direct Examination of Unstained Specimens* (Continued)

Methods and Materials	Purpose	Techniques

India Ink Preparation
1. India Ink (Pelikan brand)
    or
    Nigrosin (granular)*
2. Microscope slides, 3″ × 1″
3. Coverslips
	India ink or Nigrosin preparations are used for the direct microscopic examination of the capsules of many microorganisms. The fine granules of the india ink or Nigrosin give a semiopaque background against which the clear capsules can be easily seen. This technique is particularly useful in visualizing the large capsules of *Cryptococcus neoformans* in spinal fluid, sputum, or other secretions.	Centrifuge the spinal fluid or other fluid specimens lightly to concentrate any microorganisms in the sediment. Emulsify a small quantity of the sediment into a drop of india ink or Nigrosin on a microscope slide and overlay with a coverslip. Do not make the contrast emulsion too thick or the transmitted light may be completely blocked. Examine the mount directly under a microscope, using the 10× objective for screening and the 40× objective for confirmation of suspicious encapsulated microorganisms.

Dark-Field Examination 1. Compound microscope equipped with a dark-field condenser 2. Microscope slides, 3″ × 1″ 3. Coverslips 4. Physiologic saline 5. Applicator sticks or curet 6. Paraffin-Vaseline mixture	Dark-field examinations are used to visualize certain delicate microorganisms that are invisible by bright-field optics and stain only with great difficulty. This method is particularly useful in demonstrating spirochetes in biologic materials, particularly in urine suspected of containing Leptospira species, or from suspicious syphilitic chancres for *Treponema pallidum*.	The secretion to be examined is obtained from the patient. In the case of a chancre, the top crust is scraped away with a scalpel blade and a small quantity of serous material is placed on a microscope slide. Ring a coverslip with Vaseline-paraffin mixture and place over the drop of material. Examine the mount directly under a microscope fitted with a dark-field condenser with a 40× or 100× objective. Spirochetes will appear as motile, brightly appearing "corkscrews" against a black background.
Neufeld's Quellung Reaction 1. Homologous anticapsular serum 2. Physiologic saline 3. Microscope slides, 3″ × 1″ 4. Coverslips	When species of encapsulated bacteria are brought into contact with serum containing homologous anticapsular antibody, their capsules undergo swelling that is visible by microscopic examination. This serologic procedure is useful in identifying the various types of *Streptococcus pneumoniae* and *Hemophilus influenzae* in biologic fluids or in cultures.	A loopful of material, such as emulsified sputum, body fluid, or broth culture, is spread over a 1-cm area in two places on opposite ends of a microscope slide. A loopful of specific anticapsular typing serum is spread over the area of one of the dried preparations; the opposite area is overlaid with a loopful of saline to serve as a control. Each area is overlaid with a coverslip and examined under the 100× (oil immersion) objective of the microscope. Organisms showing a positive reaction appear surrounded with a ground-glass, refractile halo due to capsular swelling. Compare the test preparation with the saline control where no capsular swelling should be observed.

*Available from Harleco Co., Philadephia, Pa.

Most stains used in microbiology are derived from aniline and in the older literature were called aniline dyes.

Dyes are generally composed of two or more benzene rings connected by well-defined chemical bonds (chromophores) that are associated with color production. Although the underlying mechanism of the color development is not totally understood, it is theorized that certain chemical radicals have the property of absorbing light of different wave lengths, acting as chemical prisms. Some of the more common chromophore groupings found in dyes are: $C{=}C$, $C{=}O$, $C{=}S$, $C{=}N$, $N{=}N$, $N{=}O$, and NO_2. (Note the presence of these groups in the chemical formulas of the stains shown in Table 1-4.) The depth of color of a dye is proportional to the number of chromophore radicals in the compound.

The classification of dyes is somewhat complex and confusing, but is generally based on the chromophores present. In broad terms, dyes are referred to as "acidic" or "basic," designations not necessarily indicating their *p*H reactions in solution but rather whether a significant part of the molecule is anionic or cationic. From a practical standpoint, basic dyes stain structures that are acidic in nature, such as the nuclear chromatin in cells; acidic dyes react with basic substances, such as cytoplasmic structures. If both nuclear and cytoplasmic structures are to be stained in a given preparation, combinations of acidic and basic dyes may be used. A common example is hematoxylin (basic) and eosin (acidic) stains used in the examination of tissue sections.

All biologic dyes have a high affinity for hydrogen. When all the molecular sites that can bind hydrogen are filled, the dye is in its reduced state and generally is colorless. In the colorless state, the dye is called a "leuko compound." Looking at this concept from the opposite view, a dye retains its color only as long as its affinities for hydrogen are not completely satisfied. Since oxygen generally has a higher affinity for hydrogen than many dyes, color is retained in the presence of air. This allows certain dyes, such as methylene blue, to be used as a redox indicator in an anaerobic environment such as in a Gas-Pak jar, since the indicator becomes colorless in the absence of oxygen.

Stains Commonly Used in Microbiology. The more commonly used stains in microbiology, their chemical formulas and specific applications are outlined in Table 1-4. It is important that all stains be prepared in the laboratory following the specific instructions of the manufacturer. Exact details must be followed when formulating mixtures of multiple dye compounds, such as trichrome or polychrome stains. Wright's-Giemsa stain is one example, composed of methylene blue and a variety of azure blue degradation products. The *p*H of the buffer and the conditions and time of storage may be critical factors in determining the composition and staining quality of the final solution. It is essential that controls of known staining characteristics be used to test each new batch of stain to ensure that the intensity and hue of the color reactions are appropriate.

Of the stains listed in Table 1-4, the Gram's and acid-fast techniques are most commonly employed.

Gram's Stain Procedure. The formula for the stains used in the Gram's stain procedure are shown in Table 1-4. Crystal violet (or gentian violet) serves as the primary stain, which binds to the bacterial cell wall after treatment with a weak solution of iodine (the mordant). Some bacterial species, because of the chemical nature of their cell walls, have the ability to retain the crystal violet even after treatment with an organic decolorizer such as a mixture of acetone and alcohol. Such bacteria are called gram-positive. The gram-negative bacteria, presumably because of a higher lipid content of their cell wall, lose the crystal violet primary stain when treated with the decolorizer. Safranin is the secondary, or counterstain, used in the Gram's stain technique. Gram-negative bacteria, which have lost the crystal violet stain, appear red or pink when observed through the microscope, having affixed the safranin counterstain to their cell

(Text continues on p. 31)

Table 1-4. Common Biologic Stains Used in Bacteriology

Stain	Chemical Formula	Ingredients	Purpose
Loeffler's Methylene blue	Tetramethyl thionin	Methylene blue 0.3 g. Ethyl alcohol, 95% 30 ml. Distilled water 100 ml.	This is a simple direct stain used to stain a variety of microorganisms, specifically used in the identification of *Corynebacterium diphtheriae* by differentiating the deeply staining metachromatic granules from the lighter blue cytoplasm.
Gram's stain	Crystal Violet (Hexamethylpararosanilin) Dimethyl Phenosafranin	Crystal violet: Crystal violet 2.0 g. Ethyl alcohol, 95% 20 ml. NH₄ oxalate 0.8 g. Distilled water 100 ml. Gram's iodine: Potassium iodide 2.0 g. Iodine crystals 1.0 g. Distilled water 100 ml. Decolorizer: Acetone 50 ml. Ethyl alcohol, 95% 50 ml. Counterstain: Safranin 0 2.5 g. Ethyl alcohol, 95% 100 ml. Add 10 ml. to Distilled water 100 ml.	This is a differential stain used to demonstrate the staining properties of bacteria of all types. Gram-positive bacteria retain the crystal violet dye after decolorization and appear deep blue. Gram-negative bacteria are not capable of retaining the crystal violet dye after decolorization and are counterstained red by the safranin dye. Gram-staining characteristics may be atypical in very young, old, dead, or degenerating cultures.

Ziehl-Neel-sen Acid-fast stain

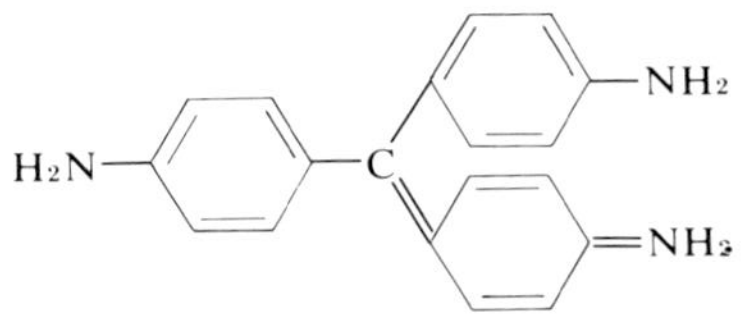

Carbolfuchsin
(Triaminotriphenylmethane)

Carbolfuchsin:	
Phenol crystals	2.5 ml.
Alcohol, 95%	5.0 ml.
Basic fuchsin	0.5 g.
Distilled water	100 ml.
Acid alcohol, 3%:	
HCl, concen-	
trated	3.0 ml.
Alcohol, 70%	100 ml.
Methylene blue:	
Methylene blue	0.5 g.
Glacial acetic	0.5 ml.
Distilled water	100 ml.

Acid-fast bacilli are so called because they are surrounded by a waxy envelope that is resistant to staining. Either heat or a detergent (Tergitol) is required to allow the stain to penetrate the capsule. Once stained, acid-fast bacteria resist decolorization, while other bacteria are destained with the acid alcohol.

Fluoro-chrome

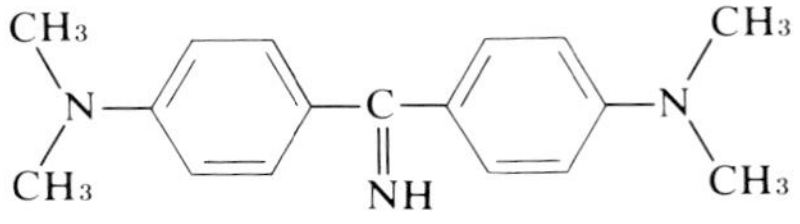

Auramine O

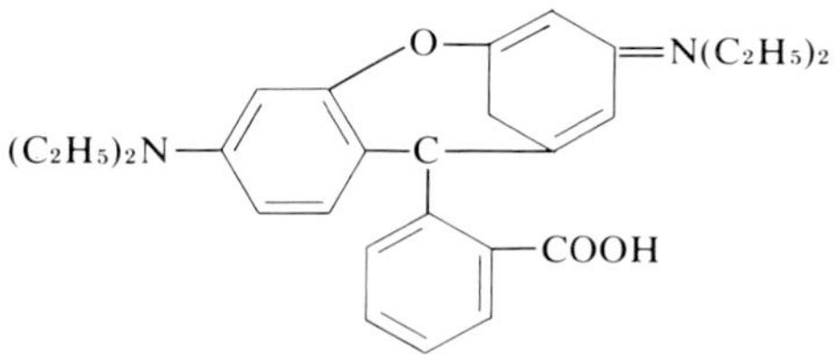

Rhodamine B

Auramine O	1.5 g.
Rhodamine B	0.75 g.
Glycerol	75 ml.
Phenol	10 ml.
Distilled water	50 ml.

The fluorochrome dye stains mycobacteria selectively by binding to the mycolic acid in the cell wall. This stain demonstrates mycobacteria better than conventional acid-fast stains and allows for screening of smears at lower magnification because organisms are more easily seen.

(Continued on p. 31)

Table 1-4. Common Biologic Stains Used in Bacteriology (Continued)

Stain	Chemical Formula	Ingredients	Purpose
Wright's-Giemsa	Polychrome methylene blue: Methylene blue Methylene azure Eosin Methylene azure B	Powdered Wright's stain 9 g. Powdered Giemsa stain 1 g. Glycerin 90 ml. Absolute methyl alcohol 2910 ml. Mix in brown bottle and let stand 1 month before using.	Wright's-Giemsa is commonly used for staining the cellular elements of the peripheral blood smear. It is useful in bacteriology for the demonstration of intracellular organisms such as *Histoplasma capsulatum* and *Leishmania* species. The stain is also useful in demonstrating intracellular inclusions in direct smears of skin or mucous membranes, such as corneal scrapings for trachoma.
Lactophenol cotton blue	Cotton blue	Phenol crystals 20 g. Lactic acid 20 g. Glycerol 40 ml. Distilled water 20 ml. Dissolve ingredients, then add: Cotton blue 0.05 g.	Because of the sulfonic groups, the dye is strongly acidic and has been used as a counterstain for unfixed tissues, bacteria, and protozoa, in combination with other dyes. Currently it is most commonly used for the direct staining of fungal mycelium and fruiting structures, which take on a delicate light blue color.

walls. The technique for performing the Gram's stain is as follows:

Gram's Stain Technique

1. Make a thin smear of the material for study and allow to air dry.
2. Fix the material to the slide so that it does not wash off during the staining procedure by passing the slide 3 or 4 times through the flame of a Bunsen burner.
3. Place the smear on a staining rack and overlay the surface with crystal violet solution.
4. After 1 minute (shorter times may be used with some solutions) of exposure to the crystal violet stain, wash thoroughly with distilled water or buffer.
5. Next, overlay the smear with Gram's iodine for 1 minute. Again wash with water.
6. Hold the smear between the thumb and forefinger and flood the surface with a few drops of the acetone-alcohol decolorizer until no violet color washes off. This usually requires 10 seconds or less.
7. Wash with running water and again place the smear on the staining rack. Overlay the surface with safranin counterstain for 1 minute. Wash with running water.
8. Place the smear in an upright position in a staining rack, allowing the excess water to drain off and the smear to dry.
9. Examine the stained smear under the 100× (oil) immersion objective of the microscope. Gram-positive bacteria stain dark blue; gram-negative bacteria appear pink-red.

Acid-Fast Stains. Mycobacteria possess a thick, waxy capsule that resists staining; however, once stained, this capsule withstands decolorization when treated with strong organic solvents such as acid alcohol. For this reason, organisms having this property are called acid-fast.

In order for the primary stain, carbolfuchsin, to penetrate the waxy capsules of acid-fast bacilli, some type of physical treatment is required. Heat is used in the conventional Ziehl-Neelsen technique. After the carbolfuchsin is overlaid on the surface of the smear to be stained, the flame of a Bunsen burner is passed back and forth through the stain until the solution begins to steam. Boiling of the stain must be avoided.

The Kenyoun modification of the acid-fast stain is called the "cold method" because a surface-active detergent, Tergitol, is added to the carbolfuchsin to facilitate staining in the place of heat. Either of these methods is satisfactory.

The procedure for the Ziehl-Neelsen acid-fast stain is discussed in detail in Chapter 12.

Fluorochrome dyes, such as auramine and rhodamine, have the property of selectively staining the capsules of mycobacteria, producing a green or green-yellow glow when observed under a fluorescent microscope. The fluorochrome dyes make the screening of smears much more rapid, because the technique which uses them is more sensitive and requires that the viewer search only for the bright pencils of light against a dark background. The use of a 25× objective is recommended for screening. This gives a magnification low enough to include a wide field, which allows rapid screening, but high enough contrast to permit morphologic differentiation of fluorescing objects. (Plate 1-3, *L*).

One precaution must be observed when using the acid-fast staining techniques: nonviable organisms that may be present in the smear preparations from patients on drug therapy may stain with the technique. Therefore, the presence of acid-fast organisms does not necessarily indicate drug failure and persistence of active infection. Cultures are required in these instances to determine the viability of any acid-fast organisms.

A variety of specific stains and direct mount procedures can be used to identify microorganisms in direct examinations. These are reviewed in Table 1-5.

Selection of Primary Plating Media

For optimal recovery of microorganisms, it is essential to inoculate the appropriate primary isolation medium with the specimen as soon as possible after the specimen arrives in the laboratory. From the several hundred media commercially available it is necessary for the clinical microbiologist to select a relatively small number of selective and nonselective media for daily use. A nonselective medium is a medium such as blood agar which will support the growth of most commonly encountered bacteria. A

Table 1-5. *Diagnosis of Infectious Disease by Direct Examination of Culture Specimens*

Specimen	Suspected Disease	Laboratory Procedure	Positive Findings
Throat culture	Diphtheria	Gram's stain	Delicate pleomorphic gram-positive bacilli arranged in "Chinese letters"
		Methylene blue stain	Light-blue-staining bacilli with prominent metachromatic granules
	Acute streptococcal pharyngitis	Direct fluorescent antibody technique (after 4 to 6 hours incubation in Todd-Hewett broth)	Fluorescent cocci in chains; use positive and negative controls with each stain.
Oropharyngeal ulcers	Vincent's angina	Gram's stain	Presence of gram-negative bacilli and spirochetes
Sputum Transtracheal aspirates Bronchial washings	Bacterial pneumonia	Gram's stain	Variety of bacterial types; *Streptococcus pneumoniae* with capsules particularly diagnostic
	Tuberculosis	Acid-fast stain	Acid-fast bacilli
	Pulmonary mycosis	Gram's stain or lactophenol cotton blue mount	Budding yeasts, pseudohyphae, true hyphae, or fruiting bodies
Cutaneous wounds or purulent drainage from subcutaneous sinus	Bacterial cellulitis	Gram's stain	Variety of bacterial types; suspect anaerobic species.
	Gas gangrene (myonecrosis)	Gram's stain	Gram-positive bacilli suspicious for *Clostridium perfringens;* spores usually not seen.
	Actinomycotic mycetoma	Direct saline mount	"Sulfur granules"
		Gram's stain or modified acid-fast stain	Delicate, branching gram-positive filaments; *Nocardia* species may be weakly acid-fast.
	Eumycotic mycetoma	Direct saline mount	White, grayish, or black "grains"
		Gram's stain or lactophenol cotton blue mount	True hyphae with focal swellings or chlamydospores
Cerebrospinal fluid	Bacterial meningitis	Gram's stain	Gram-negative pleomorphic bacilli, small (Hemophilus species) Gram-negative diplococci *(Neisseria meningitidis)* Gram-positive diplococci *(Streptococcus pneumoniae)*
		Quellung reaction (Type-specific antisera)	Swelling and ground-glass appearance of bacterial capsules
	Cryptococcal meningitis	India ink or Nigrosin mount	Encapsulated yeast cells with buds attached by thin thread
	Listeriosis	Gram's stain	Delicate gram-positive bacilli
		Hanging-drop mount	Bacteria with "tumbling" motility
Urine	Yeast infection	Gram's stain or lactophenol cotton blue mount	Pseudohyphae or budding yeasts
	Bacterial infection	Gram's stain	Variety of bacterial types
	Leptospirosis	Dark-field examination	Loosely coiled motile spirochetes
Purulent urethral or cervical discharge	Gonorrhea	Gram's stain	Intracellular gram-negative diplococci
	Chlamydial infection	Giemsa stain or smear	Minute intracellular coccobacillary forms

Table 1-5. *Diagnosis of Infectious Disease by Direct Examination of Culture Specimens*
(Continued)

Specimen	Suspected Disease	Laboratory Procedure	Positive Findings
Purulent vaginal discharge	Yeast infection	Direct mount or Gram's stain	Pseudohyphae or budding yeasts
	Trichomonas infection	Direct mount	Flagellates with darting motility
Penile or vulvar painless ulcer (chancre)	Primary syphilis	Dark-field mount of chancre secretion	Tightly coiled motile spirochetes
	Chancroid	Gram's stain of ulcer secretion or aspirate of inguinal bubo	Intracellular and extracellular small gram-negative bacilli
Eye	Purulent conjunctivitis	Gram's stain	Variety of bacterial species
	Trachoma	Giemsa stain of corneal scrapings	Intracellular perinuclear inclusion clusters
Feces	Purulent enterocolitis	Gram's stain	Neutrophils and aggregates of staphylococci
	Cholera	Direct mount of alkaline peptone water enrichment	Bacilli with characteristic darting motility
	Parasitic disease	Direct saline or iodine mounts. Examine purged specimens.	Adult parasites or parasite fragments; protozoa or ova
Skin scrapings, nail fragments, or plucked hairs	Dermatophytosis	40% KOH mount	Delicate hyphae or clusters of spores
	Taenia versicolor	40% KOH mount or lactophenol cotton blue mount	Hyphae and spores resembling "spaghetti and meatballs"
Blood	Relapsing fever (Borrelia)	Wright's or Giemsa stain Dark-field examination	Spirochetes with typical morphology
	Blood parasites: malaria, trypanosomiasis, filariasis	Wright's or Giemsa stain	Intracellular parasites (malaria, babesia)
		Direct examination of anticoagulated blood for the presence of microfilaria	Extracellular forms: trypanosomes or microfilaria

selective medium is one which is designed to support the growth of certain bacteria, inhibiting the growth of others.

The primary culture media to be inoculated should be selected on the basis of the anatomical source of the clinical material and knowledge of the bacteria species commonly encountered in specimens from various sources.

Table 1-6 lists types of specimens received in clinical laboratories, the species of clinically significant bacteria to be expected in each specimen, and primary culture media for optimal recovery of these bacterial species. This list is not complete and each microbiologist should develop similar charts which are appropriate for his or her own laboratory.

Physicians must inform the laboratory if they clinically suspect an unusual infectious disease so that special media other than those routinely used can be inoculated with the specimen to allow recovery of the less commonly encountered microorganisms. Table 1-7 lists some of these uncommon microorganisms, the diseases they may cause, and the media required for their recovery.

The media listed in Tables 1-6 and 1-7 can be prepared in the laboratory from dehydrated products that are commercially available. The formulation and instructions for preparation of each medium are usually found on the label of the bottle. Preparation of these dehydrated media is usually simple, requiring only that the designated quantity

Table 1-6. *Selection of Primary Isolation Media for Specimens from Different Body Sites*

Specimen Source	Potential Bacterial Pathogens	Blood Agar, 5% Sheep	Chocolate Agar	Lester-Martin (Transglow)	MacConkey Agar	HE, XLD, or SS Agar*	Selenite (GN) Broth†	BAGG (SF) Broth‡	Thioglycollate or Chopped Meat Broth
Throat	*Streptococcus pyogenes*	X							
Sputum Transtracheal aspirate	*Streptococcus pneumoniae* *Hemophilus influenzae* *Staphylococcus aureus* *Klebsiella pneumoniae* Anaerobic bacteria	X	X		X				X
Urine	*Escherichia coli* *Klebsiella* species *Proteus mirabilis* *Pseudomonas aeruginosa* Streptococcus, Group D	X			X				
Wounds and abscesses	*Staphylococcus aureus* *Escherichia coli* and other "enteric" bacilli *Streptococcus pyogenes* *Bacteroides fragilis, Clostridium perfringens* and other anaerobic bacteria Streptococcus, Group D	X			X			X	X
Stool	*Salmonella* species *Shigella* species *Escherichia coli* (enterotoxigenic) *Staphylococcus aureus* *Yersinia enterocolitica*	X			X	X	X		
Urethra cervix	*Neisseria gonorrhoeae*			X					
Spinal Fluid	*Neisseria meningitidis* *Streptococcus pneumoniae* *Hemophilus influenzae*	X	X		X				
Eye and ear	*Neisseria gonorrhoeae* *Hemophilus* species *Staphylococcus aureus* *Streptococcus pyogenes* *Pseudomonas aeruginosa* *Moraxella* species	X	X		X				X
Joint fluids	*Neisseria gonorrhoeae* *Staphylococcus aureus* *Escherichia coli* and other species of Enterobacteriaceae *Moraxella* species	X	X		X				X

*HE = Hektoen Enteric, XLD = Xylose, Lysine Desoxycholate, SS = Salmonella Shigella
†GN = Gram Negative
‡BAGG = Buffered Azide Glucose Glycerol, SF = Streptococcus faecalis

Table 1-7. *Specialized Media Required for Recovery of Pathogens in Unusual Disease States*

Culture Source	Presumptive Diagnosis	Bacterial Pathogens	Media Required for Recovery
Throat	Acute obstructive epiglottiditis and laryngitis	*Hemophilus influenzae*	Chocolate blood agar, or Casman's agar with horse blood. Place 10 µg. bacitracin disk in area of inoculation.
	Mebranous pharyngitis	*Corynebacterium diphtheriae*	Loeffler's medium Tinsdale agar Sodium tellurite medium
Nasopharyngeal swab	Whooping cough	*Bordetella pertussis*	Bordet-Gengou potato medium
Sputum	Tuberculosis	*Mycobacterium tuberculosis*	Lowenstein-Jensen medium
		Other species of mycobacteria	Middlebrook 7H-11 medium
	Pulmonary mycosis	*Blastomyces dermatitidis* or other dimorphous fungi	Brain-heart infusion agar with and without cycloheximide and chloramphenicol (C & C)
Blood, bone marrow, or tissue biopsy	Brucellosis	Brucella species	Brucella agar (Hold cultures for three weeks.)
Urine	Leptospirosis	Leptospira species	Fletcher's semisolid medium
Stool	Cholera	*Vibrio comma* *Vibrio parahaemolyticus*	Alkaline peptone medium Thiosulfate citrate bile salts sucrose (TCBS) agar
Any source for anaerobic bacteria	Paramucosal or deep abscesses	*Bacteroides fragilis* and other species of anaerobes	Blood agar Blood agar containing antibiotics: neomycin, vancomycin, kanamycin, gentamicin Phenylethyl alcohol blood agar

of powder be weighed out on a balance and dissolved in the appropriate amount of distilled water in a flask. The mixture is brought to a boil to completely dissolve all ingredients, followed by sterilization in an autoclave at 121°C. and 15 pounds pressure for 15 minutes. The sterilized media is poured into tubes or plates, allowed to cool, and stored in the refrigerator.

Inoculation Techniques

The equipment needed for the primary inoculation of specimens is relatively simple.

Nichrome or platinum wire is recommended, fashioned either into a loop or straight wire (Fig. 1-19). One end of the wire is inserted into a cylindrical handle for ease of use.

The agar surface of Petri plates may be inoculated with the specimen by several methods, one of which is shown in Figure 1-20. The primary inoculum can be made with a loop, with a swab, or with other suitable devices. Once the primary inoculum is made, a loop or straight wire can be used to spread the material into the four quadrants of the plate, as illustrated in Figure 1-21. The inoculum is successively streaked with a

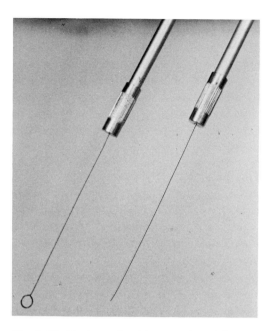

Fig. 1-19. Photograph of loop and straight wire commonly used for inoculation and transfer of cultures.

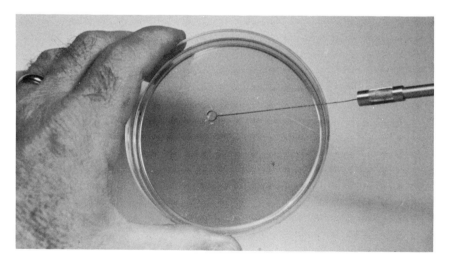

Fig. 1-20. Photograph of an agar plate illustrating inoculation of the surface with an inoculating loop. Actual inoculation is accomplished by streaking the agar surface with a back-and-forth sweeping motion of the loop.

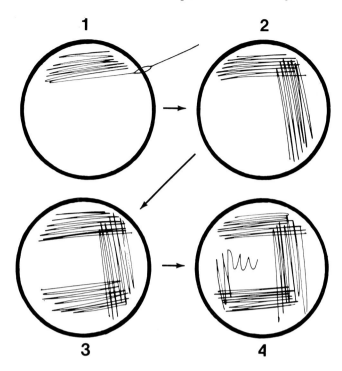

Fig. 1-21. Inoculation of culture plates for isolation of bacterial colonies.

back-and-forth motion into each quadrant by turning the plate at 90-degree angles. The loop or wire should be sterilized between each successive quadrant streak.

The purpose of this technique is to dilute the inoculum sufficiently on the surface of the agar medium so that well-isolated colonies of bacteria can be obtained from colony-forming units. The isolated colonies can then be individually subcultured to other media to obtain pure culture isolates which can be studied on differential media.

The streaking technique used for inoculation of agar media for semiquantitative colony counts is illustrated in Figure 1-22. Platinum inoculating loops, calibrated to contain either 0.01 ml. or 0.001 ml. of fluid, are immersed into an uncentrifuged urine sample, and a single streak is made across the center of an agar plate. The inoculum is spread evenly at right angles to the primary streak; then the plate is turned at 90 degrees and the inoculum is spread to cover the entire agar surface.

After 18 to 24 hours of incubation, the number of bacteria in the urine sample is estimated by counting the number of colonies that appear on the surface of the media. As illustrated in Figure 1-23, approximately 50 colonies can be counted. If a 0.001-ml. loop had been used to inoculate the medium, the number of colonies would be multiplied by one thousand; according to this, the count in this illustration is 50,000 colonies per milliliter.

Media may be dispensed in tubes, either as broth or in the form of agar. The agar may be semisolid, used for motility testing, or solid, usually poured on a slant. Broth media in a tube can be inoculated by the method shown in Figure 1-24. Tip the tube at approximately a 30-degree angle and touch an inoculating loop to the inner surface of the glass, just above the point where the surface of the agar makes an acute angle. When the culture tube is returned to its upright position, the area of inoculation is submerged beneath the surface.

Slants of agar medium are inoculated by
(Text continues on p. 40)

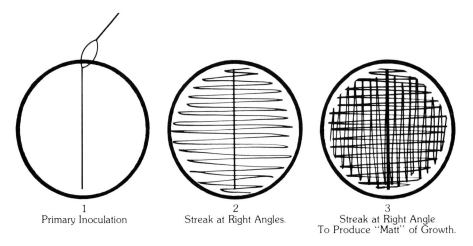

| 1 | 2 | 3 |
| Primary Inoculation | Streak at Right Angles. | Streak at Right Angle
To Produce "Matt" of Growth. |

Fig. 1-22. Procedure for inoculation of media for semiquantitative bacterial colony counts.

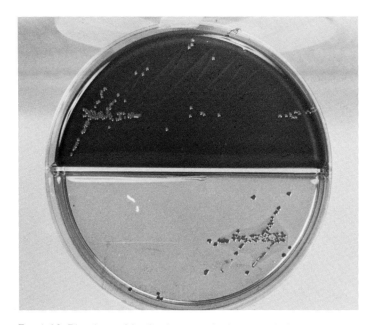

Fig. 1-23. Blood agar/MacConkey agar biplate on which are growing approximately 50 colonies of gram-negative bacteria. If a 0.001 calibrated semiquantitative loop of urine had been used to streak each medium, a colony count of $50 \times 1000 = 50{,}000$/ml. would be calculated.

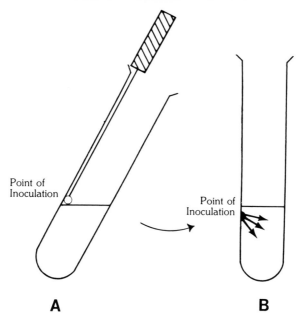

Point of
Inoculation

Point of
Inoculation

A   B

Fig. 1-24. Inoculating a tube of broth medium. *A.* Slant
and inoculate side of tube as shown. *B.* Replace tube
upright. This submerges the inoculation site under the
surface.

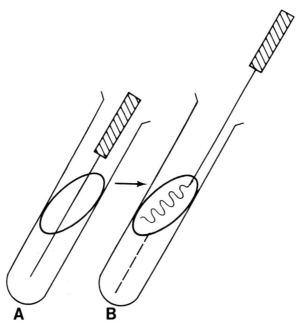

A   B

Fig. 1-25. The inoculation of an agar slant is performed
with a straight inoculating wire. *A.* The wire is first
"stabbed" into the deep of the tube to within 2 to 3 mm.
of the glass bottom. *B.* After the wire is removed from the
deep, it is streaked over the agar surface with a back-and-
forth "S" motion.

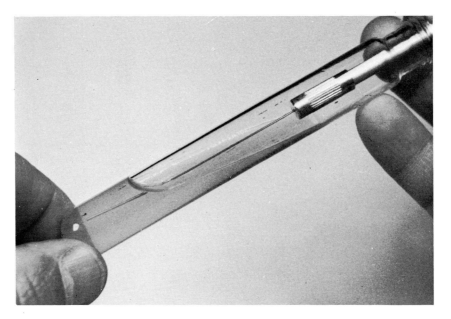

Fig. 1-26. Photograph of an agar slant, illustrating inoculation with a straight wire. The deep of the medium is stabbed with the wire and the slant is then inoculated by streaking back and forth over the agar surface. With some agar slant media only the slant surface is inoculated. Some agar tube media are inoculated by streaking the surface of the slant only.

first stabbing the "deep" of the agar, followed by streaking the slant from bottom to top with an "S" motion as the inoculating wire is removed (Figs. 1-25 and 1-26). When inoculating semisolid tubed agar for motility testing, it is important that the inoculating wire is removed exactly along the same tract used to stab the medium. A fanning motion during inoculation of this medium can result in a growth pattern along the stab line which may be falsely interpreted as bacterial motility.

THE INTERPRETATION OF CULTURES

Interpretation of primary cultures after 24 to 48 hours of incubation requires considerable skill. From initial observations the microbiologist must assess the colonial growth and decide whether additional procedures are required. This assessment is made by:

1. Noting the characteristics and relative number of each type of colony recovered on agar media

2. Determining the purity, gram reaction, and morphology of the bacteria in each type of colony

3. Observing changes in the media surrounding the colonies, which reflect specific metabolic activity of the bacteria recovered.

Gross Colony Characteristics

The assessment of gross colony characteristics is usually performed by visually inspecting growth on the surface of agar plates. Primary tubed media are not commonly used for assessment of colonial morphology, but rather are used to enrich the growth of bacteria so that they can be recovered in sufficient quantity for study. One exception to this is the use of the roll-streak tube for the recovery and direct study of anaerobic bacteria, as currently used by the

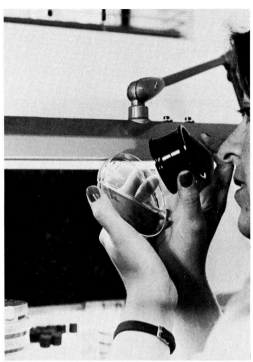

Fig. 1-27. Examination of a plate culture by reflected light.

Fig. 1-28. Examination of a plate culture using a hand lens.

Virginia Polytechnic institute Anaerobic Laboratory.[10] This system is well suited for the study of anaerobes since assessment of colonial morphology can be made directly through the glass of the tubes without exposing the organisms to ambient air during inspection.

The interpretation of primary cultures is a specialized skill which should be acquired by working with a well-trained microbiologist and is usually mastered only after many months or years of experience. Inspection of cultures is carried out by holding the plate in one hand and observing the surface of the agar for the presence of bacterial growth (see Fig. 1-27). Standard culture plates are 100 mm. in diameter and are convenient to hold in one hand. Each plate must be carefully studied because the bacteria initially recovered from specimens are often in mixed culture and a variety of colonial types may be present. Pinpoint colonies of slow-growing bacteria may be overlooked among larger colonies, particularly if there is any tendency for growth to spread over the surface of the plate.

During examination, plates should be tilted in various directions under bright, direct illumination so that light is reflected from various angles. Some microbiologists utilize a hand lens or a dissecting microscope to assist in the detection of tiny or immature colonies and to better observe their characteristics (see Fig. 1-28). Blood agar plates also should be examined when transilluminated by bright light from behind the plate in order to detect hemolytic reactions in the agar (see Fig. 1-29).

The following outline lists the commonly observed colonial characteristics that are helpful in making a preliminary bacterial identification. Observations of the colonies should be made under a dissecting microscope and by macroscopic examination.

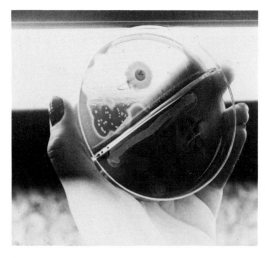

Fig. 1-29. Examination of a plate culture using transmitted light. This technique is most useful in studying the hemolytic patterns of colonies on blood agar.

Characteristics of Colonies Used in the Identification of Bacteria. Following are a number of terms used to describe gross colony characteristics. These are outlined in Figure 1-30.

1. *Size:* diameter in mm.
2. *Form:* punctiform, circular, filamentous, irregular, rhizoid, spindle
3. *Elevation:* flat, raised, convex, pulvinate, umbonate, umbilicate
4. *Margin* (edge of colony): entire, undulant, lobate, erose, filamentous, curled
5. *Color:* white, yellow, black, buff, orange, *etc.*
6. *Surface:* glistening, dull, other
7. *Density:* opaque, translucent, transparent, other
8. *Consistency:* butyryous, viscid, membranous brittle, other

Reactions in Agar Media Used in the Identification of Bacteria

1. *Hemolysis on blood agar:*
   Alpha: partial clearing of blood around colonies with green discoloration of the medium
   Beta: zone of complete clearing of blood around colonies due to lysis of the red blood cells
   Gamma: no change in the medium around the colony; no lysis of the red blood cells
   Double zone: halo of complete lysis immediately surrounding colonies with a second zone of partial hemolysis at the periphery
2. *Pigment production in agar media:*
   a. Water-soluble pigments discoloring the medium
   b. Pyocyanin
   c. Flurochrome pigments (fluorescein)
   d. Nondiffusible pigments confined to the colonies
3. *Reactions in egg-yolk agar:*
   a. Lecithinase: zone of precipitate in medium surrounding colonies
   b. Lipase: "pearly layer," an iridescent film in and immediately surrounding colonies, visible by reflected light
   c. Proteolysis: clear zone surrounding colonies
4. *Changes in differential media:* various dyes, *p*H indicators, and other ingredients are included in differential plating media to serve as indicators of enzymatic activities and aids in identification of bacterial isolates.

Odor. Although difficult to describe specifically, odors produced by the action of certain bacteria in plating media and in liquid media can be very helpful in tentative identification of the microorganisms involved. Examples of microorganisms exhibiting distinctive odors include:

Pseudomonas species (grape juice)
Proteus species (burned chocolate)
Streptomyces species (musty basement)
Clostridium species (fecal, putrid)

By assessing the described colonial characteristics and action on media, the microbiologist is able to make a preliminary identification of the different bacteria isolated by primary culture. These characteristics are helpful in the selection of other appropriate

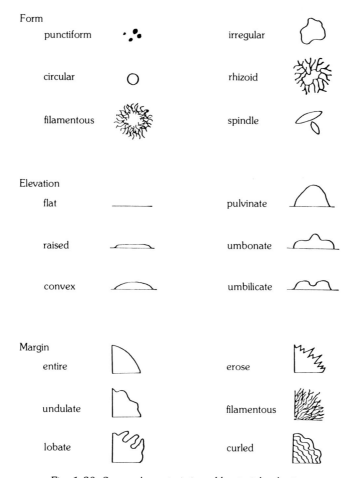

Form

punctiform

circular

filamentous

irregular

rhizoid

spindle

Elevation

flat

raised

convex

pulvinate

umbonate

umbilicate

Margin

entire

undulate

lobate

erose

filamentous

curled

Fig. 1-30. Some characteristics of bacterial colonies.

differential media and tests to complete the identification of the isolates. In order to better illustrate this approach to bacterial identification, Table 1-8 lists some of the more commonly encountered colonial types, the group of bacteria to suspect for each, additional tests required for definitive identification, and reference to the exact frame in Plate 1-2 where these colony types are illustrated.

The initial inspection of colonies for preliminary identification of bacteria is one of the cornerstones of diagnosic microbiology and will be discussed in detail in later chapters devoted to specific groups of pathogenic bacteria and other microorganisms.

Gram's Stain Reaction and Cellular Morphology

The preliminary impressions based on the observation of colony characteristics can be further confirmed by studying gram-stained smears, a technique that is relatively simple to perform. The center of the colony to be studied is first touched with the end of a straight inoculating wire (see Fig. 1-31). The portion of the colony to be sampled is emulsified in a small drop of water or physiologic saline on a microscope slide to disperse the individual bacterial cells (see Fig. 1-32). After the slide has air dried, the bacterial film is fixed to the glass surface by quickly

Table 1-8. *Preliminary Bacterial Identification by Colonial Types*

Colonial Type	Bacterial Group	Additional Tests	Illustrated in Color Plate 1-2
Convex, entire edge, 2 to 3 mm., creamy, yellowish, zone of beta hemolysis	Staphylococcus	Catalase Coagulase DNase Mannitol utilization Tellurite reduction	Frame: A, B, C
Convex or pulvinate, translucent, pinpoint in size, butyrous, wide zone of beta hemolysis	Streptococcus	Catalase "A" disk 6.5% NaCl tolerance Bile-esculin CAMP test Hippurate hydrolysis	Frame: D, E
Umbilicate or flat, translucent, butyrous or mucoid, broad zone of alpha hemolysis	Pneumococcus	"P" disk Bile solubility	Frame: F

Pulvinate, semi-opaque, gray, moist to somewhat dry. Beta hemolysis may or may not be present.	*Escherichia coli* and Enterobacteriaceae	Multiple tests:
		  Indole
		  Citrate
		  Decarboxylase
		  Urease
		(See Chap. 5)
			Frame: G

Flat, opaque, gray to greenish, margins erose or spreading, greenblue pigment, grapelike odor	Pseudomonas	Cytochrome oxidase Fluorescence Growth at 42° C	Frame: H
Flat, gray; spreading as thin film over agar surface; burned chocolate odor	Proteus	Phenylalanine deaminase Urease Lysine deaminase	Frame: I

passing the slide 4 or 5 times through the flame of a Bunsen burner. The fixed smear is placed on a staining rack and the Gram's stain is performed as described on page 31.

The stained smear should be examined microscopically using an oil immersion objective. In addition to the Gram's stain reaction of the bacterial cells (gram-positive bacteria appear blue, gram-negative bacteria appear red or pink), three other characteristics helpful in making a preliminary identification of isolates include:

1. The size and shape of the bacterial cells
2. The arrangement of the bacterial cells
3. The presence of specific structures or organelles (spores, metachromatic granules, swollen bodies, *etc.*)

The microbiologist should evaluate each of these characteristics in making a preliminary identification of bacterial isolates. A series of photomicrographs of several Gram's stains illustrating a number of the morphologic cell types and spatial arrangements of bacteria commonly encountered in clinical laboratories is shown in Plate 1-3.

From the information derived from the examination of bacterial colonies and gram-stained smears of the cells, the microbiologist is able to proceed a long way toward the identification of isolates without performing differential tests. For example, a raised, creamy, yellow hemolytic colony on blood agar that shows gram-positive cocci in clusters in the Gram's stain is suggestive of staphylococcus. A pinpoint translucent beta-hemolytic colony on blood agar showing gram-positive cocci in chains is most probably a streptococcus.

The microbiologist soon learns, however, not to rely solely on the examination of gram-stained smears, in that the staining reactions may be variable, particularly with very young or older colonies. Also, the clumping and chaining of gram-positive cocci may be less pronounced in colonies picked from an agar surface than in smears made from broth cultures. These variabilities must be taken into account when evaluating Gram's stains.

Nevertheless, microbiologists should make known to the physician as much preliminary information as possible. In selected cases, such as the observation of bacteria in a blood culture broth or directly in infected cerebrospinal fluid, this type of preliminary information can be quite useful in directing specific antibiotic therapy, prior to the time that final species identification or antimicrobial susceptibility test results are available.

Preliminary Identification Based on Metabolic Characteristics

Most tests used to assess the biochemical or metabolic activity of bacteria, by which a final species identification can be made, are usually performed by subculturing the primary isolate to a series of differential tests that can then be interpreted after one or more days of additional incubation. Preliminary observations can be made when using certain primary isolation media, or a limited number of tests may be performed directly on colonies recovered. For example, the lactose-utilizing properties of the Enterobacteriaceae can be directly evaluated from MacConkey agar by observing the red discoloration of the colony; H_2S production may be detected on salmonella-shigella (SS) agar by detecting colonies with black centers; and lysine decarboxylase activity can be suspected in the presence of red colonies on xylose, lysine, dextrose (XLD) agar.

Direct tests that can be performed on isolated colonies recovered on primary culture plates include:

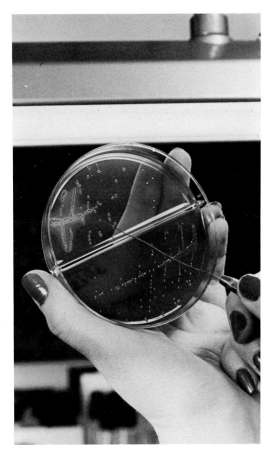

Fig. 1-31. "Picking" an isolated bacterial colony using a straight inoculating wire.

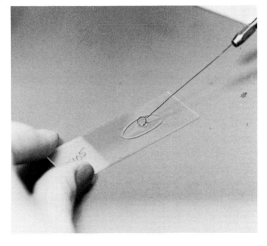

Fig. 1-32. Preparation of a smear for Gram's stain. A portion of the bacterial colony to be studied is sampled with an inoculating loop or needle and emulsified in a drop of water or saline on a 3" × 1" glass microscope slide.

Table 1-9. *Select Bacterial Groups and Commonly Measured Differential Characteristics*

Bacterial Group Suspected from Gross Colony and Gram's Stain Morphology	Differential Characteristics Commonly Measured
Staphylococci	Catalase production
	Cytochrome oxidase activity
	Oxidative or fermentative (O-F) glucose utilization
	Mannitol fermentation
	Coagulase production
	DNase activity
Streptococci including *S. pneumoniae*	Catalase production
	Bacitracin susceptibility ("A" disk)
	6.5% NaCl tolerance
	Reaction on bile-esculin
	Sodium hippurate production
	Bile solubility
	Optochin sensitivity ("P" disk)
Enterobacteriaceae	Lactose utilization
	Nitrate reduction
	Cytochrome oxidase activity
	Decarboxylase or dehydrolase: lysine, ornithine, and arginine
	Hydrogen sulfide production
	Indole production
	Citrate utilization
	Methyl red reaction
	Motility
	Orthonitrophenyl galactosidase activity (ONPG test)
	Production of acetyl-methyl carbinol (Voges-Proskauer reaction)
	Urease production
Nonfermentative gram-negative bacilli	Oxidative or fermentative utilization of glucose (O-F)
	Growth on MacConkey agar
	Cytochrome oxidase activity

1. *The Catalase Test.* A few drops of hydrogen peroxide are placed directly on the colony. Rapid effervescence indicates hydrogen gas production and a positive test. (see Chart 7-1).

2. *The Bile Solubility Test.* A few drops of sodium desoxycholate are placed on colonies suspected of being *Streptococcus pneumoniae.* Pneumococcal colonies are completely lysed and disappear after about 30 minutes (see Chart 7-7).

3. *The Slide Coagulase Test.* The colony suspected of being a staphylococcus species is emulsified in a drop of rabbit plasma on a glass slide. The presence of bacterial clumping within one minute indicates the presence of bound coagulase and is a positive test (see Chart 7-2).

4. *Direct Spot Indole Test.* A small portion of the colony to be tested is transferred to a strip of filter paper that has been saturated with Kovac's reagent. The immediate development of a red color indicates the presence of indole and a positive test (see Chart 2-4).

5. *The Cytochrome Oxidase Test.* A portion of the colony to be tested is smeared on the reagent-impregnated area of an oxidase test strip. The immediate development of a blue color in the zone of smearing indicates cytochrome oxidase activity and a positive test (see Chart 2-1).

6. *Direct Serologic Typing.* The unknown bacterial suspension is emulsified with a drop of specific antiserum on a glass slide. Bacterial agglutination indicates a positive test.

The techniques and interpretations of the

Table 1-9. *Select Bacterial Groups and Commonly Measured Differential Characteristics*
(Continued)

Bacterial Group Suspected from Gross Colony and Gram's Stain Morphology	Differential Characteristics Commonly Measured
Nonfermentative gram-negative bacilli *(Continued)*	Motility
	Nitrate reduction
	Denitrification of nitrates and nitrites
	Gluconate reduction
	Fluorescein pigment production
	Pigment production
	Lysine decarboxylase activity
	Utilization of 10% lactose and glucose
	Penicillin sensitivity
	Catalase production
Neisseria	Cytochrome oxidase activity
Branhamella	Nitrate reduction:
Moraxella	Carbohydrate utilization
	Glucose
	Maltose
	Sucrose
	Lactose
	Growth on modified Thayer-Martin medium
Hemophilus	Catalase production
	Cytochrome oxidase activity
	Growth on sheep blood agar
	Nitrate reduction
	Growth requirements for CO_2
	Indole production
	Utilization of sucrose
	Requirements for growth factors:
	X (Hemin)
	V (NAD)

above procedures will be discussed in greater detail in later chapters.

Bacterial Species Identification and Selection of Differential Identification Characteristics

Preliminary bacterial identification can be made by observing colonial characteristics and Gram's stain morphology as discussed above. However, the final characterization of an unknown bacterial isolate into genus and species is usually accomplished by the detection of certain enzyme systems that are unique to each species and serve as identification markers. In the laboratory, these enzyme systems are detected by inoculating a small portion of a well-isolated bacterial colony into a series of culture media containing specific substrates and chemical indicators which detect pH changes or the presence of specific by-products. The clinical microbiologist must select the appropriate set of characteristics that are required for the identification of each group of bacteria.

For the final identification of bacteria within any given group, several dozen characteristics are available from which to choose. Table 1-9 lists several groups of bacteria commonly encountered in the clinical laboratory and the characteristics often selected for their identification and species differentiation. The various schemas by which these characteristics are used in the identification of the several bacterial groups listed will be discussed in detail in subsequent chapters.

QUALITY CONTROL

Quality control in the laboratory is a systematic assessment of the laboratory work being performed to ensure that the final product has an acceptable degree of conformity within previously established tolerance limits.

Space does not allow for a full discussion in this text of quality control as it relates to microbiology. Rather, only a brief overview outline can be included. A number of resources dealing with this subject are listed in the references at the end of this chapter for the reader desiring a broad overview.[2,4,5,13,17]

Implementation of quality control in clinical laboratories did not occur until the late 1950's when analytical instruments of increasing complexity were becoming part of the laboratory scene. Applications to microbiology were delayed until the late 1960's, primarily because there are only a few items in the microbiology laboratory that can be accurately measured. Most micorbiology results are derived from a series of observations requiring interpretative judgments that cannot be quantitated. The calculation of standard deviations and coefficients of variation, so much a part of the strictly analytical functions of the laboratory, have precious few applications in microbiology.

Nevertheless, all microbiology laboratories, large and small, must establish some means for continually monitoring the quality of work being performed. Supervisors of low-volume laboratories may decide to limit the types of procedures or services offered; however, the procedures selected must be controlled within the same tolerance limits as similar procedures performed in large laboratories.

The following outline of a basic microbiology quality control program lists a number of specific items that must be considered when implementing the various phases of the program. Bartlett,[5] in his outline on developing a quality control program, discusses different levels of activity, ranging from basic, to more advanced, to most advanced. Using this outline, a supervisor can select the level of activity that is appropriate for the volume of work and personnel available to any given laboratory.

The Commission on Laboratory Inspection and Accreditation of the College of American Pathologists has established standards for accreditation of medical laboratories, including an inspection checklist for microbiology laboratories.[11] This checklist provides microbiology supervisors with a valuable guideline to follow in making a point-by-point assessment of the quality control needs in the laboratory being served.

Microbiology Quality Control Program

At the outset, a quality control coordinator must be selected. The duties of the coordinator must be clearly established and authority conferred to the extent that problems can be efficiently handled when they arise. It is the coordinator's responsibility to establish the minimal standards for quality control that are to be met by the laboratory and to outline the several steps to be taken for daily monitoring and surveillance of all facets of the program.

Developing a Procedure Manual

One of the most important documents for directing the day-by-day activities in the microbiology laboratory is an up-to-date procedure manual. All of the various activities of the laboratory should be clearly outlined, bound into one or more volumes, and placed in an accessible part of the laboratory for ready reference by all employees at all times.

The exact order in which the material is to appear in the manual must be determined by the microbiology supervisor to best meet the needs of the laboratory. Following are items that should appear in all procedure manuals:

1. Names, addresses, and telephone numbers of the laboratory director, staff

pathologists, supervisors, and all employees

2. List of all general policies and regulations of the microbiology laboratory

3. List of the exact locations of equipment, media, reagents, and supplies, particularly if the laboratory is covered by part-time personnel on evenings or weekends

4. Complete description of all forms, reports, and files used in the microbiology laboratory

5. Detailed descriptions of all techniques and procedures that are performed in the laboratory

6. List of all media and reagents used, including full descriptions of their formulations and instructions for preparation

7. List of all identification schemas used in the identification and classification of microorganisms

8. Names, addresses, telephone numbers, and the procedures and policies of reference laboratories pertaining to the shipment of reference samples

9. Inclusion of all quality control procedures, with specific details on the frequency and manner with which each item is to be carried out

Current laboratory inspection guidelines require that the procedure manual be revised and updated at least once a year, and the initials of the laboratory director or supervisor must appear on each page, indicating that the update has been accomplished.

Monitoring of Laboratory Equipment

All electrical or mechanical equipment items should be included in a preventative maintenance program that provides a check on all functions at a prescribed time interval. Certain working parts should be replaced after a specified time of use even though visually they may not appear worn. A brief listing of some of the equipment items, the monitoring procedure to be carried out, and the frequency and tolerance limits is shown in Table 1-10. Assignments should be made among laboratory personnel to see that these functions are carried out, and all data should be recorded on charts or in maintenance manuals in such a way that upward or downward trends can be immediately detected and appropriate corrective action taken before serious errors result.

Monitoring of Culture Media, Reagents, and Other Supplies

Even though many expendable supplies procured from commercial sources have been quality controlled by the manufacturer, in many instances a repeat assessment must also be made in the laboratory prior to use of such supplies. Culture medium is a prime example, requiring that each batch of new media be selectively tested with stock organisms so that both positive and negative reactions can be observed. A suggested list of organisms and acceptable results for the culture media most commonly used in clinical laboratories is found in Table 1-11. Quality control stock organisms may be maintained in the laboratory by subculturing bacterial isolates recovered as part of the routine work, or dried stock organisms can be obtained from the American Type Culture Collection (ATCC),* Difco Laboratories (Bactrol disks),† or Roche Diagnostics (Bact-Check disks).‡

Each culture tube, plate of medium, and reagent must bear a label that clearly indicates the content, date of preparation, and date of expiration.

All antimicrobial susceptibility disks must be tested with organisms of known susceptibility each time that a new run is made (usually daily). *Escherichia coli* (ATCC 25922), *Staphylococcus aureus* (ATCC 25923), and *Pseudomonas aeruginosa* (ATCC 27853) serve as the standard control

(Text continues on p. 52)

*American Type Culture Collection, Baltimore, Md.
†Difco Laboratories, Detroit, Mich.
‡Roche Diagnostics, Division of Hoffmann-LaRoche, Inc., Nutley, N.J.

Table 1-10. *Quality Control Surveillance Procedures of Commonly Used Microbiology Equipment*

Equipment	Procedure	Schedule	Tolerance Limits
Refrigerators	Recording of temperature*	Daily or continuous	2° to 8° C.
Freezers	Recording of temperature	Daily or continuous	−8° to −20° C.
Incubators	Recording of temperature	Daily or continuous	35.5° C. + or − 1° C.
Incubators (CO_2)	Measuring of CO_2 content. Use blood gas analyzer or Fyrite† device.	Daily or twice daily	5% to 10%
Water baths	Recording of temperature	Daily	36° to 38° C. 55° to 57° C.
Heating blocks	Recording of temperature	Daily	Within + or −1° C. of setting
Autoclaves	Test with spore strip (*Bacillus stearothermophilus*).	At least weekly	No growth of spores in subculture indicates sterile run.
pH Meter	Test with *pH*-calibrating solutions.	Each time of use	Within + or − 0.1 *pH* units of standard being used
Anaerobic jars	Methylene blue strip	With each use	Conversion of strip from blue to white indicates low O_2 tension.
	Clostridium novyi culture, Type B	Run periodically	Growth indicates very low O_2 tension. It is used only where extremely low O_2 tension is required.
Serology rotator	Count revolutions/min.	Each use	180 r.p.m. + or − 10 r.p.m.
Centrifuges	Check revolutions with tachometer.	Monthly	Within 5% of dial indicator setting
Safety hoods	Measure air velocity‡ across face opening.	Semiannual or quarterly	50 feet of air flow per minute + or − 5 feet/min.

*Each monitoring thermometer must be calibrated against a standard thermometer.
†Bacharach Instrument Co., Pittsburg, PA 15238
‡Velometer Jr., Alnor Instrument Co., Chicago, Ill.

Table 1-11. *Quality Control of Commonly Used Media; Suggested Control Organisms and Expected Reactions*

Medium	Control Organisms	Expected Reactions
Blood agar	Group A streptococcus	Good growth, beta hemolysis
	Streptococcus pneumoniae	Good growth, alpha hemolysis
Bile-esculin agar	Enterococcus species	Good growth, black color
	Alpha-hemolytic streptococcus, not Group D	No growth; no discoloration of media
Chocolate agar	*Hemophilus influenzae*	Good growth
	Neisseria gonorrhoeae	Good growth
Christensen urea agar	*Proteus mirabilis*	Pink color throughout (positive)
	Klebsiella pneumoniae	Pink slant (partial positive)
	Escherichia coli	Yellow color (negative)
Simmons citrate agar	*Klebsiella pneumoniae*	Growth or blue color (positive)
	Escherichia coli	No growth, remains green (negative)
Cysteine-trypticase agar (CTA):		
Dextrose	*Neisseria gonorrhoeae*	Yellow color (positive)
	Branhamella catarrhalis	No color change (negative)
Sucrose	*Escherichia coli*	Yellow color (positive)
	Neisseria gonorrhoeae	No color change (negative)

(Continued)

Table 1-11. *Quality Control of Commonly Used Media; Suggested Control Organisms and Expected Reactions* (Continued)

Medium	Control Organisms	Expected Reactions
Maltose	Salmonella species, or *Neisseria meningitidis*	Yellow color (positive)
	Neisseria gonorrhoeae	No color change (negative)
Lactose	*Neisseria lactamicus*	Yellow color (positive)
	Neisseria gonorrhoeae	No color change (negative)
Decarboxylases:		
Lysine	*Salmonella typhimurium*	Bluish color (positive)
	Shigella flexneri	Yellow color (negative)
Arginine (dihydrolase)	*Enterobacter cloacae*	Bluish color (positive)
	Proteus mirabilis	Yellow color (negative)
Ornithine	*Proteus mirabilis*	Bluish color (positive)
	Klebsiella pneumoniae	Yellow color (negative)
Deoxyribonuclease (DNase)	*Serratia marcescens*	Zone of clearing (Add 1 N HCl)
	Enterobacter cloacae	No zone of clearing
Eosin–methylene blue agar	*Escherichia coli*	Good growth, green metallic sheen
	Klebsiella pneumoniae	Good growth, purple colonies, no sheen
	Shigella flexneri	Good growth, transparent colonies (lactose-negative)
Hektoen enteric agar	*Salmonella typhimurium*	Green colonies with black centers
	Shigella flexneri	Green transparent colonies
	Escherichia coli	Growth slightly inhibited, orange colonies
Indole (Kovac's)	*Escherichia coli*	Red color (positive)
	Klebsiella pneumoniae	No red color (negative)
Kligler iron agar	*Escherichia coli*	Acid slant/acid deep
	Shigella flexneri	Alkaline slant/acid deep
	Pseudomonas aeruginosa	Alkaline slant/alkaline deep
	Salmonella typhimurium	Alkaline slant/black deep
Lysine iron agar	*Salmonella typhimurium*	Purple deep and slant, $+H_2S$
	Shigella flexneri	Purple slant, yellow deep
	Proteus mirabilis	Red slant, yellow deep
MacConkey agar	*Escherichia coli*	Red colonies (lactose-positive)
	Proteus mirabilis	Colorless colonies, no spreading
	Enterococcus species	No growth
Malonate	*Escherichia coli*	No growth
	Klebsiella pneumoniae	Good growth, blue color (positive)
Motility (semisolid agar)	*Proteus mirabilis*	Media cloudy (positive)
	Klebsiella pneumoniae	No feather edge on streak line (negative)
Nitrate broth or agar	*Escherichia coli*	Red color on adding reagents
	Acinetobacter lwoffi	No red color (negative)
Phenylethyl alcohol blood agar	Streptococcus species	Good growth
	Escherichia coli	No growth
ONPG	*Serratia marcescens*	Yellow color (positive)
	Salmonella typhimurium	Colorless (negative)
Phenylalanine deaminase	*Proteus mirabilis*	Green color (add 10% $FeCl_3$)
	Escherichia coli	No green color (negative)
Salmonella-shigella agar (SS)	*Salmonella typhimurium*	Colorless colonies, black centers
	Escherichia coli	No growth
Voges-Proskauer	*Klebsiella pneumoniae*	Red color (add reagents)
	Escherichia coli	No color development (negative)
Xylose, lysine, dextrose agar (XLD)	Salmonella species	Red colonies (positive lysine)
	Escherichia coli	Yellow colonies (positive sugars)
	Shigella species	Transparent colonies (negative)

Table 1-12. *Quality Control of the Bauer-Kirby Antimicrobial Susceptibility Test; Acceptable Ranges of Zone Diameters in Millimeters**

Antimicrobial Agent	Disk Content	E. coli	S. aureus	P. aeruginosa
Penicillins:				
Ampicillin	10 μg.	15–20	24–35	—
Carbenicillin	100 μg.	24–29	—	19–25
Methicillin	5 μg.	—	17–22	
Penicillin G	10 units	—	26–37	
Aminoglycosides:				
Gentamicin	10 μg.	19–26	19–27	16–22
Kanamycin	30 μg.	17–25	19–26	
Tobramycin	10 μg.	18–26	19–29	19–25
Cephalothin	30 μg.	18–23	25–37	
Chloramphenicol	30 μg.	21–27	19–26	
Clindamycin	2 μg.	—	23–29	
Erythromycin	15 μg.	—	23–30	
Neomycin	30 μg.	17–23	18–26	
Polymyxin B	300 units	12–16	—	13–18
Streptomycin	10 μg.	12–20	14–22	
Tetracycline	30 μg.	18–25	19–28	
Trimethoprim-	1.25 μg.	24–32	24–32	
Sulfamethoxazole	23.75 μg.			

*Area Committee for Microbiology Performance Standards, established by the National Committee for Clinical Laboratory Standards (NCCLS), Villanova, Pa.

organisms. The acceptable zone diameter ranges against these organisms for the more commonly used antibiotics is shown in Table 1-12.

Laboratory Safety

Laboratory safety is not often thought of as a part of quality control, yet maintenance of good safety practices has an important influence on the overall performance and productivity of the laboratory. If equipment or instruments are damaged because of fire or electrical accidents, or if personnel are injured, there may be a serious interruption in work flow and a delay in producing laboratory results.

One individual in the laboratory should be designated as the safety officer whose duty it shall be to see that the laboratory is in compliance with all electrical and fire codes and who shall instruct the laboratory staff in safety principles and in eliminating potential hazards.

Safety items that directly involve laboratory personnel include rules against mouth pipetting of reagents and serum, smoking, eating, drinking, or applying cosmetics in the laboratory area, and other practices known to be unsafe. Face shields or eye protectors should be worn when handling or mixing caustic chemicals, and asbestos gloves should be worn when handling hot implements or glassware. Eye baths should be installed in accessible locations in the laboratory.

Fire drills should be conducted frequently. All laboratory personnel should know the location and proper use of fire extinguishers and fire blankets. All flammable liquids must be stored in explosion-proof cabinets and kept away from open flames during use.

One of the hospital electricians should be invited to inspect the electrical wiring of the laboratory. All electrical equipment must be properly grounded and care must be taken not to overload circuits. Extension cords and the use of household appliances such as coffee pots or electrical heaters should bring to the attention of the supervisor any electrical heaters should be strictly forbidden.

Laboratory personnel should bring to the attention of the supervisor any electrical shocks incurred in the use of any piece of equipment. Instrument repair should never be attempted while any device is connected to the electrical circuit.

All tanks of compressed gases must be securely chained to the wall and clearly marked with the name of their contents. All corrosive or caustic chemicals should be designated with a brightly colored label and stored in closed cabinets near the floor. Corrosive solutions that are poured in the sink must be flushed with copious amounts of water, both before and after actual disposal. Rules must be established for the disposal of solid wastes, including hypodermic needles, which should be placed into sealed containers labeled with a "danger" or "hazardous materials" sign. All spent culture media or serum samples suspected of harboring an infectious disease, such as serum hepatitis, must be autoclaved before final disposal.

The discussion of many other safety items for the laboratory is beyond the scope of this text and the reader is referred to a number of safety pamphlets that have been prepared by the Commission of Laboratory Inspection and Accreditation of the College of American Pathologists in Skokie, Illinois.

Laboratory Personnel and Quality Control

Accredited laboratories must be directed by a pathologist, physician, or an individual with a doctoral degree in a specific area of laboratory science. Laboratories must be staffed at all times with a qualified supervisor with at least 4 years of laboratory experience.

Quality control of personnel requires an effective continuing education program. In-service training must be an ongoing activity. Personnel should be encouraged to participate as often as possible in local, regional, and national seminars and workshops. All laboratories should participate in one or more of the available proficiency test services, and these should be used for teaching

exercises. Blind, unknown samples for laboratory testing should be periodically circulated with test runs and any discrepancies in the results that are discovered should be openly discussed and the sources of error pinpointed and corrected. Proficiency-testing programs should be particularly made available to personnel who work on evenings or weekends, and provisions must be made that their test results are reviewed by an individual on the regular shift to see that errors are not being made. Supervisory personnel should check all results for accuracy, reproducibility, and compliance with all quality control standards.

In a broad sense, quality control in microbiology is more of an art than a science. It involves intangible items such as common sense, good judgment, and paying constant attention to details. Programs should be organized with well-defined objectives in mind. In the end, high-level laboratory performance requires an alert, interested, and well-motivated laboratory staff.

REFERENCES

1. Attebery, H. R., and Finegold, S. M.: A miniature anaerobic jar for tissue transport or for cultivation of anaerobes. Am. J. Pathol. Clin. *53:*383–388, 1970.
2. Barry, A. L., Moronde, G. R., and Beckley, E. A.: Methods for quality control in the clinical bacteriology laboratory. Nutley, N. J., Roche Diagnostics, Division of Hoffmann-Laroche, Inc. 1973.
3. Bartlett, J. G., *et al.:* Should fiberoptic bronchoscopy aspirates be cultured? Am. Rev. Respir. Dis., *114:*73–78, 1976.
4. Bartlett, R. C.: A plea for clinical relevance in microbiology. Am. J. Clin. Pathol., *61:*867–872, 1974.
5. ——— Medical Microbiology: Quality Cost and Clinical Relevance. New York, John Wiley & Sons, 1974.
6. Bartlett, R. C., Ellner, P. D., and Washington, J. A.: Blood cultures. *In* Sherris, J. C. (ed.): Cumulative Techniques and Procedures in Clinical Microbiology. Washington, D.C., American Society for Microbiology, 1974.
7. Dowell, V. R., Jr., and Hawkins, T. M.: Laboratory methods in anaerobic bacteriology. CDC laboratory manual. DHEW publi-

cation No. (CDC) 74-8272, Center for Disease Control, Atlanta, 1974.

8. Finegold, S. M.: Anaerobic bacteria in human disease. New York, Academic Press, 1977.

9. Hoeprich, P. D.: Etiologic diagnosis of lower respiratory tract infections. Calif. Med., *112*:108, 1970.

10. Holdeman, L. V., and Moore, W. E. C. (eds.): Anaerobe Laboratory Manual. ed. 3. Blacksburg, Va., Virginia Polytechnic Institute and State University, 1975.

11. Inspection Checklist: Section IV, Microbiology. Skokie, Ill., Commission on Laboratory Inspection and Accreditation, College of American Pathologists, 1974.

12. Kaye, D.: Antibacterial activity of human urine. J. Clin. Invest., *47*:2374–2390, 1968.

13. Koneman, E. W.: Quality control in microbiology. *In* Listen, Look and Learn: Clinical Microbiology. Chap. 3. Bethesda, Md., Health and Education Resources, 1977.

14. Kunin, C. M.: Detection, Prevention and Management of Urinary Tract Infections: A Manual for the Physician, Nurse and Allied Health Worker. ed. 2. Philadelphia, Lea & Febiger, 1974.

15. Lorian, V. (ed.): Medical Microbiology in Care of Patients. Baltimore, Williams & Wilkins, 1977.

16. Murray, P. R., and Washington, J. A., II: Microscopic and bacteriologic analysis of expectorated sputum. Mayo Clin. Proc., *50*:339–344, 1975.

17. Prier, J. E., Bartola, J., and Friedman, H.: Quality Control in Microbiology. Baltimore, University Park Press, 1975.

18. Quality Control of Culture Media. Technical Information Bulletin, Difco Laboratories, Detroit, 1974.

19. Rein, M. F., and Mandell, G. L.: Bacterial killing by bacteriostatic saline solutions–potential for diagnostic error. N. Engl. J. Med., *289*:794–795, 1973.

20. Reller, L. B.: Blood Culture Techniques: Critical Factors in Detection of Bacteremia in Cultures. Research Study, University of Colorado School of Medicine, Department of Internal Medicine. *Personal Communication.*

21. Sutter, V. L.; Vargo, V. L., and Finegold, S. M.: Wadsworth Anaerobic Bacteriology Manual. ed. 2. Los Angeles, Wadsworth Hospital Center, Veterans Administration and the Department of Medicine, U.C.L.A. School of Medicine, 1975.

22. Van Scoy, R. E.: Bacterial sputum cultures: a clinician's viewpoint. Mayo Clin. Proc., *52*:39–41, 1977.

23. Washington, J. A., II: Anaerobic blood cultures. *In* Lennette, E. H., Spaulding, E. H., and Truant, J. P., (eds.): Manual of Clinical Microbiology. ed. 2, Chap. 43. Washington, D.C., American Society for Microbiology, 1974.

2 The Enterobacteriaceae

Gram-negative bacilli belonging to the family Enterobacteriaceae are the most frequently encountered microorganisms in the clinical microbiology laboratory. As the family name implies, many members of this group are indigenous to the gastrointestinal tract. Prior to the advent of antibiotics, chemotherapy, and immunosuppressive measures, the members of this group were essentially limited to causing diseases of the gastrointestinal and urinary tracts. Today members of the Enterobacteriaceae may be recovered from infections of virtually every anatomical site, and the microbiologist must be prepared to isolate these organisms from any culture sent to the laboratory.

Gram's stain morphology is neither helpful in separating members of the Enterobacteriaceae from other gram-negative bacilli nor in making species identifications. The morphology of colonies growing on blood agar is also of limited diagnostic usefulness because most species appear as dull gray, dry to mucoid colonies (Plate 2-1, A and B). Some species of Proteus may grow diffusely over the agar surface as a thin film, a phenomenon called "swarming" (Plate 2-1, C).

BIOCHEMICAL SPECIES IDENTIFICATION

Classification of the Enterobacteriaceae is based primarily on the determination of the presence or absence of different enzymes coded by the genetic material of the bacte-rial chromosome. These enzymes direct the metabolism of bacteria along one of several pathways that can be detected by special media used in *in vitro* culture techniques. Substrates upon which these enzymes can react are incorporated into the culture medium, together with an indicator system that can detect either the decay of the substrate or the presence of specific metabolic products. By selecting a series of media that measure different metabolic characteristics of the microorganism to be tested, a biochemical "fingerprint" can be determined for making a species identification.

SCREENING PRELIMINARY CHARACTERISTICS FOR THE FAMILY ENTEROBACTERIACEAE

Since the identification of the members of the Enterobacteriaceae may require a somewhat complex series of biochemical tests, considerable time and a potential misidentification can be avoided if a few preliminary observations are made to ensure that the organism under test belongs to this group. If the organism is a gram-negative bacillus of another group, it may be necessary to measure a totally separate set of characteristics, and those commonly used for the identification of the Enterobacteriaceae may be inappropriate or misleading. With few exceptions, all members of the Enterobacteriaceae show the following characteristics:

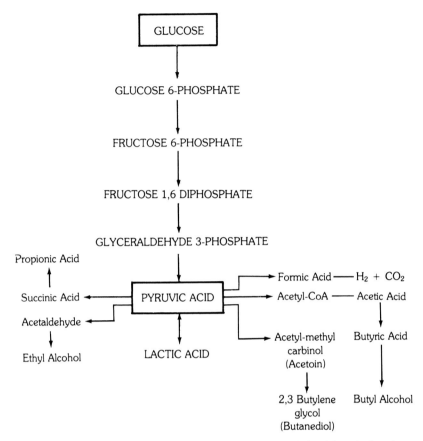

Fig. 2-1. Mixed acid fermentation of glucose via the Embden-Meyerhof pathway.

- Glucose is metabolized fermentatively (Plate 2-1, *D, E, F,* and *G*).
- Cytochrome oxidase activity is lacking (Plate 2-1, *H*).
- Nitrates are reduced to nitrites (Plate 2-1, *I*).

Fermentative Metabolism of Glucose

A variety of different growth media, either in broth or in agar, can be used to measure the capability of a test organism to utilize carbohydrates fermentatively. The principle of carbohydrate fermentation is based on Pasteur's yeast studies of over 100 years ago—that the action of many species of microorganisms on a carbohydrate substrate results in acidification of the medium.

The formulation of a typical fermentation medium is:

Trypticase (BBL)	10.000 g.
Sodium chloride	5.000 g.
Phenol red	0.018 g.
Distilled water to:	1000 ml.

The carbohydrate to be tested, glucose for instance, is filter-sterilized and added to the basal medium in a 0.5 to 1.0 per cent final concentration. Trypticase is a protein hydrolysate that serves as a source for carbon and nitrogen, sodium chloride is added as an osmotic stabilizer, and phenol red is a *p*H indicator that turns yellow when the *p*H of the medium drops below 6.8 (Plate 2-1, *F* and *G*).

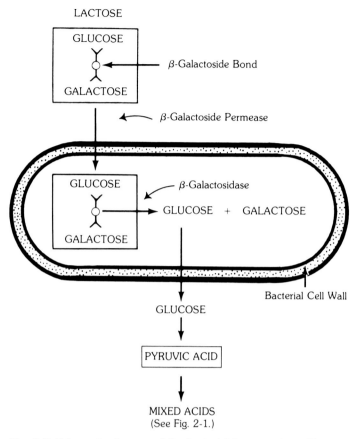

Fig. 2-2. Schematic diagram of the bacterial fermentation of lactose. Lactose, a disaccharide composed of molecules of glucose and galactose joined by a β-galactoside bond, diffuses through the bacterial cell wall under the action of β-galactoside permease. If the bacterium produces β-galactosidase, the lactose is hydrolyzed to produce glucose and galactose. The glucose is then metabolized as illustrated in Fig. 2-1.

All of the Enterobacteriaceae grow well in this type of medium, and the base formulation used is a matter of personal preference. Briefly, glucose fermentation follows the anaerobic Embden-Meyerhof-Parnas (EMP) pathway leading to the formation of pyruvic acid, from which a variety of organic acids are derived (Fig. 2-1). All Enterobacteriaceae, by definition, ferment glucose via this pathway, producing a mixed acid fermentation and a yellow color in a medium using phenol red as the *p*H indicator.

In practice, microorgansims that are incapable of fermenting glucose are most commonly detected by observing the reactions they produce when growing on Kligler iron agar or triple sugar iron agar. An alkaline-slant/alkaline-deep reaction (Fig. 2-3) on either of these media indicates the absence of acid production and the inability of the test organism to ferment the glucose and other carbohydrates present. This reaction alone is sufficient to exclude an organism from the family Enterobacteriaceae.

Chart 2-1. *Cytochrome Oxidase*

Introduction	The cytochromes are iron-containing hemoproteins that act as the last link in the chain of aerobic respiration by transferring electrons (hydrogen) to oxygen, with the formation of water. The cytochrome system is possessed of aerobic or facultatively anaerobic organisms, so that the oxidase test is important in identifying those organisms that either lack the enzyme or are obligate anaerobes. The test is most helpful in screening colonies suspected of being one of the Enterobacteriaceae (all negative) and in identifying colonies suspected of being a pseudomonas species or a neisseria species (positive).
Principle	The cytochrome oxidase test utilizes certain reagent dyes, such as p-phenylenediamine dihydrochloride, that substitute for oxygen as artificial electron acceptors. In the reduced state the dye is colorless; however, in the presence of cytochrome oxidase and atmospheric oxygen, p-phenylenediamine is oxidized, forming indophenol blue.

Media and Reagents	1. Tetramethyl-p-phenylenediamine dihydrochloride, 1% (Kovac's reagent)
	2. Dimethyl-p-phenylenediamine dihydrochloride, 1% (Gordon and McLeod's reagent)
	3. Commercial disks and strips:
	    A. Difco Laboratories: Bacto differentiation disks, oxidase
	    B. BBL: Taxo N disks
	    C. General Diagnostics: PathoTec oxidase strips

Procedure	The test is commonly performed by one of two methods: (1) the direct plate technique, in which 2 to 3 drops of reagent are directly added to isolated bacterial colonies growing on plate medium; and (2) the indirect paper strip

Cytochrome Oxidase Activity

Any organism that displays cytochrome oxidase activity when subjected to a test for cytochrome oxidase is *also excluded* from the family Enterobacteriaceae. The details of cytochrome oxidase and the test used for its detection are presented in Chart 2-1.

The commercial cytochrome oxidase disks or strips are most commonly used because of their convenience. The reaction colors are clearly visible within a few seconds and the detection of cytochrome oxidase is both selective and sensitive for the bacterial species most commonly encountered in clinical microbiology. Platinum inoculating loops or wires should be used instead of stainless steel because trace amounts of iron oxide on the flamed surface of stainless steel may produce false positive reactions. Since the oxidase test is so simple to perform, it is recommended that any colony suspected of being one of the Enterobacteriaceae, presenting as a nonlactose fermenter on a selective agar plate, be tested for cytochrome oxidase activity before choosing the set of differential test media to be used.

Nitrate Reduction

All Enterobacteriaceae with the exception of certain biotypes of *Enterobacter*

Chart 2-1. *Cytochrome Oxidase* (Continued)

Procedure *(Continued)*	procedure, in which a few drops of the reagent are either added to a filter paper strip, or commercial disks or strips impregnated with dried reagent are used. In either method, a loopful of suspected colony is smeared into the reagent zone of the filter paper.
Interpretation	Bacterial colonies having cytochrome oxidase activity will develop a deep blue color at the inoculation site within seconds (Plate 2-1, *H*). Stainless steel inoculating loops or wires should not be used for this test because surface oxidation products formed when flame sterilizing may result in false positive reactions. The tetramethyl derivative of p-phenylenediamine is recommended in that the reagent is more stable in storage and it is more sensitive to the detection of cytochrome oxidase and less toxic than the dimethyl derivative.
Controls	Bacterial species showing positive and negative reactions should be run as controls at frequent intervals. The following can be suggested: Positive control: *Pseudomonas aeruginosa* Negative control: *Escherichia coli*
Bibliography	Gordon, J., and McLeod, J. W.: The practical application of the direct oxidase reaction in bacteriology. J. Pathol. Bacteriol., *31*:185–190, 1928. MacFaddin, J. F.: Biochemical Tests for Identification of Medical Bacteria. Baltimore, Williams & Wilkins, 1975; pp. 154–162, 1976. Steel, K. J.: The oxidase reaction as a taxonomic tool. J. Gen. Microbiol., *25*:297–306, 1961. Weaver, D. K., Lee, E. K. H., and Leahy, M. S.: Comparison of reagent impregnated paper strips and conventional methods for identification of *Enterobacteriaceae*. Am. J. Clin. Pathol., *49*:494–499, 1968.

agglomerans and the Erwinia reduce nitrates to nitrites. The details of the nitrate reduction test are presented in Chart 2-2.

Because it requires 18 to 24 hours to perform the nitrate reduction test, the test is not commonly used to pre-screen unknown bacterial isolates for this characteristic. A more rapid test using PathoTec* reagent-impregnated nitrate test strips has alleviated this problem to some degree. However, the nitrate reduction test is used in most laboratories either to confirm the correct classification of an unknown microorganism or as an aid in arbitrating the identification of a bacterial species showing atypical reactions in tests measuring other characteristics.

Any basal medium that supports the growth of the organism under test and contains a 0.1 per cent concentration of potassium nitrate (KNO_3) is suitable for performing this test. Nitrate broth and solid nitrate agar poured on a slant are the media forms most commonly used in most clinical laboratories. Since the enzyme nitratase is activated only under anaerobic conditions, ZoBell[18] has recommended the use of semisolid agar. Semisolid media enhance the growth of many bacterial species and provide the anaerobic environment needed for enzyme activation.

---

*General Diagnostics, Morris Plains, N.Y.

Chart 2-2.   *Nitrate Reduction*

Introduction

The capability of an organism to reduce nitrates to nitrites is an important characteristic used in the identification and species differentiation of many groups of microorganisms. All Enterobacteriaceae except certain biotypes of *Enterobacter agglomerans* and Erwinia demonstrate nitrate reduction. The test is also helpful in identifying members of the Hemophilus, Neisseria, and Branhamella genera.

Principle

Organsims demonstrating nitrate reduction have the capability of deriving oxygen from nitrates to form nitrites and other reduction products. The chemical equation is:

$$NO_3^- + 2e^- + 2H \longrightarrow NO_2^- + H_2O$$
Nitrate   Nitrite

The presence of nitrites in the test medium is detected by the addition of α-naphthylamine and sulfanilic acid, with the formation of a red diazonium dye, p-sulfobenzene-azo-α-naphthylamine.

Media and Reagents

1. Nitrate broth; or nitrate agar (slant):

Beef extract	3.0 g.
Peptone	5.0 g.
Potassium nitrate (KNO_3)	1.0 g.
Agar (nitrite-free)	12.0 g.
Distilled water to	1000 ml.

2. Reagent A:

α-Naphthylamine	5.0 g.
Acetic acid (5 N), 30%	1000 ml.

3. Reagent B:

Sulfanilic acid	8.0 g.
Acetic acid (5 N), 30%	1000 ml.

Procedure

Inoculate the nitrate medium with a loopful of the test organism isolated in pure culture on agar medium and incubate at 35° C. for 18 to 24 hours. At the end of incubation, add 1 ml. each of reagents A and B to the test medium, in that order.

Interpretation

The development of a red color within 30 seconds after adding the test reagents indicates the presence of nitrites and represents a positive reaction for nitrate reduction (Plate 2-1, *I*). If no color develops after adding the test reagents, this may indicate either that nitrates have not been reduced (a true negative reaction), or that they have been reduced to products other than nitrites, such as ammonia, molecular nitrogen (denitrification), nitric oxide (NO) or nitrous oxide (N_2O), and hydroxylamine. Since the test reagents detect only nitrites, the latter process would lead to a false negative reading. Thus, it is necessary to add a small quantity of zinc dust to all negative reactions. Zinc ions reduce nitrates to nitrites, and the development of a red color after adding zinc dust indicates the presence of residual nitrates and confirms a true negative reaction.

Chart 2-2. *Nitrate Reduction (Continued)*

Controls	It is important to test each new batch of media and each new formulation of test reagents for positive and negative reactions. The following organisms are suggested: Positive control: *Escherichia coli* Negative control: *Acinetobacter calcoaceticus* (Herellea)
Bibliography	MacFaddin, J. F.: Biochemical Tests for Identification of Medical Bacteria. pp. 142–149. Baltimore, Williams & Wilkins, 1976. Wallace, G. I., and Neave, S. L.: The nitrite test as applied to bacterial cultures. J. Bacteriol., *14*:377–384, 1927.

The addition of zinc dust to all negative reactions, as discussed in Chart 2-2, should be a routine procedure.

Most organisms capable of reducing nitrates will do so within 24 hours; some may produce detectable quantities within 8 to 12 hours. Both α-naphthylamine and sulfanilic acid are relatively unstable, and their reactivity should be determined at frequent intervals by testing with positive and negative control organisms. Since the diazonium compound formed by nitrate-reducing organisms is unstable, the color reactions should be read within a short period of time before they fade.

GUIDE TO THE SELECTION OF MEDIA FOR THE PRIMARY RECOVERY OF THE ENTEROBACTERIACEAE

Since most specimens submitted to the microbiology laboratory contain several species of bacteria, often mixed with other microorganisms, a selective medium must be used to recover those species that may be of medical importance. In order to make rational selections, microbiologists must know the composition of each medium's formula and the purpose and relative concentration of each chemical or compound that is included. For instance, it is not sufficient to know that bile salts are included in the formulas of a number of selective media to inhibit the growth of gram-positive and some of the more fastidious gram-negative bacterial species. It is the concentration of bile salts that often determines which bacteria will be inhibited and how selective a given medium might be.

For the recovery of the Enterobacteriaceae from clinical specimens that potentially harbor mixed bacteria, three general types of media are available: (1) primary isolation agars; (2) selective isolation agars; and (3) enrichment broths. These will each be discussed in detail below, and Tables 2-1, 2-2, and 2-3 compare a number of different media commonly used in clinical practice, listing their formulations, purpose of the selective ingredients, and some of the final reactions and their interpretations. The formulas are somewhat complex, including ingredients that not only inhibit the growth of certain bacterial species but also detect a variety of biochemical characteristics that are important in making a preliminary identification of the microorganisms that are selected.

General Types of Chemicals and Compounds Used in Selective Media

Following is a list of the general types of chemicals and compounds used in selective media, including brief comments on the function of each.

1. *Protein Hydrolysates* (Peptones, Meat Infusion, Tryptones, Casein). Basic nu-

trients supporting the growth of most bacteria, specifically, providing the sources of carbon and nitrogen needed for bacterial metabolism.

2. *Carbohydrates.* A variety of disaccharides (lactose, sucrose, and maltose), hexoses (dextrose), and pentoses (xylose) are included in selective media for two purposes: (1) to provide a ready source of carbon for energy production; and (2) to serve as substrates in biochemical tests to determine the ability of an unknown microorganism to utilize different sugars with the formation of acid and/or gas.

3. *Buffers.* Balanced mono- and disodium or potassium phosphates are most commonly used. Buffers serve two primary purposes: (1) to provide a stable *p*H best suited for the growth of microorganisms; and (2) to provide a standard reference *p*H for those media in which a shift in *p*H is used as an end point to detect metabolic products used in the identification of microorganisms.

4. *Enrichments* (Blood, Serum, Vitamin Supplements, and Yeast Extracts). Compounds added to media to support the growth of more fastidious organisms. Enrichments are less commonly used in the media for the recovery of the Enterobacteriaceae since most members of this group grow quite well.

5. *Inhibitors.* A number of different compounds may serve to inhibit the growth of certain undesired bacterial species: (1) dyes (brilliant green, eosin, azide); (2) heavy metals (bismuth); (3) chemicals (citrate, desoxycholate, selenite, phenylethyl alcohol); and (4) antimicrobial agents (neomycin, vancomycin, chloramphenicol). Their relative concentration is important in determining the exact selectivity of the medium in which they are contained.

6. *pH Indicators.* Fushsin, methylene blue, neutral red, phenol red, and bromcresol purple are the more commonly used indicators to measure *p*H shifts in test media resulting from bacterial action on a given substrate.

7. *Miscellaneous Indicators.* Other indicators may be included for the detection of specific chemicals, such as ferric and ferrous ions or sulfide precursors for the detection of H_2S gas.

8. *Miscellaneous Compounds and Chemicals.* Agar, a gelatinous extract of red seaweed, is commonly added to media in varying concentrations as a solidifying agent. Concentrations of 1 to 2 per cent are used for plating media, 0.05 to 0.3 per cent for motility media, and trace amounts are added to broths to prevent convection currents and penetration of oxygen into the deeper anaerobic portions. Sodium thiosulfate is a chemical commonly added to provide a source of sulfur for the production of H_2S gas.

Primary Isolation Media

Table 2-1 compares the formulas, inhibitory ingredients, and key biochemical reactions of the following four primary isolation media used for the recovery of the Enterobacteriaceae:

MacConkey agar
Eosin methylene blue agar (EMB)
Desoxycholate-citrate agar (DCA)
Endo agar

MacConkey first described a primary isolation medium in 1905, which he called neutral red-bile salt agar, formulated to select out the gram-negative enteric bacilli from specimens containing mixed bacterial species. Although at that time all non-spore-forming gram-negative bacilli were still referred to as enteric organisms, microbiologists had already recognized the importance of identifying certain species that seemed to be more pathogenic to man than others. The carbohydrate-utilization patterns for several species of bacteria were already known by the turn of the century, and the fermentation of lactose in particular was recognized as an important marker in differentiating these enteric pathogens. MacConkey incorporated lactose in the neutral red-bile salt agar together with the indicator neutral red which provides a direct visual means for

Table 2-1. *Primary Isolation Media for Recovery of Enterobacteriaceae*

Medium	Formulation		Purpose and Differential Ingredients	Reactions and Interpretation
MacConkey agar (Plate 2-2, A, B, and C)	Peptone	17.000 g.	MacConkey agar is a differential plating medium for the selection and recovery of the Enterobacteriaceae and related enteric gram-negative bacilli.	Typical strong lactose fermenters, such as species of Escherichia, Klebsiella, and Enterobacter, produce red colonies surrounded by a zone of precipitated bile.
	Polypeptone	3.000 g.		
	Lactose	10.000 g.		
	Bile salts	1.500 g.		
	Sodium chloride	5.000 g.		
	Agar	13.500 g.		Slow or weak lactose fermenters, such as Citrobacter, Providencia, Serratia, and Hafnia, may appear colorless after 24 hours, or slightly pink in 24 to 48 hours.
	Neutral red	0.030 g.	The bile salts and crystal violet inhibit the growth of gram-positive bacteria and some fastidious gram-negative bacteria.	
	Crystal violet	0.001 g.		
	Distilled water to:	1.000 liter		
	Final pH = 7.1		Lactose is the sole carbohydrate. Lactose-fermenting bacteria produce colonies that are varying shades of red, due to the conversion of the neutral red indicator dye (red below pH 6.8) from the production of mixed acids. Colonies of non-lactose-fermenting bacteria appear colorless or transparent.	Proteus, Edwardsiella, Salmonella, and Shigella, with rare exceptions, produce colorless or transparent colonies. Representative colonies, showing these various reactions, are shown in Plate 2-2.
Eosin methylene blue agar (EMB) (Plate 2-2, D, E, and F)	Peptone	10.000 g.	Eosin methylene blue (EMB) agar is a differential plating medium that can be used in place of MacConkey agar in the isolation and detection of the Enterobacteriaceae or related coliform bacilli from specimens with mixed bacteria.	Typical strong lactose-fermenting colonies, notably *Escherichia coli,* produce colonies that are green-black with a metallic sheen.
	Lactose	5.000 g.		
	Sucrose*	5.000 g.		
	Dipostassium, PO₄	2.000 g.		Weaker acid formers, including Klebsiella, Enterobacter, Serratia, and Hafnia, produce purple colonies within 24 to 48 hours.
	Agar	13.500 g.		
	Eosin y	0.400 g.		
	Methylene blue	0.065 g.	The aniline dyes (eosin and methylene blue) inhibit gram-positive and fastidious gram-negative bacteria. They also combine to form a precipitate at acid pH, thus also serving as indicators of acid production.	Nonlactose fermenters, including Proteus, Salmonella and Shigella, produce transparent colonies.
	Distilled water to:	1.000 l.		
	Final pH = 7.2			*Yersinia enterocolitica,* a non lactose, sucrose fermenter, produces transparent colonies on Levine EMB, purple to black colonies on the modified formula. See Plate 2-2.
			Levine EMB, with only lactose, gives reactions more in parallel with MacConkey agar; the modified formula also detects sucrose fermenters.	

(Continued)

Table 2-1. *Primary Isolation Media for Recovery of Enterobacteriaceae (Continued)*

Medium	Formulation	Purpose and Differential Ingredients	Reactions and Interpretation
Desoxycho-late-citrate agar (DCA) (Plate 2-2, G, H, and I)	Meat, infusion from 375.00 g. Peptone 10.00 g. Lactose 10.00 g. Sodium citrate 20.00 g. Ferric citrate 1.00 g. Sodium desoxycholate 5.00 g. Agar 17.00 g. Neutral red 0.02 g. Distilled water to: 1.00 liter Final pH = 7.3	Desoxycholate-citrate agar is a differential plating medium used for the isolation of members of the Enterobacteriaceae from mixed cultures. Since the medium contains about 3 times the concentration of bile salts (sodium desoxycholate) as MacConkey agar, it is most useful in selecting species of Salmonella from specimens overgrown or heavily contaminated with coliform bacilli or gram-positive organisms. The sodium and ferric citrate salts retard the growth of *Escherichia coli*. Lactose is the sole carbohydrate and neutral red is the pH indicator and detector of acid production.	All gram-positive bacteria are inhibited. Colonies of *Escherichia coli* appear small and deep red. Weaker acid-producing organisms such as Klebsiella and Enterobacter show somewhat mucoid, colorless colonies with light pink centers. Non-lactose-fermenting organisms such as Proteus, Salmonella, and Shigella produce large, colorless colonies. See Plate 2-2.
Endo agar	Potassium phosphate 3.5 g. Peptone 10.0 g. Agar 15.0 g. Lactose 10.0 g. Sodium sulfite 2.5 g. Basic fuchsin 0.5 g. Distilled water to: 1.0 liter Final pH = 7.4	Endo agar is a solid plating medium used to recover coliform and other enteric organisms from clinical specimens or materials of sanitary importance such as water, milk, and other foodstuffs. The sodium sulfite and basic fuchsin serve to inhibit the growth of gram-positive bacteria. Acid production from lactose is not detected by a pH change, rather from the reaction of the intermediate product, acetaldehyde, which is fixed by the sodium sulfite.	Lactose-fermenting colonies appear pink to rose red. Strong acid producers, such as *Escherichia coli*, may color the medium surrounding the colonies or produce a metallic sheen from reaction with the basic fuchsin. Non-lactose-fermenting organisms, including Salmonella, Shigella, and Proteus, produce colonies about the same color as the medium, being almost colorless to faint pink.

*Modified Holt-Harris Teague formula. Not contained in Levine EMB.

detecting lactose utilization by the test organism. Any organism capable of degrading lactose with the production of acids produces red colonies on neutral red-bile salt agar (MacConkey agar) because of a color change in the indicator at an acid pH (Plate 2-2, *A, B,* and *C*). Thus, MacConkey was among the first to devise a medium that is capable of not only selecting a certain group of bacteria, but also provides for the preliminary identification of metabolic or biochemical characteristics as well.

Primary isolation media are only moderately inhibitory and designed to recover many different species of bacteria within a broad group. The four media listed in Table 2-1 inhibit the growth of almost all species of gram-positive bacteria and many species of fastidious gram-negative organisms as well; however, they permit the growth of all Enterobacteriaceae and other gram-negative bacilli.

The choice of which of these four media to use is largely one of personal preference, although there are differences in their inhibitory properties and in the visual appearance of the gross colonies. All four media have the capability of differentiating those bacterial species that utilize lactose from those that do not. On MacConkey and desoxycholate agars, both of which contain neutral red as the *p*H indicator, lactose-utilizing colonies appear red from the production of acids. Strong acid producers such as *Escherichia coli* form deep red colonies with diffusion of pigment into the surrounding agar. Weaker acid producers form lighter pink colonies, or colonies that are clear at the periphery with pink centers.

On EMB and Endo agars, strong acid-producing bacteria form colonies with a metallic sheen. Many microbiologists prefer these media because this sheen assists in the identification of *Escherichia coli;* however, this characteristic is nonspecific in that other species of enteric bacilli may produce a sheen and not all strains of *Escherichia coli* have this classic appearance. EMB agar has an advantage in demonstrating the mucoid nature of some strains of *Escherichia coli* and species of Klebsiella because of the enhanced production of polysaccharide substances on this medium.

Because of the somewhat higher concentration of bile salts in desoxycholate-citrate agar, many strains of *Escherichia coli* and other species of enteric bacilli grow poorly if at all on this medium (Plate 2-2, *G*). Since most strains of Shigella and all of the salmonellae are not inhibited by bile salts, desoxycholate-citrate agar is preferred by many to recover these potential pathogens from heavily contaminated specimens such as feces or sewage (Plate 2-2, *H* and *I*).

Endo agar is not commonly used in clinical laboratories, but rather it is widely employed by sanitarians and epidemiologists for the recovery and preliminary identification of enteric organisms from food and water supplies.

Selective Isolation Media

Media are made selective by the addition of a variety of inhibitors to their formulas generally in higher concentration than found in primary isolation media. With the use of these media, certain unwanted bacteria are inhibited, allowing only those species of potential medical significance to grow out in culture. For instance, the selective media used for the recovery of the Enterobacteriaceae from mixed cultures are designed to inhibit the growth of gram-positive bacteria and to varying degrees retard the development of *Escherichia coli* and other coliform bacilli. This permits the recovery of the salmonellae and shigellae from specimens where they are few in number in comparison to the massive concentration of other enteric organisms.

Although a number of selective media have been formulated for use in clinical laboratories, only the four most commonly employed will be discussed here. The formulas and other characteristics of these media are listed in Table 2-2. These media are:

Salmonella-Shigella (SS) agar
Hektoen enteric (HE) agar
Xylose, lysine, desoxycholate (XLD) agar
Bismuth sulfite agar

The choice of which of these selective media to use for the recovery of Enterobacteriaceae is dependent both on personal preference and on the species to be selected. In general, selective media are used for the recovery of Salmonella and Shigella from the feces of patients suffering from diarrhea or from food and water supplies where there is a suspicion of fecal contamination. Virtually all species of Salmonella grow well in the presence of bile salts, explaining why this organism is commonly recovered from

(Text continues on p. 68)

Table 2-2. *Selective Media for Recovery of Enterobacteriaceae*

Medium	Formulation		Purpose and Differential Ingredients	Reactions and Interpretation
Salmonella-Shigella agar (SS) (Plate 2-2, J, K, and L)	Beef extract Peptone Lactose Bile salts Sodium citrate Sodium thiosulfate Ferric citrate Agar Neutral red Brilliant green Distilled water to: Final pH = 7.4	5.000 g. 5.000 g. 10.000 g. 8.500 g. 8.500 g. 8.500 g. 1.000 g. 13.500 g. 0.025 g. 0.0330 g. 1.000 liter	SS Agar is a highly selective medium formulated to inhibit the growth of most coliform organisms and permit the growth species of Salmonella and Shigella from environmental and clinical specimens. The high bile salts concentration and sodium citrate inhibit all gram-positive bacteria and many gram-negative organisms, including coliforms. Lactose is the sole carbohydrate and neutral red the indicator for acid detection. Sodium thiosulfate is a source for sulfur. Any bacteria that produce H_2S gas are detected by the black precipitate formed with ferric citrate (relatively insensitive). High selectivity of SS agar permits use of heavy inoculum.	Any lactose-fermenting colonies that appear are colored red by the neutral red. Rare strains of Salmonella (*Arizona* strains) are lactose-fermenting and colonies may simulate *Escherichia coli.* Growth of species of Salmonella is uninhibited and colonies appear colorless with black centers due to H_2S gas production. Species of Shigella show varying inhibition and colorless colonies with no blackening. Motile strains of Proteus that appear on SS agar do not swarm. See Plate 2-2.
Hektoen enteric agar (HE) (Plate 2-3, A, B, and C)	Peptone Yeast extract Bile salts Lactose Sucrose Salicin Sodium chloride Sodium thiosulfate Ferric ammonium citrate Acid fuchsin Thymol blue Agar Distilled water to: Final pH = 7.6	12.00 g. 3.00 g. 9.00 g. 12.00 g. 12.00 g. 2.00 g. 5.00 g. 5.00 g. 1.50 g. 0.10 g. 0.04 g. 14.00 g. 1.00 liter	HE Agar is a recent formulation devised as a direct plating medium for fecal specimens to increase the yield of species of Salmonella and Shigella from the heavy numbers of normal flora. The high bile salt concentration inhibits growth of all gram-positive bacteria and retards the growth of many strains of coliforms. Acids may be produced from three carbohydrates, and acid fuchsin reacting with thymol blue produces a yellow color when the pH is lowered. Sodium thiosulfate is a sulfur source, and H_2S gas is detected by ferric ammonium citrate (relatively sensitive).	Rapid lactose fermenters (*Escherichia coli*) are moderately inhibited and produce bright orange to salmon pink colonies. Salmonella colonies are blue-green, typically with black centers from H_2S gas. Shigella appear more green than Salmonella, with the color fading to the periphery of the colony. Proteus strains are somewhat inhibited; colonies that develop are small, transparent, and more glistening or watery in appearance than species of Salmonella or Shigella. See Plate 2-3.

Table 2-2. *Selective Media for Recovery of Enterobacteriaceae (Continued)*

Medium	Formulation		Purpose and Differential Ingredients	Reactions and Interpretation
Xylose lysine desoxycholate agar (XLD) (Plate 2-3, D through J)	Xylose Lysine Lactose Sucrose Sodium chloride Yeast extract Phenol red Agar Sodium Desoxycholate Sodium thiosulfate Ferric ammonium citrate Distilled water to: Final pH = 7.4	3.50 g. 5.00 g. 7.50 g. 7.50 g. 5.00 g. 3.00 g. 0.08 g. 13.50 g. 2.50 g. 6.80 g. 0.80 g. 1.00 liter	XLD agar is less inhibitory to growth of coliform bacilli than HE and was designed to detect shigellae in feces after enrichment in GN broth. Bile salts in relatively low concentration make this medium less selective than the other three included in this chart. Three carbohydrates are available for acid production and phenol red is the pH indicator. Lysine-positive organisms, such as most *Salmonella enteriditis* strains, produce initial yellow colonies from xylose utilization, and delayed red colonies from lysine decarboxylation. H_2S detection system is similar to HE agar.	Organisms such as *E. coli* and Klebsiella-Enterobacter species may utilize more than one carbohydrate and produce bright yellow colonies. Colonies of many species of Proteus are also yellow. Most species of Salmonella and Arizona produce red colonies, most with black centers from H_2S gas. Shigella, Providencia, and many Proteus species utilize none of the carbohydrates and produce translucent colonies. Citrobacter colonies are yellow with black centers; many Proteus species are yellow or translucent with black centers; salmonellae are red with black centers. See Plate 2-3.
Bismuth sulfite agar (Plate 2-3, K and L)	Beef extract Peptone Glucose Disodium phosphate Ferrous sulfate Bismuth sulfite Brilliant green Agar Distilled water to: Final pH = 7.5	5.000 g. 10.000 g. 5.000 g. 4.000 g. 0.300 g. 8.000 g. 0.025 g. 20.000 g. 1.000 liter	Bismuth sulfite agar is highly selective for *Salmonella typhi*. Note that glucose is included in the formula. In the presence of glucose fermentation, the sulfite is reduced with the production of iron sulfide and black colonies. The heavy metal bismuth and brilliant green are inhibitory to the growth of all gram-positive bacteria and most species of gram-negative bacteria except the salmonellae. The medium must be used on the day that it is prepared, which limits its general use in many laboratories.	Most lactose-fermenting coliforms and the shigellae are completely inhibited. Colonies of *Salmonella typhi* appear black with a metallic sheen. Most species of *Salmonella enteriditis* are black without a sheen. *S. gallinarum, S. cholerae-suis,* and *S. paratyphi* produce greenish colonies. A brownish discoloration of the agar many times the size of the colony may be observed. See Plate 2-3.

the gallbladder, particularly in subjects that are salmonella carriers. Bile salts are commonly added to selective media because other species of enteric bacilli, including some of the more fastidious strains of Shigella, grow poorly if at all in such media. SS and HE agars, both containing relatively high concentrations of bile salts, are well adapted for the recovery of salmonellae from specimens heavily contaminated with other coliform bacilli. Most of the shigellae also grow on these media, although some strains are inhibited. XLD agar, containing lower concentrations of bile salts, is more effective in the recovery of all shigellae, particularly after the specimen has been enriched for a few hours in gram-negative (GN) broth. However, because other species of coliform bacilli, particularly *Escherichia coli,* are less inhibited on XLD, either SS or HE is better adapted for the recovery of Salmonella and Shigella from fecal specimens plated directly to the agar.

HE and XLD agars are now widely used in clinical laboratories, not only because they enhance the recovery of Salmonella and Shigella from clinical specimens, but also because a variety of biochemical characteristics can be detected to assist in making a preliminary species identification.

The carbohydrates contained in HE agar are lactose, sucrose, and salicin. Microorganisms that utilize these carbohydrates with the production of acid grow as yellow colonies on HE agar (Plate 2-3, A and B). On the other hand, bacterial species such as Shigella that fail to utilize any of these carbohydrates grow as translucent or light green (due to the background color of the medium) colonies (Plate 2-3, A). *Klebsiella rhinoscleromatis,* which utilizes only salicin, produces only a small amount of acid and can be suspected on HE agar by the yellow-orange appearance of the colonies.

HE agar also contains ferric salts. Therefore, any H_2S-producing colony will assume a dark brown or black pigmentation (Plate 2-3, B and C).

XLD agar contains lactose, sucrose, and xylose; therefore, any microorganism utilizing these with formation of acid will produce yellow colonies (Plate 2-3, D). Bacteria not utilizing these carbohydrates will produce colorless colonies. Salmonella species can be presumptively identified on XLD agar by observing certain colonial features. Most strains produce a small amount of acid from xylose, forming colonies with a yellowish tinge (Plate 2-3, E). This differs from the reaction on selective media that contain only lactose or sucrose, on which colonies appear colorless. However, since XLD agar also contains lysine and since most species of Salmonella can decarboxylate this amino acid with the production of alkaline amines, the colonies often have a somewhat orange tinge (Plate 2-3, F). This feature is often of value in distinguishing Salmonella species from Proteus species, many strains of which also appear as "non lactose fermenters" on other selective media and also produce H_2S. On XLD agar, Proteus species produce translucent colonies the centers of which may or may not turn black (Plate 2-3, G and H).

Colonies with an intense black pigmentation suggest strong acid and H_2S producers such as Citrobacter species (Plate 2-3, I and J).

Bismuth sulfite agar is used almost exclusively for the recovery of *Salmonella typhi* from feces. This medium is particularly useful when screening a large number of individuals who have potentially been exposed in localized or regional epidemics. The bismuth sulfite and brilliant green in the media inhibit almost all organisms except the salmonellae, and S. typhi in particular can be suspected because of the production of colonies with a black sheen (Plate 2-3, K and L). Bismuth sulfite agar is also useful in demonstrating those rare strains of lactose-positive Salmonella species that may be overlooked or discounted as unimportant when observed on other selective media. This medium is not widely used in clinical laboratories because of its high selectivity and its very short shelf life of only one day.

Enrichment Broths

As the name indicates, enrichment broths are used to enhance the growth of certain bacterial species while inhibiting the development of unwanted microorganisms. Enrichment broths are most commonly used in clinical laboratories for the recovery of Salmonella and Shigella from fecal specimens. This is particularly necessary in the case of salmonella carriers or in some cases of shigella infection where the number of organisms may be as low as 200 per gram of feces, in comparison to the massive numbers of *Escherichia coli* or other enteric bacilli that may reach 10^9/g. of feces or more. Enrichment broths work on the principle that the normal enteric flora are held in a lag phase of growth, while Salmonella and Shigella are uninhibited and enter into the normal log phase of growth. However, the normal fecal flora is only temporarily inhibited and in time will also enter into a log phase of growth. Therefore, subculture from the enrichment broth to one of the selective media within 6 to 12 hours is recommended.

The formulas and salient characteristics of the three most commonly used enrichment broths are summarized in Table 2-3. These broths are:

Selenite broth
Gram-negative (GN) broth
Tetrathionate broth

Because of differences in the inhibitors used, these broths differ in their selectivity and specific applications. Selenite and tetrathionate broths are more inhibitory to the growth of *Escherichia coli* and other enteric gram-negative bacilli than is GN broth. Thus, they are best adapted for the recovery of Salmonella or Shigella from heavily contaminated specimens such as feces or sewage where a heavy inoculum may be desirable. However, GN broth is used with greater frequency in clinical laboratories because it is less inhibitory to the growth of many of the more fastidious strains of Shigella. Enrichment of fecal specimens in GN broth for

4 to 6 hours and subculture to XLD agar is the optimal technique for the recovery of shigellae from suspected cases of bacillary dysentery.

Suggested Guide to the Choice of Selective Media for the Recovery of the Enterobacteriaceae

The relatively large number of media listed in Tables 1-1, 1-2, and 1-3 and the several combinations in which they can be used may leave the beginning microbiologist somewhat confused. Following is a guide to the selection of media that may be helpful in the recovery of the Enterobacteriaceae from clinical specimens.

For specimens other than feces or rectal swabs, a combination of MacConkey or EMB agar together with a blood agar plate is usually sufficient. Media with greater inhibitory properties are not routinely required since the concentration of commensal flora or contaminating organisms is relatively low in most nonenteric specimens. Subculturing to a more inhibitory medium can be done in those instances where it appears necessary.

For fecal specimens or rectal swabs, it is necessary to select only one medium from each of the three groups listed in Tables 1-1, 1-2, and 1-3. The following can be suggested:

1. *Inoculate the specimen directly to a MacConkey or EMB agar plate* for the primary isolation of all species of enteric gram-negative bacilli.
2. *Directly inoculate either an SS or an HE agar plate* for the selective screening of Salmonella and Shigella. Use bismuth sulfite agar if typhoid fever is clinically suspected.
3. *"Enrich" the specimen* by placing a relatively heavy inoculum in either selenite, tetrathionate, or GN broth for 6 to 12 hours. Subculture to SS, HE, or XLD after 6 to 12 hours of incubation at 35° C. The combination of GN broth to XLD agar is optimal for the recovery of Shigella species.
4. *Incubate all plate cultures* at 35° C. for

Table 2-3. *Enrichment Broths for Recovery of Enterobacteriaceae*

Broth	Formulation		Purpose and Differential Ingredients	Reactions and Interpretation
Selenite broth	Peptone Lactose Sodium selenite Sodium phosphate Distilled water to: pH = 7.0	5.0 g. 4.0 g. 4.0 g. 10.0 g. 1.0 liter	Selenite broth is recommended for the isolation of salmonellae from specimens such as feces, urine, or sewage that have heavy concentrations of mixed bacteria. Sodium selenite is inhibitory to *E. coli* and many other coliform bacilli, including many strains of Shigella. The medium functions best under anaerobic conditions and a pour depth of at least two inches is recommended.	Within a few hours the broth becomes cloudy after inoculation with the specimen. Because coliforms or other intestinal flora may overgrow the pathogens within a few hours, subculture to SS agar or bismuth sulfite is recommended within 8 to 12 hours. Overheating of the broth during preparation may produce a visible precipitate, making it unsatisfactory for use.
Gram-negative (GN) broth	Polypeptone peptone Glucose D-mannitol Sodium citrate Sodium desoxycholate Dipotassium phosphate Monopotassium phosphate Sodium chloride	20.00 g. 1.00 g. 2.0 g. 5.0 g. 0.5 g. 4.0 g. 1.5 g. 5.0 g.	Because of the relatively low concentration of desoxycholate, GN broth is less inhibitory to *Escherichia coli* and other coliforms. Most strains of Shigella grow well. The desoxycho-	GN broth is designed for the recovery of Salmonella species and Shigella species when they are in small numbers in fecal specimens. The broth becomes cloudy within one or

24 to 48 hours. Based on visual observation of the colony morphology and biochemical characteristics described above, pick suspected colonies and transfer them to a series of differential media or perform serologic typing when applicable.

DIFFERENTIAL CHARACTERISTICS AND TESTS USED FOR THE SPECIES IDENTIFICATION OF THE ENTEROBACTERIACEAE

Although a preliminary identification of the Enterobacteriaceae is possible based on colonial colony characteristics and biochemical reactions on primary isolation media, further species identification requires the determination of additional metabolic characteristics that reflect the genetic code and unique identity of the organism being tested. It is the purpose of this discussion to review the salient features of the tests most commonly used in clinical laboratories to measure these metabolic characteristics. This is necessary so that the laboratory worker can develop a fundamental understanding of the principles behind these procedures, so that any biochemical inconsistencies, problems with mixed cultures, or faulty techniques can be quickly recognized and corrected.

A variety of differential tests and numerous schemas for the final species identification of the Enterobacteriaceae are available. It will not be possible to discuss all of these here. However, the following set of tests is widely used in clinical laboratories and also

Table 2-3. *Enrichment Broths for Recovery of Enterobacteriaceae (Continued)*

Broth	Formulation		Purpose and Differential Ingredients	Reactions and Interpretation
	Distilled water to:	1.0 liter	late and citrate are inhibitory to gram-positive bacteria. The increased concentration of mannitol over glucose limits the growth of Proteus species while encouraging growth of Salmonella and Shigella species, both of which are capable of fermenting mannitol.	two hours of inoculation, and subculture to HE agar or XLD agar within 1 to 6 hours is recommended.
	pH = 7.0			
Tetrathionate broth	Polypeptone peptone	18.00 g.	Tetrathionate broth is used primarily for the recovery of the salmonellae from feces. To supplement the inhibitory action of the tetrathionate broth base, 20 ml. of a solution containing 6 g. of iodine and 5 g. of potassium iodine is added to each liter of base. The inclusion of brilliant green is optional, making the broth more inhibitory to the growth of coliforms.	Subculture to either SS agar or bismuth sulfite agar within 12 to 24 hours is recommended for optimal recovery of salmonellae. The tetrathionate broth base cannot be heated after the iodine solution has been added. The medium must be used within 24 hours after adding the iodine solution.
	Yeast extract	2.00 g.		
	Sodium chloride	5.00 g.		
	D-mannitol	0.50 g.		
	Glucose	0.50 g.		
	Sodium desoxycholate	0.50 g.		
	Sodium thiosulfate	38.00 g.		
	Calcium carbonate	25.00 g.		
	Brilliant green	0.01 g.		
	Distilled water to:	1.00 liter		
	pH = 7.0			

provides for the measurement of those metabolic characteristics by which all but a few rare or atypical species of the Enterobacteriaceae can be identified. These characteristics are:

Utilization of carbohydrates
Orthonitrophenyl galactosidase activity (ONPG)
The IMViC reactions:
  Production of indole
  Methyl red reaction
  Production of acetyl-methyl carbinol (Voges-Proskauer test)
  Utilization of citrate
Production of urease
Decarboxylation of lysine, ornithine, and arginine
Production of phenylalanine deaminase
Production of hydrogen sulfide gas
Motility

Utilization of Carbohydrates

The terms "sugars" and "fermentation" require a brief discussion.

It is common for laboratory microbiologists to refer to all carbohydrates as sugars. This is convenient in an operational sense, with the understanding that polyhedral alcohols such as dulcitol and mannitol or cationic salts of acetate or tartrate are not truly sugars in a chemical sense.

The term fermentation is also somewhat loosely used in reference to the utilization of carbohydrates by bacteria; that is, we speak of lactose fermenters and nonlactose fermenters. By definition fermentation is an oxidation-reduction metabolic process that takes place in an anaerobic environment, with an organic substrate serving as the final hydrogen (electron) acceptor in place of oxygen. In bacteriologic test systems this process is detected by visually observing color changes of pH indicators as acid products are formed. Acidification of a test medium may occur through the degradation of carbohydrates by pathways other than strictly fermentative, or there may be ingredients included in some media other than carbohydrates that result in acid end products. Although most bacteria that utilize carbohydrates are facultative anaerobes, the utilization of carbohydrates may not always be under strictly anaerobic conditions, as is witnessed by the production of acid products by bacterial colonies growing on the surface of agar media. Even though all tests used to measure an organism's ability to enzymatically degrade a "sugar" into acid products may not always be "fermentative," we will use these terms in this text for convenience, acknowledging some degree of chemical license.

Carbohydrate-Fermentation Media Bases

A number of broth and agar media bases have been formulated to detect the carbohydrate-fermenting properties of bacteria. Since the Enterobacteriaceae as a group grow well on all of these media, the selection is largely one of personal preference. Following is a representative formulation of a fermentation base medium:

Trypticase (BBL)	10.000 g.
Sodium chloride	5.000 g.
Phenol red	0.018 g.
Distilled water to:	1.000 liter

Trypticase is a casein hydrolysate that is commonly included in media formulas to supply carbon and nitrogen. Sodium chloride serves both as an osmotic stabilizer and as an inhibitor to the growth of many bacterial species that are common environmental contaminants. The phenol red is a pH indicator, turning the media yellow if acid production lowers the pH below 6.8.

The carbohydrate to be tested is added to this base to achieve a final concentration of 0.5 to 1.0 per cent. It is recommended that the fermentation base be autoclaved first, adding a filter-sterilized solution of the carbohydrate afterward, to avoid hydrolytic degradation of the sugar and potentially false negative test results.

The Basic Principle of Carbohydrate Fermentation

Pasteur's mid-19th century studies of the action of yeasts on wine form the basis of our present understanding of carbohydrate fermentation. Pasteur observed that certain contaminating bacterial species produce a pH drop in wine (a carbohydrate substrate) from the production of a variety of acids. Full descriptions of the fermentative pathways, by which a monosaccharide such as glucose is degraded, evolved soon thereafter. Through a series of enzymatic glycolytic cleavages and transformations the glucose molecule is split into a series of three carbon compounds, the most important of which is pyruvic acid. This chemical sequence is known as the anaerobic Embden-Meyerhof pathway. Many bacteria, including all of the Enterobacteriaceae, utilize carbohydrates by what is called a "mixed acid fermentation," in which a variety of organic acids are ultimately derived from pyruvic acid. The mixed acid fermentation scheme for the degradation of glucose is shown in Figure 2-1.

Bacteria differ in the carbohydrates that they can utilize and in the types and quantities of mixed acids produced. These differences in enzymatic activity serve as one of the important characteristics by which the different species are recognized. For instance, glucose fermentation by *Escherichia coli* results in the production of large quantities of acetic and lactic acid, with a marked drop in the pH of the test medium. This is detected by a positive methyl red test, to be discussed below. On the other hand, the Klebsiella-Enterobacter group of coliform bacilli metabolize pyruvic acid primarily via the butylene glycol pathway, producing acetyl-methyl carbinol and a positive Voges-Proskauer test (see Fig. 2-5). Mixed acids are also produced, but they are produced in smaller quantities.

Many of the anaerobic bacteria, particularly species of Clostridium, Peptococcus, and Fusobacterium, produce relatively large quantities of butyric acid, accounting for the offensive foul odor they produce in clinical infections and in culture media. The gas formed by fermenting bacteria is primarily a mixture of hydrogen and carbon dioxide through the cleavage of formic acid. It is an accepted rule of thumb that any bacterium that forms gas in carbohydrate test medium must first form gas, which is self evident after studying the EMP scheme in Figure 2-1. Gas is best detected by using a broth carbohydrate-fermentation medium into which have been placed small inverted Durham tubes (see Plate 2-1, *G*). Even trace amounts of gas can be detected, which collect as bubbles under the inverted bottoms of the Durham tubes.

The formation of ethyl alcohol by microorganisms is of utmost commercial importance in the manufacture of alcoholic beverages and organic reagents; however, it is of limited usefulness in the laboratory identification of bacteria.

Lactose Fermentation

The bacterial fermentation of lactose is more complex than that of glucose. Lactose is a disaccharide composed of glucose and galactose connected through an oxygen linkage known as a glactoside bond. Upon hydrolysis, this bond is severed, with the release of glucose and galactose. In order for a bacterium to utilize lactose, two enzymes must be present: β-*galactosidase permease,* permitting the transmigration of β-galactosides (such as lactose) across the bacterial cell wall; and β-galactosidase, required to hydrolyze the β-galactoside bond once the disaccharide has entered the cell. The final acid reaction is from the degradation of glucose as shown in Figure 2-1. Lactose degradation is schematically illustrated in Figure 2-2.

Since lactose fermentation ultimately proceeds by way of glucose degradation via the EMP pathway, it follows that any organism incapable of utilizing glucose cannot form acid from lactose. This explains why glucose is omitted from the formulas of primary isolation media such as MacConkey or EMB; otherwise the ability to detect the lactose-fermenting capability of the test bacteria would be lost. Since the end point of lactose fermentation in the test medium is the detection of acid production, a non-lactose-fermenting organism is one that lacks β-galactosidase or cannot attack glucose. Since the Enterobacteriaceae by definition ferment glucose, only lack of the enzyme applies. So-called "late lactose fermenters" are thought to be organisms that possess β-galactosidase activity, but show sluggish β-galactosidase permease activity.

ONPG Test for the Detection of β-Galactosidase

Orthonitrophenyl galactoside (ONPG) is a compound structurally similar to lactose except that the glucose has been substituted by an orthonitrophenyl radical. This rather ingenous manipulation of the molecule forms the basis for the ONPG test which is outlined in Chart 2-3. This test allows for the detection of the enzyme β-galactosidase far more quickly than the test for lactose fermentation described above. This is helpful in identifying those late lactose-fermenting organisms that are deficient in β-galactoside permease. ONPG is more permeable through the bacterial cell wall than is lactose, and under the action of β-galactosidase ONPG is hydrolyzed into galactose and orthonitrophenol. The ONPG test medium is buffered at an alkaline *p*H, allowing any orthonitrophenol that is released to be detected by a visible pale yellow color change in the medium (Plate 2-4, *F*).

ONPG test tablets than can be easily reconstituted by adding a small amount of water are commercially available for convenient use in the laboratory. Organisms with strong β-galactosidase activity may produce a positive test within a few minutes after

Chart 2-3.   *Orthonitrophenyl Galactoside (ONPG)*

Introduction   Orthonitrophenyl galactoside (ONPG) is structurally similar to lactose, except that orthonitrophenyl has been substituted for glucose as shown in the following chemical reaction:

Orthonitrophenyl Galactoside (ONPG)	Galactose	Orthonitrophenol

Upon hydrolysis, through the action of the enzyme β-galactosidase, ONPG cleaves into two residues, galactose and orthonitrophenol. ONPG is a colorless compound; orthonitrophenol is yellow, providing visual evidence for hydrolysis.

Principle   Lactose-fermenting bacteria possess both lactose permease and β-galactosidase, two enzymes required for the production of acid in the lactose-fermentation test. The permease is required for the lactose molecule to penetrate the bacterial cell where the β-galactosidase can cleave the galactoside bond, producing glucose and galactose. Non lactose fermenters are devoid of both enzymes and are incapable of producing acid from lactose. Some bacterial species appear as non lactose fermenters because they lack permease but do possess β-galactosidase and give a positive ONPG test. So-called late lactose fermenters may be delayed in their production of acid from lactose because of sluggish permease activity. In these instances, a positive ONPG test may provide a rapid identification of delayed lactose fermentation.

Media and Reagents   1. Sodium phosphate buffer, 1 M, pH 7.0
2. O-nitrophenyl-β-galactoside (ONPG), 0.75M (Buffered ONPG tablets are commercially available.*)
3. Physiologic saline
4. Toluene

Procedure   Bacteria grown in medium containing lactose, such as Kligler iron agar (KIA) or triple sugar iron agar (TSI), produce optimal results in the ONPG test. A loopful of bacterial growth is emulsified in 0.5 ml. of physiologic saline to produce a heavy suspension. One drop of toluene is added to the suspension and vigorously mixed for a few seconds to release the enzyme from the bacterial cells. An equal quantity of buffered ONPG solution is added to the suspension, and the mixture is placed in a 37° C. water bath.

When using ONPG tablets, a loopful of bacterial suspension is added directly to the ONPG substrate resulting from adding 1 ml. of distilled water to a tablet in a test tube. This suspension is also placed in a 37° C. water bath.

Interpretation   The rate of hydrolysis of ONPG to orthonitrophenol may be rapid for some organisms, producing a visible yellow color reaction within 5 to 10 minutes (Plate 2-4, F). Most tests are positive within one hour; however, reactions should not be interpreted as negative before 24 hours of incubation. The

Interpretation *(Continued)*	yellow color is usually distinct, and indicates that the organism has produced orthonitrophenol from the ONPG substrate through the action of β-galactosidase.
Controls	Positive control: *Escherichia coli* Negative control: Proteus species
Bibliography	Belliveau, R. R., Grayson, J. W., Jr., and Butler, T. J.: A rapid, simple method of identifying Enterobacteriaceae. Am. J. Clin. Pathol. *50*:126–128, 1968. Blazevic, D. J., and Ederer, G. M.: Principles of Biochemical Tests in Diagnostic Microbiology. Pp. 83–85. New York, John Wiley & Sons, 1975. Lederberg, J.: The beta-*d*-galactosidase of *Escherichia coli,* strain K-12. J. Bacteriol., *60*:381–392, 1950. Lowe, G. H.: The rapid detection of lactose fermentation in paracolon organisms by the demonstration of β-galactosidase. J. Med. Lab. Technol., *19*:21–25, 1962. MacFaddin, J. F.: Biochemical Tests for Identification of Medical Bacteria. Pp. 71–78. Baltimore, Williams & Wilkins, 1976.

*Key Scientific Products, Los Angeles, Calif., 90066.

inoculation into the medium. The ONPG test is most helpful in the detection of β-galactosidase activity in late lactose fermenters such as some strains of *Escherichia coli* where differentiation from species of Shigella may otherwise be difficult. The test is also helpful in distinguishing some strains of Citrobacter and Arizona (ONPG-positive) from most salmonellae (ONPG-negative). The ONPG test is not a substitute for the determination of lactose fermentation since only the enzyme β-galactosidase is measured.

Detection of Glucose and Lactose Fermentation Using Kligler Iron Agar or Triple Sugar Iron Agar

Kligler iron agar (KIA) and triple sugar iron agar (TSI) are virtually indispensable in the identification of gram-negative bacilli recovered on primary isolation media. The reaction patterns are an integral part of many Enterobacteriaceae identification schemas and also serve as a valuable quality control confirmation of the reactions observed in other test media.

Double sugar agar was originally formulated by Russell in 1911 to detect the production of acid and gas from dextrose and lactose in a single test medium. Kligler modified the formula in 1917 by adding ferrous sulfate and sodium thiosulfate to detect the production of H_2S gas as well. The formulas for KIA and TSI are identical, except that TSI contains 10 g. per liter of sucrose in addition to dextrose and lactose (triple sugar). The formula for KIA is:

Kligler Iron Agar	
Beef extract	3.000 g.
Yeast extract	3.000 g.
Peptone	15.000 g.
Proteose peptone	5.000 g.
Lactose	10.000 g.
Dextrose	1.000 g.
Ferrous sulfate	0.200 g.
Sodium chloride	5.000 g.
Sodium thiosulfate	0.300 g.
Agar	12.000 g.
Phenol red	0.024 g.
Distilled water to:	1.000 liter
Final pH = 7.4	

The following observations are important in studying the formula of KIA. The incorporation of four protein compounds, beef extract, yeast extract, peptone, and proteose peptone, makes KIA and TSI nutritionally very rich, and the lack of inhibitors permits the growth of all but the most fastidious bacterial species (excluding the obligate

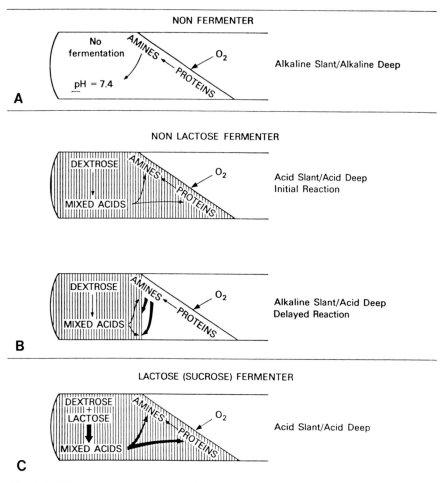

NON FERMENTER

No fermentation

AMINES ← O₂

PROTEINS

pH = 7.4

Alkaline Slant/Alkaline Deep

A

NON LACTOSE FERMENTER

DEXTROSE

AMINES ← O₂

↓

PROTEINS

MIXED ACIDS

Acid Slant/Acid Deep
Initial Reaction

DEXTROSE

AMINES ← O₂

↓

PROTEINS

MIXED ACIDS

Alkaline Slant/Acid Deep
Delayed Reaction

B

LACTOSE (SUCROSE) FERMENTER

DEXTROSE + LACTOSE

AMINES ← O₂

↓

PROTEINS

MIXED ACIDS

Acid Slant/Acid Deep

C

Fig. 2-3. Schematic illustration of the three general types of reactions produced by bacteria growing on Kligler iron agar. Diagram A shows nonfermentative bacilli that are unable to produce acids from the fermentation of glucose or lactose; there is no change in the media (represented by white). Diagram B illustrates an initial acidification of both the deep and the slant of the medium (vertical lines) by bacteria that ferment glucose, but the slant reverts back to an alkaline pH as alkaline amines are formed from the oxidative decarboxylation of proteins near the surface. Diagram C illustrates the complete permanent acidification of both the deep and the slant of the tube by lactose-fermenting bacteria.

anaerobes). For this reason, KIA and TSI can be used only in testing a single bacterial species picked from a single colony recovered on primary or selective agar plates. Lactose is present in 10 times the concentration of dextrose (the ratio of sucrose to dextrose is also 10:1 in TSI). Ferrous sulfate as an H₂S detector is somewhat less sensitive than other ferric or ferrous salts; therefore there may be discrepancies in the H₂S readings between KIA or TSI and other test media (Plate 2-4, *D* and *E*). The phenol red indicator is yellow below a pH of 6.8. Since the final pH of the medium is buffered at 7.4, relatively small quantities of acid production will result in a visible color change.

The biochemical principles underlying the reactions observed in KIA or TSI are illustrated in Figure 2-3. Note that the agar is poured on a slant. This configuration results in essentially two reaction chambers within the same tube. The slant portion, exposed throughout its surface to atmospheric oxygen, is aerobic; the lower portion, called the "butt" or the "deep," is protected from the air and is relatively anaerobic. It is important when preparing the media that the slant and the deep be kept equal in length, approximately 1½ inches or 3 cm. each, so that this two-chamber effect is preserved.

KIA and TSI tubes are inoculated with a long straight wire. The well-isolated test colony recovered on an agar plate is touched with the end of the inoculating needle which is stabbed into the deep of the tube. It is important that the stab line extend no closer than 3 to 5 mm to the glass bottom, to prevent the entrance of air into the deep and an alteration in the anaerobic environment. Upon removing the inoculating wire from the deep of the tube, the slant surface is streaked with a back-and-forth motion. Inoculated tubes are placed into an incubator at 35° C. for 18 to 24 hours.

Biochemical Principles Underlying KIA and TSI Reactions

A schematic drawing illustrating the biochemical principles underlying the KIA reactions is shown in Figure 2-3, and color photographs are presented in Plate 2-4, *A, B,* and *C.* These reactions are:

Reactions on KIA

Alkaline Slant/Alkaline Deep
No carbohydrate fermentation. Characteristic of nonfermentative bacteria such as *Pseudomonas aeruginosa.*
Alkaline Slant/Acid Deep
Dextrose fermented; lactose (or sucrose for TSI) not fermented. Characteristic of non-lactose-fermenting bacteria such as Shigella species.
Alkaline Slant/Acid (Black) Deep
Dextrose fermented; lactose not fermented, H_2S gas produced. Characteristic of non-lactose-fermenting, H_2S-producing bacteria such as Salmonella species, *Arizona* species, *Citrobacter* species and some species of Proteus.
Acid Slant/Acid Butt
Dextrose and Lactose (and/or sucrose with TSI) fermented. Characteristic of lactose-fermenting coliforms such as *Escherichia coli* and the Klebsiella-Enterobacter group.

The slant portion of the tube that is exposed to atmospheric oxygen has a propensity to turn alkaline from the oxidative decarboxylation of proteins, proteoses, and amino acids in the medium. Accelerated by the action of bacteria growing on the slant, amines are formed from these protein derivatives and the slant portion has a tendency to remain alkaline and appear red. In the deep of the tube, where oxygen is excluded, protein degradation is minimal and even small quantities of acid may be detected by the appearance of a yellow color.

It is interesting that the production of volatile amines is the process that earlier workers recognized as the putrefaction or decay that occurs in the decomposition of cadavers or animal carcasses. Collectively they referred to these alkaline by-products as "ptomaines" and theorized that "ptomaine poisoning" underlies the morbid effects that bacteria have on individuals with infections. The descriptive terms "cadaverine" and "putrescine," referring to the decarboxylation products of lysine and ornithine, respectively, reflect these early theories. Although it is a little far fetched to think in terms of putrefaction or ptomaine poisoning of a KIA or TSI slant, the underlying biochemical principles are similar to that which occurs with protein degradation in nature and indicates why alkalization occurs under bacterial action in the slant portion of protein-containing media.

Thus, as shown in Figure 2-3, *A,* in the absence of carbohydrate fermentation, no acids are formed and the amine production in the slant together with the alkaline buffers produce a red color throughout the medium. Bacteria that produce this type of reaction are known as "nonfermenters." A negative KIA or TSI is one of the important initial indications that an organism does not belong to the family Enterobacteriaceae, and that a different set of identification characteristics must be measured in order to make a species identification. The identification of nonfermentative gram-negative bacilli is discussed in Chapter 3.

If the KIA tube is inoculated with a dextrose-fermenting organism that cannot utilize lactose, only a relatively small quantity of acid can be obtained from the 0.1 per cent concentration of dextrose in the medium. Initially, during the first 8 to 12 hours of incubation, even this amount of acid may be sufficient to convert both the deep and the slant to a yellow color (see Fig. 2-3, *B*). However, within the next few hours the proteins within the slant portion of the tube, under the action of oxygen and bacteria, begin to release amines that soon counteract the small quantities of acid. By 18 to 24 hours the entire slant reverts to an alkaline *p*H and a return to a red color. In the deep of the tube, however, protein degradation is insufficient to counteract the acid formed and it remains yellow in color. Thus, the alkaline-slant/acid-deep reaction on KIA (or TSI) is an important initial indicator that the test organism is a non lactose fermenter.

Bacteria that utilize lactose (lactose and/ or sucrose in TSI) produce relatively large quantities of acid in KIA because of the higher concentration of lactose (10:1 over dextrose) in the medium. This quantity of acid is sufficient to overcome the alkaline reaction evolving in the slant and the entire tube remains yellow in color. Acid/acid reactions, therefore, are indicative of lactose-fermenting organisms (see Fig. 2-3, *C*, and Plate 2-4, *A*, *B*, and *C*).

Many microbiologists prefer TSI over KIA because the addition of sucrose to the formula helps to screen out Salmonella and Shigella species, since neither of these (except for rare strains) utilizes either lactose or sucrose. Therefore, any acid/acid reaction on TSI indicates that either lactose, sucrose, or both have been fermented, excluding Salmonella and Shigella. TSI also has an advantage in the early detection of *Yersinia enterocolitica,* an organism that ferments sucrose but not lactose. On KIA this organism appears as a nonlactose fermenter and cannot be detected in initial screening.

Since hydrogen sulfide is a colorless gas, an indicator must be included in the medium for its detection. Sodium thiosulfate is the chemical most often included in media to supply the sulfur atoms needed by the bacteria to produce H_2S gas. The iron salts ferrous sulfate and ferric ammonium citrate are most commonly used, each reacting with H_2S gas to produce the insoluble black precipitate, ferrous sulfide. Because an acid environment is required for an organism to produce H_2S gas (see p. 94), the blackening in KIA or TSI is often confined to the deep of the agar, particularly with non-lactose-fermenting bacteria (Plate 2-4, *B*, left tube). Thus it follows that a black deep should be read as acid, even if the usual yellow color is obscured by the black precipitate. KIA and TSI are less sensitive in the detection of H_2S than other iron-containing media such as sulfide-indole-motility (SIM). Thus, some organisms that only weakly produce H_2S may show only a trace or even an absence of blackening of KIA medium whereas a distinct black color may appear in SIM medium (see Plate 2-4, *D* and *E*).

Other Carbohydrates Useful in the Identification of the Enterobacteriaceae

The ability of microorganisms to utilize carbohydrates other than glucose, lactose, and sucrose provides useful additional characteristics for the identification of the Enterobacteriaceae. The differential fermentation patterns observed with mannitol, dulcitol, sorbitol, arabinose, raffinose, and rhamnose are helpful in differentiating different species within the Klebsiella and Enterobacter groups (see Table 2-7). Melibiose utilization helps to differentiate *Yersinia pseudotuberculosis* (90% positive and 10% slow positive) from *Yersinia pestis* (none positive). Sorbitol separates *Yersinia enterocolitica* (98.7% positive) from *Yersinia pseudotuberculosis* (none positive). Other examples will be discussed in the later section on the Enterobacteriaceae identification schema.

The IMViC Reactions

One of the earliest sets of tests used for the identification of enteric bacilli was the IMViC reactions. This acronym stands for

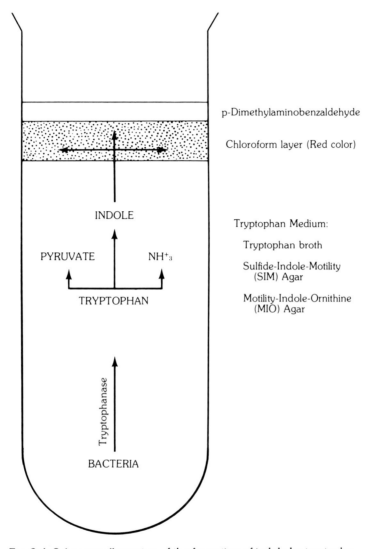

p-Dimethylaminobenzaldehyde

Chloroform layer (Red color)

INDOLE

PYRUVATE NH$^+_3$

TRYPTOPHAN

Tryptophanase

BACTERIA

Tryptophan Medium:

Tryptophan broth

Sulfide-Indole-Motility
(SIM) Agar

Motility-Indole-Ornithine
(MIO) Agar

Fig. 2-4. Schematic illustration of the formation of indole by tryptopha-
nase-producing bacteria growing on a culture medium containing tryptophan. Indole is one of the immediate degradation products (in addition to pyruvic acid and ammonia) resulting from the deamination of
tryptophan. Indole can be extracted from the aqueous phase of the
medium by chloroform and detected by the addition of Ehrlich's reagent
(dimethylaminobenzaldehyde).

79

Chart 2-4. *Indole*

Introduction	Indole, a benzyl pyrrole, is one of the metabolic degradation products of the amino acid tryptophan. Bacteria that possess the enzyme tryptophanase are capable of hydrolyzing and deaminating tryptophan with the production of indole, pyruvic acid, and ammonia. Indole production is an important characteristic in the identification of many species of microorganisms, being particularly useful in separating *Escherichia coli* (positive) from members of the Klebsiella-Enterobacter group (mostly negative).
Principle	The indole test is based on the formation of a red color complex when indole reacts with the aldehyde group of p-dimethylaminobenzaldehyde. This is the active chemical in Kovac's and Ehrlich's reagents, shown below. A medium rich in tryptophan must be used. In practice, combination media such as sulfide-indole-motility (SIM), motility-indole-ornithine (MIO), or indole-nitrate are used. Rapid spot tests, utilizing filter paper strips impregnated with Kovac's reagent, are useful in screening for bacteria that are prompt indole producers.

Media and Reagents

1. Tryptophan broth (1% tryptophan):

Peptone or pancreatic digest of casein (trypticase)	2.0 g.
Sodium chloride	0.5 g.
Distilled water	100 ml.

2. Kovac's reagent:

Pure amyl or isoamyl alcohol	150 ml.
p-Dimethylaminobenzaldehyde	10.0 g.
Concentrated HCl	50 ml.

3. Ehrlich's reagent:

p-Dimethylaminobenzaldehyde	2.0 g.
Absolute ethyl alcohol	190 ml.
Concentrated HCl	40 ml.

indole, methyl red, Voges-Proskauer, and citrate, a set of differential characteristics originally used by sanitarians and epidemiologists to detect the cross-contamination of sewage or fecal material into food and water supplies. The presence of *Escherichia coli* has been used by public health officials for many years to indicate fecal contamination. When the IMViC tests were first devised, *Aerobacter aerogenes* (now included in the genus Enterobacter) posed a problem because the colonies produced on primary isolation medium may be indistinguishable from those of *Escherichia coli*. The recovery of *Aerobacter aerogenes* was not considered to be of significance since this species is widespread in soil, grasses, and vegetative matter and does not necessarily indicate fecal contamination of water or food. The IMViC reactions were important in differentiating these two species as follows:

	Indole	Methyl Red	Voges-Proskauer	Citrate
Escherichia coli	+	+	−	−
Aerobacter aerogenes	−	−	+	+

The term "paracolon bacilli" evolved from the use of the IMViC reactions, referring to a large number of related species of enteric bacteria that produced intermediate reactions. The so-called paracolon group is no longer recognized; rather, its members

Chart 2-4. *Indole (Continued)*

Procedure	Inoculate tryptophan broth (or other indole media) with the test organism and incubate at 35° C. for 18 to 24 hours. At the end of this time, add 5 drops of reagent down the inner wall of the tube. If Ehrlich's reagent is used, this step should be preceded with the addition of 1 ml. of chloroform. This is not necessary with Kovac's reagent.
Interpretation	The development of a bright fuchsia red color at the interface of the reagent and the broth (or the chloroform layer) within seconds after adding the reagent is indicative of the presence of indole and is a positive test (Plate 2-5, *A*).
Controls	Each new batch of media or reagents should be tested for positive and negative indole reactions. The following organisms serve well as controls: Positive control: *Escherichia coli* Negative control: *Klebsiella pneumoniae* (most strains)
Bibliography	Blazevic, D. J., and Ederer, G. M.: Principles of Biochemical Tests in Diagnostic Microbiology. Pp. 63–67. New York, John Wiley & Sons, 1975. Isenberg, H. D., and Sundheim, L. H.: Indole reactions in bacteria. J. Bacteriol., *75*:682–690, 1958. McFaddin, J. F.: Biochemical Tests for Identification of Medical Bacteria. Pp. 99–108. Baltimore, Williams & Wilkins, 1976. Vracko, R., and Sherris, J. C.: Indole-spot test in bacteriology. Am. J. Clin. Pathol., *39*:429–432, 1963.

have been assigned to other well-defined genera based on additional taxonomic characteristics.

The IMViC reactions are shown in Plate 2-5. Although the characteristics included in the IMViC tests are still used for bacterial identification today, they are most commonly included as separate, unrelated reactions in most of the Enterobacteriaceae identification schemas currently being employed. Therefore, each characteristic will be discussed separately.

Indole

Indole is one of the degradation products in the metabolism of the amino acid tryptophan. Bacteria which possess the enzyme tryptophanase are capable of cleaving tryptophan with the production of indole, pyruvic acid, and ammonia (NH_3). Indole can be detected in an appropriate test medium by observing the development of a red color after adding a reagent containing p-dimethylaminobenzaldehyde (Ehrlich's or Kovac's reagents). The biochemistry of this test is schematically illustrated in Figure 2-4, and the details of the indole test are shown in Chart 2-4. A color reproduction is shown in Plate 2-5, *A*.

The choice between Ehrlich's and Kovac's reagents is one of personal preference. Ehrlich's reagent is more sensitive and is preferred when testing nonfermentative bacilli or anaerobes where indole production is often scant. Since indole is soluble in organic compounds, it is recommended that chloroform be added to the test medium prior to adding Ehrlich's reagent. This step is not necessary with Kovac's reagent since the amyl alcohol that is used for the diluent (ethyl alcohol is used with Ehrlich's reagent)

Chart 2-5. *Methyl Red*

Introduction	Methyl red is a pH indicator with a range between 6.0 (yellow) and 4.4 (red). The pH at which methyl red detects acid is considerably lower than the pH for other indicators used in bacteriologic culture media. This, in order to produce a color change, the test organism must produce large quantities of acid from the carbohydrate substrate being used.
Principle	The methyl red test is a quantitative test for acid production, requiring positive organisms to produce strong acids (lactic, acetic, formic) from glucose via the mixed acid fermentation pathway (Fig. 2-1). Since many species of the Enterobacteriaceae may produce sufficient quantities of strong acids that can be detected by methyl red indicator during the initial phases of incubation, only those organisms that can maintain this low pH after prolonged incubation (48 to 72 hours), overcoming the pH buffering system of the medium, can be called methyl-red-positive.
Media and Reagents	The medium most commonly used is methyl red-Voges-Proskauer (MR/VP) broth, as formulated by Clark and Lubs. This medium also serves for the performance of the Voges-Proskauer test.

        MR/VP broth:

Polypeptone	7.0 g.
Dextrose	5.0 g.
Dipotassium phosphate	5.0 g.
Distilled water to:	1.0 liter

Final pH = 6.9

       Methyl red pH indicator:
          Methyl red, 0.1 g., in 300 ml. of 95% ethyl alcohol
          Distilled water, 200 ml.

Procedure	Inoculate the MR/VP broth with a pure culture of the test organism. Incubate the broth at 35° C. for 48 to 72 hours (no less than 48 hours). At the end of this time, add 5 drops of the methyl red reagent directly to the broth.
Interpretation	The development of a stable red color in the surface of the medium is indicative of sufficient acid production to lower the pH to 4.4, and is a positive test (Plate 2-5, *B*). Since other organisms may produce lesser quantities of acid from the test substrate, an intermediate orange color between yellow and red may develop. This does not indicate a positive test.
Controls	Positive and negative controls should be run after preparation of each lot of medium and after making each batch of reagent. Suggested controls are: Positive control: *Escherichia coli* Negative control: *Enterobacter aerogenes*
Bibliography	Barry, A. L., *et al.*: Improved 18-hour methyl red test. Appl. Microbiol., *20*:866–870, 1970. Blazevic, D. J., and Ederer, G. M.: Principles of Biochemical Tests in Diagnostic Microbiology. Pp. 75–77. New York, John Wiley & Sons, 1975. MacFaddin, J. F.: Biochemical Tests for Identification of Medical Bacteria. Pp. 133–137. Baltimore, Williams & Wilkins, 1976.

Color Plates
2-1 *to* 2-6

Plate 2-1
Presumptive Identification of the Enterobacteriaceae

A bacterial isolate can be presumptively grouped in the family Enterobacteriaceae by assessing both colonial and biochemical characteristics. On blood agar the colonies appear gray, opaque, mucoid, or butyrous and may or may not produce hemolysis. *Proteus vulgaris* and *Proteus mirabilis* swarm on blood agar.

As a family, all members of the Enterobacteriaceae ferment glucose, reduce nitrates to nitrites, and are negative for cytochrome oxidase. These characteristics are illustrated in this plate.

A.
Blood agar plate of a 24-hour culture showing gray, opaque, somewhat mucoid colonies characteristic of one of the members of the Enterobacteriaceae.

B.
Higher-power view comparing the large, gray, mucoid colonies of a Klebsiella species with the smaller, convex, smooth, pale yellow colonies of *Staphylococcus aureus*.

C.
Blood agar plate illustrating the swarming pattern of a motile species of Proteus. The growth from the primary colony may appear in sequential waves as shown here; or, the growth may cover the agar surface with a thin, almost invisible veil.

D-G.
All members of the Enterobacteriaceae ferment glucose with the production of acid. The acid production is detected in the medium by the color conversion of a *p*H indicator.

D.
Hektoen enteric (HE) agar with a 24-hour growth of *Escherichia coli*. The yellow color of the medium indicates acid production through the fermentation of lactose or sucrose. Colonies not producing acid remain clear and do not form a yellow color in the surrounding medium.

E.
Kligler Iron Agar (KIA) slant showing an acid (yellow) slant and acid deep. This reaction indicates fermentation of both lactose and glucose.

F.
Purple broth medium showing fermentation of glucose in the right tube (yellow color from acid production) compared to the purple, uninoculated control tube on the left.

G.
Purple broth medium showing glucose fermentation in the two tubes on the left and the production of a small amount of gas that has collected under the inverted Durham tubes.

H.
Cytochrome oxidase paper test strips (Pathotec) revealing a positive purple reaction (top) compared to the colorless control (bottom). Any organism giving a positive cytochrome oxidase reaction is not a member of the Enterobacteriaceae.

I.
Nitrate test medium showing a positive reaction (red color after addition of reagents; see Chart 2-2). All members of the Enterobacteriaceae except certain biotypes of *Enterobacter agglomerans* and Erwinia species will reduce nitrates to nitrites.

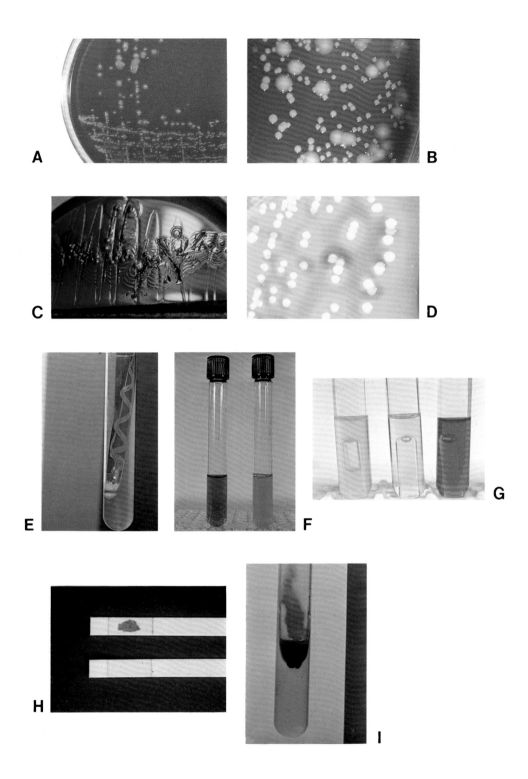

A number of selective culture media are available that help to separate members of the Enterobacteriaceae from other groups of bacteria that may be present in mixed culture. In addition to different types of growth inhibitors contained in these media, glucose and other carbohydrates, a pH indicator to detect fermentation reactions, and iron salts to detect H_2S are included in many. This plate illustrates the appearance of a number of the Enterobacteriaceae on a variety of selective media.

A.
MacConkey agar with *E. coli* after 24-hour incubation. The red pigmentation within and surrounding the colonies indicates acid production from utilization of the lactose in the medium. This reaction is characteristic of the "lactose fermenters."

B.
MacConkey agar with red, lactose-fermenting bacterial colonies and nonpigmented colonies of a nonlactose-fermenting organism. All non-lactose-fermenting bacteria must be identified because the Salmonellae and the Shigellae are included in this group.

C.
MacConkey agar showing a mixture of red lactose and transparent non-lactose-fermenting bacterial colonies. Extension of the red color into the medium indicates strong acid production from lactose fermentation.

D.
Eosin methylene blue agar (EMB) with colonies of *E. coli*. The green, metallic sheen, although not diagnostic of *E. coli*, is frequently seen with this organism. The green sheen is indicative of strong acid production with a drop in pH to 4.5 or lower.

E.
Closer view of an EMB agar plate with colonies of *E. coli*. The green sheen here is less obvious.

F.
EMB agar with a mixture of lactose- (green sheen) and non-lactose-fermenting colonies (light purple and semitranslucent).

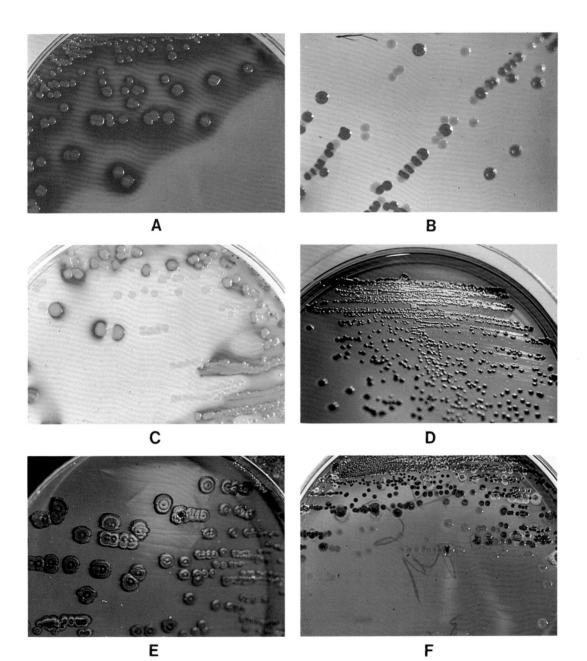

A

B

C

D

E

F

Plate 2-2 (Part II)

Appearance of Enterobacteriaceae Colonies
on Selective Isolation Media

G.
Desoxycholate citrate agar (DCA) with a few colonies of one of the species of the Enterobacteriaceae. The pink tinge of the colonies indicates weak lactose fermentation.

H.
DCA containing a few colonies of lactose-fermenting bacteria (pink) and many non-lactose-fermenting colonies. DCA agar is inhibitory to the growth of *E. coli*, allowing the selective growth of non-lactose-utilizing bacteria such as Shigella and Salmonella species.

I.
DCA with non-lactose-fermenting colonies in pure culture.

J.
Salmonella-Shigella agar (SS) with a few red colonies denoting lactose-fermenting bacteria mixed with numerous colonies of nonlactose fermenting bacteria. SS agar is more inhibitory to the growth of *E. coli* than Salmonella and Shigella species, permitting selective growth.

K.
Higher-power view of SS agar showing both lactose- (red) and non-lactose-fermenting colonies (gray-white).

L.
SS agar with non-lactose-fermenting bacterial colonies in pure culture. The growth of *E. coli* and other lactose-fermenting species has been totally inhibited.

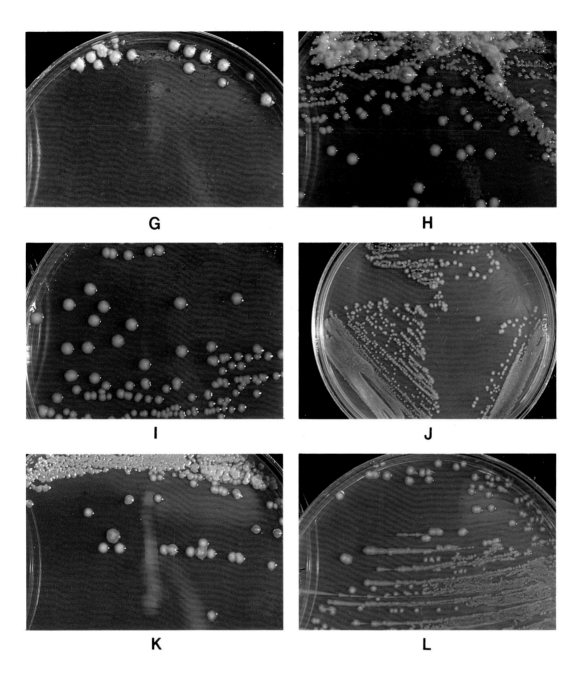

G

H

I

J

K

L

Plate 2-3
Appearance of Enterobacteriaceae on Selective Isolation Culture Media

A.

Hektoen enteric (HE) agar with a mixture of yellow and green bacterial colonies. HE agar contains three carbohydrates: lactose, sucrose, and salicin with thymol blue as a pH indicator. Bacteria fermenting one or more of these three carbohydrates appear yellow. Bacteria not fermenting one of these three carbohydrates are translucent, appearing green.

B.

High-power view of an HE agar plate containing a mixture of green, yellow and black colonies. HE agar contains ferric ammonium citrate and bacteria producing H_2S will appear black.

C.

HE agar plate with a mixture of black, H_2S-producing colonies and semitranslucent, blue-green colonies of one of the non-lactose-fermenting bacteria.

D.

Xylose lysine desoxycholate (XLD) agar with colonies of a lactose-fermenting species of the Enterobacteriaceae. The carbohydrates in XLD agar are xylose, lactose, and sucrose; phenol red is the pH indicator. Colonies fermenting one or more of these carbohydrates produce acid, will appear pale yellow, and will change the color of the adjacent agar from pink to yellow, as shown in this frame.

E.

XLD agar plate with translucent, non-lactose-fermenting bacterial colonies with black centers. XLD agar contains ferric ammonium citrate and H_2S-producing bacteria appear black. The appearance of the colonies seen in this frame are suggestive of Salmonella species or Proteus species.

F.

Higher-power view of an XLD agar plate showing in more detail the colonies seen in Frame E.

G.

XLD agar plate with bacterial colonies illustrating different patterns of H_2S production. The single,

deeply pigmented black colony indicates strong H_2S production with little or no carbohydrate utilization, as often seen with Proteus species. The small, pale yellow colonies with a black center are characteristic of bacteria capable of both carbohydrate fermentation and H_2S production, suggestive of an organism such as *Edwardsiella tarda*. Note also the pale yellow lactose-fermenting colonies of *E. coli* near the bottom of the frame.

H.

XLD agar plate with many colonies characterized by orange-yellow pigmentation. A few non-lactose-fermenting, H_2S-producing colonies are also present. The orange color indicates that only small amounts of acid have been formed from the carbohydrates. Of the carbohydrates included in the medium, *Klebiella rhinoscleromatis*, ferments only salicin. This leads to weak acid production and colonies that appear orange rather than the pale yellow of strongly fermenting bacteria.

I.

XLD agar plate with many black H_2S-producing bacterial colonies mixed with a few pale yellow lactose-fermenting colonies. Note the narrow, bright red halo around many of the black colonies. This suggests that the bacteria in the colonies are capable of decarboxylating the lysine in the medium, with an increase in the pH of the adjacent agar due to release of alkaline amines. This colonial appearance is highly suggestive of one of the Salmonella species.

J.

XLD agar plate showing the pale yellow colonies characteristic of *E. coli* mixed with black colonies typical of Salmonella species.

K.

Bismuth sulfite agar plate with a number of black colonies and a metallic sheen on the surrounding surface. Bismuth sulfite agar is highly selective for *Salmonella typhi* and is one of the most sensitive media to detect H_2S produced by this organism.

L.

Bismuth sulfite agar showing colonies characteristic of *S. typhi*.

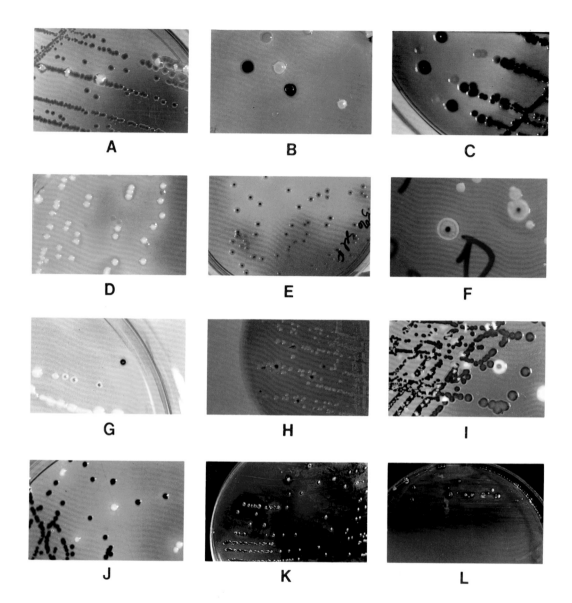

A

B

C

D

E

F

G

H

I

J

K

L

Plate 2-4
Differential Reactions of Enterobacteriaceae
on Kligler Iron Agar

Kligler iron agar (KIA) and/or triple sugar iron agar (TSI) are commonly used in the subgrouping of members of the Enterobacteriaceae based on the capacity of the test organism to ferment dextrose and/or lactose (and sucrose in TSI) and to produce H_2S. Alkaline reactions are indicated by red and acid reactions by a yellow color of the agar. The culture medium is poured in a tube and hardens in a slant. The surface of the slant is exposed to ambient air, while the agar deeper in the tube portion provides an anaerobic environment for incubation. The various reactions produced by the Enterobacteriaceae on KIA are shown in the frames in this plate.

A.
Series of four KIA agar slants. The tube to the left is an uninoculated control. The medium before use shows an alkaline (red) slant and an alkaline deep (abbreviated alk/alk). The center two tubes reveal alkaline slant/acid deep (alk/acid) reactions indicating fermentation of dextrose but the lack of lactose utilization. This type of reaction is characteristic of non-lactose-fermenting bacteria such as the Shigella and Providence species. The fourth tube on the right shows an alk/alk reaction indicating that neither dextrose nor lactose were utilized. An alk/alk reaction on KIA is characteristic of a group of gram-negative bacteria that do not ferment glucose and therefore excludes the organism from the Enterobacteriaceae.

B.
Series of four KIA agar slants. The tube to the left reveals an alkaline slant; the deep is black, reflecting H_2S production. This reaction is characteristic of several species of Proteus and most species of Salmonellae. The second tube shows an alk/acid reaction which is characteristic of H_2S negative strains of Proteus and Salmonella species. The last two tubes in this frame show acid/acid reactions characteristic of lactose fermenting bacteria. The third tube also shows the

presence of a small amount of gas, indicated by the bubble in the deep of the agar; this reaction is often produced by *E. coli* and many species of the Kelbsiella-Enterobacter group.

C.
Alk/acid reaction on a KIA agar slant (right), compared to the alk/alk appearance of an uninoculated tube on the left. The small amount of H_2S present primarily at the point of inoculation in the slant of the tube is a reaction commonly produced by *Salmonella typhi*.

D.
Sulfide-indole-motility (SIM) semisolid agar tube (left), and a KIA agar slant (right), showing an alk/acid reaction. This frame illustrates the difference in sensitivity in the detection of H_2S between the two media. The SIM tube indicates the production of H_2S not detected in the KIA slant, after both media were inoculated with the same strain of *Salmonella typhi*.

E.
SIM and KIA media inoculated with the same organism with a faint amount of H_2S detected in only the SIM tube. Note the extension of black pigment from the inoculation stab line in the SIM medium, indicating that the test organism not only produces H_2S but is motile as well.

F.
Orthonitrophenylgalactopyraniside (ONPG) test is performed in culture tubes and a positive reaction is interpreted by observing the development of a yellow color (presence of orthonitrophenol), as seen in the tube on the left, compared to the clear appearance of the uninoculated control tube (right). A positive reaction indicates that the organism is capable of producing beta galactosidase, one of the enzymes required for the initial degradation of lactose.

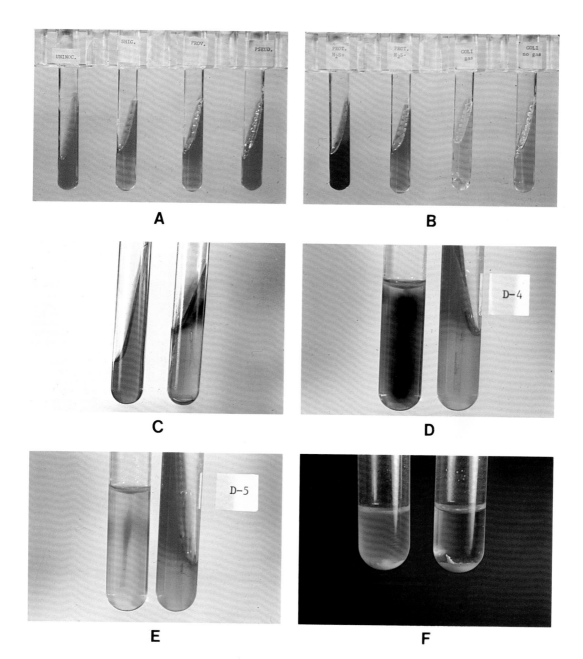

Plate 2-5
Identification of the Enterobacteriaceae
The IMViC Reactions

The acronym IMViC stands for indole, methyl red, Voges-Proskauer, and citrate, four differential characteristics that, historically, have been important in the preliminary separation of *E. coli* from other members of the Enterobacteriaceae. These reactions were used for many years by epidemiologists and sanitarians to detect possible fecal contamination of food and water supplies, indicated by the presence of *E. coli*. *E. coli* will give a positive reaction for the first two characteristics but is negative for the other two. In contrast to *E. coli*, members of the Enterobacter group will give negative reactions for indole production and the methyl red test, but are positive for the Voges-Proskauer test and citrate utilization.

A.
Two tubes of SIM medium illustrating a positive indole reaction, evidenced by the bright red ring of color near the top of the medium in the tube on the right, compared to a negative test performed on the uninoculated culture medium on the left. Indole is produced from tryptophan contained in the test medium and the red color is produced upon addition of Kovac's or Ehrlich's reagent (p-dimethylaminobenzaldehyde).

B.
Broth media illustrating the methyl red and Voges-Proskauer tests. The red color in the positive methyl red tube indicates the production of mixed acids in sufficient quantities to drop the pH below 4.4, the level at which the indicator dye turns red.

C.
Positive Voges-Proskauer test as illustrated by the light red color in the V-P tube on the right. The red color indicates the presence of acetyl methyl carbinol, detected by adding potassium hydroxide and alpha-naphthol to the inoculated medium.

D, E and F.
Citrate agar slants showing various reactions. The utilization of sodium citrate as the sole source of carbon by the organism under test is indicated by either the development of a blue color in the medium (alkaline reaction) or the visible presence of growth on the slant surface. (Both are shown in Frame *E*. Frame *D* is a negative control tube.) Microorganisms that utilize sodium citrate only weakly produce the blue color in only the slanted portion of the medium (Frame *F*).

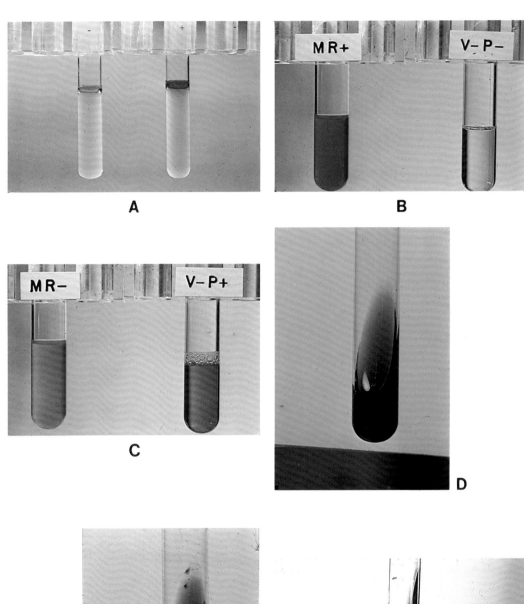

Plate 2-6
Characteristics in the Identification of the Enterobacteriaceae

A.

Christensen's urea agar slant. The bright red color seen in the medium is indicative of the hydrolysis of urea by the test organism with the release of ammonia and the development of an alkaline pH. Although many members of the Enterobacteriaceae can hydrolyze urea, the rapid and complete conversion of the tube to a red color within a few hours is characteristic of the members of the Proteus genus.

B.

Three Christensen's urea agar slants. The tube to the right shows a strong positive test (red color throughout the medium), compared to the negative control on the left (yellow color). The center tube reveals moderate urease with only the slant portion of the tube showing the red color. This latter pattern is often produced by Klebsiella species and certain members of the Enterobacter group.

C and D.

A series of tubes containing Moeller decarboxylase broth. In Frame C, the tubes, from left to right, contain lysine, arginine, ornithine, and a control tube of the same culture medium but free of amino acids. The decarboxylation of the amino acid releases alkaline amines, evidenced by a deep red or purple color, in contrast to the yellow color of the control and negatively reacting tubes.

In Frame C, the organism decarboxylated lysine and ornithine (purple color) but not arginine (yellow). Frame D compares a positive decarboxylase reaction (right) with an uninoculated tube of Moeller decarboxylase broth. Although the uninoculated medium has a shade of purple, it is not difficult to distinguish true positive reactions, because the intensity of the purple color is greater in the tube containing the test organism.

E.

Two tubes of phenylalanine agar. Upon addition of $FeCl_3$, a green color on the surface of the medium, as shown in the tube on the left, indicates the presence of phenylpyruvic acid (a degradation product of the phenylalanine in the medium) through the action of phenylalanine deaminase. Of the Enterobacteriaceae, only members of the Proteus and Providence groups produce phenylalanine deaminase to give a positive reaction on phenylalanine agar.

F.

SIM and KIA media. A small amount of H_2S is noted in the SIM tube but not in the KIA slant. The flare out of the dark discoloration from the inoculum indicates that the test organism is motile.

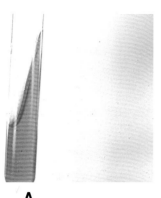

A

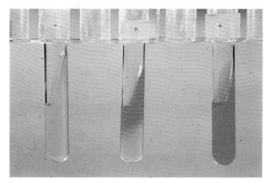

B

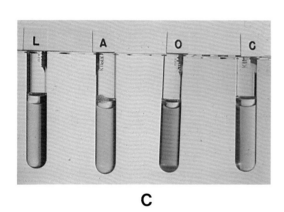

C

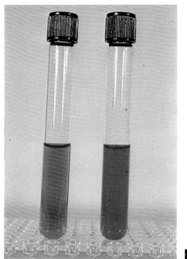

D

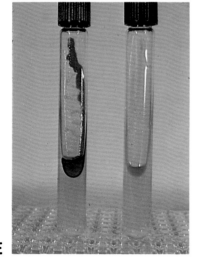

E

F

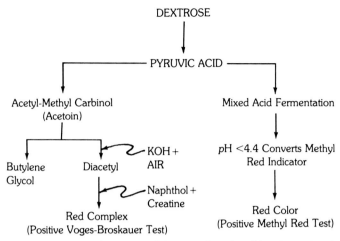

Fig. 2-5. Simplified scheme of the mixed-acid and butylene glycol pathways of dextrose fermentation.

is capable of extracting sufficient indole from the aqueous medium to produce a positive reaction.

Methyl Red

A simplified schema showing only two alternative pathways (mixed acid and butylene glycol) for the metabolism of the pyruvate formed from the fermentation of glucose is shown in Figure 2-5. Bacteria that follow primarily the mixed acid fermentation route often produce sufficient acid to maintain the pH below 4.4 (the acid color break point of the methyl red indicator) against the buffer system of the test medium. This provides a valuable characteristic for identifying those bacterial species that produce strong acids and are referred to as methyl-red-positive.

The details of the methyl red test are shown in Chart 2-5. The test as originally described requires 48 to 72 hours of incubation before a valid result can be read, a time delay unacceptable in most clinical microbiology laboratories. Barry[1] (see Chart 2-5) has described a modification that can be read in 18 to 24 hours. A 0.5-ml. aliquot of broth is used with a relatively heavy inoculum of the test organism. One or two drops of methyl red reagent are added after 18 to 24 hours of incubation at 35° C., and the development of a red color indicates a positive test. The Barry modification is as accurate as the test originally described and results in a significant saving of time.

Voges-Proskauer Test

The details of the Voges-Proskauer test are shown in Chart 2-6 and are based on the conversion of acetyl-methyl carbinol (acetoin) to diacetyl through the action of KOH and atmospheric oxygen. Diacetyl is converted into a red complex under the catalytic action of α-naphthol and creatine (Plate 2-5, *C*).

Note in Figure 2-5 that the fomation of acetoin and butylene glycol is an alternative pathway for the metabolism of pyruvic acid. Bacteria that utilize this pathway, such as the Klebsiella-Enterobacter group, produce

Chart 2-6. *Voges-Proskauer Test*

Introduction	Voges-Proskauer is a double eponym, named after two microbiologists working at the time of the turn of the 20th century. They first observed the red color reaction produced by appropriate culture media after treatment with potassium hydroxide. It was later discovered that the active product in the medium formed by bacterial metabolism is acetyl-methyl carbinol, a product of the butylene glycol pathway shown in Figure 2-5.
Principle	Pyruvic acid, the pivotal compound formed in the fermentative degradation of glucose, is further metabolized via a number of metabolic pathways depending upon the enzyme systems possessed by different bacteria. One such pathway results in the production of acetoin (acetyl-methyl carbinol), a neutral-reacting end product. Organisms such as members of the Kleb-siella-Enterobacter group produce acetoin as the chief end product of glucose metabolism, and form less quantities of "mixed acids." In the presence of atmospheric oxygen and 40% potassium hydroxide, acetoin is converted to diacetyl, and α-naphthol serves as a catalyst to bring out a red color complex.
Media and Reagents	Media: The medium is the MR/VP broth described in Chart 2-5. Reagents:

1. Alpha-naphthol (5%)	5.0 g.	
Absolute ethyl alcohol	100 ml.	
2. Potassium hydroxide (40%)	40.0 g.	
Distilled water to:	100 ml.	

lesser quantities of mixed acids, insufficient to lower the pH of the methyl red medium to produce a color change. For this reason, most species of the Enterobacteriaceae that are Voges-Proskauer-positive, with rare exceptions, are methyl-red-negative (Plate 2-5, *C*). Also, methyl-red-positive bacteria are usually Voges-Proskauer-negative, as shown by the following schema:

	Methyl Red	Voges-Proskauer
Escherichia	+	−
Shigella	+	−
Salmonella	+	−
Arizona	+	−
Citrobacter	+	−
Klebsiella	−	+
Enterobacter	−	+
Serratia	−	+
Proteus	+	−
Providencia	+	−

Citrate Utilization

The principle of the citrate utilization test, as outlined in Chart 2-7, is to determine the capability of an organism to utilize sodium citrate as the sole source of carbon for metabolism and growth.

The original formula described by Koser in 1923 was a broth medium containing sodium ammonium phosphate, monopotassium phosphate, magnesium sulfate, and sodium citrate. Proteins and carbohydrates were omitted as carbon and nitrogen sources. The end point of the Koser test is the presence or absence of visible turbidity after inoculation and incubation of the test organism. This end point is acutally a measure of the ability of the organism to utilize carbon from sodium citrate in order to produce sufficient growth to become visible. Unfortunately, it was soon recognized that

Chart 2-6.   *Voges-Proskauer Test (Continued)*

Procedure	Inoculate a MR/VP broth tube with a pure culture of the test organism. Incubate for 24 hours at 35° C. At the end of this time, aliquot 1 ml. of broth to a clean test tube. Add 0.6 ml. of 5% α- naphthol, followed by 0.2 ml. of 40% KOH. It is essential that the reagents be added in this order. Shake the tube gently to expose the medium to atmospheric oxygen and allow the tube to remain undisturbed for 10 to 15 minutes.
Interpretation	A positive test is the development of a red color after 15 minutes following addition of the reagents, indicating the presence of diacetyl, the oxidation product of acetoin (Plate 2-5, *C*). The test should not be read after standing for over one hour because negative Voges-Proskauer cultures may produce a copperlike color, potentially resulting in a false positive interpretation.
Bibliography	Barritt, M. M.: The intensification of the Voges-Proskauer reaction by the addition of α-naphthol. J. Pathol. Bacteriol., *42*:441–454, 1936.
	Barry, A. L., and Feeney, K. L.: Two quick methods for Voges-Proskauer test. Appl. Microbiol., *15*:1138–1141, 1967.
	Blazevic, D. J., and Ederer, G. M.: Principles of Biochemical Tests in Diagnostic Microbiology. Pp. 105–107. New York, John Wiley & Sons, 1975.
	MacFaddin, J. F.: Biochemical Tests for Identification of Medical Bacteria. Pp. 194–204. Baltimore, Williams & Wilkins, 1976.
	Voges, O., and Proskauer, B.: Beitrag zur Ernährungsphysiologie und zur Differential-diagnose der Bakterien der hämorrhagischen Septicamia. Z. Hyg., *28*:20–32, 1898.

turbidity in Koser's medium is not always caused by bacterial growth. Simmons[17] resolved this problem by adding agar and bromthymol blue to the Koser formula, changing the end point to a visual color change in the medium.

Simmons citrate medium is poured into a test tube on an agar slant. A light inoculum from a colony of growth of the test organism is streaked to the surface of the agar slant. If the inoculum is too heavy, performed organic compounds within the cell walls of dying bacteria may release sufficient carbon and nitrogen to produce a false positive test. When inoculating a series of tubes of differential culture media with an unknown organism, it is important that the citrate medium be streaked first in order to prevent carryover of proteins or carbohydrates from the other media.

The production of a blue color in the test medium after 24 hours of incubation at 35° C. indicates the presence of alkaline products and a positive citrate-utilization test (Plate 2-5, *E*). If carbon is utilized from sodium citrate, nitrogen is also extracted from the ammonium phosphate contained in the medium, with release of ammonium. Occasionally visible growth may be detected along the streak line before conversion of the medium to a blue color (Plate 2-5, *F*). This also may be interpreted as a positive test for reasons similar to those used in reading the turbidity of Koser's medium as a positive test. Incubation for an additional 24 hours will usually bring out the blue color and confirm the positive test.

The citrate-utilization reactions are shown in Plate 2-5, *D*, *E*, and *F*.

Malonate, acetate, and mucate are other anionic radicals that are commonly used to determine the ability of bacteria to utilize these simple compounds as a sole source of carbon.

Chart 2-7. *Citrate Utilization*

Introduction	Sodium citrate is a salt of citric acid, a simple organic compound found as one of the metabolites in the tricarboxylic acid cycle (Krebs cycle). Some bacteria can obtain energy in a manner other than the fermentation of carbohydrates by utilizing citrate as the sole source of carbon. The measurement of this characteristic is important in the identification of many members of the Enterobacteriaceae. Any medium used to detect citrate utilization by test bacteria must be devoid of protein and carbohydrates as sources of carbon.
Principle	The utilization of citrate by a test bacterium is detected in citrate medium by the production of alkaline by-products. The medium include sodium citrate, an anion, as the sole source of carbon and ammonium phosphate as the sole source of nitrogen. Bacteria that can utilize citrate also can extract nitrogen from the ammonium salt, with the production of ammonia (NH^+), leading to alkalinization of the medium from conversion of the NH_3^+ to ammonium hydroxide (NH_4OH). Bromthymol blue, yellow below pH 6.0 and blue above pH 7.6 is the indicator.
Media and Reagents	The citrate medium most commonly used is the formula of Simmons. The medium is poured into a tube on a slant. The formula of Simmons citrate medium is:

Ammonium dihydrogen phosphate	1.00 g.
Dipotassium phosphate	1.00 g.
Sodium chloride	5.00 g.
Sodium citrate	2.00 g.
Magnesium sulfate	0.20 g.
Agar	15.00 g.
Bromthymol blue	0.08 g.
Distilled water to:	1.00 liter

Final pH = 6.9

Urease Production

Microorganisms that possess the enzyme urease have the capability of hydrolyzing urea with the release of ammonia, an important characteristic for making a species identification. The details of the urease test are shown in Chart 2-8.

Important differences between Stuart's urea broth and Christensen's urea agar should be noted. Stuart's broth is heavily buffered with phosphate salts at pH 6.8. Relatively large quantities of ammonia must be formed by the test organism before the buffer system is overcome and the pH of the medium is elevated sufficiently to produce a color change of the indicator (above 8.0).

Stuart's broth, therefore, is virtually selective for species of the genus Proteus.

Christensen's urea agar[6] is far less buffered and in addition contains peptones and dextrose. This medium will support the growth of many species of bacteria that cannot grow in Stuart's broth, and the decreased buffer capacity allows for the detection of less ammonia production. Thus, many bacterial species with less active urease production, such as the Klebsiella, many of the Enterobacter species Yersinia species, Cryptococcus species, Brucella species, and Bordetella bronchiseptica, can be tested with Christensen's urea agar. With many of these species, a positive urease reaction is first detected by a pink to red color change

Chart 2-7. *Citrate Utilization (Continued)*

Procedure	A well-isolated colony is picked from the surface of a primary isolation medium and inoculated as a single streak on the slant surface of the citrate agar tube. The tube is placed into a 35° C incubator for 24 to 48 hours.
Interpretation	A positive test is the development of a deep blue color within 24 to 48 hours, indicating that the test organism has been able to utilize the citrate contained in the medium, with the production of alkaline products (Plate 2-5, *D, E,* and *F*). A positive test may also be read in the absence of a blue color if there is visible colonial growth along the inoculation streak line. This is possible because for growth to be visible the organism must enter the log phase of growth, possible only if carbon and nitrogen had been assimilated. A positive interpretation from reading the streak line can be confirmed by incubating the tube for an additional 24 hours, when a blue color usually develops.
Controls	Each new batch of medium should be tested with a positive- and a negative-reacting organism. The following species are suggested controls: Positive control: *Enterobacter aerogenes* Negative control: *Escherichia coli*
Bibliography	BBL Manual of Products and Laboratory Procedures. ed. 5., Pp. 115 and 138. Cockeysville, Md., Division of BioQuest, Division of Becton, Dickinson and Co., 1968. Blazevic, D. J., and Ederer, G. M.: Principles of Biochemical Tests in Diagnostic Microbiology. Pp. 15–18. New York, John Wiley & Sons, 1975. Koser, S. A.: Utilization of the salts of organic acids by the colon-aerogenes group. J. Bacteriol., 8:493–520, 1923. MacFaddin, J. F.: Biochemical Tests for Identification of Medical Bacteria. Pp. 35–40. Baltimore, Williams & Wilkins, 1976. Simmons, J. S.: A culture medium for differentiating organisms of typohod-colon aerogenes groups and for isolation of certain fungi. J. Infect. Dis., 39:209–214, 1926.

in the slant portion of the agar (see Plate 2-6, *A* and *B*). It is the slant that initially turns red because the alkaline reaction resulting from the splitting of small quantities of urea is augmented by the amines formed from the oxidative decarboxylation of the proteins in the media (similar to the slant reactions in KIA or TSI described on p. 76).

Decarboxylase

Many species of bacteria possess enzymes capable of decarboxylating specific amino acids in the test medium, with the release of alkaline-reacting amines and carbon dioxide as products. A number of test systems have been described to measure this property, based either on detecting an alkaline pH shift in the test medium, or by directly measuring the reaction products. An elaborate manometric method has been described for measuring the CO_2 gas that is formed, or paper chromatography can be used to detect the presence of specific amines. Neither of these methods is commonly employed in clinical microbiology laboratories. The amines can be more easily determined by having them react with Ninhydrin reagent after extracting the amines from the test medium with chloroform. This is the relatively sensitive Carlquist reaction,[5] most commonly used for detecting the weak decarboxylase activity of many of the nonfermentative gram-negative bacilli and certain

Chart 2-8. *Urease*

Introduction	Urea is a diamide of carbonic acid with the formula:

$$\underset{NH_2-C-NH_2.}{\overset{\overset{O}{\parallel}}{}}$$

All amides are easily hydrolyzed with the release of ammonia and carbon dioxide.

Principle	Urease is an enzyme possessed by many species of microorganisms that can hydrolyze urea following the chemical reaction:

$$NH_2-\overset{\overset{O}{\parallel}}{C}-NH_2 + 2HOH \xrightarrow{\text{Urease}} CO_2 + H_2O + 2NH_3 \rightleftharpoons (NH_4)_2CO_3$$

The ammonia reacts in solution to form ammonium carbonate, resulting in alkalization and an increase in the pH of the medium.

Media and Reagents	Stuart's urea broth and Christensen's urea agar are the two media most commonly used in clinical laboratories for the detection of urease activity.

Stuart's Urea Broth		*Christensen's Urea Agar*	
Yeast extract	0.1 g.	Peptone	1.000 g.
Monopotassium phosphate	9.1 g.	Dextrose	1.000 g.
Disodium phosphate	9.5 g.	Sodium chloride	5.000 g.
Urea	20.0 g.	Monopotassium phosphate	2.000 g.
Phenol red	0.01 g.	Urea	20.000 g.
Final pH = 6.8		Phenol red	0.012 g.
		Agar	15.000 g.
		Final pH = 6.8	

Procedure	The broth medium is inoculated with a loopful of the test organism previously isolated in pure culture; the surface of the agar slant is streaked

---

species of anaerobic bacteria. This method will be discussed in more detail in Chapters 3 and 10.

The decarboxylase activity of the Enterobacteriaceae is most commonly measured in clinical microbiology laboratories with Moeller decarboxylase broth.[13] The details of this test are shown in Chart 2-9. The end point of the reaction is the production of an alkaline pH shift in the medium and the development of a blue-purple color after incubation with the test organism (Plate 2-6, *C* and *D*).

Note in the Moeller formula included in Chart 2-9 that the medium is buffered at pH 6.0. This is relatively more acid than most culture media. This low pH is necessary since the decarboxylase enzymes are not optimally active until the pH of the medium drops below 5.5. The drop from 6.0 to 5.5 is accomplished by the growing bacteria that utilize the small amount of glucose in the medium, with the production of mixed acids. A control tube, devoid of amino acid, must always be included when performing the decarboxylase test to ensure that this

Chart 2-8. *Urease (Continued)*

with the test organism. Both media are incubated at 35° C. for 18 to 24 hours.

Interpretation	Organisms that hydrolyze urea rapidly may produce positive reactions within one or two hours; less active species may require three or more days. The reactions are: Stuart's broth: red color throughout the medium indicates alkalization and urea hydrolysis. Christensen's agar: 1. Rapid urea splitters (Proteus species): red color throughout medium (Plate 2-6, *A*). 2. Slow urea splitters (Klebsiella species): red color initially in slant only, gradually converting the entire tube (Plate 2-6, *B*). 3. No urea hydrolysis: media remains original yellow color (Plate 2-6, *B*).
Controls	Positively and negatively reacting control organisms should be run with each new batch of medium. Following are suggested organisms: Positive control: Proteus species Positive control (weak): Klebsiella species Negative control: *Escherichia coli*
Bibliography	BBL Manual of Products and Laboratory Procedures. ed. 5, p. 154. Cockeysville, Md., Division of BioQuest, Division of Becton, Dickinson and Co., 1968. Christensen, W. B.: Urea decomposition as a means of differentiating *Proteus* and paracolon cultures from each other and from *Salmonella* and *Shigella* types. J. Bacteriol. 52:461–466, 1946. MacFaddin, J. F.: Biochemical Tests for Identification of Medical Bacteria. Pp. 187–194. Baltimore, Williams & Wilkins, 1976. Stuart, C. A., Van Stratum, E., and Rustigian, R.: Further studies on urease production by *Proteus* and related organisms. J. Bacteriol., 49:437–444, 1945.

initial pH drop has occurred. This is detected by observing a yellow color change in the medium due to the conversion of the bromcresol purple indicator. Pyridoxal phosphate is included in the medium which acts as a coenzyme to further enhance the decarboxylase activity.

Many microbiologists prefer Falkow lysine broth[10] over the Moeller medium because the test depends only on a shift in the pH indicator and neither an anaerobic nor an acid environment is required. However, this medium cannot be used for detecting lysine decarboxylase activity of Klebsiella or Enterobacter species. Both of these produce acetyl-methyl carbinol, which interferes with the final alkaline pH shift, leading to false negative interpretations. Modifications of this medium form the basis of the motility indole ornithine (MIO) semisolid agar widely used in clinical microbiology laboratories.

Edwards and Fife[8] described a solid lysine decarboxylase medium based on the Falkow formula that includes ferric ammonium citrate and thiosulfate for the detection of H_2S

Chart 2-9. *Decarboxylases*

Introduction	Decarboxylases are a group of substrate-specific enzymes that are capable of attacking the carboxyl (COOH) portion of amino acids, with the formation of alkaline-reacting amines. This reaction, known as decarboxylation, forms carbon dioxide as a second product. Each decarboxylase enzyme is specific for an amino acid. Lysine, ornithine, and arginine are the three amino acids routinely tested in the identification of the Enterobacteriaceae. The specific amine products are:

$$\text{Lysine} \longrightarrow \text{Cadaverine}$$
$$\text{Ornithine} \longrightarrow \text{Putrescine}$$
$$\text{Arginine} \longrightarrow \text{Citrulline}$$

The conversion of arginine to citrulline is a dihydrolase rather than a decarboxylase reaction, in which an NH_2 group is removed from arginine as a first step. Citrulline is next converted to ornithine, which then undergoes decarboxylation to form putrescine.

Principle   Moeller decarboxylase medium is the base most commonly used for determining the decarboxylase capabilities of the Enterobacteriaceae. The amino acid to be tested is added to the decarboxylase base prior to inoculation with the test organism. A control tube, consisting of only the base without the amino acid, must also be set up in parallel. Both tubes are anaerobically incubated by overlying with mineral oil. During the initial stages of incubation, both tubes turn yellow due to the fermentation of the small amount of glucose in the medium. If the amino acid is decarboxylated, alkaline amines are formed and the medium reverts to its original purple color.

Media and Reagents   Moeller decarboxylase broth base:

Peptone	5.000 g.
Beef extract	5.000 g.
Bromcresol purple	0.010 g.
Cresol red	0.005 g.
Glucose	0.500 g.
Pyridoxal	0.005 g.
Distilled water to:	1.000 liter
Final pH	= 6.0

Amino acid:
    Add 10 g. (final concentration = 1%) of the l-form of the amino acid (lysine, ornithine, or arginine). Double this amount if the d-l (dextro) form is to be used since only the l (levo) form is active.

Procedure   From a well-isolated colony of the test organism previously recovered on primary isolation agar, inoculate two tubes of Moeller decarboxylase medium, one containing the amino acid to be tested, the other a control tube devoid of amino acid. Overlay both tubes with sterile mineral oil to cover about 1 cm. of the surface and place in a 35° C. incubator for 18 to 24 hours.

Interpretation   Conversion of the control tube to a yellow color indicates that the organism is viable and that the pH of the medium has been lowered sufficiently to

Chart 2-9.

Interpretation *(Continued)*	activate the decarboxylase enzymes. Reversion of the tube containing the amino acid to a blue-purple color indicates a positive test due to the release of amines from the decarboxylation reaction (Plate 2-6, *C* and *D*).
Controls	The following organisms are suggested for positive and negative controls: *Positive Control* *Negative Control* Lysine: *Enterobacter aerogenes* *Enterobacter cloacae* Ornithine: *Enterobacter cloacea* *Klebsiella* species Arginine: *Enterobacter cloacae* *Enterobacter aerogenes*
Bibliography	Blazevic, D. J., and Ederer, G. M.: Principles of Biochemical Tests in Diagnostic Microbiology. Pp. 29–36. New York, John Wiley & Sons, 1975. Carlquist, P. R.: A biochemical test for separating paracolon groups. J. Bacteriol., *71*:339–341, 1956. Falkow, S.: Activity of lysine decarboxylase as an aid in the identification of Salmonellae and Shigellae. Am. J. Clin. Pathol. 29:598–600, 1958. Gale, E. F.: The bacterial amino acid decarboxylases. *In* Nord, F. F. (ed): Advances in Enzymology and Related Subjects of Biochemistry. Vol. 6. New York, Interscience Publishers, 1946. MacFaddin, J. F.: Biochemical Tests for Identification of Medical Bacteria. Pp. 52–64. Baltimore, Williams & Wilkins, 1976. Moeller, V.: Simplified tests for some amino acid decarboxylases and for the arginine dihydrolase system. Acta Pathol. Microbiol. Scand., *36*:158–172, 1955.

gas. This medium is lysine iron agar (LIA), used in many laboratories as an aid in the identification of Salmonella species, most of which are both H_2S and lysine-decarboxylase-positive. A black deep and a purple slant with LIA are virtually diagnostic of Salmonella species. LIA also has the advantage in that Proteus and Providencia species, both of which deaminate rather than decarboxylate amino acids, can be detected by a red color in the slant of the tube.

The lysine decarboxylase test is useful in differentiating lactose-negative Citrobacter species (0% positive) from the Salmonellae (94.6% positive). Almost all strains of *Shigella sonnei* possess ornithine decarboxylase activity, while only a few strains of *Shigella boydii* (2.5%) show such activity, and neither *Shigella dysenteriae* nor *Shigella flexneri* show activity. The ornithine decarboxylase test is perhaps most useful in separating Klebsiella species (all negative) from Enterobacter species (most strains positive).

The decarboxylase reactions are shown in Plate 2-6, *C* and *D*.

Phenylalanine Deaminase

The phenylalanine deaminase determination is useful in the initial differentiation of Proteus and Providencia species from other gram-negative bacilli. Only members of these genera possess the enzyme responsible for the oxidative deamination of phenylalanine, except for 4.4 per cent of *Enterobacter agglomerans*, a relatively rare isolate.

The test is easily performed, as outlined in Chart 2-10. Phenylpyruvic acid may be detected as soon as 4 hours after a heavy inoculum, although a period of 18 to 24 hours of incubation is generally recommended. The phenylalanine test medium employs yeast extract as the source for carbon and nitrogen because meat extracts or protein hydroly-

Chart 2-10. *Phenylalanine Deaminase*

Introduction	Phenylalanine is an amino acid which upon deamination forms a keto acid, phenylpyruvic acid. Of the Enterobacteriaceae, only members of the Proteus and Providencia genera possess the deaminase enzyme necessary for this conversion.
Principle	The phenylalanine test depends upon the detection of phenylpyruvic acid in the test medium after growth of the test organism. The test is positive if a visible green color develops upon addition of a solution of 10% ferric chloride.

Media and Reagents   Phenylalanine agar: the agar is poured as a slant into a tube. Meat extracts or protein hydrolysates cannot be used because of their varying natural content of phenylalanine. Yeast extract serves as the carbon and nitrogen source. The formula is:

DL-Plenylalanine	2 g.
Yeast extract	3 g.
Sodium chloride	5 g.
Sodium phosphate	1 g.
Agar	12 g.

Final pH = 7.3

Ferric chloride:

Ferric chloride	12.0 g.
Concentrated HCl	2.5 ml.
Distilled water to:	100.0 ml.

Procedure   The agar slant of the medium is inoculated with a single colony of the test organism isolated in pure culture of primary plating agar. After incubation at 35° C. for 18 to 24 hours, 4 or 5 drops of the ferric chloride reagent are added directly to the surface of the agar. As the reagent is added, the tube is rotated to dislodge the surface colonies.

Interpretation   The immediate appearance of an intense green color indicates the presence of phenylpyruvic acid and a positive test (Plate 2-6, *E*).

Controls   Each new batch of media or reagent must be tested with positive- and negative-reacting organisms. The following are suggested:
Positive control: Proteus species
Negative control: *Escherichia coli*

Bibliography   Blazevic, D. J., and Ederer, G. M.: Principles of Biochemical Tests in Diagnostic Microbiology. Pp. 23–28. New York, John Wiley & Sons, 1975.

Hendriksen, S. D.: A comparison of the phenylpyruvic acid reaction and urease test in the differentiation of *Proteus* from other enteric organisms. J. Bacteriol. *60:*225–231, 1950.

MacFaddin, J. F.: Biochemical Tests for Identification of Medical Bacteria. Pp. 170–175. Baltimore, Williams & Wilkins, 1976.

Shaw, C., and Clarke, P. H.: Biochemical classification of *Proteus* and Providence cultures. J. Gen. Microbiol., *13:*155–161, 1955.

Hendriksen, S. D., and Closs, K.: The production of phenylpyruvic acid by bacteria. Acta Pathol. Microbiol. Scand., *15:*101–113, 1938.

Table 2-4. *Media for the Detection of H₂S*

Media	Sulfur Source	H₂S Indicator
Bismuth sulfite	Peptones plus sulfite	Ferrous sulfate
Citrate sulfide agar	Sodium thiosulfate	Ferric ammonium citrate
Deoxycholate-citrate agar	Peptones	Ferric citrate
Lysine iron agar	Sodium thiosulfate	Ferric ammonium citrate
Kligler iron agar	Sodium thiosulfate	Ferrous sulfate
Triple sugar iron agar	Sodium thiosulfate	Ferrous sulfate
Lead acetate agar	Sodium thiosulfate	Lead acetate
Salmonella-Shigella agar	Sodium thiosulfate	Ferric citrate
SIM medium	Sodium thiosulfate	Peptonized iron
XLD or HE agar	Sodium thiosulfate	Ferric ammonium citrate

sates contain varying amounts of naturally occurring phenylalanine which would lead to inconsistent results. The development of the green color after addition of the ferric chloride reagent is immediate and easy to visualize (see Plate 2-6, *E*).

Hydrogen Sulfide Production

The ability of certain bacterial species to liberate sulfur from sulfur-containing amino acids or other compounds in the form of H_2S gas is an important characteristic for their identification. H_2S gas production can be detected in a test system if the following conditions are present:

1. *There is a source of sulfur in the medium.* The various protein complexes that are included in media contain sufficient qualties of the sulfur-containing amino acids, cystine and methionine, for the production of H_2S gas. Sodium thiosulfate is an inorganic compound that is commonly added to medium as an additional source of sulfur.

2. *There is an H₂S indicator in the medium.* Ferrous sulfate, ferric citrate, ferric ammonium sulfate or citrate, peptonized iron, and lead acetate are the sulfide indicators most commonly included in media formulated to detect H_2S gas.

3. *The medium supports the growth of the bacterium being tested.*

4. *The bacterium possesses the H₂S-producing enzyme systems.*

In Table 2-4 are listed the media most commonly used for the detection of H_2S gas,

showing the sources of sulfur and the sulfide indicators.

The sequence of steps leading to the production and detection of H_2S gas in a test system is thought to be as follows:

1. Release of sulfide from cysteine or from thiosulfate by bacterial enzymatic action

2. Coupling of sulfide (S^{-2}) with hydrogen ion (H^+) to form H_2S gas

3. Detection of the H_2S gas by heavy metal salts, such as iron, bismuth, or lead, in the form of a heavy metal-sulfide, black precipitate

The differences in sensitivity to H_2S gas detection by the different media result from alterations in one or more of these conditions. H_2S gas detected in one medium may not be detected in another, and it is necessary to know the test system used when interpreting identification charts. SIM is more sensitive to H_2S detection than is KIA, presumably because of its semisolid consistency, lack of carbohydrates to suppress H_2S formation, and the use of peptonized iron as the indicator (see Plates 2-4, *D* and *E*, and 2-6, *F*). KIA, in turn, is more sensitive than TSI because sucrose in particular is thought to suppress the enzyme mechanisms responsible for H_2S production. Lead acetate is the most sensitive indicator and should be used whenever testing bacteria that produce H_2S only in trace amounts. Unfortunately, lead acetate is also inhibitory to the growth of many fastidious bacteria, precisely the ones that may require a sensitive detector system. These organisms can be tested for H_2S gas production by draping a lead-acetate-impregnanted filter paper strip under the cap of a culture tube of KIA medium (see Plate 4-1, *I*). In this way the extreme sensitivity of the lead acetate indicator can be used without incorporating it directly into the medium.

With all H_2S-detection systems, the end point is an insoluble, heavy metal-sulfide, black precipitate in the medium or on the filter paper strip. Since the availability of hydrogen ions is necessary for H_2S gas formation, the blackening is first seen in test media where acid formation is maximum, that is, along the inoculation line or within the deeps of slanted agar media or in the center of colonies growing on agar surfaces (see Plates 2-3 and 2-4).

Motility

Bacterial motility is another important characteristic in making a final species identification. Bacteria move by means of flagella, the number and location of which vary with the different species. Flagellar stains are available for this determination, but are not commonly used.

Bacterial motility can be directly observed by placing a drop of culture broth medium on a microscope slide and viewing this under a microscope. Hanging-drop chambers are available so that the preparation can be viewed under higher magnification without danger of lowering the objectives into the contaminated drop. This technique is used primarily for detecting the motility of those bacterial species that do not grow well in agar media. This is not a problem with the Enterobacteriaceae, and tubes containing semisolid agar are most commonly employed.

Motility media have agar concentrations of 0.4 per cent or less. At higher concentrations the gel is too firm to allow free spread of the organisms. Combination media, such as sulfide-indole-motility (SIM) or motility-indole-ornithine (MIO), have found wide use in clinical microbiology laboratories because more than one characteristic can be measured in the same tube. The motility test must be interpreted first since the addition of indole reagent may obscure the results. Because SIM and MIO have a slightly turbid background, interpretations may be somewhat difficult with bacterial species that grow slowly in these media. In these cases, the following test medium is recommended because it supports the growth of most fastidious bacteria and has a crystal-clear appearance:

Motility Test Medium	
(Edwards and Ewing)	
Beef extract	3 g.
Peptone	10 g.
Sodium chloride	5 g.
Agar	4 g.
Distilled water to:	1 liter
Final pH = 7.3	

The motility test is interpreted by making a macroscopic examination of the medium for a diffuse zone of growth flaring out from the line of inoculation (see Plate 2-6, *F*). The use of tetrazolium salts in motility medium has been advocated to aid in the visual detection of bacterial growth. Tetrazolium salts are colorless but are converted into insoluble red formazan complexes by the reducing properties of growing bacteria. In a motility test medium containing tetrazolium, the development of this red color helps to trace the spread of bacteria from the inoculation line. However, these salts are inhibitory to some fastidious bacteria, and these bacteria are often those that grow slowly in motility medium where the tetrazolium indicator would be most helpful.

Of the Enterobacteriaceae, species of Shigella and the Klebsiellae are uniformly nonmotile. Most motile species of the Enterobacteriaceae can be interpreted at 35° C.; however, *Yersinia enterocolitica,* in which flagellar proteins develop more rapidly at lower temperatures, is motile only at 22° C. (room temperature). *Listeria monocyto-*

genes is another bacterial species that requires room-temperature incubation before motility becomes active. *Pseudomonas aeruginosa,* an organism that grows well only in the presence of oxygen, produces a spreading film on the surface of motility agar and does not show the characteristic fanning out from the inoculation line because it does not grow in the deeper oxygen-deficient portions of the tube.

IDENTIFICATION SYSTEMS USED FOR THE SPECIES IDENTIFICATION OF THE ENTEROBACTERIACEAE

To this point the discussion in this chapter has been devoted to developing a working understanding of the culture media and tests used to measure the characteristics by which the various members of the Enterobacteriaceae can be identified. This is necessary so that the various schemas by which these test results are grouped into identification patterns can be better understood.

Identification schemas are systems that provide a means for identifying all members of a given group of microorganisms by listing them in some form together with notations of all their positive and negative characteristics. Although there are numerous such systems in use, most are based on one of the following four general approaches:

The crosshatch or checkerboard matrix
The grouping system of Edwards and Ewing
The branching flow diagrams
The numerical coding systems

The first three of these will be discussed here. The numerical coding system is presented in detail in Chapter 6.

The Crosshatch or Checkerboard Matrix

One of the earliest versions of the crosshatch or checkerboard matrix system is the chart showing the determinations of the IMViC reactions shown on page 80. Although this chart is quite simple, limited to only two species of bacteria, it follows the general format of the checkerboard matrix by listing the biochemical tests along one coordinate and the organisms that could be separated by these reactions on the opposite coordinate. An expanded checkerboard matrix is shown in Table 2-5. Lines are drawn to produce a crosshatch grid with the reactions listed in the squares of intersection. For instance, in Table 2-5 the appropriate lines of intersection indicate that Salmonella is positive for the methyl red test while Klebsiella is negative for this characteristic.

Although not commonly done in the past, current authors of identification schemas should define the confidence levels for all of the reactions listed. Thus, at the bottom of Table 2-5 all of the positive reactions indicated with a $(+)$ signify that at least 90 per cent of the bacterial species tested are positive for that characteristic. For instance, 90 per cent or more of the salmonellae are positive for methyl red. Similarly, at least 90 per cent of salmonellae are negative for the production of acetyl-methyl carbinol, as indicated by the $(-)$ notation in the Voges-Proskauer square of intersection. The notations $(+/-)$ and $(-/+)$ indicate that more than 50 per cent but less than 90 per cent of the species tested for that character are either positive or negative, respectively. For instance, in Table 2-5, the $(-/+)$ in the Shigella-Indole intersection indicates that more than 50 per cent of shigellae are indole-negative, but not more than 90 per cent indole-negative. The actual percentages listed by Edwards and Ewing[7] based on their data base of cultures submitted to the Center for Disease Control in Atlanta are that 72.2 per cent of shigellae are indole-negative and 37.8 per cent are indole-positive. If one decides to accept identification characteristics only at a 90 per cent or greater confidence level, the indole reaction must be excluded from the list of characteristics indicative of Shigella species.

The checkerboard matrix system shown in

**Table 2-5. Sample of a Simplified Checkerboard Matrix
for Identification of the Enterobacteriaceae**

	Escherichia	Shigella	Salmonella	Arizona	Citrobacter	Klebsiella	Enterobacter cloacae	Enterobacter aerogenes	Hafnia	Serratia	Proteus vulgaris	Proteus mirabilis	Proteus morganii	Proteus retgeri	Providencia
Indole	+	−/+	−	−	−	−/+	−	−	−	−	+	−	+	+	+
Methyl red	+	+	+	+	+	−	−	−	+/−	−/+	+	+	+	+	+
Voges-Proskauer	−	−	−	−	−	+	+	+	+/−	+	−/+	−/+	−	−	−
Simmons citrate	−	−	d	+	+	+	+	+	+/−	+	d	+	−	+	+
H₂S gas (TSI)	−	−	+	+	+/−	−	−	−	−	−	+	+	−	−	−
Urease	−	−	−	−	dʷ	+ʷ	+ʷ−	−	−	dʷ	+	+	+	+	−
Motility	+/−	−	+	+	+	−	+	+	+	+	+	+	+	+	+
Lysine	d	−	+	+	−	+	−	+	+	+	−	−	−	−	−
Ornithine	d	d	+	+	d	−	+	+	+	+	−	+	+	−	−
Phenylalanine	−	−	−	−	−	−	−	−	−	−	+	+	+	+	+
Gas from glucose	+	−	+	+	+	+	+	+	+	+/−	+/−	+	d	−/+	−
Lactose	+	−	−	d	d	+	+	+	−/+	−/+	−	−	−	−	−
Sucrose	d	−	−	−	d	+	+	+	d	+	+	d	−	d	d

Key: (+) = 90% + of isolates positive; (−) = 90% + of isolates negative; (d) = delayed positive (3–5 days); (w) = weak reaction; (+/−) = majority positive, but less than 90% positive; (−/+) = majority negative, but less than 90% negative.

Table 2-5 has the advantage that all of the organisms of a given group and their characteristics are listed in the chart. Since all possible reactions of the tests included are listed, the chart can be read with great accuracy since positive, negative, and variable characters can be crosschecked. However, this makes the identification of any given bacterial species somewhat tedious in that each identification characteristic must be checked against all other organisms listed. In the event that the identification is in doubt because a 90 per cent confidence level has not been reached from the information available on the chart, there is no listing of additional identification characteristics that might be helpful. In addition, it is not possible from such a schema to easily spot identification patterns that may be suggestive of a certain organism. For example, over 90 per cent of *Escherichia coli* can be identified from the reactions on KIA, SIM (or MIO), and citrate. This information cannot be determined from the checkerboard matrix;

Table 2-6. *Tribes, Genera, and Identification Characteristics of the Enterobacteriaceae (Edwards and Ewing)*

Tribe	Genus		Genus Characteristics	Reactions
Tribe I: Escherichieae	Genus I:	Escherichia	Methyl Red	+
	Genus II:	Shigella	Voges-Proskauer	−
			Citrate	−
			Hydrogen sulfide	−
			Urease	−
			Phenylalanine deaminase	−
			Indole	+/−
Tribe II: Edwardsielleae	Genus I:	Edwardsiella	Indole	+
			Hydrogen sulfide	+
			Lysine decarboxylase	+
			Urease	−
			Phenylalanine deaminase	−
			ONPG	−
Tribe III: Salmonelleae	Genus I:	Salmonella	Indole	−
	Genus II:	Arizona	Methyl red	+
	Genus III:	Citrobacter	Voges-Proskauer	−
			Citrate	+
			Urease	−
			Phenylalanine deaminase	−
			Hydrogen sulfide	+
Tribe IV: Klebsielleae	Genus I:	Klebsiella	Indole	−
	Genus II:	Enterobacter	Methyl red	−
	Genus III:	Pectobacterium	Voges-Proskeuer	+
	Genus IV:	Serratia	Citrate	+
			Potassium cyanide (KCN)	+
			Hydrogen sulfide	−
			Urease	−*
			Phenylalanine deaminase	−
Tribe V: Proteeae	Genus I:	Proteus	Phenylalanine deaminase	+
	Genus II:	Providencia	Methyl red	+

*Some strains are weakly positive, turning the slant of Christensen's urea agar a light pink.

rather, all of the reactions must be determined in order to derive full value from the chart.

The Grouping System of Edwards and Ewing

In order to overcome some of the disadvantages of the checkerboard chart as cited above, Edwards and Ewing,[7] from a study of a large number of isolates received for many years at the Center for Disease Control in Atlanta, have divided the Enterobacteriaceae into five tribes. This division is based on the selection of a key group of positive and negative characteristics by which each tribe can be easily identified. The five tribes of Edwards and Ewing, the genera of bacteria within each tribe, and the determining characteristics for each tribe are listed in Table 2-6. For example, the information shown in Table 2-6 indicates that the genus Salmonella can be suspected by observing positive reactions for methyl red, citrate, and hydrogen sulfide gas production, and negative reactions for indole, Voges-Proskauer, urease, and phenylalanine deaminase. Note that the determination of only two characteristics, phenylalanine deaminase and methyl red, is necessary to suspect Proteeae. In order to differentiate the species within each genus, additional characteristics must be determined, as discussed on the following page.

This subgrouping is particularly helpful in facilitating identification. Often an unknown bacterium can be readily placed into one of the five tribes, based on the recognition of a few select identification characteristics. These recognition patterns are important in providing the direction for selecting the appropriate additional characteristics that must be determined in order to most quickly arrive at a final identification. Often it is helpful to a physician to know to which tribe an unknown bacterium belongs so that treatment may be initiated before a final species identification is possible. In the following sections are brief discussions of each of these tribes and the bacterial species included in each.

Tribe I: The Escherichia-Shigella Group. As shown in Table 2-6, the Escherichia-Shigella group should be suspected if a test organism is positive-reacting for methyl red, but is negative-reacting for Voges-Proskauer, citrate, H₂S, urease, and phenylalanine. The old IMViC formula, indole +, methyl red +, Voges-Proskauer −, and citrate −, still serves as a good clue for *Escherichia coli.* Generally the differentiation of *Escherichia coli* from Shigella species is not difficult because the colonies of *Escherichia coli* appear as lactose fermenters on primary isolation media, turn KIA or TSI acid throughout, and are motile, in contract to the nonlactose, nonmotile characteristics of the shigellae. However, there are strains of late lactose-fermenting, nonmotile *Escherichia coli* that can closely mimic the shigellae in primary culture, and the following set of differential tests may be required for their differentiation:

	Escherichia coli	Shigellae
Motility	+	−
Lysine	+	−
Arginine	+	−
Ornithine	+	−
Acetate	+	−
Mucate	+	−
Glucose	+	−
Lactose	+	−
Salicin	+	−

Escherichia coli is the currently accepted name for the common coliform bacillus originally called *Bacillus coli commune* by Escherich in 1885, *Bacillus coli* by Migula in 1895, and *Bacterium coli* by Lehmann in 1896. Most strains are nonpathogenic in the bowel, and entereotoxin-producing strains are no longer felt to be clinically significant. *Escherichia coli* is currently of most clinical significance in humans because of its role as an opportunistic pathogen causing urinary tract, wound, and blood infections in man.

Shigella is a group of four well-defined species and a number of serotypes which regularly cause infectious dysentery in man. The genus is named after K. Shiga, who termed the bacillus he discovered *Bacillus dysenteriae* in 1898. This group represents one of the pathogenic nonlactose fermenters that are screened for in cultures in outbreaks of infectious diarrhea. Shigella can be suspected in differential tests by its negative reactions in various test media, as shown in the above chart.

Tribe II: Edwardsielleae. The tribe Edwardsielleae is the most recently described, and was initially called the Asakusa group by Sakazaki and Murata[16] in 1962 and the Bartholomew group by King and Adler in 1964.[12] Ewing and McWhorter suggested the name Edwardsielleae in 1965[9], in honor of the prominent American microbiologist, P. R. Edwards. This bacterium resembles *Escherichia coli* in its IMViC reactions and failure to produce urease or deaminate phenylalanine, but simulates Citrobacter species or the Salmonelleae by its production of H₂S gas in TSI agar and failure to utilize lactose. *Edwardsiella tarda* is the single species, and has been described associated with cases of human dysentery, meningitis, and septicemia.

Tribe III: Salmonelleae. Of all the Enterobacteriaceae, the Salmonelleae are by far the most complex, with over 2200 serotypes currently described in the Kauffmann-White scheme. The Salmonelleae cause varying degrees of gastroenteritis in man, with *Salmonella typhi* causing the most severe disease, typhoid fever. Along with the Shigelleae, the Salmonelleae constitute the other group of pathogenic nonlactose fermenters which are routinely screened for in cases of diarrheal disease.

Salmonella species can be suspected in screening tests for any organism that is non-lactose-fermenting, produces H_2S in the deep of KIA or TSI agars, utilizes citrate as a carbon source, and does not produce urease. Serologic grouping can be easily performed for early confirmation of suspicious colonies.

Arizona and Citrobacter species are culture "look-alikes" in primary cultures that can be separated from Salmonella by the characteristics shown in the following scheme:

	Salmonella	Arizona	Citrobacter
Malonate	−	+	+ / −
Lysine	+ *	+	−
Arginine	+ †	+	d
Ornithine	+ †	+	d
KCN	−	−	+
Geletin	−	d	−
Lactose	−	d	d
Dulcitol	+	−	d

Salmonella paratyphi A is negative.
†*Salmonella typhi* and *Salmonella gallinarum* are negative.
d = delayed

Salmonella is named after the American microbiologist; D. E. Salmon. The more than 2200 serotypes have been divided into three major clinical groups: *Salmonella typhosa,* causing typhoid fever; *Salmonella choleraesuis,* causing enteric fever with septicemia; and *Salmonella enteritidis;* the remaining strains cause various degrees of gastroenteritis in man, with considerable individual variation in severity. Salmonella can be easily subgrouped into six O groups, A, B, C_1, C_2, D, and E; however, further species differentiation requires the services of a reference laboratory since more than 30 antisera are required to derive the exact subspecies name.

The Arizona group, given a separate ge-

nus designation by Ewing and Fife *(Arizona hinshawii)* because of certain differences in characteristics as listed in the chart above, has been included as a subspecies of *Salmonella enteritidis* in the edition 8 of *Bergey's Manual of Determinative Bacteriology. Arizona hinshawii* has been recovered from cases of gastroenteritis and a variety of localized infections in man and lower animals.

The genus Citrobacter has had a variety of prior designations, including *Bacterium freundii* (Braak, 1928), *Escherichia freundii* (Yale, 1939), and *Bathesda Ballerup* (West and Edwards, 1949). *Citrobacter freundii* is the current type species, closely resembling the Salmonella and Arizona species in preliminary screening. The negative lysine decarboxylation reaction and ability to grow in potassium cyanide (KCN) broth are the important differential characteristics. *Citrobacter freundii* is considered an opportunistic and not a true pathogen in man.

Tribe IV: Klebsielleae. The tribe Klebsielleae includes a heterogeneous group of bacteria separated into four genera: Klebsiella, Enterobacter, Pectobacterium, and Serratia. The taxonomy is still unsettled, and continuing shifts in nomenclature are confusing. For example, *Enterobacter liquefaciens* has recently been shifted to the genus Serratia; *Bacterium herbicola,* formerly in the Erwinia group, is now *Enterobacter agglomerans,* and the pectobacteria are now included in the genus Erwinia.

As shown in Table 2-6, a bacterium can be suspected of belonging to the tribe Klebsielleae if the IMViC reactions are − − + +, if growth is observed in KCN broth, and if H_2S gas, urease, and phenylalanine deaminase are not produced. These reactions, however, may be variable for any given species and additional characteristics often must be determined before a final identification can be made. Because of its short shelf life and inconsistent reactions, KCN is not commonly used in clinical laboratories. The differential characteristics of the genera and species of the tribe Klebsielleae are listed in Table 2-7.

Table 2-7. *Differential Characteristics of Klebsiella, Enterobacter, and Serratia*

	Klebsiella	Enterobacter				Serratia		
		cloacae	aerogenes	hafniae	agglomerans	marcescens	liquefaciens	rubidaea
Motility	−	+	+	+	+/−	+	+	+/−
Lysine	+	−	+	+	−	+	+	+
Arginine	−	+	−	d	−	−	−	−
Ornithine	−	+	+	+	−	+	+	−
Inositol	+	d	+	−	d	d	+	d
Sorbitol	+	+	+	−	d	+	+	−
Adonitol	+	+/−	+	−	−	d	d	+
Raffinose	+	+	+	−	d	−	+	+
Arabinose	+	+	+	+	−	−	+	+
Rhamnose	+	+	+	+	−	−	−	−

Genus Klebsiella. Members of the genus Klebsiella are important pathogens for man, causing enteritis in children and upper respiratory tract infections or pneumonia, meningitis, and urinary tract infections in both children and adults. Klebsiella is suspected when a large colony with a mucoid consistency is recovered on primary isolation agar plates. Because of lactose fermentation, red colonies are produced on MacConkey agar and acid slant/acid deep reactions are seen on KIA or TSI. Many species of Enterobacter can closely simulate the klebsiellae in screening tests; however, the lack of motility and inability to decarboxylate ornithine separate the latter species. Many strains of Klebsiella can slowly split urea, producing a light pink color change in the slant of Christensen's urea agar.

The genus Klebsiella is named after Edwin Klebs, a late 19th-century German microbiologist. The Klebsiella bacillus was described by Carl Friedlander and for many years the Friedlander bacillus was well known as a cause of severe, often fatal, pneumonia. *Klebsiella pneumoniae* is the type species and the classic reactions are shown in the above chart. Although the IMViC reactions are − − + + for most *Klebsiella pneumoniae*, 6 per cent of strains are indole-positive, 13.3 per cent are methyl-red-positive, and 8.9 per cent are Voges-Proskauer-negative in the Edwards and Ewing data base.[7] *Klebsiella ozaenae*

and *Klebsiella rhinoscleromatis*, two less frequently encountered species, are more likely to yield aberrant biochemical reactions (almost 100% are methyl-red-positive and none are Voges-Proskauer-negative).

Genus Enterobacter. The members of the genus Enterobacter differ from Klebsiella as shown in the chart above. In particular, the members of Enterobacter are motile and capable of decarboxylating ornithine. There are four species in the Ewing classification: *E. aerogenes, E. cloacae, E. hafniae,* and *E. agglomerans.* These bacteria are common commensals in the intestinal tract of man and animals, but are being recovered with increasing frequency from patients with urinary tract infections, wound infections, and septicemia.

Because of the large amount of gas produced by many strains of the Enterobacter group, the type species for many years was called *Aerobacter aerogenes.* The genus designation was changed to Enterobacter by Edwards and Ewing in 1962. The many ill-defined paracolon bacilli, producing intermediate reactions in the IMViC tests between *Escherichia coli* and *Aerobacter aerogenes,* have also been classified within the Enterobacter group. *Enterobacter hafniae* and *Enterobacter agglomerans* are the more recent additions to the genus, important in clinical medicine as opportunistic pathogens in immunosuppressed hosts. These organisms were recently isolated with frequency from septicemic patients receiving contaminated intravenous solutions.[11]

Genus Serratia. The emergence of certain strains of Serratia as important opportunistic pathogens causing severe pulmonary infections and/or septicemia in immunosuppressed hosts is a classic example of how a previously innocuous organism can assume spontaneous virulence. The type-specific species, *Serratia marcescens,* is readily identified in cultures because of the production of deep red pigment. This organism is a free-living commensal in natural bodies of water and soil and has been considered nonpathogenic to man. Recently, however, nonpigmented strains have emerged that are not only virulent, but also resistant to many of the antibiotics currently used in clinical practice.

The genus is separated into three species, *Serratia marcescens, Serratia liquefaciens,* and *Serratia rubidaea* under the classification proposed by Edwards and Ewing.[7]

Because of the potential virulence of Serratia species, it is important that this organism be separated from the Enterobacter group. The production of detectable extracellular DNase by Serratia is one reliable characteristic by which this separation can be made. Of the Enterobacter group, only occasional strains of *E. liquefaciens* (now included in the genus Serratia) are DNase-positive. Differences in the utilization of carbohydrates such as dulcitol, adonitol, inositol, sorbitol, arabinose, and raffinoe also aid in the identification of Serratia species.

Tribe V: Proteeae

Ewing proposes two genera, Proteus and Providencia, in the tribe Proteeae. However, the classification of Providencia is still unsettled. In the edition 8 of *Bergey's Manual of Determinative Bacteriology,* Providencia is not recognized as a genus; rather, it is designated *Proteus inconstans* in the genus Proteus. It is common usage in most clinical laboratories to consider Providencia a separate bacterial genus with two species. Clinically this controversy is of minor relevance in that both Proteus species and Providencia species are most commonly associated with urinary tract infections and wound infections. Species of Proteus are recovered far more frequently in cases of human infection than are Providencia species.

Rustigan and Stuart[15] have divided the genus Proteus into four species: *Proteus vulgaris, Proteus mirabilis, Proteus morganii,* and *Proteus rettgeri.* There are species differences in the production of indole, utilization of citrate, and decarboxylation of ornithine by which the identification can be made. Species differentiation of the Proteus group is of clinical importance in that *Proteus mirabilis* is usually susceptible to penicillin while *Proteus vulgaris* commonly displays resistance to a wide spectrum of antibiotics.

Two species of Providencia are recognized: *Providence alcalifaciens* and *Providence stuartii.* Differentiation is based primarily on the production of gas from glucose (*P. alcalifaciens,* positive), utilization of adonitol (*P. alcalifaciens,* positive), and utilization of inositol (*P. stuartii,* positive).

The Proteus and Providencia bacilli can be separated from the rest of the Enterobacteriaceae by their ability to deaminate phenylalanine. They also have the unique property of oxidatively deaminating lysine, which can be detected by the appearance of a red color in the slant portion of a lysine iron agar (LIA) tube.

Most of the species of Proteus are rapid hydrolyzers of urea; the species of Providencia are urease-negative. *Proteus mirabilis* and *Proteus vulgaris* actively swarm on blood agar, covering the surface with a thin film. Only *Proteus mirabilis* is indole-negative, an important characteristic in screening tests to derive a rapid diagnosis.

Except for rare exceptions, all species of Proteus and Providencia are unable to utilize lactose and appear as transparent colonies on primary isolation agars such as

(Text continues on p. 104)

Table 2-8.   *Biochemical Reactions for Identification of*

| Tribe | Escherichieae | | Edwardsielleae | Salmonelleae | | | Klebsielleae | Enterobacter | |
Test or Substrate	Escherichia	Shigella	Edwardsiella	Salmonella	Arizona	Citrobacter	Klebsiella	cloacae	aerogenes
Indol	+	- or +	+	-	-	-	- or +	-	-
Methyl red	+	+	+	+	+	+	-	-	-
Voges-Proskauer	-	-	-	-	-	-	+	+	+
Simmons's citrate	-	-	-	d	+	+	+	+	+
Hydrogen sulfide (TSI)	-	-	+	+	+	+ or -	-	-	-
Urease	-	-	-	-	-	d^w	+	+ or -	-
KCN	-	-	-	-	-	+	+	+	+
Motility	+ or -	-	+	+	+	+	-	+	+
Gelatin (22 C.)	-	-	-	-	(+)	-	-	(+) or -	- or (+)
Lysine decarboxylase	d	-	+	+	+	-	+	-	+
Arginine dihydrolase	d	- or (+)	-	(+) or +	+ or (+)	d	-	+	-
Ornithine decarboxylase	d	d*	+	+	+	d	-	+	+
Phenylalanine deaminase	-	-	-	-	-	-	-	-	-
Malonate	-	-	-	-	+	d	+	+ or -	+ or -
Gas from glucose	+	- *	+	+	+	+	+	+	+
Lactose	+	- *	-	-	d	d	+	+	+
Sucrose	d	- *	-	-	-	d	+	+	+
Mannitol	+	+ or -	-	+	+	+	+	+	+
Dulcitol	d	d	-	d[†]	-	d	- or +	- or +	-
Salicin	d	-	-	-	-	d	+	+ or (+)	+
Adonitol	-	-	-	-	-	-	+ or -	- or +	+
Inositol	-	-	-	d	-	-	+	d	+
Sorbitol	+	d	-	+	+	+	+	+	+
Arabinose	+	d	-	+[†]	+	+	+	+	+
Raffinose	d	d	-	-	-	d	+	+	+
Rhamnose	d	d	-	+	+	+	+	+	+

* Certain biotypes of *S. flexneri* produce gas; *S. sonnei* cultures ferment lactose and sucrose slowly and decarboxylate ornithine.

† *S. typhosa, S. choleraesuis, S. enteritidis* bioser. Paratyphi A and Pullorum, and a few others ordinarily do not ferment dulcitol promptly. *S. choleraesuis* does not ferment arabinose.

‡ Gas volumes produced by cultures of Serratio, Proteus, and Providencia are small.

Klebsielleae						Proteeae					
Enterobacter				Serratia	Pectobacterium	Proteus				Providencia	
hafniae		liquefaciens				vulgaris	mirabilis	morganii	rettgeri	alcalifaciens	stuartii
37 C.	22 C.	37 C.	22 C.		25 C.						
−	−	−	−	−	− or +	+	−	+	+	+	+
+ or −	−	+ or −	− or +	− or +	+ or −	+	+	+	+	+	+
+ or −	+	− or +	+ or −	+	− or +	−	− or +	−	−	−	−
(+) or −	d	+	+	+	d	d	+ or (+)	−	+	+	+
−	−	−	−	−	−	+	+	−	−	−	−
−	−	d	−	dw	dw	+	+	+	+	−	−
+	+	+	+	+	+ or −	+	+	+	+	+	+
+	+	d	+	+	+ or −	+	+	+	+	+	+
	−	+	+	+ or (+)	+ or (+)	+	−	−	−	−	−
+	+	+ or −	+	+	−	−	−	−	−	−	−
−	−	−	−	−	− or +	−	−	−	−	−	−
+	+	+	+	+	−	−	+	+	−	−	−
−	−	−	−	−	−	+	+	+	+	+	+
+ or −	+ or −	−	−	−	− or +	−	−	−	−	−	−
+	+	+	+	+ or −(‡)	− or +	+ or −	+	d	− or +	+ or −	−
− or (+)	− or (+)	d	(+)	− or (+)	d	−	−	−	−	−	−
d	d	+	+	+	+	+	d	−	d	d	d
+	+	+	+	+	+	−	−	−	+ or −	−	d
−	−	−	−	−	−	−	−	−	−	−	−
d	d	+	+	+	+	d	d	−	d	−	−
−	−	d	d	d	−	−	−	−	d	+	−
−	−	+	+	d	−	−	−	−	+	−	+
−	−	+	+	+	−	−	−	−	d	−	d
+	+	+	+	−	+	−	−	−	−	−	−
−	−	+	+	−	+ or (+)	−	−	−	−	−	−
+	+	−	−	−	d	−	−	−	+ or −	−	−

+, 90 per cent or more positive in 1 or 2 days. −, 90 per cent or more negative. d, different biochemical types [+, (+), −]. (+), delayed positive. + or −, majority of cultures positive.
− or +, majority negative. w, weakly positive reaction.

Courtesy, Enteric Section, Enterobacteriology Branch, Bacteriology Division, CDC, Atlanta, Ga.

Table 2-9.   *Classification of the Enterobacteriaceae*
(Bergey's Manual of Determinative Bacteriology, ed. 8)

	Genus	Species
Genus I	Escherichia	Escherichia coli
Genus II	Edwardsiella	Edwardsiella tarda
Genus III	Citrobacter	Citrobacter freundii
		Citrobacter intermedius
Genus IV	Salmonella	Salmonella choleraesuis
		Salmonella typhi
		Salmonella enteritidis
		(numerous serotypes)
Genus IV	Shigella	Shigella dysenteriae
		Shigella flexneri
		Shigella boydii
		Shigella sonnei
Genus VI	Klebsiella	Klebsiella pneumoniae
		Klebsiella ozaenae
		Klebsiella rhinoscleromatis
Genus VII	Enterobacter	Enterobacter cloacae
		Enterobacter aerogenes
Genus VIII	Hafnia	Hafnei alvei
Genus IX	Serratia	Serratia marcescens
Genus X	Proteus	Proteus vulgaris
		Proteus morganii
		Proteus mirabilis
		Proteus rettgeri
		Proteus inconstans
Genus XI	Yersinia	Yersinia pestis
		Yersinia pseudotuberculosis
		Yersinia enterocolitica
Genus XII	Erwinia	Erwinia amylovora
		Erwinia salicis
		Erwinia tracheiphila

MacConkey or EMB. Since many species of Proteus also produce H_2S gas, these non-lactose-fermenting colonies are often confused with Salmonella.

Table 2-8 is a recent chart published by the Enteric Bacteriology Laboratory at the Center for Disease Control in Atlanta, listing all of the differential characters for the identification of the Enterobacteriaceae using the Edwards and Ewing classification.

Bergey's Classification of the Enterobacteriaceae

The current classification of the Enterobacteriaceae proposed in the 1975 edition 8 of *Bergey's Manual of Determinative Bacte-*

riology[3] is shown in Table 2-9. Note the following changes compared to the Edwards and Ewing classification discussed above:

1. The family Enterobacteriaceae is now divided into 12 genera. Escherichia, Shigella, Citrobacter, Klebsiella, Hafnia, and Serratia, previously sharing a tribe designation with other organisms, now have separate genus designations.

2. The Arizona group is no longer recognized as a separate entity, but rather has been regrouped with Salmonella as the subspecies *Salmonella arizonae.*

3. The Providencia group, including *Providence alcalifaciens* and *Providence stuartii,* is no longer recognized, but is included in

the genus Proteus with the designation *Proteus inconstans.*

4. The Yersinia group, only recently separated from the genus Pasteurella, has now been incorporated into the Enterobacteriaceae. In 1954, Thal proposed this reclassification based on the utilization of carbohydrates by Yersinia by fermentative rather than oxidative pathways and the lack of cytochrome oxidase activity.

Yersinia is named after the French bacteriologist Alexander Yersin, who in 1894 first identified this bacterium as the causative organism of plague. *Yersinia pestis,* formerly *Pasteurella pestis*, is the type species and presently is endemic in rodents, with sporadic cases of human infections reported annually, particularly in the southwestern part of the United States.[14] Two other species, *Yersinia pseudotuberculosis* and *Yersinia enterocolitica,* that cause tuberculosis-like lesions of the lymph glands of the intestine (mesenteric lymphadinitis) belong to the genus.

The bacteria appear in Gram's stains as small, pleomorphic, coccobacillary gram-negative forms with bipolar staining (safety-pin effect). They grow well on blood agar and bile media, such as MacConkey and EMB agars. Colonial growth is better at 25° to 30° C. than at 35° to 37° C. The colonies appear initially as smooth, gray-yellow translucent forms, but with prolonged incubation may become rough with serrated margins.

Since carbohydrate fermentation is weak, one initial clue to the Yersinea is the yellow-orange conversion of the butt of KIA agar. *Yersinia enterocolitica* is the exception on TSI agar since it ferments sucrose and converts the deep and the slant to a yellow color. The distinguishing characteristics between the species of Yersinia, Salmonella, and Pasteurella are given in Table 2-10.

5. The Erwinia, a group of organisms better known as plant pathogens causing plant wilt and fireblight, have recently been included in the Enterobacteriaceae as a distinct genus.

Table 2-10. *Distinguishing Characteristics Between the Species of Yersinia, Salmonella, and Pasteurella*

Characteristics	*Y. pestis*	*Y. pseudotuberculosis*	*Y. enterocolitica*	*Salmonella species*	*Pasteurella multocida*
Motility					
22° C.	–	+	+	+	–
37° C.	–	–	–	+	–
Urease	–	+	+	–	–
Rhamnose	d	+	–	+	–
Salacin	d	+	–	–	–
Sucrose	–	–	+	–	+
Gas from glucose	–	–	–	+	–
ONPG	+	+	+	d	–
Ornithine	–	–	+	+	–
Citrate	–	–	–	+	–
Indole	–	–	d	–	d

The organisms of the Erwinia group are opportunistic pathogens for man, causing infections in hosts with a depressed immune response. Members of the herbicola group, currently classified as *Enterobacter agglomerans*, were recently incriminated in an outbreak of septicemia in hospitalized patients receiving contaminated intravenous fluids.[11]

Members of the Erwinia group ferment glucose with the formation of acid but not gas and are oxidase-negative. There are six species within the genus; however, only *E. amylovora, E. salicis,* and *E. tracheiphila* are considered of importance in causing human infections.

The initial clue that an organism may belong to the Erwinia genus is the formation of a somewhat dry colony with a yellow pigmentation on primary isolation agars. The second clue is from the biochemical reactions, because, with the exception of *Enterobacter agglomerans* (previously *Erwinia herbicola*), the Erwinia are unique

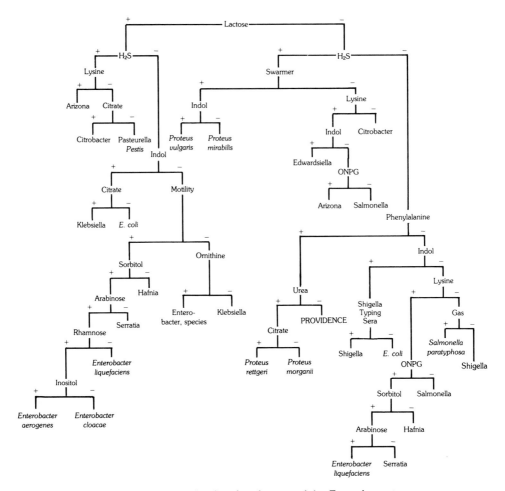

Fig. 2-6. Flow diagram for the identification of the Enterobacteriaceae.

among the Enterobacteriaceae in their inability to decarboxylate lysine, ornithine, and arginine.

Identification of the Enterobacteriaceae Using Branching Flow Diagrams

Flow diagrams have been designed to simplify the interpretation of a series of reactions used in the identification of bacteria.[2] Figure 2-6 is a representative example of one such diagram used for the identification of the Enterobacteriaceae. For optimal use of this type of diagram, it is recommended that all of the identification characteristics included in the chart be determined, even though with some bacteria only a limited number of determinations are re-

quired to derive an answer. If some of the characteristics are omitted, a branch point may be reached that requires the result of that determination before progressing further down the diagram.

The author of a flow chart must establish the confidence levels of the reactions at each point. A major disadvantage of a flow diagram such as the one illustrated in Figure 2-6 is the potential for making an erroneous identification if all branch points along the schema do not approach 100 per cent confidence. A 90 per cent confidence level is inadequate in that there is a 10 per cent chance of error at each branch point, which becomes compounded when more than one characteristic is used in making the final

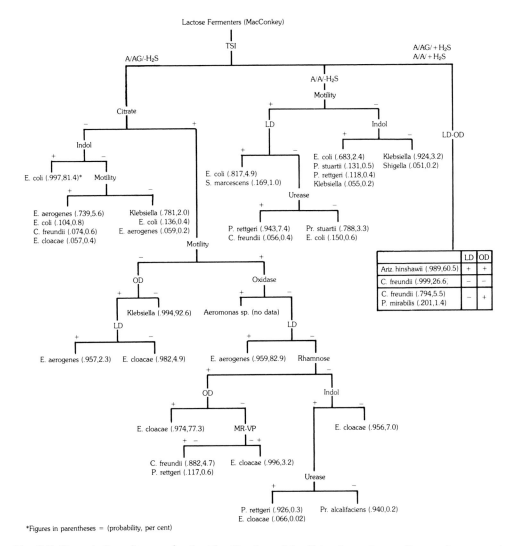

Fig. 2-7. Gavan's flow diagram for the identification of the Enterobacteriaceae (lactose fermentens).

identification. Even a 10 per cent error can conceivably lead to a 100 per cent wrong answer. Figure 2-6 is based on a 97.5 per cent confidence level, allowing the correct identification of about 97 per cent of the Enterobacteriaceae.

The inability to easily assess other positive and negative results for a given isolate is a second shortcoming of the flow diagrams. For example, according to Figure 2-6, a bacterium with the characteristics lactose-positive, H_2S-positive, and lysine-positive is identified as an "Arizona species." It would be more reassuring in establishing this iden-

tification to assess other positive and negative characters of the Arizona group before issuing a final answer based on only three reactions.

Flow charts have been constructed by Thomas Gavan that partially overcome these shortcomings. The Gavan charts are illustrated in Figures 2-7 and 2-8. The basic construction of these charts is similar to Figure 2-6 except that the choice of the identification characters and their sequence are different. An added feature of these modified flow diagrams is a notation after each bacterial species of the percentage probabil-

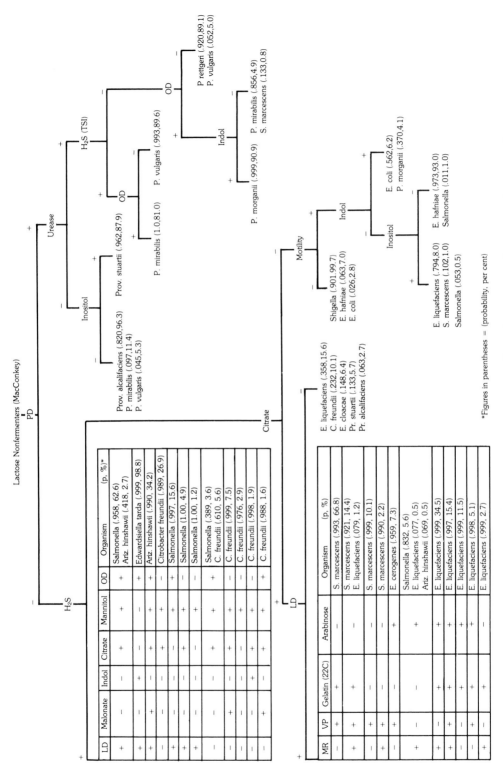

Fig. 2-8. Gavan's flow diagram for the identification of the Enterobacteriaceae (lactose nonfermentens).

ity that an identification is correct, together with a listing of alternative possibilities. For instance, the example of the lactose-positive, H₂S-positive, and lysine-positive organism used above can also be read from Figure 2-7. These reactions are designated along the right branch, leading to a box that includes *Arizona hinshawii, Citrobacter freundii,* and *Proteus mirabilis* as potential answers. An additional characteristic, ornithine decarboxylase, is also used in this schema. Note that after the name *Arizona hinshawii* the numbers .989 and 60.5 appear in parentheses. The first figure, .989, indicates that any organism that is lactose-positive, H₂S-positive, lysine-positive, and ornithine-positive has a 98.9 per cent chance of being *Arizona hinshawii.* This is sufficient for most diagnostic purposes. In this same box, note that an organism that is lactose-positive, H₂S-positive, lysine-negative, and ornithine-positive is *Citrobacter freundii* with a probability of 79.4 per cent; however, based on these four reactions, there is a 20.1 per cent chance that the correct identification is *Proteus mirabilis.* Since 79.4 per cent is less than an acceptable 90 per cent confidence level, the answer of *Citrobacter freundii* cannot be derived without determining other characteristics (urease and phenylalanine activity, for example).

The second figure in the notations behind the species name indicates the percentage of strains within a given species that give the positive and negative reactions leading to that identification. For example, the number 60.5 listed after *Arizona hinshawii* in Figure 2-7 indicates that 60.5 per cent of *Arizona hinshawii* strains are lactose-positive, H₂S-positive, lysine-positive, and ornithine-positive. This relatively low percentage results because only 61.3 per cent of *Arizona hinshawii* strains are lactose-positive according to the data base established by Edwards and Ewing. Note that *Arizona hinshawii* also appears in the left upper box of Figure 2-8, a schema designed to identify the non-lactose-fermenting members of the Enterobacteriaceae.

These modified flow diagrams have the

advantage that the user can determine from the answers derived whether or not the confidence level is acceptable, and, if not, the other organisms that must be considered in the differential diagnosis and the additional characteristics that must be determined in order to reach an acceptable confidence level.

Numerical Coding Systems for the Identification of the Enterobacteriaceae

A numerical code is a system by which the several identifying characteristics of bacteria are translated into a sequence of numbers that represent one or more bacterial species. The binary number is the easiest to derive, in that all positive reactions are designated "1" and all negative reactions "0." For example, the binary number for *Arizona hinshawii,* based on the lactose-positive, H₂S-positive, lysine-positive, and ornithine-positive, is "1111." Similarly, *Citrobacter freundii* is "1101," the "0" indicating that the lysine reaction is negative. Binary numbers are convenient for input into computers that have been programmed to derive bacterial identifications based on the data available. However, since ten or more characteristics may be required to identify a given bacterial species, binary numbers are too cumbersome to use without computers. Therefore, other numerical codes have been derived that are easier to translate by people. A full discussion of numerical codes, the derivation of biotype numbers, and their applications in clinical microbiology is presented in detail in Chapter 6.

REFERENCES

1. Barry, A. L., *et al.:* Improved 18-hour methyl red test. Appl. Microbiol., *20:*866–870, 1970.
2. Belliveau, R. R., Grayson, J. W., Jr., and Butler, T. J.: A rapid, simple method of identifying *Enterobacteriaceae.* Am. J. Clin. Pathol., *50:*126–128, 1968.
3. Breed, R. S., Murray, E. G. D., and Smith, N. R.: Bergey's Manual of Determinative Bacteriology. ed. 7. Baltimore, Williams & Wilkins, 1957.

4. Buchanan, R. E., and Gibbons, N. E.: Bergey's Manual of Determinative Bacteriology. ed. 8. Baltimore, Williams & Wilkins, 1974.

5. Carlquist, P. R.: A biochemical test for separating paracolon groups. J. Bacteriol., *71*:339–341, 1956.

6. Christensen, W. B.: Urea decomposition as a means of differentiating Proteus and paracolon cultures from each other and from Salmonella and Shigella types. J. Bacteriol., *52*:461–466, 1946.

7. Edwards, P. R., and Ewing, W. H.: Identification of *Enterobacteriaceae*. ed. 3. Minneapolis, Burgess Publishing Co., 1972.

8. Edwards, P. R., and Fife, M. A.: Lysine-iron agar in the detection of *Arizona* cultures. Appl. Microbiol., *9*:478–480, 1961.

9. Ewing, W. H., *et al.: Edwardsiella*, a new genus of *Enterobacteriaceae* based on a new species of *E. tarda*. Int. Bull. Bact. Nomencl. Taxon., *15*:33–38, 1965.

10. Falkow, S.: Activity of lysine decarboxylase as an aid in the identification of Salmonella and Shigellae. Am. J. Clin. Pathol., *29*:598–600, 1958.

11. Felts, S. K., *et al.:* Sepsis caused by contaminated intravenous fluids: epidemilogical,

clinical and laboratory investigation of an outbreak in one hospital. Ann. Intern. Med., *77*:881–890, 1972.

12. King, B. M., and Adler, D. L.: A previously unclassified group of *Enterobacteriaceae*. Am. J. Clin. Pathol., *41*:230–232, 1964.

13. Moller, V.: Simplified tests for some amino acid decarboxylases and for the arginine-dehydrolase system. Acta Pathol. Microbiol. Scand., *36*:158–172, 1955.

14. Reed, W. P., *et al.:* Bubonic plague in the Southwestern United States. Medicine, *49*:465–486, 1970.

15. Rustigan, R., and Stuart, C. A.: Taxonomic relationships in the genus *Proteus*. Proc. Soc. Exp. Biol. Med., *53*:241–243, 1943.

16. Sakazaki, R., and Murata, Y.: The new group of *Enterobacteriaceae*. The Asakusa group. Jap. J. Bacteriol., *17*:616–617, 1963.

17. Simmons, J. S.: A culture medium for differentiating organisms of typhoid-colon aerogenes groups and for isolation of certain fungi. J. Infect. Dis., *39*:209–214, 1926.

18. ZoBell, C. E.: Factors influencing the reduction of nitrates and nitrites by bacteria in semisolid media. J. Bacteriol., *24*:273–281, 1932.

3 The Nonfermentative Gram-Negative Bacilli

The term "nonfermenters" refers to a group of aerobic, non-spore-forming, gram-negative bacilli that are either incapable of utilizing carbohydrates as a source of energy or degrade them via "oxidative" rather than "fermentative" metabolic pathways.

It is not possible in this text to present in detail the biochemistry of carbohydrate metabolism and the complex pathways used by various groups of bacteria. Texts by Doelle[2] and Thimann[13] should be consulted for details. However, a brief summary of bacterial metabolism is necessary to gain a clear working understanding of such terms as "aerobic," "anaerobic," "fermentation," and "oxidation," since these refer to characteristics not only relating to the taxonomy of bacteria but also apply to procedures useful in the identification of microorganisms in the clinical laboratory.

DERIVATION OF ENERGY BY BACTERIA

Bacteria that derive their energy from organic compounds are known as chemoorganotropes. Most of the bacteria encountered in clinical medicine utilize carbohydrates by one of several metabolic pathways to achieve their energy needs. Some bacteria, such as members of the genus Moraxella, to be discussed below, do not utilize carbohydrates but can derive their energy from other types of organic compounds, such as amino acids, alcohols, or organic acids. Neutral or slightly alkaline reactions are produced when these bacteria are grown in carbohydrate test media. Some free-living bacteria, such as the nitrogen-fixing groups or those capable of oxidizing sulfur or iron, can derive their energy from simple inorganic chemicals. These chemolithotropes are seldom implicated as the cause of disease in humans.

FERMENTATIVE AND OXIDATIVE METABOLISM

The bacterial degradation of carbohydrates proceeds by several metabolic pathways involving a series of steps in which hydrogen ions (electrons) are successively transferred to compounds of higher redox potential, with the ultimate release of energy in the form of adenosine triphosphate (ATP). All six-, five-, and four-carbon carbohydrates are initially degraded to pyruvic acid, a key intermediate. Glucose serves as the main carbohydrate source of carbon for bacteria, and degradation proceeds by three major pathways; namely, the Embden-Meyerhof-Parnas (EMP), the Entner-Doudoroff (ED), and the Warburg-Dickens (hexose monophosphate or HMP) pathways. As shown in Figure 3-1, glucose is converted to pyruvic acid in each of these three pathways via a different set of degradation steps. Bacteria utilize one or more of these pathways for glucose metabolism, depending upon their enzymatic composition and the presence or absence of oxygen.

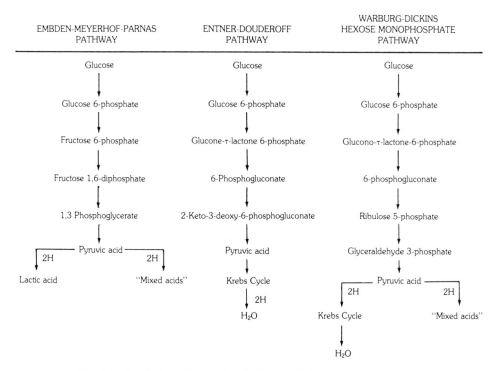

Fig. 3-1. Metabolic pathways for the bacterial degradation of glucose.

The Embden-Meyerhof-Parnas (EMP) Pathway

The Embden-Meyerhof-Parnas (EMP) pathway has been discussed previously in Chapter 2. Since glucose is degraded in the absence of oxygen, this pathway has also been known as the glycolytic or the anaerobic pathway, utilized exclusively by anaerobic bacteria and to some degree by facultatively anaerobic bacteria as well. The intermediate steps in the EMP pathway include the initial phosphorylation of glucose, conversion to fructose-phosphate, cleavage to form two molecules of glyceraldehyde phosphate, which through a series of intermediate steps not shown in Figure 3-1 forms pyruvic acid.

Historically the EMP pathway has also been termed the "fermentative" pathway. Fermentation and anaerobic metabolism have been considered synonymous ever since Pasteur demonstrated that acids and alcohols are the major end products of carbohydrate degradation when oxygen is excluded from the system. Under a more current concept, fermentative metabolism is said to exist in a glycolytic system when organic compounds serve as the final hydrogen (electron) acceptor. Thus, as shown in Figure 3-1, via the EMP pathway outlined in the left column, pyruvic acid acts as an intermediate hydrogen acceptor but is then oxidized by giving up its hydrogen ions to sodium lactate to form lactic acid, or to other organic salts to form so-called "mixed acids." These acids are the end products of glucose metabolism via the EMP pathway, accounting for the drop in pH in fermentation tests used in the identification of bacteria. Bacteria that possess the appropriate enzyme systems can further degrade these mixed acids into alcohols, CO_2 gas, or other organic compounds.

Although these biochemical principles seem somewhat removed from the daily work in the laboratory, microbiologists must have a basic understanding of bacterial metabolism when designing or interpreting test procedures comparing fermentation with oxidation. Fermentation must be determined in test systems that exclude oxygen. The glycolytic products formed by fermentation have relatively strong acidity, easily detected by pH indicators, and significant amounts of gas may be produced. This is not true for the Entner-Douderoff pathway, as will be discussed below.

The Entner-Douderoff (ED) Pathway

The Entner-Douderoff pathway is also called the "aerobic" pathway in that oxygen is required for glycolysis to occur. Note in the center column of Figure 3-1 that glucose is not broken into two triose carbon molecules as in the EMP pathway; rather, it is oxidized to 6-phosphogluconate and 2-keto-3-deoxy-6-phosphogluconate, prior to the formation of pyruvic acid. MacFaddin[8] points out that some bacteria utilize shunt pathways through which glucose is oxidized directly into glucuronic and ketoglucuronic acid without initial phosphorylation. In either event, the intermediate pyruvic acid is formed, and "oxidation" refers more to the manner in which pyruvic acid transfers its hydrogen ions than to the pathway by which it is formed. Lacking the dehydrogenase enzymes necessary to oxidize pyruvic acid to lactic acid or other "mixed acids," oxidative bacteria rather transfer the available hydrogen ions from pyruvic acid into the Krebs cycle, where they ultimately link with elemental oxygen to form water. Thus, the oxidative metabolism of carbohydrates is presently defined as those energy-yielding reactions that require molecular oxygen (or other nonorganic chemicals) to serve as the terminal hydrogen (electron) acceptor.

This difference in metabolism necessitates an alternate practical approach in the laboratory identification of oxidative and fermentative bacteria. The acids that are formed in the Entner-Douderoff pathway (glucuronic acid and its derivatives), or that may be produced in the Krebs cycle (citric acid and its derivatives), are extremely weak

compared to the "mixed acids" resulting from fermentation. Since the end product of oxidative metabolism is water, gas is not formed from carbohydrates by oxidative organisms. Therefore, test systems with more sensitive detectors of acid production must be used when studying oxidative bacteria, to be discussed in detail below. Test systems designed to detect acid production from fermentative bacteria often cannot be applied to oxidative organisms which produce insufficient acids to convert the pH indicator system used.

The Warburg-Dickins Hexose Monophosphate (HMP) Pathway

Facultatively anaerobic bacteria have the capacity to grow in the presence of oxygen on the surface of an agar plate, or they can reproduce equally as well in an anaerobic environment. It must be understood that just because a bacterial species can grow in an aerobic environment does not necessarily mean that oxygen is metabolically utilized. That is, not all aerobes are "oxidative," and the term "aerotolerant" is more appropriate for nonoxidative bacteria that are capable of growing in the presence of oxygen.

Many of the facultative anaerobes can utilize either the EMP or the ED pathway, depending upon the environmental conditions in which they are growing. The hexose monophosphate pathway (HMP), as shown in the right-hand column of Figure 3-1, is actually a hybrid between the EMP and the ED pathways. Note that the initial steps in the degradation of glucose in the HMP pathway parallel those of the ED pathway; however, later in the HMP schema, the triose glyceraldehyde 3-phosphate is formed as the precursor of pyruvic acid, similar to the EMP pathway. Therefore, the HMP pathway provides a means for nonoxidative bacteria (those incapable of utilizing the Krebs cycle and passing hydrogen on to molecular oxygen, that lack the isomerase or aldolase enzymes necessary to carry out the initial steps in the EMP fermentative pathway) to degradate glucose, with the ultimate formation of pyruvic acid. These organisms would appear fermentative in test systems, even though the EMP pathway was not strictly utilized.

Note in Figure 3-1 that the precursor to the formation of glyceraldehyde 3-phosphate in the HMP pathway is ribulose 5-phosphate. Ribulose is a pentose, and for this reason the HMP pathway has also been referred to as the pentose cycle. It provides the major avenue by which pentoses can be metabolized by a number of bacterial species.

THE NONFERMENTATIVE GRAM-NEGATIVE BACILLI: IDENTIFICATION SCHEMAS

Historically, the identification and characterization of the nonfermenting bacteria have been somewhat difficult for microbiologists working in clinical laboratories. Many members of the nonfermenter group are slow growing or have metabolic requirements necessitating the use of special cultivation media. As discussed above, because some of these bacteria produce only weakly acid metabolites they often cannot be detected with test systems routinely used with other groups of bacteria. Until recently, a number of the nonfermenters were looked upon as nonpathogenic commensals of little clinical importance. Their relatively low rate of recovery in most clinical settings has made it difficult for personnel, particularly those working in low-volume laboratories, to maintain the necessary level of competence for the identification of the nonfermenting bacilli. The almost endless shifting of nomenclature and reclassification of these bacteria have made it difficult for all except those working in a research capacity to keep up. For example, the organism now called *Acinetobacter calcoaceticus* has undergone at least seven name changes and reclassifications since the Morax-Axenfeld bacillus was first described in the late 1800's.

Nevertheless, during the past two decades, through the work of Elizabeth King and Robert Weaver and associates[6] at the Center for Disease Control, G. L. Gilardi,[3] M. J. Pickett,[9,10,11] and others working in clinical laboratories and research centers, the characterization and medical significance of the nonfermenting bacilli have been substantially clarified and a number of practical identification schemas have been formulated for use in clinical laboratories.

Classification of the Nonfermenting Bacilli

The nonfermentative gram-negative bacilli do not represent a well-defined taxonomic group of bacteria, but rather include more than 30 species in seven major genera, some of uncertain affiliation. There is some unavoidable overlap with the genera of "unusual bacilli," to be discussed in Chapter 4. However, as common denominators, the nonfermentative bacilli are not particularly fastidious in their growth requirements and are unable to utilize glucose by fermentative pathways. Some species are oxidative in their metabolism and others are "nonsaccharolytic"; that is, they use compounds other than carbohydrates for their energy needs, as outlined above.

The primary characteristics used for differentiating genera of the nonfermenting bacilli are: (1) glucose utilization; (2) motility; (3) cytochrome oxidase activity; (4) ability to grow on MacConkey agar; and (5) cellular morphology. These differential characteristics, in addition to others that are useful in identifying an unknown bacterium in one of the seven major genera of nonfermentative bacilli, are listed in Table 3-1.

Initial Clues to the Recognition of Nonfermentative Bacilli

The microbiologist observing the growth of colonies on a primary isolation medium may suspect the presence of nonfermentative gram-negative bacilli in one of three ways:

1. *Alkaline-Slant/Alkaline-Deep Reaction in Kligler Iron Agar or Triple Sugar Iron Agar.* The absence of acid production in either of these media suggests one of the

Color Plates
3-1 *to* 3-3

Plate 3-1
The Identification of the Nonfermentative Gram-Negative Bacilli

The important characteristics distinguishing the nonfermentative gram-negative bacilli from the Enterobacteriaceae and other groups of bacteria are illustrated in Frames *A, B,* and *C.*

A.
No reaction on KIA agar (left tube). This indicates the inability of nonfermenting bacteria to utilize either the lactose or the dextrose in KIA, in contrast to the Enterobacteriaceae, which produce acid in KIA (right tube).

B.
Failure to grow on MacConkey agar. A blood agar/MacConkey agar biplate with small colonies of a nonfermentative bacillus are present on the blood agar plate but not on the MacConkey agar. Although many species of nonfermentative bacilli are capable of growing on MacConkey agar, the lack of growth on this medium excludes the Enterobacteriaceae, where all species will show positive growth.

C.
Oxidative utilization of glucose. Oxidative-Fermentative (OF) medium of Hugh-Leifson inoculated with a test organism. Note that acid production (yellow color) is evident in only the open tube, where the medium is exposed to ambient air (left), while no acid is produced in the closed, fermentative tube (right), which is overlayed with mineral oil.

Other characteristics often determined in different schema for the identification of nonfermentative bacilli include motility, indole production, pigment production, fluorescence, denitrification (formation of gas in nitrate/nitrite medium), and gluconate reduction.

D.
Semisolid motility test agar inoculated with an oxidative nonfermenting bacterium. The organism is motile only near the top of the medium, covering the surface with a thin film. Lack of readily available oxygen below the surface of the medium prevents significant growth in the deep portion of the tube.

E.
Indole medium showing a positive reaction for indole formation, evidenced by the distinct ring of red color at the interface between the surface of the medium and the added reagent.

F.
A blood agar/MacConkey agar biplate showing growth on the blood agar but not on the MacConkey agar. Note distinct yellow pigmentation of the colonies, a helpful identification feature in the identification of many members of the nonfermenting bacilli.

G.
Tube of fluorescence-denitrification medium showing a blue fluorescent glow when viewed under the Wood's lamp. Note that the colonies do not fluoresce in the agar plate. This is also included in the frame.

H.
Denitrification agar. The presence of gas bubbles in the medium indicates that the nitrates and/or nitrites in the medium have been completely reduced to nitrogen gas (denitrification).

I.
Tubes containing gluconate substrate, to which Benedicts reagent (Clinitest tablets) have been added after a 24-hour incubation with the test organism. Upon addition of a Clinitest tablet, the appearance of a yellow-orange precipitate indicates the presence of reducing substances from the gluconate (right tube), compared to the blue color of the negative control (left tube). None of the Enterobacteriaceae have the enzymes necessary to reduce gluconate; many nonfermentative bacteria have this capability and the test can be a valuable identification characteristic to distinguish between the two groups of bacteria.

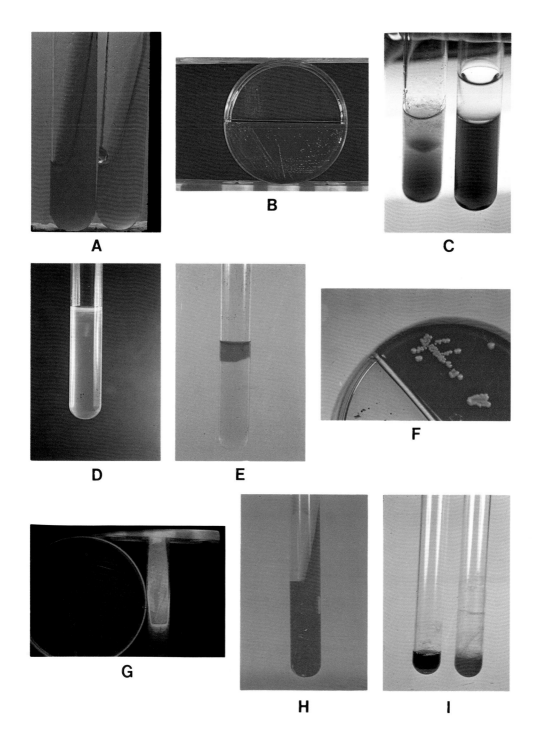

Plate 3-2
The Identification of *Pseudomonas aeruginosa*

Pseudomonas aeruginosa, the nonfermentative gram-negative bacillus most frequently recovered in clinical specimens, is a common cause of human infections. It is important to learn its characteristics so that it will be possible to identify this organism. A number of useful tests are illustrated in this plate.

A.
Desoxycholate-citrate agar with slightly spreading, dull gray, non-lactose utilizing colonies of *P. aeruginosa*. The green background pigmentation of the medium indicates the production of pyocyanin by the organism.

B.
MacConkey agar with dull gray, non-lactose fermenting colonies of *P. aeruginosa* mixed with red lactose-utilizing colonies of *E. coli*.

C.
Pseudosel agar with colonies of *P. aeruginosa*. This medium contains a strong detergent, cetrimide, and is selective for *P. aeruginosa*. The production of pyocyanin is enhanced (observe blue-green color) and the medium can be used for the demonstration of fluorescein pigment. Note that the blue-green pigment diffuses from the areas of bacterial growth, while the uninoculated medium remains clear.

D.
Cytochrome oxidase test paper strip showing the blue color of a positive reaction (right), compared to the negative control (left). All strains of *P. aeruginosa* produce cytochrome oxidase.

E.
Oxidative-fermentative (OF) media inoculated with *P. aeruginosa*, illustrating acid production (yellow color) in the open tube but not in the medium overlayed with mineral oil. All strains of *P. aeruginosa* utilize carbohydrates only oxidatively.

F.
Semisolid motility medium inoculated with *P. aeruginosa*. All strains of *P. aeruginosa* are motile and produce a thick film of growth on the surface of semisolid media. Little growth or evidence of motility is present below the surface of the medium.

G.
Fluorescent-denitrification (FN) medium inoculated with *P. aeruginosa*. All strains of *P. aeruginosa* produce nitrogen gas from nitrates and/or nitrites. Free nitrogen is indicated by the presence of gas bubbles in the denitrification medium.

H.
FN medium inoculated with *P. aeruginosa*. All strains produce fluorescein, which can be detected in the medium by observing bright blue fluorescence with a Wood's lamp.

I.
Flagellar stain of *P. aeruginosa*. Note the single polar flagellum characteristic of *P. aeruginosa*.

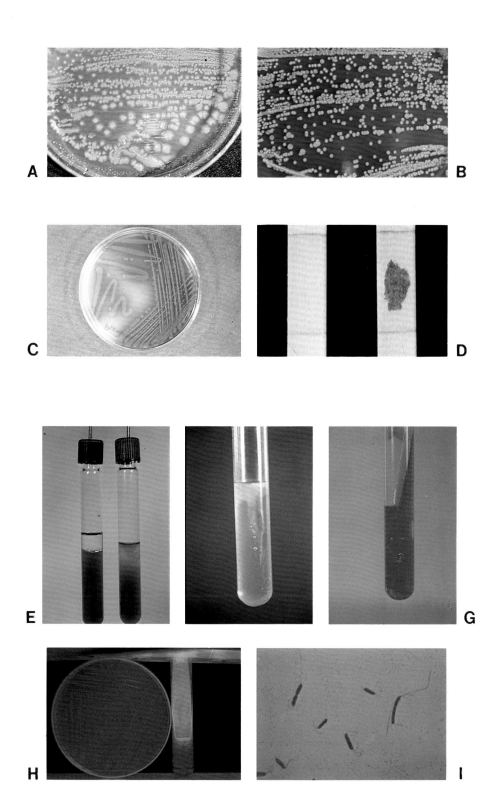

Plate 3-3

Identification of Nonfermentative Bacilli
Other Than Pseudomonas Species

A.

Blood/MacConkey agar biplate illustrating the slightly mucoid, dull gray colonies of the Acinetobacter group, which grow equally as well on MacConkey agar as on blood agar.

B.

Illustration of a Gram's stain of Acinetobacter species illustrating the characteristic, short, gram-negative coccobacillary forms. These organisms have a tendency to form diplococci similar to Neisseria species (thus the derivation of the now illegitimate genus name "mima").

C.

Identification characteristics useful for non-fermentative bacteria. The tubes, from left to right, show the following:

Tubes 1 and 2. O-F medium with acid production in the open tube, indicating only oxidative utilization of carbohydrates. (The closed tube is overlayed with Vaspar, a mixture of vaseline and paraffin oil.)

Tube 3. Alk/alk reaction on KIA (no carbohydrate utilization).

Tube 4. Lack of motility on semisolid agar.

Tube 5. Yellow band in the center of Seller's agar, indicating utilization of 10 per cent dextrose. Seller's agar, which permits detection of fluorescein, denitrification; and carbohydrate utilization in the same tube, is not commonly used in clinical laboratories today.

Tube 6. One per cent lactose broth showing a red, alkaline reaction, indicating the inability of the organism to utilize lactose at this concentration.

Tube 7. Ten per cent lactose agar slant, showing a yellow, acid reaction. The ability to utilize lactose in a 10 per cent, but not a 1 per cent concentration, is characteristic of *A. anitratus.*

Tube 8. Indole test medium illustrating the inability of Acinetobacter species to produce indole.

D.

Blood agar plate with the characteristic yellow pigmented colonies of Flavobacterium species.

E.

Agar medium containing tryptophan, illustrating a positive indole reaction (red ring at interface of agar surface and reagent). Although yellow pigment may be produced by other species of the nonfermenting bacteria, only members of the Flavobacterium group produce indole.

F.

Blood agar/MacConkey agar biplate illustrating growth of characteristic mucoid, yellow, beta-hemolytic colonies of Xanthomonas species. Xanthomonas species will not grow on MacConkey agar. Although Xanthomonas colonies may appear similar to those of Flavobacterium species, the indole reaction is negative.

G.

Semisolid motility medium illustrating surface motility. Xanthomonas species are motile, a helpful characteristic in differentiating them from the nonmotile Flavobacterium species.

H.

Blood/MacConkey agar biplate illustrating the growth of tiny colonies of Moraxella species on blood agar and inhibited growth on MacConkey agar.

I.

Identification characteristics useful for Moraxella species. The tubes, from left to right, show the following:

Tubes 1 and 2. O-F medium illustrating a lack of dextrose utilization in both open and closed tubes, both tubes thus remain blue. Most Moraxella species are nonsaccharolytic.

Tube 3. KIA showing an alk/alk reaction.

Tube 4. Semisolid agar showing lack of motility.

Tube 5. Seller's medium showing no denitrification (lack of gas bubbles) and no utilization of carbohydrates (slant shows no yellow color).

Tube 6. One per cent lactose broth, showing a red, alkaline reaction and the inability of the test organism to utilize lactose.

Tube 7. Ten per cent lactose agar slant showing a red, alkaline reaction and the inability of the organism to utilize lactose, even at this increased concentration.

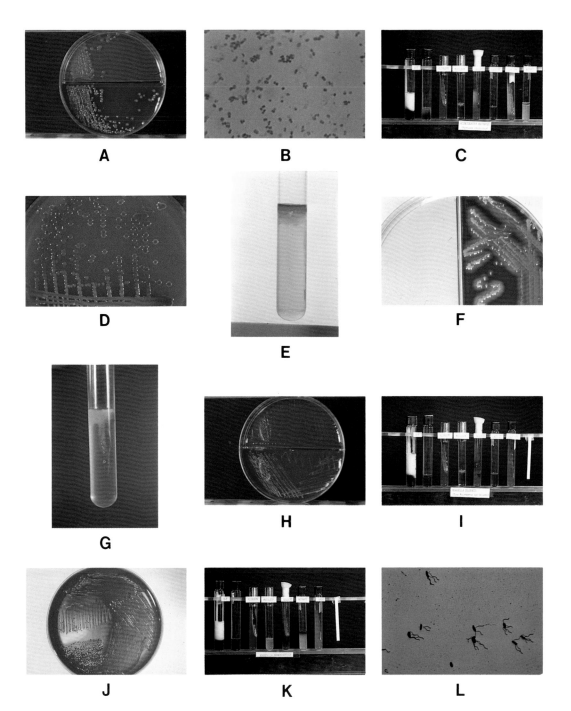

A

B

C

D

E

F

G

H

I

J

K

L

Plate 3-3 (Continued)

J.
Blood agar plate illustrating small, mucoid, yellow-pigmented colonies characteristic of *Bordetella bronchiseptica*.

K.
Identification characteristics for *B. bronchiseptica*. The tubes, from left to right, show the following:

Tubes 1 and 2. O-F medium showing no evidence of dextrose utilization in either the open or the closed tube, both of which remain blue.

Tube 3. Alk/alk reaction on KIA.

Tube 4. Seller's medium revealing no denitrification or evidence of carbohydrate utilization (lack of gas bubbles and yellow color in medium).

Tube 5. Indole test medium showing a positive reaction (band of red in the reagent zone at the top of the medium).

Tube 6. Urease slant showing strongly positive reaction (diffuse red color).

L.
Flagellar stain of *P. bronchiseptica*, illustrating characteristic multiple, lateral, peritrichous flagellae.

Table 3-1. *Identifying Characteristics of Seven Genera of Nonfermentative Bacilli*

Genus	Metabolism	Motility	Oxidase	Growth on MacConkey	Additional Characteristics
Pseudomonas	Oxidative	Positive via polar flagella	+	Positive	Monotrichous and multitrichous polar flagella. Special growth factors not required. Pyocyanin and fluorescein pigments produced by some species. All species denitrify nitrates to nitrogen gas.
Acinetobacter (Herrellea and Mima)	Oxidative or nonsaccharolytic	Negative	−	Positive	Special growth factors not required. Acid production from glucose weak (var. *calcoaceticus*) or absent (var. *lwoffi*). Cells appear as diplococci on Gram's stain. All strains are penicillin-resistant.
Flavobacterium	Oxidative (some strains questionable fermenters)	Variable. Some species have peritrichous flagella.	+	Poor or negative	Supplemental nitrogen and B-complex vitamins required for growth of many strains. Yellow pigments often produced. No denitrification of nitrates. Growth optimal at 30° C. All species are resistant to polymyxin B. Some species are weakly indole-positive.
Bordetella	Oxidative	Variable. Some strains have lateral polytrichous flagella.	+	Variable; poor or negative	Nicotinic acid, cystine, and methionine required by some strains for growth. Potato glycerol blood agar (Bordet-Gengou) needed for growth of *B. pertussis*. *B. bronchiseptica* rapidly splits urea.
Moraxella	Oxidative and nonsaccharolytic	Negative	+	Scant to negative	Most strains fastidious in growth requirements, some requiring serum supplement. Strict aerobes. May appear as coccobacilli on Gram's stain. Highly sensitive to penicillin.
Xanthomonas	Oxidative	Motile via single polar flagellum	−/+ weak	Negative	Methionine, glutamic acid, and nicotinic acid growth supplements needed for growth by some strains. Yellow pigment produced. Catalase-positive. Nitrates not reduced to nitrites. Primarily a plant pathogen.
Alcaligenes	Oxidative and nonsaccharolytic	Negative	+	Negative	Strict aerobes, although some strains utilize nitrate instead of oxygen as the final electron acceptor. Simple nitrogenous growth requirements.

Chart 3-1.   *Oxidative-Fermentative (O-F) Test (Hugh and Leifson)*

Introduction	Saccharolytic microorganisms degrade glucose either fermentatively or oxidatively, as shown in Figure 3-1. The end products of ferementation are relatively strong "mixed acids" that can be detected in a conventional fermentation test medium. However, the acids formed in oxidative degrad- ation of glucose are extremely weak, and the more sensitive oxidation- fermentation medium of Hugh and Leifson (O-F medium) is required for their detection.
Principle	The O-F medium of Hugh and Leifson differs from carbohydrate fermenta- tion media as follows: 1. The concentration of protein (peptones) has been decreased from 1% to 0.2%. 2. The concentration of carbohydrate is increased from 0.5% to 1.0%. 3. The concentration of agar is decreased to 0.2% from 0.3%, making it semisolid in consistency. The lower protein to carbohydrate ratio reduces the formation of alkaline amines that can neutralize the small quantities of weak acids that may form from oxidative metabolism. The relatively larger amount of carbohydrate serves to increase the amount of acid that can potentially be formed. The semisolid consistency of the agar permits acids that form on the surface of the agar to permeate throughout the medium, making interpretation of the pH shift of the indicator easier to visualize. Motility can also be determined with this medium.
Media and Reagents	For comparison, the formulas for a conventional carbohydrate fermenta- tion medium and O-F medium are as follows:

Carbohydrate Fermentation Medium		*O-F Medium of Hugh and Leifson*	
Peptone	10.00 g.	Peptone	2.00 g.
Dextrose	5.00 g.	Dextrose	10.00 g.
Bromcresol purple	0.04 g.	Bromthymol blue	0.03 g.
Agar*	15.00 g.	Agar	2.50 g.
Distilled water to:	1000.00 ml.	Sodium chloride	5.00 g.
pH = 7.0		Dipotassium phosphate	0.30 g.
		Distilled water to:	1000.00 ml.
		pH = 7.1	

	O-F medium should be poured without a slant into tubes with an inner diameter of 15 to 20 mm. to increase surface area.
Procedure	Two tubes are required for the O-F test, each inoculated with the unknown organism, using a straight needle, stabbing the medium almost to the bottom of the tube. One tube of each pair is covered with a 1-cm. layer of sterile mineral oil or melted paraffin, leaving the other tube open to the air. Incubate both tubes at 35° C. for 48 hours or longer.
Interpretation	Acid production is detected in the medium by the appearance of a yellow color. In the case of oxidative organisms, color production may be first

Chart 3-1. *Oxidative-Fermentative (O-F) Test (Hugh and Leifson)* *(Continued)*

Interpretation *(Continued)*	noted near the surface of the medium. Following is a chart listing the reaction patterns:

Open Tube	Covered Tubes	Type of Metabolism
Acid (Yellow)	Alkaline (Green)	Oxidative
Acid (Yellow)	Acid (Yellow)	Fermentative
Alkaline (Green)	Alkaline (Green)	Nonsaccharolytic

These color reactions are shown in Plate 3-1, *C.* For slower growing species, incubation for three days or longer may be required to detect positive reactions.

Controls	Glucose fermenter: *Escherichia coli* Glucose oxidizer: *Pseudomonas aeruginosa* Nonsaccharolytic: Moraxella species

Bibliography	*BBL Manual of Products and Laboratory Procedures.* ed. 5, pp. 129–130. Cockeysville, Maryland: Division of BioQuest, Division of Becton, Dickinson and Co, 1973. Hugh, R., and Leifson, E.: The taxonomic significance of fermentative versus oxidative metabolism of carbohydrates by various gram-negative bacilli. J. Bacteriol., *66:*24–26, 1953. MacFaddin, J. F.: Biochemical Tests for Identification of Medical Bacteria. Pp. 162–170. Baltimore, Williams & Wilkins, 1976.

*Agar is omitted from broth medium and Durham tubes added to detect gas.

nonfermenting gram-negative bacilli (see Fig. 2-3, Chap. 2, and Plate 3-1, *A*).

2. *Positive Cytochrome Oxidase Reaction.* Any colony composed of gram-negative bacteria, growing on blood agar or other primary isolation medium, that is oxidase-positive should be suspected of belonging to the nonfermenter group. Not all nonfermenters are oxidase-positive; however, the demonstration of its production excludes the microorganism from the Enterobacteriaceae. (See Plate 2-1, *H.*)

3. *Failure of the Bacterium to Grow on MacConkey Agar.* Any gram-negative bacillus that grows on blood agar but not on MacConkey agar should be suspected of belonging to the nonfermenter group. Microbiologists must learn by experience the appearance of colonies of gram-negative bacilli on blood agar if those species not growing on MacConkey agar are to be recognized and identified (Plate 3-1, *B*).

When one or more of these initial characteristics are observed, the possibility of a nonfermenting gram-negative bacillus must be considered and some schema selected by which a final identification can be made. Of several schemas that have been proposed for identification of nonfermenters, three have stood the test of time and are now commonly used in many clinical laboratories, including those of relatively small size. These are the King-Weaver schema, the Gilardi schema, and the Pickett schema.

The King-Weaver Schema

One of the first schemas designed for the identification of aerobic gram-negative bacilli was developed at the Center for Disease

Closed   Open   Closed   Open   Closed   Open

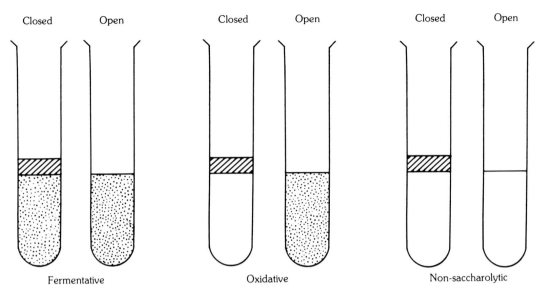

Fermentative   Oxidative   Non-saccharolytic

Fig. 3-2 The Oxidative-Fermentative (O-F) Test Fermentative organisms produce acid in both the closed and the open tubes (stippled effect); oxidative organisms produce acid only in the open tube. Organisms that do not utilize carbohydrates produce no change in either tube.

Control by Elizabeth O. King.[6] For well over a decade the King charts have been an integral part of most diagnostic laboratories throughout the world, and modern clinical microbiology is indebted to Miss King and her pioneering efforts. This schema utilizes the following four characteristics for preliminary differentiation of gram-negative bacilli:

1. Utilization of glucose (fermentative, oxidative, or nonsaccharolytic)
2. Ability to grow on MacConkey agar
3. Cytochrome oxidase production
4. Motility

Following is a brief discussion of these key characteristics:

1. *Utilization of Glucose.* It should be emphasized that conventional culture media designed to detect acid production from fermentative bacteria such as the Enterobacteriaceae are not suitable for the study of nonfermentative bacilli, most species of which grow slowly and produce extremely weak acids. Hugh and Leifson[4] were the first to design an oxidative-fermentative (O-F) medium to accommodate the metabolic properties of the nonfermentative bacilli, as outlined in Chart 3-1.

Note that the Hugh and Leifson medium contains 0.2 per cent peptone and 1.0 per cent carbohydrate, so that the ratio of peptone to carbohydrate is 0.2:1, in contrast to the 2:1 ratio found in media designed for carbohydrate fermentation. The decrease in peptone minimizes the formation of oxidative products from peptone which tend to raise the *p*H of the medium and may neutralize the weak acids produced by the nonfermentative bacilli. The increase in carbohydrate concentration, on the other hand, enhances acid production by the microorganism. The semisolid consistency of the agar, the use of bromthymol blue as the *p*H indicator, and the inclusion of a small quantity of diphosphate buffer are all designed to increase the detection of acid production.

Two tubes of media are required when performing the test, one exposed to the air, the other having the agar surface covered with mineral oil or a wax seal (illustrated in Fig. 3-2). Oxidative microorganisms produce acid only in the open tube exposed to atmospheric oxygen; fermenting organisms produce acid in both tubes; nonsaccharolytic bacteria are inert in this medium, which remains at an alkaline *p*H after incubation. Plate 3-1, *C* shows these O-F reactions.

The O-F test has the limitations that slower-growing nonfermentative bacilli may not produce color changes for several days, and species that are especially proteolytic may cause reversions of weak acid reactions with time, thus confusing the final interpretation. It is extremely important that the Hugh-Leifson formula be strictly followed when performing the O-F test. Any alteration of the formula of the media produced by commercial companies must be carefully evaluated with appropriate control cultures within each laboratory before the reactions produced can be accepted.

2. *Growth on MacConkey Agar.* The ability of bacteria to grow on MacConkey agar is determined by inspecting by reflected light the surface of plates that have been inoculated and incubated for 24 to 48 hours. The visualization of tiny colonies may be aided by the use of a hand lens or a dissecting microscope.

3. *Cytochrome Oxidase Activity.* The tests for determining cytochrome oxidase activity were discussed in Chapter 2 (Chart 2-1, p. 58). The more sensitive tetramethyl derivative of p-phenylenediamine (Kovac's reagent) is recommended for determining the cytochrome oxidase activity of the nonfermentative bacilli, many of which are weak. It is important when using interpretive charts to use the methodology and differential tests employed by the author(s) who prepared the charts if unnecessary errors are to be avoided.

4. *Motility.* Semisolid agar, used for the detection of motility by fermentative organisms, may not be suitable for nonfermenting species that grow only on the surface of the agar. If semisolid agar is used for nonfermentative bacilli, stab inoculate only the upper 4 mm. of the medium and make an initial reading at 4 to 6 hours. Many motile strains of nonfermentative bacilli will show only an early, faint haziness near the surface of the agar, which tends to disappear with prolonged incubation (Plate 3-1, *D*). Also, readings should again be made at 24 and 48 hours to detect slow-growing motile strains.

The direct microscopic examination of a hanging-drop preparation is not only more rapid but also more accurate in the detection of motility of many species of nonfermentative bacilli. The use of flagellar stains, discussed below, is also helpful in differentiating certain motile species (Plate 3-2, *I*).

After these characteristics of a culture have been assessed for preliminary grouping, additional characteristics must be determined before a final species identification can be made. One criticism of the original King charts has been the large number of additional characteristics that must be determined before the charts can be used. Only larger reference laboratories can afford the time and effort necessary for final identification. However, the latest revision of the King charts by Weaver has provided a more practical algorithm for the identification of nonfermentative bacilli, well within the capabilities of most diagnostic laboratories, even those of smaller volume. An adaptation of the Weaver scheme is included in the following pages of this text.

Utilizing the four preliminary characteristics discussed above, Weaver has separated the aerobic gram-negative bacilli into 13 separate groups. The group numbers used in this text are for discussion purposes only and do not reflect any official designation given by Dr. Weaver or by the Center for Disease Control. The 13 groups are as follows:

Group 1: Gram-Negative Fermenter, MacConkey-Positive, Oxidase-Negative

Group 2: Gram-Negative Fermenter, MacConkey-Positive, Oxidase-Positive (Motile Via Polar or Polar + Lateral Flagella)

Group 3: Gram-Negative Fermenter, MacConkey-Positive, Oxidase-Positive (Nonmotile)

Group 4: Gram-Negative Fermenter, MacConkey-Negative, Oxidase-Negative

Group 5: Gram-Negative Fermenter, MacConkey-Negative, Oxidase-Positive

Group 6: Gram-Negative Glucose Oxidizer, MacConkey-Positive, Oxidase-Negative

Group 7: Gram-Negative Glucose Oxidizer, MacConkey-Positive, Oxidase-Positive

Group 8: Gram-Negative Glucose Oxidizer, MacConkey-Negative, Oxidase-Negative

Table 3-2. Group 1: Gram-Negative Fermenter, MacConkey-Positive, Oxidase-Negative*; Differential Characteristics (King-Weaver Schema)

Group or Species	Gas from Glucose	Urea	Lysine Decarboxylase	Arginine Dihydrolase	Ornithine Decarboxylase	Lactose	Sucrose	Rhamnose	Motility 35°	Motility 25°	Motility 5°	Indol
Enterobacteriaceae	Variable; may be large volume	Variable	Variable	Variable	Variable	Variable	Variable	Variable	Variable			Variable
Yersinia enterocolitica	+ or –; not large volume	+	–	–	+	– (delayed)	+	–	–	+	+	Variable
Yersinia pseudotuberculosis	–	+	–	–	–	–	–	+	–	+ or –	+	–
Yersinia pestis	–	–	–	–	–	–	–	–	–	–	–	–
Chromobacterium violaceum	–	– (delayed)	–	+	–	–	Variable	–	+	–	–	– (? +)
HB-5	+ or –; not large volume	–	–	–	–	–	–	–	–	–	–	+, weak

Key: + = 90% or more strains are positive; (+) = 51–89% of strains are positive; – = less than 10% of strains are positive.
*Organisms to consider: Enterobacteriaceae, Yersinia pestis, Yersinia pseudotuberculosis, Yersinia enterocolitica, Chromobacterium violaceum (oxidase variable), HB-5 (oxidase and MacConkey variable).
(Courtesy of the Enteric Section, Entereobacteriology Branch, Bacteriology Division, Bureau of Laboratories, Center for Disease Control, Atlanta, Georgia, 1976)

Table 3-3. *Group 2: Gram-Negative Fermenter, MacConkey-Positive, Oxidase-Positive (Motile Via Polar or Polar and Lateral Flagella)*; Differential Characteristics (King-Weaver Schema)*

Group or Species	Gas from Glucose	Arginine Dihydrolase	Ornithine Decarboxylase	NaCl Required	Mannitol
Aeromonas hydrophilia	+ or −	+	−	−	+
Aeromonas (Plesiomonas) shigelloides	−	+(−)	+ or −	−	−
Vibrio cholerae and related species	−	−	+	−	+
Vibrio parahaemolyticus and related species	−	−	+	+	+
Chromobacterium violaceum (oxidase variable)	−	+	−	−	−
Providencia (oxidase weak, if positive)	+ or −	−	−	−	− or +

Key: + = 90% or more of strains are positive; (−) = 10–50% of strains are positive; − = less than 10% of strains are positive.
*Organisms to consider: *Aeromonas hydrophilia, Aeromonas (Plesiomonas) shigelloides, Vibrio cholerae* and related species, *Vibrio parahaemolyticus* and related species, *Chromobacterium violaceum,* Providencia.
(Courtesy of the Enteric Section, Entereobacteriology Branch, Bacteriology Division, Bureau of Laboratories, Center for Disease Control, Atlanta, Georgia, 1976)

Group 9: Gram-Negative Glucose Oxidizer, MacConkey-Negative, Oxidase-Positive

Group 10: Gram-Negative Glucose Nonoxidizer (Nonfermenter), MacConkey-Positive, Oxidase-Negative

Group 11: Gram-Negative Glucose Nonoxidizer (Nonfermenter), MacConkey-Positive, Oxidase-Positive

Group 12: Gram-Negative Glucose Nonoxidizer (Nonfermenter), MacConkey-Negative, Oxidase-Negative

Group 13: Gram-Negative Glucose Nonoxidizer (Nonfermenter), MacConkey-Negative, Oxidase-Positive

The organisms to be considered within each of these groups and the differential characteristics required to make a final identification are shown in Tables 3-2 through 3-14. (Courtesy of the Enteric Section, Entereobacteriology Branch, Bacteriology Division, Bureau of Laboratories, Center for Disease Control, Atlanta, GA). Each microbiologist must determine the extent to which species identification within each of these groups is to be carried out. The additional characteristics required for some groups are minimal and the procedures are not difficult; in other groups the additional media required may be within the capabilities of only reference laboratories. For smaller-volume laboratories, it may be sufficient to report only the major group to which an organism belongs, and to refer the culture to a larger laboratory where a final identification can be made.

One additional difficulty in the use and interpretation of the King charts or the Weaver revision is in understanding the taxonomic position of a number of lettered and numbered groups of bacteria currently accepted by the Center for Disease Control (CDC). Designations such as EO-1, HB-1, M-1, IIa, IVa, Va-1, Ve-1, *etc.* are quite confusing to microbiologists who have either not encountered these bacterial species or who know them by other designations. Table 3-15 lists a number of these CDC lettered and numbered groups and some of their more familiar synonyms.

Table 3-4. Group 3: Gram-Negative Fermenter, MacConkey-Positive, Oxidase-Positive (Nonmotile)*; Differential Characteristics (King-Weaver Scheme)

Group or Species	Glucose	Xylose; Mannitol, Lactose	Gas from Nitrate	Urea	Indol
EF-4	+	−	+ or −	−	−
HB-5	+	−	−	−	+, weak
Pasteurella haemolytica	+	At least one +	−	−	−
Actinobacillus lignieresi and *Actinobaccillus equuli*	+	At least one +	−	+	−

Key: + = 90% or more strains are positive; − = 10% or less of strains are positive.

*Organisms to consider: EF-4; HB-5; *Pasteurella haemolytica, Actinobacillus lignieresii,* and *Actinobacillus equuli.*

(Courtesy of the Enteric Section, Entereobacteriology Branch, Bacteriology Division, Bureau of Laboratories, Center for Disease Control, Atlanta, Georgia, 1976)

Table 3-5. Group 4: Gram-Negative Fermenter, MacConkey-Negative, Oxidase-Negative*; Differential Characteristics (King-Weaver Schema)

Group or Species	Catalase	Nitrate Reduction	Indol	Lactose	Maltose	Other
Hemophilus aphrophilus	weakly + or −	+	−	+	+	Few strains oxidase positive
Hemophilus vaginalis	−	−	−	−	+	
Actinobacillus actinomycetemcomitans	+	+	−	−	+	
HB-5	−	+	+, weak	−	−	

Key: + = 90% or more strains are positive; (+) = 51–89% of strains are positive; − = 10% or less of strains are positive.

*Organisms to consider: *Hemophilus aphrophilus, Hemophilus (Corynebacterium) vaginalis, Actinobacillus actinomycetemcomitans,* HB-5, Bacillus species (May stain weakley gram-positive or gram-negative).

(Courtesy of the Enteric Section, Entereobacteriology Branch, Bacteriology Division, Bureau of Laboratories, Center for Disease Control, Atlanta, Georgia, 1976)

Gilardi's Schema for Identification of Nonfermentative Gram-Negative Bacilli

Gilardi's[3] schema was designed as a practical approach for the identification of nonfermentative gram-negative bacilli in diagnostic laboratories with only a limited number of testing procedures available. This schema utilizes biochemical and nutritional tests that are commonly used for the identification of other groups of bacteria, omitting those that have special requirements or require long incubation periods.

Although the Gilardi schema may have insufficient characteristics to identify some of the less commonly encountered nonfermentative bacilli or atypical mutants with exacting nutritional requirements, the majority of the commonly encountered species can be accurately identified within 24 to 48 hours. Most members of the three major groups recognized by Gilardi can be readily identified:

1. The motile, oxidase-positive bacillus group (Pseudomonas species and Alcaligenes species)
2. The nonmotile, oxidase-negative diplococcus (Acinetobacter species) and oxidase-positive diplobacillus group (Moraxella species)
3. The nonmotile, oxidase-positive, yellow-pigmented bacillus group (Flavobacterium species) (See Plate 3-2, C, for an illustration of a pigmented colony.)

Table 3-6. *Group 5: Gram-Negative Glucose Fermenter, MacConkey-Negative, Oxidase-Positive*; Differential Characteristics (King-Weaver Schema)*

Species or Group	Nitrate Reduction	Indol	Urea	Xylose	Catalase	Other
Pasteurella multocida	+	+	−	Variable	+	
Pasteurella pneumotropica	+	+	+	+	+	
Pasteurella ureae	+	−	+	−	+ weak (−)	
Pateurella gallinarum	+	−	−	−(+)	+	
Pasteurella "gas"	+	+	+	−	+	
Cardiobacterium hominis	−	+	−	−	−	
EF-4	+ or +, gas	−	−	−	+	Glucose the only carbohydrate fermented
HB-5 (oxidase and Mac-Conkey variable)	+	+, weak	−	−	−	

Key: + = 90% or more of strains are positive; (+) = 51–89% of strains are positive; (−) = 10–50% of strains are positive; − = less than 10% of strains are positive.
*Organisms to consider: *Pasteurella multocida, Pasteurella pneumotropica, Pasteurella ureae,* Pasteurella "gas," *Pasteurella gallinarum, Cardiobactetrium hominis,* EF-4, HB-5 (Oxidase and MacConkey variable), Neisseria species.
(Neisseria species separated from above on basis of morphology.)
(Courtesy of the Enteric Section, Entereobacteriology Branch, Bacteriology Division, Bureau of Laboratories, Center for Disease Control, Atlanta, Georgia, 1976)

Table 3-7. *Group 6: Gram-Negative Glucose Oxidizers, MacConkey-Positive, Oxidase-Negative*; Differential Characteristics (King-Weaver Schema)*

Species or Group	Motility	Reduction of Nitrate	Lysine Decarboxylase	Arginine Dihydrolase	Ornithine Decarboxylase	O-F Lactose
Pseudomonas cepacia	+	Variable	+ or −	−	− or +	Acid
Herellea (Acinetobacter)	−	−	−	−	−	Acid
Ve-1	+	Variable	−	+	−	Alkaline
Ve-2	+	−	−	−	−	Alkaline
Pseudomonas mallei	−	+	−	+	−	− or acid > 7 days

Key: + = 90% or more strains are positive; − = 10% or less of strains are positive.
*Organisms to consider: *Pseudomonas cepacia,* Herellea (Acinetobacter), Ve-1, Ve-2, *Pseudomonas mallei.*
(Courtesy of the Enteric Section, Entereobacteriology Branch, Bacteriology Division, Bureau of Laboratories, Center for Disease Control, Atlanta, Georgia, 1976)

Table 3-16, modified from Gilardi's paper,[3] includes identification tests more commonly performed in diagnostic microbiology laboratories, and shows the salient features by which members of the Pseudomonas and Alcaligenes genera can be recognized. Similarly, Table 3-17 lists the salient features for identifying members of the Acinetobacter-Moraxella and Flavobacterium groups. Since the species of bacteria listed in these two tables comprise well over 95 per cent of the nonfermenting bacilli encountered in diagnostic clinical laboratories, this is a realistic and practical approach. Only rarely will it be necessary to perform additional biochemical or nutritional tests for definitive identification of an isolate.

Microbiologists who are interested in implementing the Gilardi schema in their laboratories should consult Gilardi's paper (cited in the references at the end of this chapter) *(Text continues on p. 127)*

Table 3-8. *Group 7: Gram-Negative Glucose Oxidizers, MacConkey-Positive, Oxidase-Positive*; Differential Characteristics (King-Weaver Schema)*

Species or Group	Indol	Gas from Nitrate	Test for Nitrite Formation	Complete Reduction of Nitrate and Nitrite	Growth on SS	Lysine Decarboxylase	Arginine Dihydrolase	Motility	Flagella	Fluorescin (Pyoverdin)
IIk, type 1	–	–	–(+)	–	–	–	–	–(weak +)	1 polar	–
IIk, type 2	–	–	– or +	–	–	–	–	–(weak +)	?1 polar	–
Flavobacterium meningosepticum	+	–	–	–	–	–	–	–	–	–
IIb, Flavobacterium species	+	–	–(+)	– or +	–	–	–	–	–	–
P. aeruginosa	–	+	– or +	+ or –	+(–)	–	+	+	<3 polar	+
P. fluorescens group	–	–(+)	– or +	–(+)	+	–	+	+	>2 polar	+
P. pseudomallei	–	+	+	–	–	–	+	+	>2 polar	–
P. stutzeri group	–	+	– or +	+ or –	+(–)	–	+ or –	+	<3 polar	–
P. vesiculare	–	–	–	–	–	–	–	+	1 polar, short wave, low amplitude	–
P. cepacia	–	–	variable	–	–(poor)	+(–)	–	+	>2 polar	–
Va	–	+	+	–	–	nt	nt	+	<3 polar	–
Vd	–	+	– or +	+ or –	+	nt	nt	+	Pentrichous	–

Key: + = 90% or more of strains are positive; (+) = 51–89% of strains are positive; (–) = 10–50% of strains are positive; – = less than 10% of strains are positive; nt = test not done.
*Organisms to consider: IIk, type 1, IIk, type 2, Flavobacterium meningosepticum, IIb, Flavobacterium species; Pseudomonas aeruginosa, Pseudomonas fluorescens group; P. pseudomallei, P. Stutzeri, P. vesiculare, P. Cepacia, Va, Vd.
(Courtesy of the Enteric Section, Enterobacteriology Branch, Bacteriology Division, Bureau of Laboratories, Center for Disease Control, Atlanta, Georgia, 1976)

Table 3-8A. *Group 7: Gram-Negative Glucose Oxidizers, MacConkey-Positive, Oxidase-Positive; Carbohydrate Reactions (King-Weaver Schema)*

Species or Group	Glucose	Xylose	Mannitol	Lactose	Sucrose	Maltose
IIk, types 1 and 2	A	A	Alk	A	A	A
Flavobacterium meningo-septicum	A	Alk	A	Var	Alk	A
IIb, Flavobacterium species	A	Alk	Alk	Alk	Alk	A
Pseudomonas aeruginosa	A	A	Var	Alk	Alk	Alk
Pseudomonas fluorescens group	A	A	Var	Var	Var	Var
Pseudomonas pseudo-mallei	A	A	A	A	Var	A
Pseudomonas stutzeri group	A	A	A (weak)	Alk	Alk	Var
Pseudomonas cepacia	A	A	A (weak)	A	Var	A
Pseudomonas vesiculare	A	A (Alk)	Alk	Alk	Alk	A
Va	A	A	Alk	Var	Alk	Var
Vd	A	A	Var	Alk	Var	Var

Key: A = acid; Alk = alkaline; Var = variable.

(Courtesy of the Enteric Section, Entereobacteriology Branch, Bacteriology Division, Bureau of Laboratories, Center for Disease Control, Atlanta, Georgia, 1976)

Table 3-9. *Group 8: Gram-Negative Glucose Oxidizer, MacConkey-Negative, Oxidase-Negative*; Differential Characteristics (King-Weaver Schema)*

Species or Group	O-F Base		
	Glucose	Xylose	Maltose
IIk, type 1	A	A	A
Pseudomonas mallei	A	Alk	Alk

Key: A = acid; Alk = alkaline.

*Organisms to consider: *Pseudomonas (Actinobacillus) mallei;* IIk, type 1 (most strains).

(Courtesy of the Enteric Section, Entereobacteriology Branch, Bacteriology Division, Bureau of Laboratories, Center for Disease Control, Atlanta, Georgia, 1976)

Table 3-10. **Group 9: Gram-Negative Glucose Oxidizers, MacConkey-Negative, Oxidase-Positive*; Differential Characteristics (King-Weaver Schema)**

Species or Group	Indol	Nonsoluble Pigment	Glucose	Xylose	Mannitol	Lactose	Sucrose	Maltose	Urea	Other
IIk, type 1	–	Yellow	A	A	Alk	A (O-F Base)	A	A	–(+ delayed)	
Flavobacterium meningosepticum	+	± Yellow	A	Alk	A(–)	Alk(A) (O-F Base)	Alk	A	–(+ delayed)	
IIb, Flavobacterium species	+	Yellow	A	Alk(A)	Alk(A)	Alk (O-F Base)	Alk(A)	A	–(+ delayed)	
Moraxella kingii	–	–	A	–	–	Broth Base	–	A	–	
Vibrio extorquens	–	Pink	– or A, weak, 2–7	A 2–7 (–)–	Alk	Alk (O-F Base)	Alk	Alk	+(– delayed)	
Brucella	–	–	N A, weak	N A, weak	Alk	Alk (O-F Base)	Alk	Alk	+	Tiny, coccoid oxidase-positive, weak, or negative

Key: A = acid; Alk = alkaline; N = neutral; + = 90% or more of strains are positive; (+) = 51–89% of strains are positive; (–) = 10–50% of strains are positive; – = less than 10% of strains are positive.

*Organisms to consider: IIk, type 1: *Flavobacterium meningosepticum*; Flavobacterium species, group IIb; *Moraxella kingii*; *Vibrio extorquens*, Brucella species, Neisseria species. (Neisseria distinguished by morphology).

(Courtesy of the Enteric Section, Entereobacteriology Branch, Bacteriology Division, Bureau of Laboratories, Center for Disease Control, Atlanta, Georgia, 1976)

Table 3-11. *Group 10: Gram-Negative Glucose Nonoxidizers (Nonfermenters), MacConkey-Positive, Oxidase-Negative*; Differential Characteristics (King-Weaver Schema)*

Species	Urea	Soluble Pigment	O-F Maltose	Gelatin	Motility	Beta Hemolysis
Mima (Acinetobacter) polymorpha	− (+)	− (Brown)	−	− (+)	−	− (+)
Pseudomonas malto- philia	− (weak + , delayed)	Tan	Acid	+	+	−
Bordetella parapertussis	+	Brown	−	−	−	+

Key: + = 90% or more of strains are positive; (+) = 51–89% of strains are positive; − = less than 10% of strains are positive.
*Organisms to consider: *Mima (Acinetobacter) polymorpha; Pseudomonas maltophilia; Bordetella parapertussis.*
(Courtesy of the Enteric Section, Entereobacteriology Branch, Bacteriology Division, Bureau of Laboratories, Center for Disease Control, Atlanta, Georgia, 1976)

for a complete description of procedures so that proper interpretations can be made which are consistent with the data presented in Tables 3-16 and 3-17.

Pickett's Schema for Detection and Identification of Nonfermentative Bacilli

The approach used by Pickett[10] for the detection and identification of nonfermentative bacilli employs the following unique and useful features:

1. Preliminary subculture of the unknown bacterial colony recovered on primary isolation medium to an enriched medium where luxurient growth can take place
2. Use of a heavy inoculum of this preliminary growth for all secondary tests of nutritional and biochemical characteristics
3. Preliminary screening with a minimal number of characteristics to quickly identify the more commonly encountered nonfermenter species
4. Utilization of buffered single-substrate (BSS) media for the determination of the bulk of all secondary characteristics. Each medium contains only a single substrate (carbohydrate, alcohol, amine, *etc.*) which specifically reacts with the preformed products of the pregrown organisms in the heavy inoculum.

The step-by-step sequence of the Pickett approach to the identification of the nonfermentative gram-negative bacilli is outlined in the following paragraphs.

By early or mid morning, a well-isolated colony of the unknown bacterial species to be studied that has been recovered on primary isolation medium is spot inoculated to the surface of a Kligler iron agar slant, covering an area approximately that of a dime. Kligler iron agar is suggested as the enrichment medium because it supports the growth of most nonfermenters and contains ample quantities of carbohydrates. Blood agar or chocolate agar could also be used for this preliminary growth; however, most blood agar bases are deficient in carbohydrates, which may compromise the development of enzyme systems that are required for the organism to produce positive reactions in the differential test media.

Since the KIA slant will be used to harvest the massive inocula for the buffered single-substrate tests, a larger, 20×150-mm. screwcap tube containing 14 to 15 ml. of medium is recommended over the conventional 13-mm. tubes. This provides a butt of about 3 cm. and a slope of 10 cm., giving as large a surface area as possible.

A tube of semisolid motility-nitrate medium should also be inoculated along with the KIA spot inoculum. Motility-nitrate medium consists of 4 ml. of medium poured without a slant into a 13-mm. screwcap tube, and is composed of 1.0 g. per cent tryptose (Difco), 0.8 g. per cent infusion agar (Difco), and 0.1 g. per cent of KNO_3. The medium is inoculated by stabbing only the top 3 to 5 mm. because nonfermenters will
(Text continues on p. 134)

Table 3-12. Group 11: Gram-Negative Glucose Nonoxidizers (Nonfermenters), MacConkey-Positive, Oxidase-Positive*; Differential Characteristics (King-Weaver Schema)

Species or Group	Flagella	Nitrate Reduction	Gas from Nitrate	Gas from Nitrite (Nitrate Not Reduced)	H_2S, Butt of TSI	O-F Xylose	Growth on SS	Urea
Pseudomonas acidovorans-testosteroni (Comamonas)	Polar >2	+ (rare −)	−	−	−	−	Variable	− or +
Pseudomonas alcaligenes-pseudoalcaligenes	Polar 1–2	+ (−)	− (?+)	−	−	−	Variable	−
Pseudomonas denitrificans	Polar <3?	+	+	−	−	−	− (+)	−
Pseudomonas diminuta	Polar 1–2; short wave-length and amplitude	− (+)	−	−	−	−	− (+)	− (+)
Pseudomonas putrefaciens (Ib)	Polar 1–2	+	−	−	+	Neutral or Alkaline	Variable	− or +
Bordetella bronchiseptica	Peritrichous	+	−	−	−	−	+	strong +
Alcaligenes faecalis	Peritrichous	− or +	−	−	−	−	+	−
Alcaligenes odorans	Peritrichous	−	−	+	−	−	+	−
Alcaligenes denitrificans	Peritrichous	+	+	−	−	−	Variable	−
IIIa and IIIb	Peritrichous	+	+ or −	− (IIIa)	−	Acid	+	−
IVc-2	Peritrichous	−	−	−	−	−	−	Moderate +
IVe	Peritrichous	+	+	−	−	−	−	strong +

Key: + = 90% or more strains are positive; (+) = 51–89% of strains are positive; (−) = 10–50% of strains are positive; − = less than 10% of strains are positive.
*Organisms to consider: Pseudomonas acidovorans-testosteroni (Comamonas); P. alcaligenes-pseudoalcaligenes; P. denitrificans; P. diminuta; P. putrefaciens; Bordetella bronchiseptica; Alcaligenes faecalis; A. odorans; A. denitrificans, IIIa and IIIb; IVc-2; IVe.
(Courtesy of the Enteric Section, Enterobacteriology Branch, Bacteriology Division, Bureau of Laboratories, Center for Disease Control, Atlanta, Georgia, 1976)

Table 3-13. *Group 12: Gram-Negative Glucose Nonoxidizers (Nonfermenters), MacConkey-Negative, Oxidase-Negative*; Differential Characteristics (King-Weaver Schema)*

Species	Urea	Soluble Pigment	O-F Maltose	Gelatin	Motility	Beta Hemolysis
Mima (Acinetobacter) polymorpha	−(+)	−	−	−(+)	−	−(+)
Pseudomonas maltophilia	−(weak +, delayed)	Tan	Acid	+	+	−
Bordetella parapertussis	+	Brown	−	−	−	+

Key: + = 90% or more of strains are positive; (+) = 51–89% of strains are positive; − = 10% or less of strains are positive.
*Organisms to consider: *Mima (Acinetobacter) polymorpha; Pseudomonas maltophilia; Bordetella parapertussis.*
(Courtesy of the Enteric Section, Entereobacteriology Branch, Bacteriology Division, Bureau of Laboratories, Center for Disease Control, Atlanta, Georgia, 1976)

Table 3-14. *Group 13: Gram-Negative Glucose Nonoxidizers (Nonfermenters), MacConkey-Negative, Oxidase-Positive*; Differential Characteristics (King-Weaver Schema)*

Species or Group	Indol	Urea	O-F Xylose	Reduction of Nitrate	Other
Moraxella	−	− (rare +)	−	+ or −	Frequently typical diplobacilli
Neisseria	−	−	−	− or + (gas)	Typical diplococci. Some Neisseria may grow on MacConkey.
IIf	+	−	−	−	Moist growth (capsules)
IIj	+	+	−	−	
Brucella	−	+	Acid or Acid weak (−)	+ or + with gas	Tiny, coccoid
HB-1	−	−	−	+	Lysine and ornithine decarboxylase +
Vibrio extorquens	−	+ delayed or −	Acid weak or acid delayed (−)	−	Pink pigment; nonsoluble

Key: + = 90% or more strains are positive; (+) = 51–89% of strains are positive; (−) = 10–50% of strains are positive; − = less than 10% of strains are positive.
*Organisms to consider: Moraxella; Neisseria, IIf, IIj, Brucella, HB-1, *Vibrio extorquens.*
(Courtesy of the Enteric Section, Entereobacteriology Branch, Bacteriology Division, Bureau of Laboratories, Center for Disease Control, Atlanta, Georgia, 1976)

Table 3-15. *CDC Lettered and Numbered Bacterial Groups: Synonyms*

CDC Letters and Numbers	Synonyms
DF-1	None
DF/0-2	None
EF-4	None (Pasteurella-like)
EO-1	*Pseudomonas cepacia; Pseudomonas multivorans*
HB-1	*Eikenella corrodens*
HB-2	*Hemophilus aphrophilus*
HB-3 & 4	*Actinobacillus actinomycetemcomitans*
HB-5	None (Hemophilus-like)
M-1	*Moraxella kingii; Kingella kingae*
M-3	*Moraxella atlantae*
M-4	*Moraxella urethralis*
M-4f	None (Moraxella-like)
M-5	None (Moraxella-like)
M-6	None (Moraxella-like)
I	*Pseudomonas maltophilia*
Ia	*Pseudomonas diminuta*
Ib-1	*Pseudomonas putrefaciens*
Ib-2	None (Putrefaciens-like)
IIa	*Flavobacterium meningosepticum*
IIb	Flavobacterium species
IIc	None (saccharolytic flavobacterium)
IId	*Cardiobacterium hominis*
IIe	None (saccharolytic flavobacterium)
IIf	None (nonsaccharolytic flavobacterium)
IIh, IIi	None (saccharolytic flavobacterium)
IIj	None (nonsaccharolytic flavobacterium)
IIk-1	Xanthomonas species; *Pseudomonas paucimobilis*
IIk-2	Xanthomonas species
IIIa	*Achromobacter xylosoxidans*
IIIb	*Achromobacter xylosoxidans*
IVa	*Bordetella bronchicanis*
IVb	*Bordetella parapertussis*
IVc	None (Alcaligenes-like)
IVd	None (Bronchicanis-like)
IVe	None
Va-1	None
Va-2	*Pseudomonas pickettii*
Vb-1	*Pseudomonas stutzeri*
Vb-2	*Pseudomonas mendocina*
Vb-3	None
Vc	*Alcaligenes denitrificans*
Vd-1	Achromobacter species
Vd-2	Achromobacter species
Vd-3	*Agrobacterium radiobacter*
Ve-1	*Chromobacterium typhiflavum*
Ve-2	*Chromobacterium typhiflavum*
VI	*Alcaligenes faecalis*

Table 3-16.   *Salient Features for Identification of the Pseudomonas-Alcaligenes Group* (Gilardi Schema)*

Characteristics	P. aeruginosa (51)[†]	P. fluorescens (26)[†]	P. putida (46)[†]	P. pseudomallei (5)[†]	P. cepacia (22)[†]	P. acidovorans (11)[†]	P. alcaligenes (7)[†]	P. pseudoalcaligenes (10)[†]	P. stutzeri (31)[†]	P. mendocina (5)[†]	P. putrefaciens (12)[†]	P. maltophilia (117)[†]	P. dimunita (4)[†]	A. faecalis (9)[†]	A. oderans (23)[†]	A. bronchiseptica (3)[†]
Acid: O-F glucose	100	100	100	100	100	100	0	100	100	100	100	100	0	0	0	0
Acid: O-F maltose	0	84	26	100	100	0	0	10	100	100	25	100	0	0	0	0
ONPG	0	0	0	0	100	0	0	0	0	0	0	98	0	0	0	0
Fluorescence	90	88	90	0	0	0	0	0	0	0	0	0	0	0	0	0
Hydrogen sulfide	0	0	0	0	0	0	0	0	0	0	100	0	0	0	0	0
Urea	86	54	63	60	27	0	14	0	19	60	25	0	0	0	0	100
Nitrate reduction	26	0	0	80	27	100	57	100	26	100	100	38	0	22	0	33
Denitrification (N_2)	61	0	0	100	0	0	0	0	100	100	0	0	0	0	100	0
Oxidase	100	100	100	100	91	100	100	100	100	100	100	0	100	100	100	100
Decarboxylases:																
Arginine	98	100	98	100	0	0	0	0	100	0	0	0	0	0	0	0
Lysine	0	0	0	0	100	0	0	0	0	0	0	0	100	0	0	0
Ornithine	0	0	0	0	0	0	0	0	0	0	100	0	0	0	0	0
Gelatin liquefaction	59	100	0	100	45	0	14	0	0	20	92	100	100	0	0	0
Growth on SS agar	94	96	100	0	0	82	0	80	94	100	58	0	0	44	100	33
Growth at 42° C	100	0	0	100	77	0	100	100	100	40	58	63	100	56	91	66
Penicillin-susceptible	0	0	0	0	0	0	43	0	16	0	9	4	0	33	26	33
Polymyxin-susceptible	100	100	100	0	0	73	86	90	100	100	100	98	0	100	100	100
Phenylalanine deaminase	2	0	0	0	0	0	72	50	60	0	0	0	0	0	0	0

*Modified from Gilardi, G. L.: Practical schema for the identification of nonfermentative Gram-negative bacteria encountered in medical bacteriology. Am. J. Med. Technol., *38*:65–72, 1972.

[†]The number in parentheses is the number of strains tested. The numbers in the columns below indicate the percentage of strains positive for the characteristic tested.

Table 3-17. *Salient Features for Identification of the Acinetobacter-Moraxella Group and Flavobacterium* (Gilardi Schema)*

Test or Substrate	A. anitratum (377)[+]	A. haemolyticus, subsp. haemolyticus (30)[+]	A. haemolyticus, subsp. alcaligenes (7)[+]	A. lwoffi (46)[+]	M. osloensis (7)[+]	M. liquefaciens (2)[+]	M. nonliquefaciens (27)[+]	Flavobacterium (17)[+]
Acid: O-F glucose	100	100	0	0	0	0	0	100
Nitrate reduction	0	0	0	0	29	100	100	0
Cytochrome oxidase	0	0	0	0	100	100	100	100
Esculin hydrolysis	0	0	0	0	0	0	0	100
Lipase activity	100	100	100	96	43	0	7	100
Starch hydrolysis	0	0	0	0	0	0	0	100
Gelatin liquefaction	0	0	100	100	0	100	0	100
Hemolysis (5% sheep blood)	0	97	100	2	0	0	0	41
Growth on SS agar	0	97	100	2	14	0	22	0
Growth on MacConkey agar	100	100	100	100	43	50	41	0
Acetate assimilation (basal mineral medium)	99	100	100	100	100	0	0	0
Penicillin-susceptible	0	3	1	17	100	100	100	0
Polymyxin-susceptible	100	100	100	100	100	100	100	0

Table 3-18. *Identification of Pseudomonas aeruginosa and Acinetobacter calcoaceticus (anitratus) Using Pickett's Preliminary Screening Tests*

Characteristics	Pseudomonas Aeruginosa	Acinetobacter calcoaceticus (anitratus)
Pigmented growth	90% +	90% −
Oxidase	+	−
Motility-nitrate medium:		
Motility	+	−
Gas	+	−
Fluorescence-nitrate medium:		
Fluorescence	+	−
Acidification of slant	−	+
Gas formation (denitrification)	+	−
Growth at 42° C.	+	−
Penicillin-sensitive	−	−

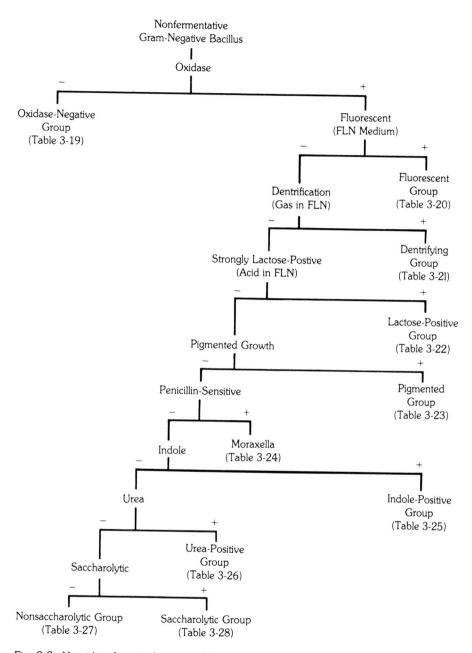

Fig. 3-3. Algorithm for initial group differentiation of the nonfermentative gram-negative bacilli. (Modified from Pickett.)

Table 3-19.   *Oxidase-Negative Group* (Pickett Schema)*

Differential Characteristics	Acinetobacter anitratus	Acinetobacter lwoffi	Pseudomonas cepacia	Pseudomonas maltophilia	CDC's IIk	CDC's Ve
Primary Features:						
Diameter of colonies, mm.	1–2	0.5–1	0.1–0.5	0.2–1	0.1–0.5	0.5–2
MacConkey medium, growth	+	+	(−)	+	(−)	+
Pigmented growth	−	−	(−)	(−)	(+)	+
Fluorescence-lactose-denitrification (FLN) medium						
Slant, acid	+	−	(+)	−	(+)	−
Motility-nitrate medium						
Motility	−	−	+	+	(−)	+
Nitrite	−	−	(+)	(+)	(−)	(+)
Secondary Features:						
Arabinose	+	+	+	−	+	+
Glucose	+	(+)	+	+	+	+
Lactose	+	−	+	+	+	(−)
Acetamide	−	−	(+)	−	−	−
Gluconate	−	−	−	−	−	(+)
Lysine decarboxylase (LDC)	−	−	+	+	−	−
Urea	(−)	−	−	−	(−)	−
Additional Features:						
Fructose	(+)	(+)	+	+	+	+
Mannitol	−	−	−	−	−	+
Rhamnose	+	−	−	−	(−)	(−)
Sucrose	−	−	(+)	+	+	(−)
Loeffler's slant, liquefied	−	−	−	+	−	−

Key: + = 90% or more strains are positive; (+) = 51–89% of strains are positive; (−) = 10–50% of strains are positive; − = less than 10% of strains are positive.
*Organisms to consider: *Acinetobacter anitratus, Acinetobacter lwoffi, Pseudomonas cepacia, Pseudomonas maltophilia,* CDC's IIk, CDC's Ve.

not grow in the deeper portions of the medium where atmospheric oxygen is excluded.

By mid to late afternoon, sufficient growth will have taken place in the area of spot inoculation on the KIA slant to perform a cytochrome oxidase test and a Gram's stain. An initial interpretation of the motility tubes should also be made in 6 to 8 hours. Motile nonfermentative bacilli often produce far less opacity in semisolid agar than do the fermentative organisms, and the opacity may be so slight that it could be missed if tubes are interpreted only at 24 hours.

If the organism is motile and oxidase-positive (suspicious for Pseudomonas species), inoculate a tube of brain-heart infusion broth for incubation at 42° C.

If the organism is nonmotile and oxidase-positive (suspicious for Moraxella species), then prepare a two-unit penicillin disk susceptibility test.

Heavily inoculate a tube of fluorescence-nitrate (FN) medium, to be discussed below. Then stab the KIA agar slant through the

Table 3-20.   *Fluorescent Group* (Pickett Schema)*

Differential Characteristics	Species		
	Pseudomonas aeruginosa	Pseudomonas fluorescens	Pseudomonas putida
Primary Features:			
Diameter of colonies, mm.	0.5–>2	0.1–>1	0.4–>2
Pigmented growth	(+)	−	−
Growth at 42° C.	+	−	−
FLN medium			
Gas	+	(−)	−
Motility-nitrate medium			
Gas	+	(−)	−
Secondary Features:			
Lactose	−	(+)	−
Acetamide	+	−	−
Gluconate	(+)	(+)	+
Additional Features:			
Maltose	−	+	−
Rhamnose	(−)	+	−
Tartrate	−	−	+

Key: + = 90% or more strains are positive; (+) = 51–89% of strains are positive; (−) = 10–50% of strains are positive; − = less than 10% of strains are positive.
*Organisms to consider: *Pseudomonas aeruginosa, Pseudomonas fluorescens, Pseudomonas putida.*

area of the spot inoculation and completely streak the slant using an inoculating loop so that a lawn of growth covering the entire surface will be produced after 24 hours of incubation at 35° C.

On the following morning, first determine if the isolate is *Pseudomonas aeruginosa* or *Acinetobacter calcoaceticus* (Herellea-Mima group), the two species accounting for more than 50 per cent of nonfermenter isolates in most clinical laboratories. The identification of the bacteria can be made with the reactions listed in Table 3-18. If either of these species can be identified, a final report can be issued and no further work is needed.

If these initial observations fail to identify either *P. aeruginosa* or *A. calcoaceticus*, a heavy inoculum is prepared from the lawn of growth on the surface of the KIA slant. Overlay the surface growth with 2 to 3 ml. of sterile water and, with an applicator stick or glass rod, emulsify the organisms to produce a slurry having the consistency of skim milk.

Three drops of this heavy suspension are then added to each tube of buffered single-substrate medium to be used in determining the final identification of the organism. The following "secondary" set of substrates should be used. These allow identification of well over 95 per cent of the nonfermenters encountered in a clinical laboratory:

> Tryptophan (indole)
> Gluconate
> Lysine decarboxylase
> Lactose
> Fructose
> Dextrose
> Arabinose
> Acetamide
> Urea

Figure 3-3 is the algorithm developed by Pickett by which the majority of the nonfermentative gram-negative bacilli can be identified in the clinical laboratory. According to

Table 3-21. Denitrifying Group* (Pickett Schema)

Differential Characteristics	Achromobacter	Pseudomonas aeruginosa	Pseudomonas pickettii	Pseudomonas pseudomallei	Pseudomonas stutzeri	CDC's Va	Alcaligenes denitrificans	Alcaligenes odorans	Pseudomonas denitrificans
Species									
Primary Features:									
Diameter of colonies, mm.	1	0.5–2.0	0.1–0.5	0.8–1.5	0.2–1.5	0.5–1.0	0.2–2	0.2–1.2	0.7–1.0
Pigmented growth	–	(+)	–	(–)	(–)	–	+	–	–
Growth at 42° C.	(–)	+	(+)	+	+	(–)	+	+	(–)
FLN medium									
Fluorescence	–	+	–	(–)	–	–	–	–	–
Slant, acid	–	–	–	(–)	–	(–)	–	–	–
Gas	ND	+	(–)	(+)	(+)	(–)	+	+	+
Motility-nitrate medium									
Gas	ND	+	(–)	(+)	(+)	(–)	+	–	+
Secondary Features:									
Arabinose	(+)	+	+	+	+	+	–	–	–
Lactose	–	–	–	+	–	+	–	–	–
Acetamide	(+)	+	–	(+)	(–)	(–)	+	+	–
Gluconate	(+)	(+)	–	–	–	–	–	–	–
LDC	ND	–	–	–	–	–	–	–	–
Urea	(+)	–	–	–	–	+	–	–	–
Additional Features:									
Fructose	ND	+	+	+	+	+	–	–	–
Sucrose	(–)	–	–	(+)	–	–	–	–	–
Gelatin	–	+	(–)	+	–	–	–	–	–
Amylase (Mueller-Hinton)	–	–	–	+	(+)	–	–	–	–
Esculin (in KIA)	(–)	–	–	(+)	–	–	–	–	–

Key: + = 90% or more strains are positive; (+) = 51–89% of strains are positive; (–) = 10–50% of strains are positive; – = less than 10% of strains are positive; ND = no data (not tested).

*Organisms to consider: *Pseudomonas aeruginosa, Pseudomonas pickettii, Pseudomonas pseudomallei, Pseudomonas stutzeri, CDC's Va, Alcaligenes denitrificans, Alcaligenes odorans, Pseudomonas denitrificans.*

Table 3-22. Strongly Lactose-Positive Group* (Pickett Schema)

Differential Characteristics	Pseudomonas cepacia	Pseudomonas pseudomallei	CDC's IIk	CDC's Va
Primary Features:				
Diameter of colonies, mm.	0.1–0.5	0.8–1.5	0.1–0.5	0.5–1.0
MacConkey medium, growth	(−)	+	(−)	+
Pigmented growth	(−)	(−)	(+)	−
Growth at 42° C.	(+)	+	−	(−)
FLN medium				
Slant, acid	(+)	(−)	(+)	(−)
Gas	−	(+)	−	(−)
Motility-nitrate medium				
Motility	+	+	(−)	+
Gas	−	(+)	−	(−)
Nitrite	(+)	+	(−)	+
Secondary Features:				
Acetamide	(+)	(+)	−	(−)
LDC	+	−	−	−
Urea	−	−	(−)	+
Additional Features:				
Mannitol	+	+	−	−
Sucrose	(+)	(+)	+	−
Amylase (Mueller-Hinton medium)	−	+	(−)	−
Esculin (in KIA)	(+)	(+)	(+)	−

Key: + = 90% or more strains are positive; (+) = 51–89% of strains are positive; (−) = 10–50% of strains are positive; − = less than 10% of strains are positive.
*Organisms to consider: *Pseudomonas cepacia, Pseudomonas pseudomallei,* CDC's IIk, CDC's Va.

this schema, the nonfermenters are divided into the following subgroups:

1. The oxidase-negative group
2. The fluorescent group
3. The denitrifying group (oxidase-positive and nonfluorescent)
4. The strongly lactose-positive group (oxidase-positive, nonfluorescent, and nondenitrifying)
5. The pigmented group (oxidase-positive, nonfluorescent, and nondenitrifying)
6. The penicillin-sensitive group (oxidase-positive, nonfluorescent, and nondenitrifying)
7. The indole-positive group
8. The urea-positive group (oxidase-positive, nonfluorescent, nondenitrifying, and indole-negative)
9. The nonsaccharolytic group (oxidase-positive, nondenitrifying, nonpigmented, and urea-negative)
10. The saccharolytic group (oxidase-positive, nondenitrifying, nonpigmented, and urea-negative)

A list of the organisms to consider within each of these groups and their differential characteristics for identification are included in Tables 3-19 through 3-28, which follow Pickett's schema. These tables were adapted from charts made at a workshop at the Colorado Association for Continuing Medical Laboratory Education, Denver, Colorado, in December, 1977.

A brief review of the several nutritional and biochemical characteristics listed in Tables 3-19 through 3-28 and some of their specific applications in the use of the Pickett schema are included in the following paragraphs.

Evaluation of Colonial Size. There may be instances where the estimation of the diameter of colonies of an unknown isolate after 24 and 48 hours of incubation may be helpful in placing it within one of the Pickett groups. The range in diameter of colonies is therefore included in the charts for this reason.

(Text continues on p. 142)

Table 3-23. *Pigmented Group* (Pickett Schema)*

Differential Characteristics	Species				
	Flavobacterium, Group 2	*Flavobacterium,* Group 3	*Pseudomonas Maltophilia*	*Pseudomonas stutzeri*	CDC's IIk
Primary Features:					
Diameter of colonies, mm.	0.5–1.5	0.2–0.8	0.2–1.0	0.2–1.5	0.1–0.5
MacConkey medium, growth	(−)	−	+	+	(−)
Pigmented growth	(+)	(+)	(−)	(−)	(+)
Growth at 42° C.	(−)	+	(−)	+	−
Penicillin sensitivity	−	+	−	−	(+)
FLN medium					
Slant, acid	−	−	−	−	(+)
Gas	−	−	−	(+)	−
Motility-nitrate medium					
Motility	−	−	+	+	(−)
Gas	−	−	−	(+)	−
Nitrite	(−)	−	(+)	+	(−)
Secondary Features:					
Arabinose	(−)	−	−	+	+
Glucose	+	−	+	+	+
Lactose	−	−	+	−	+
Acetamide	−	−	−	(−)	−
Indole	+	+	−	−	−
LDC	−	−	+	−	−
Urea	−	−	−	−	(−)
Additional Features:					
Fructose	+	−	+	+	+
Sucrose	−	−	+	−	+
Indol pyruvic acid (IPA) and phenyl pyruvic acid (PPA)	−	+	−	−	−
Amylase (Mueller-Hinton medium)	+	−	−	(+)	(−)
Esculin (in KIA)	+	−	(+)	−	(+)

Key: + = 90% or more strains positive; (+) = 51–89% strains positive; (−) = 10–50% strains positive; − = less than 10% strains positive.

*Organisms to consider: Flavobacterium, group 2, Flavobacterium, group 3, *Pseudomonas maltophilia, Pseudomonas stutzeri,* CDC's IIk.

Table 3-24. *Penicillin-Sensitive Group—Moraxella* (Pickett Schema)*

Differential Characteristics	Species				
	Flavobacterium, Group 3	*Moraxella nonliquefaciens*	*Moraxella osloensis*	*Moraxella phenylpyruvica*	CDC's IIk
Primary Features:					
Diameter of colonies, mm.	0.2–0.8	0.1–0.5	0.1–0.8	0.1–0.8	0.1–0.5
MacConkey medium, growth	−	(−)	(+)	+	(−)
Pigmented growth	(+)	−	−	−	(+)
Growth at 42° C.	+	−	−	+	−
Penicillin sensitivity	+	+	+	+	(+)
FLN medium					
Slant, acid	−	−	−	−	(+)

Table 3-24. *Penicillin-Sensitive Group—Moraxella* (Pickett Schema) (Continued)*

	Species				
Differential Characteristics	Flavobacterium, Group 3	Moraxella nonliquefaciens	Moraxella osloensis	Moraxella phenylpyruvica	CDC's IIk
Motility-nitrate medium					
Motility	−	−	−	−	(−)
Nitrite	−	−	−	−	(−)
Secondary Features:					
Arabinose	−	−	(+)	−	+
Glucose	−	−	−	−	+
Lactose	−	−	−	−	+
Indole	+	−	−	−	−
Urea	−	−	−	−	(−)
Additional Features:					
Fructose	−	−	−	−	+
Ribose	−	−	+	−	+
Asparagine	+	−	−	+	(+)
Glutamine	+	−	−	+	(+)
Formate	−	−	−	+	(+)
IPA and PPA	+	−	−	+	−

Key: + = 90% or more strains positive; (+) = 51–89% strains positive; (−) = 10–50% strains positive; − = less than 10% strains positive.
*Organisms to consider: Flavobacterium, group 3, *Moraxella nonliquefaciens, Moraxella osloensis, Moraxella phenylpyruvica,* CDC's IIk.

Table 3-25. *Indole-Positive Group* (Pickett Schema)*

	Species		
Differential Characteristics	Flavobacterium meningosepticum	Flavobacterium, Group 2	Flavobacterium, Group 3
Primary Features:			
Size of colonies, mm.	0.5–1.5	0.5–1.5	0.2–0.8
MacConkey medium, growth	(−)	(−)	−
Pigmented growth	−	(+)	(+)
Growth at 42° C.	(−)	(−)	+
Penicillin sensitivity	−	−	+
Motility-nitrate medium			
Nitrite	−	(−)	−
Secondary Features:			
Arabinose	−	(−)	−
Glucose	+	+	−
Lactose	+	−	−
Additional Features:			
Fructose	+	+	−
Mannitol	+	−	−
Amylase (Mueller-Hinton medium)	−	+	−
Esculin (in KIA)	+	+	−
IPA and PPA	−	−	+

Key: + = 90% or more strains positive; (+) = 51–89% strains positive; (−) = 10–50% strains positive; − = less than 10% strains positive.
*Organisms to consider: *Flavobacterium meningosepticum,* Flavobacterium, group 2, Flavobacterium, group 3.

Table 3-26. *Urea-Positive Group* (Pickett Schema)*

Differential Characteristics	Species		
	Bordetella bronchiseptica	CDC's Ilk, Biotype 2	CDC's Va
Primary Features:			
Diameter of colonies, mm.	0.2–1.5	0.1–0.5	0.5–1.0
MacConkey medium, growth	+	(−)	+
Pigmented growth	−	(+)	−
Growth at 42° C	+	(−)	(−)
Penicillin sensitivity	−	(−)	−
FLN medium			
Slant, acid	−	(+)	(−)
Gas	−	−	(−)
Motility-nitrate medium			
Motility	+	(−)	+
Gas	−	−	(−)
Nitrite	(−)	(−)	+
Secondary Features:			
Arabinose	−	+	+
Glucose	−	+	+
Lactose	−	+	+
Acetamide	−	−	(−)
Urea	+	(+)	+
Additional Features:			
Fructose	−	+	+
Sucrose	−	+	−
Amylase (Mueller-Hinton medium)	−	(+)	−
Esculin (in KIA)	−	(+)	−

Key: + = 90% or more of strains are positive; (+) = 51–89% of strains are positive; (−) = 10–50% of strains are positive; − = 10% or less of strains are positive.
*Organisms to consider: *Bordetella bronchiseptica,* CDC's Ilk (biotype 2), CDC's Va.

Table 3-27. *Nonsaccharolytic Group* (Pickett Schema)*

Differential Characteristics	Species					
	Alcaligenes faecalis	Pseudomonas acidovorans	Pseudomonas alcaligenes	Pseudomonas diminuta	Pseudomonas pseudoalcaligenes	Pseudomonas testosteroni
Primary Features:						
Diameter of colonies, mm.	ND	0.4–1.0	0.5–1.8	0.1–0.4	0.8–1.0	ND
Growth at 42° C.	+	−	+	(−)	+	−
Motility-nitrate medium						
Nitrite	(−)	+	+	−	+	+
Secondary Features:						
Arabinose	−	−	−	(−)	(−)	−
Glucose	−	(−)	−	−	(−)	−
Acetamide	+	+	−	−	−	−
Additional Features:						
Fructose	−	+	−	−	+	−
Mannitol	−	(+)	−	−	−	−
Allantoin	−	(+)	−	−	−	+
Citrate	+	+	+	−	+	+
Gelatin	−	+	(−)	+	(−)	−
Tartrate	+	+	−	−	−	−

Key: + = 90% or more of strains are positive; (+) = 51–89% of strains are positive; (−) = 10–50% of strains are positive; − = 10% or less of strains are positive; ND = no data (not tested).
*Organisms to consider: *Alcaligenes faecalis, Pseudomonas acidovorans, Pseudomonas alcaligenes, Pseudomonas diminuta, Pseudomonas pseudoalcaligenes, Pseudomonas testosteroni.*

Table 3-28. Saccharolytic Group* (Pickett Schema)

Differential Characteristics	Achromobacter	Pseudomonas acidovorans	Pseudomonas cepacia	Pseudomonas maltophilia	Pseudomonas pickettii	Pseudomonas pseudoalcaligenes	Pseudomonas vesicularis
				Species			
Primary Features:							
Diameter of colonies, mm.	<1	0.4–1.0	0.1–0.5	0.2–1	0.1–0.5	0.8–1.0	<0.2
Pigmented growth	—	—	(—)	(—)	—	—	(—)
Growth at 42° C.	(+)	—	(+)	(—)	(+)	+	(—)
MacConkey medium, growth	+	+	(—)	+	+	+	(+)
FLN medium							
Slant, acid	—	—	(+)	—	—	—	—
Gas	ND	—	—	—	(—)	—	—
Motility-nitrate medium							
Gas	ND	—	—	—	(—)	—	—
Nitrite	+	+	(+)	(+)	+	+	—
Secondary Features:							
Arabinose	(+)	—	+	—	+	(—)	+
Glucose	+	(—)	+	+	+	(—)	+
Lactose	—	—	+	+	—	—	—
Acetamide	(+)	+	(+)	—	—	—	—
Gluconate	(+)	—	+	—	—	—	—
LDC	ND	—	+	+	—	—	—
Urea	(+)	—	—	—	—	—	—
Additional Features:							
Fructose	ND	+	+	+	+	+	(—)
Mannitol	(—)	(+)	+	—	—	—	—
Sucrose	(—)	—	(+)	+	—	—	—
Esculin (in KIA)	(—)	—	(+)	(+)	—	—	(—)

Key: + = 90% or more of strains are positive; (+) = 51–89% of strains are positive; (—) = 10–50% of strains are positive; — = 10% or less of strains are positive; ND = no data (not tested).

*Organisms to consider: Achromobacter, Pseudomonas acidovorans, Pseudomonas cepacia, Pseudomonas maltophilia, Pseudomonas pickettii, Pseudomonas pseudoalcaligenes, Pseudomonas vesicularis.

Chart 3-2.   *Fluorescence-Denitrification*

Introduction   The ability to produce fluorescein pigment and to reduce nitrates and/or nitrites completely to nitrogen gas are two important characteristics in the identification of the pseudomonads and other nonfermentative bacilli. Fluorescence-denitrification (FN) medium is formulated to detect these two characteristics. Fluorescence-lactose-nitrate (FLN) medium is a modification in which lactose and phenol red indicator are added to permit the detection of acid formed from utilization of lactose which is helpful in identifying the strongly lactose-positive group of nonfermenters.

Principle   Fluorescein is an organic luminescent dye that upon excitation with ultraviolet light emits a green-yellow fluorescence. A few of the nonfermentative bacilli, notably *Pseudomonas aeruginosa,* are capable of producing fluorescein, the detection of which is helpful in their identification. Fluorescence of colonies may not be detected on an ordinary isolation medium such as blood agar or MacConkey agar; rather, media containing cationic salts such as magnesium sulfate (included in FN medium), which act as activators or coactivators to intensify luminescence, often must be used.
The reduction of nitrate to nitrogen gas is shown by the following chemical equation:

$$2\ NO_3^- + 10\ e^- + 12\ H^+ \longrightarrow N_2 \uparrow + 6\ H_2O$$

In this reduction process, 5 electrons are accepted by the nitrate radical, resulting in formation of nitrogen gas and 6 molecules of water. This process is known as denitrification and is helpful in the separation of Pseudomonas species (most strains are positive) from other nonfermentative bacilli.

Media and Reagents   The formula for fluorescence-denitrification medium is:

Proteose peptone #3 (Difco)	1.00 g.
Magnesium sulfate · 7 H$_2$O	0.15 g.
Dipotassium hydrogen phosphate	0.15 g.
Potassium nitrate	0.20 g.
Sodium nitrite	0.05 g.
Agar	1.50 g.
Distilled water	100.00 ml.

Pigment Production. Colonial pigmentation is assessed visually. Pigmented strains of bacteria may not produce a visible color within the initial 24-hour incubation period; however, an additional 24 hours of incubation at room temperature usually is sufficient to show pigment development. (See Plate 3-1, *F.*)

Growth on MacConkey Agar. MacConkey agar contains substances inhibitory to the growth of some strains of bacteria as discussed in Table 2-1 on page 63. MacConkey agar plates are visually examined with strong reflected light for the development of colonies after 24 to 48 hours incubation at 35° C. The use of a hand lens or a dissecting microscope may be helpful for detection of small colonies.

Fluorescence-Denitrification Medium. Fluorescence-denitrification medium is discussed in Chart 3-2. Note in the formula that the medium contains no carbohydrates; therefore, the presence of gas indicates that the test organism has reduced the nitrate and/or nitrite to nitrogen gas (denitrification; see Plate 3-1, *H*). In some formulations, phenol red and lactose are added (FLN medium). This is helpful in detecting

Chart 3-2.   *Fluorescence-Denitrification (Continued)*

Media and Reagents *(Continued)*	If FLN medium is to be prepared, add 2.00 g. of lactose and 0.002 g. of phenol red indicator. Dispense 4 ml. of medium into 13-mm. screwcap tubes and let solidify to give a deep and a slant of approximately equal length. Sellers medium, available commercially, is also suitable for the determination of fluorescence and denitrification by nonfermentative bacteria.
Procedure	Inoculate the medium by stabbing the deep with a heavy suspension of the culture and then streaking the slant. Incubate at 35° C. for 24 to 48 hours.
Interpretation	Examine the tube for fluorescence with an ultraviolet light source (Wood's lamp). A bright yellow-green glow indicates a positive test (Plate 3-1, *G*). The presence of gas bubbles in the deep of the medium indicates that nitrogen gas was produced from denitrification and a yellow slant with FLN medium indicates acid was produced from the utilization of lactose by the microorganism (Plate 3-1, *H*).
Controls	The following organisms are appropriate controls: Positive fluorescence/positive denitrification: *Pseudomonas aeruginosa* Negative fluorescence/positive denitrification: *Pseudomonas denitrificans* Negative fluorescence/negative denitrification: *Escherichia coli*
Bibliography	Pickett, M. J., and Petersen, M. M.: Characterization of saccharolytic nonfermentative bacilli associated with man. Can. J. Microbiol., *16*:351–362, 1970.

the strongly lactose-positive group of nonfermentative bacilli.

    Fluorescence can be detected by comparing under ultraviolet light (Wood's lamp) an inoculated tube after incubation with one that has not been inoculated. An apple-green glow is indicative of fluorescein dye and a positive test (Plate 3-1, *G*). Magnesium sulfate is a necessary ingredient in the formula to provide the necessary cations to activate fluorescein production.

Motility-Nitrate Medium. This combination medium is used to detect bacterial motility and the ability of the organism to reduce nitrate to nitrite. As mentioned above, only the upper 5 mm. of the agar is stabbed and observations for motility are made at 6, 24, and 48 hours (Plate 3-2, *D*). Nitrate reduc-

tion is detected by adding 5 drops each of alpha-naphthylamine and sulfanilic acid reagent and observing for the development of a red color within 30 seconds (Plate 3-1, *E*). The appearance of gas in this medium also indicates denitrification.

*Buffered Single-Substrate Media.** Most of the secondary and additional characteristics listed in the Pickett charts are determined by the use of buffered single-substrate media as described in Chart 3-3. Each of these media contains only a single substance, usually a carbohydrate, alcohol, or amine, a pH indicator, and phosphate buffers to provide a

(Text continues on p. 146)

—————————

*Buffered single-substrate media are available in tablet form from Key Scientific Products Co., Los Angeles, California

Chart 3-3. *Buffered Single-Substrate Media*

Introduction   Many of the differential media commonly used for the identification of fermentative organisms are not suitable for the nonfermentative bacilli because some of these grow slowly and produce weak reactions in the media. Buffered single-substrate media designed by Pickett allow the results of biochemical tests to be determined quickly (usually within 24 hours) through the use of heavy bacterial suspensions of pregrown cultures.

Principle   Buffered single-substrate media include a single substrate (a carbohydrate, alcohol, amine, or other chemical), a pH indicator (phenol red), and phosphate buffers of a pH appropriate for the test being performed. Three to five drops of a turbid cell suspension prepared from the surface growth of the unknown bacterium on KIA slant are added to the buffered substrate. The reactions can usually be read within 24 hours indicating specific products or changes in the pH of the medium.

Media and Reagents   All single-substrate media are dispensed in 1-ml. quantities into 13- × -100-mm. tubes. Gluconate, indole, indolepyruvic acid, lysine decarboxylase, and phenylpyruvic acid substrates are steam-sterilized; carbohydrate-, amide-, organic salt-, and urea-buffered single-substrate media need not be autoclaved.

Carbohydrates and Alcohols
1. Prepare all stock carbohydrates as 10% solutions in 20- × -150-mm. screwcap tubes.
2. Add 0.5 M KH_2PO_4 and 0.5 M K_2HPO_4.
   A. Stock solutions (All are stable for months at room temperature.):
      1. 0.5M KH_2PO_4
      2. 0.5M K_2HPO_4
      3. Solution of phenol red and crystal violet

Phenol red	2.0 g.
Crystal violet	0.2 g.
Distilled water	200.0 ml.

      4. Carbohydrate substrates: prepare as 10% solutions in 20- × -150-mm. screwcap tubes. A few drops of chloroform can be added as a preservative.
      5. Stock basal medium:

K_2HPO_4, 0.5 M	5.0 ml.
Phenol red-crystal violet	1.0 ml.
Agar	0.5 g.
Distilled water	400.0 ml.

   B. Preparation of working media and use:

1. Basal medium (13-mm. tube)	0.8 ml.
2. Steam 10 minutes.	
3. Carbohydrate substrate	0.2 ml.
4. Add inoculum.	0.1 ml.
5. Incubate at 35° C. for 4 to 6 days and read daily.	

Amides and Organic Salts
1. Stock substrates:

Acetamide	1.0 g.%
Formate	1.0 g.%
Nicotinamide	2.5 g.%

Chart 3-3. *Buffered Single-Substrate Media* (Continued)

Media and Reagents	Tartrate	1.0 g.%
(Continued)		

Stocks should be in the range of *p*H 6.5 to 7.0. Store in 20-mm. screwcap tubes over a few drops of chloroform as a preservative.

2. Stock basal medium:

KH_2PO_4, 0.5 M	14 ml.
K_2HPO_4, 0.5 M	6 ml.
Phenol red-crystal violet	1 ml.
Agar	0.5 g.
Distilled water	400 ml.

3. Preparation and use of working medium:

Basal medium (13-mm. tube)	0.8 ml.
Substrate	0.2 ml.
Steam 10 minutes.	
Add inoculum.	0.1 ml.

Incubate at 35° C. for 4 to 6 days with daily reading.

Specific Media

1. Indole:

L-tryptophan	1.0 g.
NaCl	1.0 g.
K_2HPO_4, 0.5 M	7.2 ml.
KH_2PO_4, 0.5 M	0.8 ml.
Distilled water	192.0 ml.

2. Gluconate:

Potassium gluconate	4.0 g.
KNO_3	0.4 g.
KH_2PO_4, 0.5 M	16.0 ml.
$NaHCO_3$	0.2 g.
Distilled water	184.0 ml.

3. Lysine decarboxylase:

L-lysine-HCl	1.0 g.
Glucose	1.0 g.
KH_2PO_4	1.0 g.
Distilled water	200 ml.

4. Phenylpyruvic acid:

DL-phenylalanine	0.8 g.
KH_2PO_4, 0.5 M	4.0 ml.
K_2HPO_4, 0.5 M	4.0 ml.
Distilled water	192.0 ml.

5. Urea:

Urea	4.0 g.
KH_2PO_4, 0.5 M	25 ml.
K_2HPO_4, 0.5 M	25 ml.
Phenol red-crystal violet	0.4 ml.
Distilled water	200.0 ml.

Interpretation

1. Carbohydrate substrates: positive = yellow; negative = red
2. Amide substrates: positive = blue; negative = yellow
3. Indole: positive = development of red ring upon addition of Kovac's reagent.
4. Gluconate: positive = development of green or brown color upon addition of Benedict's reagent

(Continued)

145

Chart 3-3. *Buffered Single-Substrate Media* (Continued)

Interpretation *(Continued)*	5. Phenylpyruvic acid: positive = development of a green color on addition of 10% $FeCl_3$ 6. Urea: positive = development of a red color
Bibliography	Pickett, M. J.: Nonfermentative Bacilli: Detection and Identification. *Personal Communication*, 1973. Pickett, M. J., and Pedersen, M. M.: Nonfermentative bacilli associated with man: II. Detection and Identification. Am. J. Clin. Pathol., *54*:164–177, 1970.

*p*H appropriate for the substrate contained. Growing the bacteria on KIA agar initially helps to circumvent a prolonged lag phase of growth that may result when the unknown bacterium is inoculated directly to secondary media. The use of a heavy suspension charges the single-substrate media with sufficient preformed enzymes so that the reaction is sufficiently rapid to permit an interpretation usually within 24 hours.

Amylase (Starch Hydrolysis). Brucella agar (BBL) supplemented with 0.2 g. per cent of potato starch is recommended by Pickett.

One ml. of molten medium is added per tube and allowed to solidify as a long slant. Inoculate with 1 loop of a heavy bacterial suspension in a single streak along the center of the slant and incubate for 72 hours. Test for residual starch adjacent to the streak by flooding the surface of the agar with 0.5 ml. of Gram's iodine. A clear zone adjacent to the streak is indicative of a positive test; nonhydrolyzed starch turns a dark blue color when complexed with iodine.

Gelatin Hydrolysis. Many species of nonfermentative gram-negative bacilli produce

Chart 3-4. *Gelatin Liquefaction*

Introduction	Gelatin is a complex protein, derived from animal collagen, that initially served as a solidifying agent for culture media. Because gelatin changes from a gel to a fluid at 28° C., agar has replaced it since most solid media are incubated at 37° C. Gelatin has poor nutritive value and currently is used in culture media almost exclusively to detect the presence of proteolytic enzymes.
Principle	Gelatinases are proteolytic enzymes capable of breaking down gelatin and other proteins into peptides and amino acids. Those bacteria that secrete gelatinase can be detected by observing the liquefaction of culture media or substrates containing gelatin, following inoculation of the test organism and incubation for the appropriate period of time. Three general substrates are used in the test: (1) nutrient gelatin medium; (2) cellulose strips coated with gelatin; and (3) Kohn gelatin-charcoal particle substrate.
Media and Reagents	1. Nutrient gelatin medium:

Beef Extract	3 g.
Peptone	5 g.
Gelatin	120 g.
Distilled water	1000 ml.

Final *p*H 6.8

2. *Key gelatin strips

Chart 3-4. *Gelatin Liquefaction (Continued)*

Media and Reagents *(Continued)*	3. Kohn gelatin-charcoal medium: denatured gelatin-charcoal disks: prepared by adding powdered charcoal to a slurry of nutrient gelatin (see MacFaddin, reference 3, for instructions).

Procedure

1. *Nutrient Gelatin Medium:* gelatin tubes must be kept in the refrigerator until just prior to use. With an inoculating wire containing a heavy inoculum from an 18- to 24-hour pure culture of the organism to be tested stab the medium to a depth of ½ to 1 inch. Set up an uninoculated control tube to be run along with the bacterium being tested. Incubate at 35° C. for 24 hours to 14 days.
2. *Cellulose Gelatin Strip:* heavily inoculate about 1 ml. of distilled water or physiologic saline with the organism to be tested. Place a Key gelatin strip into the tube so that at least one half of the strip is immersed beneath the surface of the bacterial suspension. Incubate at 35° C. for 41 hours or more.
3. *Kohn Gelatin Test:* Place a denatured gelatin-charcoal disk into about 3 ml. of sterile liquid growth medium (trypticase soy broth or brain-heart infusion broth). Heavily inoculate from an 18- to 24-hour pure culture of the organism to be tested. Set up an uninoculated control to run along with the organism to be tested. Incubate both tubes for 24 hours or longer at 35° C.

Interpretation

1. *Nutrient Gelatin Medium:* check tubes daily for 2 weeks. At the end of each 24-hour period, place both tubes (test bacterium and control) in the refrigerator for approximately 2 hours. Tilt the tubes to see if liquefaction has occurred. Compare the test bacterium with the control.
2. *Key Gelatin Strip:* observe daily for clearing of the opaque gelatin coating in the submerged zone of the strip. If the test is positive, the reaction zone will appear transparent.
3. *Kohn Gelatin Test:* if the test bacterium has liquified the gelatin, the charcoal particles are released, producing a diffuse blackening of the medium. Because the Kohn gelatin-charcoal disks require only small amounts of gelatinase to release the charcoal particles, positive results often can be detected within 24 hours, whereas as many as 7 days may be required by the nutrient gelatin test.

Controls

The following organisms serve as controls:
Positive: rapid, *Pseudomonas aeruginosa* or *Staphylococcus aureus;* delayed, *Staphylococcus epidermidis*
Negative: *Escherichia coli*

Bibliography

Blazevic, D. J., and Ederer, G. M.: Principles of Biochemical Tests in Diagnostic Microbiology. Pp. 45–49. New York, John Wiley & Sons, 1975.

Kohn, J.: A preliminary report of a new gelatin liquefaction method. J. Clin. Pathol., *6:*249–250, 1953.

Lautrop, H. A.: Modified Kohn's test for the demonstration of bacterial gelatin liquefaction. Acta Pathol. Microbiol. Scand., *39:*357–369, 1956.

Levine, M. and Carpenter, D. C.: Gelatin liquefaction by bacteria. J. Bacteriol., *8:*297–306, 1923.

MacFaddin, J. F.: Biochemical Tests for Identification of Medical Bacteria. Pp. 78–84. Baltimore, Williams & Wilkins, 1976.

*Key Scientific Products Co., Los Angeles, California

Chart 3-5. *Gluconate Test*

Introduction	Gluconic acid is one of the oxidation products formed from glucose by aerobic microorganisms that utilize carbohydrates via the Entner-Doude-roff pathway, as shown in Figure 3-1. Gluconate test medium contains potassium gluconate to serve as a substrate for those bacterial species which can oxidize gluconate to 2-ketogluconic acid and other metabolites.
Principle	Potassium gluconate is not a reducing substance; however, 2-ketogluconic acid and other intermediate metabolites formed from gluconate degradation are reducing substances that can be detected by adding cupric hydroxide (Benedict's reagent) to the test medium. If reducing substances are present, the cupric hydroxide is reduced to cuprous oxide which can be detected visually as a yellow-orange precipitate in the medium.

Media and Reagents

1. Gluconate substrate:
   Gluconate peptone broth:

Peptone	1.5 g.
Yeast extract	1.0 g.
Dipotassium phosphate	40.0 g.
Potassium gluconate	37.2 g.
Distilled water	1000.0 ml.

   Final pH = 7.0
   Key gluconate substrate test tablet (Key Scientific Products Co.)
2. Reducing reagent:
   Benedict's qualitative solution:

Copper sulfate ($CuSO_4$ 5 H_2O)	1.73 g.
Sodium carbonate (Na_2CO_3)	10.00 g.
Sodium citrate ($C_3H_4OH(COO)_3Na_3$)	17.30 g.
Distilled water	100.00 ml.

   Clinitest tablets (Ames Company, Inc.)

Procedure	From an 18- to 24-hour pure culture of the organism to be tested, heavily inoculate the gluconate culture medium and incubate at 35° C. for 24 to 48 hours. At the end of the incubation period, add 1.0 ml. of Benedict's reagent or ½ of a Clinitest tablet directly to the tube of culture medium.
Interpretation	A positive test is indicated by the development of a yellow-orange precipitate in the medium, indicating the presence of reducing substances (Plate 3-1, *I*). In a negative test, the color of the medium after addition of Benedict's reagent (or the Clinitest tablet) remains a bluish-green.
Controls	The following microorganisms may serve as controls: Positive: *Pseudomonas aeruginosa* or *Enterobacter aerogenes* Negative: *Escherichia coli*

Bibliography	MacFaddin, J. F.: Biochemical Tests for Identification of Medical Bacteria. Pp. 85–89. Baltimore, Williams & Wilkins, 1976. Pease, M., *et al.*: An approach to the problem of differentiating pseudomonads in the clinical laboratory. Am. J. Med. Technol., *34*:35–40, 1968.

proteolytic enzymes that liquify gelatin and other proteins. The three methods most commonly employed for detection of gelatin liquefaction are outlined in Chart 3-4. The Kohn test is the most sensitive of the three. A gelatin-charcoal medium is incorporated in the API 20 E strip as discussed below. Nutrient gelatin medium or the cellulose gelatin strips are sufficiently sensitive to detect strong gelatinase-producing strains of Pseudomonas; however, they may give false negative or equivocal results with nonfermenting bacilli having only weak activity.

The Gluconate Test. Gluconate is not a reducing substance. However, when incorporated into culture medium, some bacteria have the property of oxidizing gluconate into products such as ketogluconate which have reducing properties and will give a positive reaction when tested with Benedict's reagent. This test is most useful in the differentiation of the "fluorescent" group of pseudomonads (see Chart 3-1).

Decarboxylase Activity. The decarboxylase test was discussed in detail in Chart 2-9 on page 90. However, the Moeller test, in which the detection of decarboxylase activity is dependent on an alkaline pH shift in the medium as the substrate amino acid is converted to its analogous amine, is insufficiently sensitive to detect the small amounts of decarboxylase produced by some strains of the nonfermentative bacilli. Therefore, when assessing the decarboxylase activity of the nonfermentative bacilli use of ninhydrin reagent to test for the amines produced is recommended. In the Pickett schema, buffered decarboxylase substrate medium is inoculated with a heavy suspension of the bacteria to be tested and incubated at 35° C. for 24 hours. One drop of 40 per cent KOH is then added, followed by 1 ml. of 0.1 g. per cent of Ninhydrin in chloroform. A purple color appearing in the chloroform phase within 3 to 5 minutes indicates the presence of amines in the medium and indicates a positive test.

Once the user becomes familiar with the Pickett schema, most nonfermentative bacilli can be identified within 24 hours after isolation in pure culture on primary medium. The reaction patterns of various microorganisms soon become familiar and identification of most isolates is not difficult. Again it should be reiterated that secondary or additional testing is not required if *Pseudomonas aeruginosa* (positive fluorescence, gas in FN medium, and growth at 42° C.) or *Acinetobacter calcoaceticus* (oxidase-negative, nonmotile, and nitrite-negative) can be identified by the initial screening tests.

Commercial Kits for the Identification of Nonfermentative Bacilli

Several commercial kits are currently available for the identification of nonfermentative bacilli. These are discussed in detail in Chapter 6.

FLAGELLAR STAINS

Although usually not required, flagellar stains may on occasion be useful in the identification of certain motile nonfermentative bacilli, particularly when biochemical reactions are weak or equivocal.

Reliable results may be obtained by using Leifson's[7] staining technique as described in Chart 3-6 if the following items in the protocol are given strict attention:

1. The slides must be scrupulously clean. Slides should be soaked in acid dichromate or acid alcohol (3% concentrated HCL in 95% ethyl alcohol) for 3 or 4 days. Final cleaning can be done immediately prior to use by heating the slides in the blue flame of a Bunsen burner.

2. Bacteria must be grown in carbohydrate-free media. A low pH may inhibit the formation of flagella and any acid formation in the medium may be detrimental. The pH of the staining solution should be maintained at 5.0 or above.

3. Bacteria should be stained during the log phase of growth, usually within 24 or 48 hours. Room-temperature incubation for 24 to 48 hours may be required to promote full flagellar development in some species.

4. Care should be taken not to transfer

Chart 3-6.   *Flagellar Stain*

Introduction   All motile bacteria possess flagella, the shape, number, and position of
                      which are important characteristics in the differentiation of genera and
                      species identification, particularly when biochemical reactions are weak or
                      equivocal. The Leifson staining technique (or modification) is most com-
                      monly employed in clinical laboratories and is not difficult to perform
                      providing exact details are followed in each step of the procedure.

Principle   Bacterial flagella can be stained by alcoholic solutions of rosaniline dyes
                      that form a precipitate as the alcohol evaporates in the staining procedure.
                      Basic fuchsin (pararosaniline acetate) serves as the primary stain with
                      tannic acid added to the solution as a mordant. A counterstain, such as
                      methylene blue, may be used to better visualize the bacteria in instances
                      where the primary stain is weak or does not react at all with the bacterial
                      cell wall.

Media and Reagents A. Basic fuchsin, 1.2% in 95% alcohol:
                         1. Commericial product must be certified for flagellar stain.
                         2. Pararosaniline acetate is preferred. If the dye solution is a mixture of
                            pararosaniline acetate and pararosaniline hydrochloride, the hy-
                            drochloride compound must not comprise more than two thirds of
                            the solution.
                         3. The stain must have the odor of acetic acid to be satisfactory.
                         4. When new stain is prepared, at least one day must be allowed for the
                            dye to enter completely into solution.
                      B. Tannic acid, 3% in water:
                         1. It must have a light yellow color to be satisfactory.
                         2. Add phenol in a 1:2000 concentration to discourage emergence of
                            molds during storage.
                      C. Sodium chloride, 1.5% in water: to prepare staining solution, combine
                         the three stock solutions in equal volumes. The stain is ready for use
                         immediately after preparation and should be stored in a tightly stop-
                         pered bottle. A precipitate will form during storage. This should not be
                         disturbed; rather remove the staining solution from the top with a
                         pipette. Shelf life of the staining solution is 1 week at room temperature,
                         1 month in the refrigerator, and indefinite if frozen.

Procedure   1. Using a young culture on an appropriate agar medium, prepare a light
                         suspension of the bacteria in water.
                      2. Place 2 drops of the bacterial suspension toward one end of an acid-
                         cleaned microscope slide. Allow to air dry. With a wax pencil, draw a
                         perpendicular line on the glass surface toward the opposite end from the
                         dried suspension.
                      3. Place the slide on a tilted rack and overlay the dried bacterial suspension
                         with a thin film of stain. The wax pencil mark will prevent the stain from
                         running off the end of the slide.
                      4. Stain for 5 to 15 minutes, allowing a precipitate to form as the alcohol
                         evaporates. The staining time is decreased if the stain is fresh, if the
                         room temperature is high, if there are air currents in the laboratory, and
                         if the layer of stain is thin. The opposite effects increase the staining
                         time.
                      5. When a precipitate forms, rinse the slide gently with distilled water,
                         drain off excess water, and allow to air dry.

Chart 3-6. *Flagellar Stain (Continued)*

Interpretation	Observe the stained slide under the oil immersion (100 ×) objective of the microscope. Dark-staining red to blue black flagella should be observed. Compare morphology with the photomicrographs provided in Leifson's *Atlas of Bacterial Flagellation*, cited in the bibliography below, and see also Plate 3-2, *I*, and 3-3, *L*.
Controls	The following bacterial species may serve as controls: Positive: peritrichous: *Escherichia coli;* polar: : monotrichous, *Pseudomonas aeruginosa;* multitrichous; *Pseudomonas cepacia* Negative: *Acinetobacter calcoaceticus*
Bibliography	Clark, W. A.: A simplified Leifson flagella stain. J. Clin. Microbiol., pp. 732–634, June, 1976. Leifson, E.: *Atlas of Bacterial Flagellation.* New York, Academic Press, 1960.

agar to the slide because it may interfere with the staining reaction. Washing the bacteria to be stained 2 or 3 times in water (lightly centrifuging between washes) prior to adding to the slides may be helpful in removing surface staining inhibitors.

Photomicrographs of representative flagellar stains are shown in Figure 3-4 and Plate 3-2, *I*, and 3-3, *L*.

REVIEW OF GENUS CHARACTERISTICS OF THE NONFERMENTATIVE GRAM-NEGATIVE BACILLI

Since various approaches are possible, it is necessary for each microbiologist to select the identification schema to use in the identification and species differentiation of the nonfermenting bacteria. This selection should be based on personal experience, the needs of the laboratory being served, and the extent to which bacterial species identification is required. It is important that students of microbiology learn to recognize key characteristics of nonfermentative gram-negative bacilli in primary culture which aid in presumptive identification of the nonfermenting group.

In review, the following are some of the more important characteristics by which gram-negative bacilli can be recognized:

1. *Enterobacteriaceae.* As a family, members of the Enterobacteriacae have the following characteristics: (1) they utilize carbohydrates fermentatively. This is detected by observing an acid reaction in the deep of a KIA or TSI agar slant. (2) They reduce nitrates to nitrites. (3) They are cytochrome-oxidase-negative. With rare exceptions, a bacterial isolate not exhibiting these characteristics can be excluded from the family Enterobacteriaceae.

2. *Pseudomonas.* Since none of the pseudomonads ferment carbohydrates, they produce no change in either the deep or the slant portions of KIA or TSI media. Except for *P. maltophilia* and *P. cepacia*, all other Pseudomonas species are cytochrome-oxidase-positive. Almost all strains are motile via monotrichous or multitrichous polar flagella. The detection of pyocyanin pigment in the culture medium is virtually diagnostic of *Pseudomonas aeruginosa*. Most *P. aeruginosa* isolates also produce fluorescein pigment and grow well at 42° C. Most Pseudomonas species also utilize carbohydrates oxidatively forming acid (yellow color) only in the O-F tube exposed to the air. The characteristics for the laboratory identifica-

Table 3-29. Important Characteristics of Commonly Encountered Pseudomonads*

Pseudomonas Species	NO$_3$→N$_{20}$ ↑ Denitrify	Cytochrome Oxidase	Pigments		O-F Reactions		Motility	Growth 42°C.	Gluconate Oxidation
			Pyocyanin	Fluorescein	O-F Glucose	O-F Maltose			
P. aeruginosa	+	+	+	+	+	-	+	+	
P. fluorescens	-/(+)	+	-	+	+	+/-	+/-	-	
P. putida	-	+/-	-	+	+	+	+	-	
P. cepacia	-	-/(+)	-	-	+	WK	+	+	
P. stutzeri	+	+	-	-	+	+	+	+	
P. maltophilia	-	-	-	-	+	+	+	-/-	
P. pseudomallei	+/-	+	-	-	+	+	+	+	

Key: + = 90% or more strains positive; (+) = 51–89% of strains positive; − = less than 10% strains negative; wk = weakly reactive.
*See Plates 3-3 and 3-4.

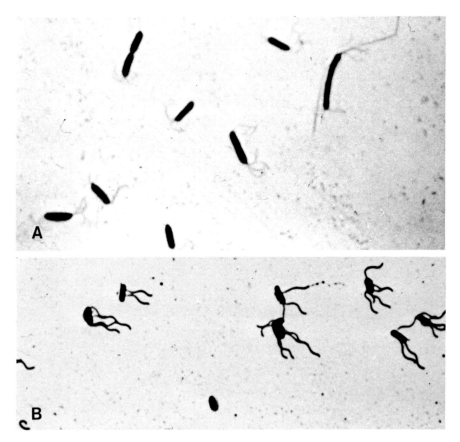

Fig. 3-4. Photomicrographs of flagellar stains. *A.* Positive flagellar stain of bacilli with polar flagella (×900). *B.* Positive flagellar stain of bacilli with peritrichous flagella (×900).

tion of *Pseudomonas aeruginosa* are shown in Plate 3-2.

There is a difference of opinion among microbiologists whether species of Pseudomonas other than *P. aeruginosa* should be identified. Although other species of Pseudomonas usually have a relatively low degree of virulence, in compromised patients some may cause life-threatening or fatal infections. Therefore, a report "Pseudomonas species, not *P. aeruginosa*" may not be sufficient for a physician treating a patient with septicemia or disseminated infection. Pseudomonas species other than *P. aeruginosa* have been recovered from a variety of human sources, including wounds (particularly burns), urine, blood, eye, bronchial washings, and sputum, joints, and the genital tract of females. *Pseudomonas*

cepacia, for example, has been frequently recovered from army troops suffering from "foot rot" or "jungle rot."[12] For those interested in differentiating some of the more commonly encountered species of Pseudomonas, the additional characteristics listed in Table 3-29 will be helpful. For a complete set of identification charts for the pseudomonads, see edition 8 of *Bergey's Manual of Determinative Bacteriology.*[1] The characteristics for the laboratory identification of Pseudomonas species other than *P. aeruginosa* are shown in Plate 3-4.

3. *Acinetobacter. Acinetobacter calcoaceticus* is the only species designation, and *A. anitratus* (formerly *Herellea vaginicola*) and *A. lwoffi* (formerly *Mima polymorpha*) are considered variant strains.

Members of the Acinetobacter group occur naturally in soil and water. Virulence for humans is low, although infrequent cases of meningitis and urinary tract infections have been reported. *A. calcoaceticus* may cause infectious disease in debilitated or immuno-suppressed patients, and septicemias have been reported. The organism is a commensal in the vagina, and the small gram-negative coccobacillus can be confused with *N. gonorrhoeae* in direct Gram's stains of vaginal secretions (Plate 3-3, *B*).

A. calcoaceticus produces no change in tubes of KIA or TSI media. The strain variant *A. anitratus* is strongly saccharolytic and will oxidatively degrade glucose in both 1 per cent and 10 per cent concentrations to produce an acid reaction in the O-F tube exposed to air. The variant *A. lwoffi* is nonsaccharolytic and produces no change in either the open or closed O-F tubes. Both variants grow well on MacConkey agar and are nonmotile. *A. calcoaceticus* is cytochrome-oxidase-positive and resistant to penicillin in the disk-diffusion agar test, two helpful characteristics in differentiating this species from Moraxella. *A. calcoaceticus* does not reduce nitrates either to nitrites or to nitrogen gas (negative denitrification). The characteristics discussed above are sufficient for a positive laboratory identification of *A. calcoaceticus*. (Plate 3-3, *A, B,* and *C* for illustrations of colonial morphology and biochemical characteristics of the genus Acinetobacter.)

4. *Flavobacterium.* A nonfermenting gram-negative bacillus that produces a colony with a bright yellow pigment is suspicious for one of the flavobacteria (Plate 3-3, *D* and *E*). Xanthomonas species and *Pseudomonas stutzeri* are two other bacterial species producing abundant yellow pigment. All species of Flavobacteria produce indole, a unique characteristic differentiating them from other nonfermenting bacteria. Indole production may be weak by some strains but can usually be detected if the culture medium is first extracted with Xylene after 48 hours of incubation at 35° C. and if Ehrlich's reagent rather than Kovac's reagent is used for the final color development.

Most flavobacteria grow well on MacConkey agar and utilize glucose oxidatively, although some strains can also ferment glucose. All species are cytochrome-oxidase-positive and nonmotile in semisolid medium. Colonies growing on blood agar for 48 or more hours often produce a lavender-green discoloration (indicative of proteolytic activity), another feature helpful in the presumptive identification of a flavobacterium.

Flavobacteria are generally of low virulence for man and opportunistic, except for groups 3 and 2b *Flavobacterium meningosepticum* which have been incriminated in cases of rapidly fatal meningitis and septicemia in newborn infants, particularly in premature infants.

5. *Moraxella.* A microorganism can be suspected of belonging to the genus Moraxella if the following characteristics are observed; (1) tiny gram-negative coccobacilli on Gram's stain; (2) the inability to grow on MacConkey agar; (3) a positive cytochrome oxidase reaction; (4) no change in KIA, TSI, or either the open or closed tubes of O-F medium (nonsaccharolytic); and (5) a high degree of susceptibility to penicillin when tested with low-concentration disks by the agar-diffusion technique.

Moraxella lacunata was originally described in the late 1890's as the Morax-Axenfeld bacillus, the cause of chronic infectious conjunctivitis. *M. osloensis* and *M. kingii* are two other species commonly encountered in the clinical laboratory; however, their pathogenicity for man appears to be minimal. Because the organisms are oxidase-positive and appear coccoid in Gram's stain, they can be confused with *N. gonorrhoeae*. The taxonomic status of some species is still uncertain and fluctuates between the genera Moraxella, Neisseria, and Branhamella. (See Plate 3-5,[3] *H* and *I*, for illustrations of colonial morphology and biochemical characteristics of the genus Moraxella.)

The Alcaligenes group includes bacteria closely related to the Moraxella and which have an uncertain affiliation. *Alcaligenes faecalis* is a common saprophyte of the intestinal tract of man, not thought to cause any known infectious disease. A tiny gram-negative bacillus that is incapable of utilizing carbohydrates either fermentatively or oxidatively can be suspected of belonging to the Alcaligenes group.

The groups of nonfermentative gram-negative bacilli discussed in this summary are those most commonly encountered in the clinical microbiology laboratory. Their identification need not be difficult when experience is gained with the use of one of the schemas described above. For rare or atypical reacting strains that may occasionally be encountered, the detailed King-Weaver charts[6] or the information included in edition 8 of *Bergey's Manual of Determinative Bacteriology*[1] should be consulted.

REFERENCES

1. Buchanan, R. E., and Gibbons, N. E.: Bergey's Manual of Determinative Bacteriology. ed. 8. Baltimore, Williams & Wilkins, 1974.
2. Doelle, H.: Bacterial Metabolism. ed. 2. New York, Academic Press, 1975.
3. Gilardi, G. L.: Practical schema for the identification of nonfermentative Gram-negative bacteria encountered in medical bacteriology. Am. J. Med. Technol., *38:*65–72, 1972.
4. Hugh, R., and Leifson, E.: The taxonomic significance of fermentative versus oxidative metabolism of carbohydrates by various Gram-negative bacteria. J. Bacteriol. *66:*24–26, 1953.
5. Isenberg, H. D., and Sampson-Scherer, J.: Clinical laboratory evaluation of a system approach to the recognition of nonfermentative or oxidase producing Gram-negative, rod-shaped bacteria. J. Clin. Microbiol., *5:*336–340, 1977.
6. King, E. O., *et al.:* The Identification of Unusual Pathogenic Gram-Negative Bacteria. Center for Disease Control, Bacteriology Division, 1972 (revised 1976 by R. E. Weaver).
7. Leifson, E.: Atlas of Bacterial Flagellation. New York, Academic Press, 1960.
8. MacFaddin, J. E.: Biochemical Tests for Identification of Medical Bacteria. Pp. 163, 164. Baltimore, Williams & Wilkins, 1976.
9. Otto, L. A., and Pickett, M. J.: Rapid method for identification of Gram-negative nonfermentative bacilli. J. Clin. Microbiol., *3:*566–575, 1976.
10. Pickett, M. J., and Pedersen, M. M.: Nonfermentative bacilli associated with man. II. Detection and identification. Am. J. Clin. Pathol., *54:*164–177, 1970.
11. ———: Characterization of saccharolytic nonfermentative bacteria associated with man. Can. J. Microbiol., *16:*351–362, 1970.
12. Taplan, D., Bassett, D. C. J., and Merz, P. M.: Foot lesions associated with *Pseudomonas cepacia.* Lancet, *2:*568–271, 1971.
13. Thimann, K. V.: The Life of Bacteria: Their Growth, Metabolism and Relationships. ed. 2. New York, Macmillan, 1963.

4 The Unusual Gram-Negative Bacilli

The "unusual gram-negative bacilli" (also called unusual bacilli or simply UB) is a designation given by Blachman[3] to a relatively large number of taxonomically unrelated species of bacteria that are rarely encountered in clinical laboratories as agents of human disease. The bacterial species currently included in the UB group are listed below.

The Unusual Gram-negative Bacilli (UB) Which May be Encountered in Clinical Materials

---

Actinobacillus actinomycetemcomitans
Actinobacillus species
Aeromonas hydrophila
Bordetella pertussis
Brucella abortus
Brucella melitensis
Brucella suis
Campylobacter fetus
Cardiobacterium hominis
CDC group DF-1
CDC group DF/0-2
CDC group EF-4
CDC group HB-5
CDC group TM-1
Chromobacterium violaceum
Eikenella corrodens
Francisella tularensis
Hemophilus aphrophilus
Hemophilus vaginalis
    (Corynebacterium vaginale)
Pasteurella multocida
Pleisomonas shigelloides
Streptobacillus moniliformis
Vibrio alginolyticus
Vibrio extorquens
Vibrio cholerae
Vibrio (lactose-positive)
Vibrio parahaemolyticus

Many of the UB are primarily endemic in mammals other than man, and may or may not cause infectious disease in these animals. These diseases may represent classic examples of the zoonoses, or those infections in which the causative microorganisms are transmitted from animals to man. In the past, human zoonoses were relatively frequent; however, through a better understanding of the modes of disease transmission and appropriate public health investigations during the past three or four decades, human infections are now quite rare. The average clinical laboratory may encounter one of these bacterial species only once or twice a year, if it receives one that often.

Based on a study of 768 strains of gram-negative bacilli encountered in his hospital's microbiology laboratory, Blachman[3] has tabulated the following distribution:

Enterobacteriaceae	78 per cent
Nonfermentative bacilli	12 per cent
Haemophilus species	9 per cent
Unusual bacilli (UB)	1 per cent

Following is a list of the total number of the infectious diseases caused by unusual bacilli encountered in the United States during 1975 which were reported to the Center for Disease Control.[21]

Brucellosis	310 cases
Tularemia	129 cases
Plague	20 cases
Whooping cough	1738 cases
Cholera	None

The low incidence of these diseases and others caused by the unusual bacilli tends to make their diagnosis somewhat difficult. It is essential that the physician inform the laboratory if he suspects one of these unusual diseases. The clinical history, including recent direct exposure to potentially diseased animals, travel to an area of the world where these diseases are more prevalent, as well as leading signs and symptoms must be carefully evaluated. Special selective media and processing techniques are required for isolating some of the unusual bacteria associated with such diseases.

Some species of the unusual bacilli, such as Brucella species, *Bordetella pertussis, Francisella tularensis, Streptobacillus moniliformis,* and *Vibrio cholerae,* produce well-known clinical syndromes. In other instances, it is more difficult to implicate an isolate of one of the unusual bacilli as the etiologic agent of an infectious disease. Nevertheless, several species of unusual bacilli have been recovered as the only potentially pathogenic organism in cases of septicemia and endocarditis, gastroenteritis, cellulitis (particularly localized skin or mucous membrane infections secondary to dog or cat bites), granulomatous processes of various organs, and, less commonly, pulmonary disease, meningitis, and osteomyelitis. *Hemophilus vaginalis (Corynebacterium vaginale)* and CDC group DF-1 are specifically thought to cause vaginitis. These clinical manifestations will be discussed in more detail in the remainder of this chapter.

LABORATORY APPROACH TO THE IDENTIFICATION OF THE UNUSUAL BACILLI

Taxonomy

There have been a number of changes in the taxonomy of many of the unusual bacilli since edition 7 of *Bergey's Manual of Determinative Bacteriology*[6] was published in 1957. This necessitates a brief update of the current genus and species designations as listed in edition 8 of *Bergey's Manual*[7], which was published in 1974:

The family Brucellaceae formerly included the following genera: (1) Pasteurella, including *P. multocida, P. pestis, P. pseudotuberculosis,* and *P. tularensis,* among others; (2) Bordetella; (3) Brucella; (4) Hemophilus; (5) Actinobacillus; (6) Calymmatobacterium; (7) Moraxella; and (8) *Noguchia.* The family Brucellaceae is no longer listed in edition 8 of *Bergey's Manual,*[7] and the genera mentioned above have either been reassigned to other genera or given separate designations.

P. pestis and *P. pseudotuberculosis* have been reassigned to the genus Yersinia, which is currently included in the family Enterobacteriaceae. *P. tularensis* is now *Francisella tularensis,* leaving the genus Pasteurella with only *P. multocida* and the infrequently encountered species *P. pneumotropica, P. haemolytica,* and *P. ureae.*

The genus Actinobacillis is now a separate designation, although its members have nutritional and biochemical characteristics closely related to the Pasteurelleae.

The genus Bordetella is now distinct, and includes *B. pertussis, B. parapertussis,* and *B. bronchiseptica.* The latter species is usually included with the nonfermentative bacilli and has been discussed in Chapter 30. The genus Moraxella is also included with the nonfermentative bacilli.

Vibrio cholerae and several species of Aeromonas are included in the family Vibrionaceae, together with two less frequently encountered genera, Plesiomonas and Photobacterium. *Plesiomonas shigelloides* is the current designation for the species formerly called *Aeromonas shigelloides. Vibrio fetus* is now included in the genus Campylobacter ("campylo," curved; "bacter," rod).

Eikenella corrodens is the current designation for the facultatively anaerobic bacteria formerly included with *Bacteroides corrodens.* It is now recognized that these bacteria are distinctly different from *B. corrodens* in several ways.

A number of CDC groups currently carry only alphanumeric designations, but some have been given species names as outlined in Table 3-15, Chapter 3.

Preliminary Identification of the Unusual Bacilli

A member of the unusual bacillus group can be suspected when an isolate has one or more of the following features:

1. *A slow growth rate*

A. On blood or chocolate agar, the unusual bacilli usually exhibit colonies less than 0.4 mm. in diameter, in contrast to the nonfermentative bacilli which produce colonies greater than 0.4 mm. after 24 hours incubation at 37° C. (Plate 4-1, *A* and *B*).

B. Unusual bacilli rarely grow at room temperature; many species of nonfermentative bacilli grow well.

C. Most species of the unusual bacilli are enhanced in their growth rate by increasing the CO_2 and humidity in the incubation chamber, provisions that have little effect on the growth of nonfermentative bacilli. For laboratories not having a moisturized CO_2 incubator, a moistened gauze placed in the bottom of a candle jar serves this purpose very well.

2. *Fastidious properties on selective media*

A. The unusual bacilli, with the exception of a few species, do not produce visible colonies on MacConkey agar within 24 hours; the majority of nonfermentative bacilli grow well on MacConkey agar.

B. Many species of unusual bacilli do not produce visible growth on the slant of Kligler iron agar within 24 hours whereas all nonfermentative bacilli do.

3. *Fermentative utilization of glucose* is not uncommon for the unusual bacilli (acidification of KIA). By definition, all nonfermentative bacilli lack this property.

Therefore, the initial laboratory assessment of an isolate that is suspected of belonging to an unusual bacillus group is to observe its growth on blood agar and MacConkey agar and whether it ferments carbohydrates in Kligler iron agar (acidification of the deep). Additionally, the evaluation of

the organism's capability of producing catalase and cytochrome oxidase can be helpful. Figure 4-1 is an algorithm by which the unusual bacilli listed on page 157 can be separated into subgroups prior to making a definitive species identification.

Definitive Identification of the Unusual Bacilli

The definitive identification of the unusual bacilli requires laboratory procedures similar to those described for the nonfermentative bacilli (Chapter 3). Assessment of motility, nitrate reduction, indole production, urease activity, lysine and ornithine decarboxylase production, and utilization of various carbohydrates are some of the features for the definitive identification of these organisms.

Although species and strain variations may occur, the following subgroups can be used to aid in laboratory identification on the basis of preliminary observations:

Group 1:

No growth on MacConkey agar at 24 hours and no growth on blood agar at 48 hours
Brucella species, *Campylobacter fetus*, Hemophilus species, and *Streptobacillus moniliformis* (Table 4-1)

Group 2:

No growth on MacConkey agar at 24 hours, growth on blood agar at 48 hours, KIA-positive or negative, and oxidase-positive or negative
1. *Actinobacillus lignieresii, Actinobacillus equuli, Cardiobacterium hominis,* EF-4, HB-5, and *Pasteurella multocida* (Table 4-2)
2. Actinobacillus actinomycetemcomitans, Campylobacter fetus, DF-1, DF/0-2, *Eikenella corrodens, Hemophilus aphrophilus, Hemophilus vaginalis,* and TM-1 (Table 4-3)

Group 3:

Growth on MacConkey agar at 24 hours, growth on blood agar at 48 hours, KIA-positive (acidified deep) and oxidase-positive
Aeromonas hydrophila, Plesiomonas shigelloides, Chromobacterium violaceum, Vibrio alginolyticus, Vibrio cholerae, Vibrio (lactose-positive) and *Vibrio parahaemolyticus* (Table 4-4)

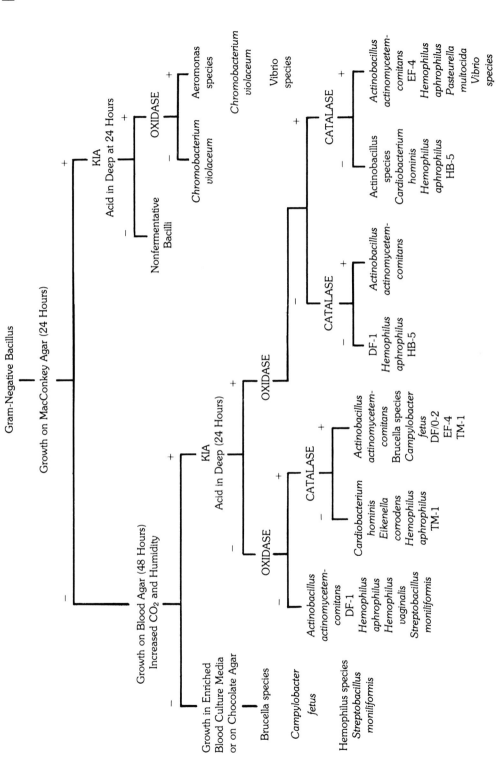

Fig. 4-1. Algorithm for identification of unusual bacilli (Modified from Blachman, U.: Distribution of aerobic gram-negative bacilli, Olive View Medical Center. Personal Communication, CACMLE Workshop Handout, December, 1977.)

The characteristics for definitive identification of these bacteria are given in Tables 4-1 through 4-4.

The clinical manifestations, epidemiology, and techniques for laboratory identification of the unusual bacilli listed are discussed in the following paragraphs.

ACTINOBACILLUS

Actinobacillus actinomycetemcomitans, a tiny gram-negative bacillus currently classified in edition 8 of *Bergey's manual*[7] as an organism of uncertain affiliation has been isolated from "sulfur granules" in association with the anaerobe *Actinomyces israelii.* Its pathogenicity has been questioned in this situation; however, recent reports indicate that *A. actinomycetemcomitans* can cause primary suppurative infection and has been implicated in cases of subacute bacterial endocarditis.[4,14,23,29]

Growth on culture media is enhanced in an atmosphere with increased CO_2. Small, nonhemolytic colonies appear on the surface of blood agar, which on prolonged incubation enlarge with filamentous extensions, giving a star or cross-cigar appearance. The colonies characteristically adhere firmly to the agar surface and are difficult to remove.

A. actinomycetemcomitans grows poorly on MacConkey agar. Many strains will not grow in O-F medium but are fermentative when growth occurs. A weakly acid reaction is produced in the deep of KIA and TSI, and a weak "orange" reaction is often produced in the slant. The reaction in the slant is not due to lactose or sucrose fermentation but from insufficient growth in the medium to prevent an alkaline reversion. The cytochrome oxidase test is negative, or weakly positive with some strains. Other reactions are listed in Table 4-3.

The organism may be recovered in blood culture bottles containing trypticase soy broth enriched with yeast extract or in thioglycollate broth, where the colonies appear as "bread crumbs" along the sides of the bottle, near the surface.

Actinobacillus lignieresii and *Actinobacillus equuli* commonly cause granulomatous lesions in the upper alimentary tract of animals ("woody tongue" in cattle) or suppurative lesions of the skin and lungs, primarily in sheep. Human infections of the skin may occur through direct contamination from contact with infected animal tissue.

AEROMONAS

Aeromonas hydrophila is the most common species of this group encountered in clinical practice. This organism may cause transient diarrhea, wound infections sometimes complicated with cellulitis and rarely septicemia, particularly in immunosuppressed hosts.[12] *Aeromonas punctata* and *Aeromonas salmonicida,* two rarely encountered species, have not been incriminated in human disease.

As the species name *hydro* ("water") *phila* ("to love") indicates, the aeromonads have their natural habitat in fresh or sea water where they commonly cause infectious diseases in aquatic cold-blooded animals. The bacteria also reside in sink traps and drainpipes, and may be recovered from tap water faucets and distilled water supplies, potential sources of organisms involved in nosocomial infections.

Aeromonas hydrophila may grow on "enteric" isolation media such as MacConkey agar, desoxycholate agar, eosin-methylene blue (EMB) agar, and salmonella-shigella (SS) agar, producing either lactose-positive or lactose-negative colonies similar to members of the Enterobacteriaceae. However, the aeromonads can be quickly excluded from the family Enterobacteriaceae because they produce cytochrome oxidase and are motile with polar rather than peritrichous flagella. Fermentation of glucose and production of indole help to differentiate the aeromonads from members of the genus Pseudomonas. (See Table 4-4 for a complete listing of biochemical characteristics of the genus Aeromonas.)

Aeromonas hydrophila has morphologic

Table 4-1. *Differential Characteristics of Brucella, Campylobacter fetus, Hemophilus Species and Streptobacillus moniliformis*

Characteristic	Brucella melitensis	Brucella abortus	Brucella suis	Campylobacter fetus	Hemophilus species	Streptobacillus moniliformis
KIA or TSI	No growth	No growth	No growth	No growth	No growth	No growth
Oxidase	+	+	+	+		-
Catalase	+	+	+	+	+	-
Motility	-	-	-	+		-
Nitrite	+	+	+	+	+	-
Lysine decarboxylase	-	-	+	-	-	-
Urease	+	+	+	-		-
Carbohydrates						
Glucose	+	+	+	-	+†	+
Xylose	-	-	+	-		
Mannitol	-	-	-	-		
Lactose	-	-	-	-		
Sucrose	-	-	-	-		
Maltose	-	-	-	-		+
Salicin	-	-	+	-		+
Ribose	-	+	+	-		
H₂S (lead acetate)	-	+	V	-	-	+w
Growth enhanced by CO₂ and humidity	+	+	+	+		+

Key: + = 90% or more of strains are positive; (+) = 51–89% of strains are positive; − = less than 10% of strains are positive; (−) = 10–50% of strains are positive; V = variable; w = weakly reactive.

* Identification made on basis of Gram's stain morphology, X and V growth requirements.

† Acid production from carbohydrates too variable to be of diagnostic value.

Color Plates
4-1 *to* 4-2

Plate 4-1
Identification of the Unusual Bacilli

As a group, the unusual bacilli are fastidious, develop tiny colonies less than 0.5 mm. in diameter in 24 to 48 hours on blood agar, and do not grow on selective media such as MacConkey agar. Microscopically, the bacterial cells are small, coccobacillary in nature, and many show bipolar staining. Biochemical reactivity in general tends to be weak or delayed.

A.
Blood/MacConkey agar biplate, illustrating a 24-hour growth of Actinobacillus species. The tiny colonies are characteristic of an unusual bacillus. Note the lack of growth on MacConkey agar, another characteristic feature.

B.
Higher-power view of blood agar plate, illustrating the tiny colonies of Actinobacillus species.

C.
Gram's stain from a 24-hour culture of Actinobacillus species, showing the short, coccobacillary nature of the cells. This is characteristic of the unusual bacilli in general.

D.
Specific fluorescent stain illustrating tiny coccobacilli of *Bordetella pertussis*. When reacted with specific fluorescent-tagged antiserum, *B. pertussis* has an apple-green glow when observed under a fluorescent microscope.

E.
Blood agar plate with tiny, mucoid, semitranslucent colonies of *Pasteurella multocida*.

F.
Identification characteristic of *P. multocida*. The tubes, from left to right, show the following:
Tube 1. Uninoculated KIA slant.
Tube 2. Alk/alk KIA reaction, characteristic of the nonfermenting bacteria.

Tube 3. KIA slant revealing a slight orange/orange reaction characteristically produced by *P. multocida*.
Tube 4. Nitrate reduction medium showing a positive reaction (red color after addition of reagents).
Tube 5. Indole test medium showing a positive reaction (red band at the top of the medium in the reagent zone).
Tube 6. Paper strip with a cytochrome oxidase tablet showing a positive reaction (blue color).

G.
Blood agar plate with the dark, purple-black, mucoid colonies of *Chromobacterium violaceum*.

H.
Select identification reactions of *Chromobacterium violaceum*. The tubes, from left to right, show the following:
Tube 1. KIA tube showing an alk/acid reaction and the diffusion of purple-black pigment in the slant of the tube. This does not represent H_2S production.
Tube 2. Simmons citrate slant showing the blue color of the medium. This indicates citrate utilization and a positive test.
Tubes 3 and 4. O-F medium showing acid production in only the open tube, indicating that *C. violaceum* utilizes dextrose only oxidatively.

I.
KIA tube in which a filter-paper strip saturated with lead acetate is draped from the lip of the tube. Lead acetate is a highly sensitive detector of H_2S and is employed to detect those organisms that produce only trace amounts. Note small amount of smoky-black pigmentation at the bottom of the filter-paper strip, indicating that the test organism growing on the medium is producing trace amounts of H_2S.

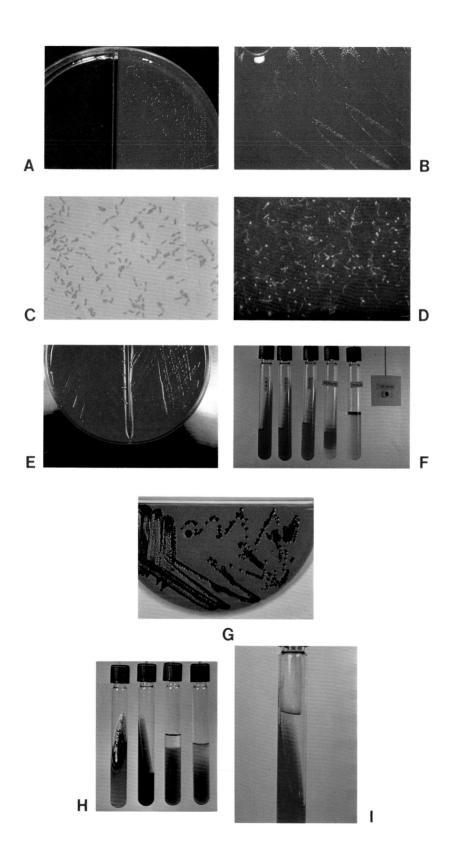

Plate 4-2 (Part I)
Identification of Vibrio Species

A.
Thiosulfate citrate bile salts sucrose (TCBS), agar illustrating the smooth, yellow colonies characteristic of *Vibrio cholerae*. The yellow colony results from citrate utilization, and acid formation, from the sucrose in the medium.

B.
Gelatin agar with white, opaque colonies of *V. cholerae*. Note the opalescence of the agar surrounding the colonies, indicating hydrolysis and denaturation of the gelatin.

C and D.
Both frames illustrate a positive "string test." Colonies of *V. cholerae*, when mixed in a drop of 0.5 per cent sodium deoxycholate, produce an extremely viscid suspension that can be drawn into a "string" with an inoculating loop.

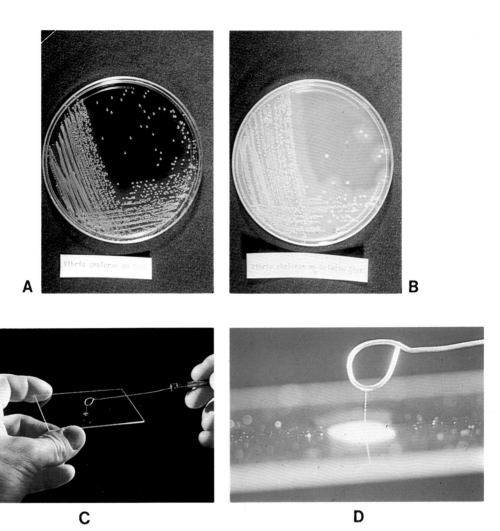

A

B

C

D

Plate 4-2 (Part II)
Identification of Vibrio Species

E.
Positive slide agglutination test for *Vibrio cholerae* using polyvalent O antiserum.

F.
Chicken erythrocyte agglutination test. Classic strains of *Vibrio cholerae* do not agglutinate chicken erythrocytes (top) in contrast to *Vibrio El Tor*, which has this property (bottom).

G, H and I.
Select identification characteristics of *Vibrio cholerae.*

G.
Reading from left to right, the reactions are: Alk/acid reaction on KIA (dextrose fermentation), red slant on lysine iron agar (LIA) indicating lysine deamination, positive indole, positive methyl red, and negative Voges-Proskauer tests.

H.
Reading from left to right, the reactions are: Simmons citrate positive, urease negative, positive motility, gelatin liquefaction, and positive cytochrome oxidase (blue color).

I.
Reading from left to right, the decarboxylase reactions are: lysine decarboxylase positive, arginine dihydrolase negative, and ornithine decarboxylase positive (positive reactions have a purple-red color; negative reactions are yellow-orange). The last tube is a growth control that does not contain an amino acid.

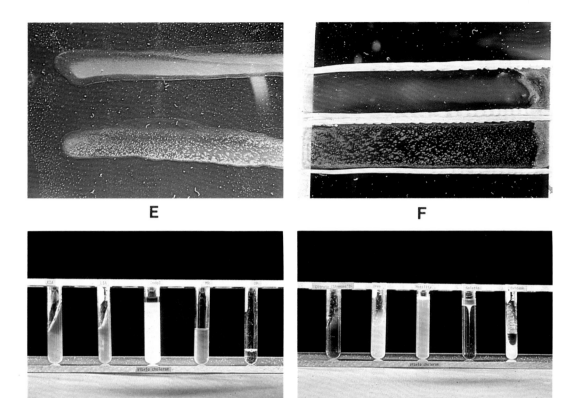

E

F

G

H

I

Table 4-2. *Differential Characteristics of Actinobacillus, Cardiobacterium, EF-4, HB-5, and Pasteurella*

Characteristic	Actinobacillus lignieresii and A. equuli	Cardiobacterium hominis	EF-4	HB-5	Pasteurella multocida
KIA (slant/deep/H$_2$S)	A/A/−; or no growth	K/A/−; or no growth	K/A−N/−	K/A/−	K−A/A/−
Oxidase	+	+	+	V	+
Catalase	+$_w$	−	+	−	+
Motility	−	−	−	−	−
Nitrite	+	−	+	+	+
Esculin	−	−	−	−	−
Indole	−	+$_w$	−	+$_w$	(+)
Urease	+	−	−	−	−
ONPG	+	−	−	−	(−)
Carbohydrates					
Lactose	+	−	−	−	V
Xylose	+	−	−	−	V
Sucrose	+	V	−	−	+
Maltose	+	V	−	−	V
Growth enhanced by increased CO$_2$ and humidity	+	+	−	+	−
Pigment	−	−	V(yellow-tan)	−	−

Key: + = 90% or more strains are positive; (+) = 51–89% of strains are positive; (−) = 10–50% of strains are positive; − = less than 10% of strains are positive; V = variable; w = weakly reactive; A/A = acid/acid deep; K/A = alkaline/acid deep; K−A/A = alkaline to acid/acid deep; N = neutral.

and biochemical characteristics most closely resembling *Vibrio cholerae,* and is included within the family vibrionaceae. *A. hydrophila* differs from a number of Vibrio species in that it grows well on all commonly used isolation culture media and is distinct from *V. cholerae* because it does not agglutinate in vibrio polyvalent antiserum.

Plesiomonas shigelloides, formerly included in the genus Aeromonas, possesses lophotrichous (2 to 7) polar flagella rather than a single flagellum and does not produce lipase, DNase, gelatinase, or caseinase, enzymes which are produced by the aeromonads. *P. shigelloides,* generally considered nonpathogenic, is suspected of causing outbreaks of gastroenteritis in Japan.[28]

BORDETELLA

Bordetella pertussis is the etiologic agent of whooping cough in man. Bordet-Gengou potato-glycerol-blood agar is required for the laboratory recovery of this organism; therefore, the laboratory must be notified in advance when this disease is suspected in order to allow for the preparation of this medium.

Cultures are commonly obtained by having the patient cough directly onto the surface of a Bordet-Gengou agar plate, but it is recommended that the organism be recovered from the nasopharynx by using a thin wire swab for collecting the sample. Samples should be plated at the bedside and direct smears made for fluorescent antibody (FA) studies. If direct plating is not feasible, the swab should be placed in 0.5 ml. of sterile casamino acid solution (ph = 7.2–7.4) for transport to the laboratory. Specimens must be processed within one to two hours. Kendrick[18] has described a charcoal agar slant for transport of nasopharyngeal swabs to reference laboratories.

Within 72 hours the colonies that appear on Bordet-Gengou plates present a smooth, transparent, convex surface with a pearly
(Text continues on p. 166)

Table 4-3. Differential Characteristics of Actinobacillus actinomycetemcomitans, Campylobacter, DF-1, DF/0-2, Eikenella, Hemophilus, and TM-1

Characteristic	Actinobacillus actinomycetemcomitans	Campylobacter fetus	DF-1	DF/0-2	Eikenella corrodens	Hemophilus aphrophilus	Hemophilus vaginalis	TM-1
KIA (slant/deep/H_2S)	K–A/A/–	K/N/–	K–A/N–A/–	N/N/– or no growth	N–K/N/–	A/A/(G)/–	N/N/– or no growth	K–N/N/–
Oxidase	–	+	–	+	+	–	–	+
Catalase	+	+	–	+	–	–	–	(–)
Motility	–	–	V	–	–	–	–	–
Esculin	–	–	V	+	–	–	–	–
Nitrite	+	+	–	–	+	+	–	–
Lysine decarboxylase	–	–	–	–	+	–	–	–
Ornithine decarboxylase	–	–	–	–	+	–	–	–
Urease	–	–	–	–	–	–	–	–
Indole	–	–	–	–	–	–	–	–
Carbohydrates								
Glucose	+	–	+	+	–	+	+	–
Lactose	–	–	V	+	–	+	V	–
Maltose	+	–	+	+	–	+	+	–
Mannitol	V	–	–	–	–	–	+	–
Starch	+	–	+	–	–	+	+	–
Beta hemolysis on blood agar	–	–	–	–	–	–	+ (human blood)	–
Pigment	–	–	(+) (yellow)	–	+ (Pale yellow)	–	–	–

Key: + = 90% or more of strains are positive; (+) = 51–89% of strains are positive; (–) = 10–50% of strains are positive; – = less than 10% of strains are positive; V = variable; K–A/A = alkaline to acid/acid; K/N = alkaline/neutral; N/N = neutral/neutral; N–K/N = neutral to alkaline/neutral; A/A = acid/acid; K–N/N = alkaline to neutral/neutral.

Table 4-4. *Differential Characteristics of Aeromonas, Chromobacterium, and Vibrio*

Characteristic	Aeromonas hydrophila	Aeromonas shigelloides	Chromobacterium violaceum	Vibrio alginolyticus	Vibrio cholerae	Vibrio (Lactose-Positive)	Vibrio parahaemolyticus
KIA (slant/deep/H$_2$S)	K/A/−	K − A/A/−	K/A/−	K/A/−	K/A/−	A/A/−	K/A/−
Catalase	+	+	+	+	+	+	+
Esculin	(+)	−	−	−	−		−
Motility	+	(+)	+	+	+	+	+
ONPG	+	+	−	−	+	+	+
Indole	(+)	+	−	+	+	+	+
Voges-Proskauer	(−)	−	−	+	(−)	−	−
Lysine decarboxylase	−	+	−	+	+	+	+
Ornithine decarboxylase	−	V	−	(−)	+	(+)	(+)
Carbohydrates							
Lactose	−	(+)	−	−	−	+	−
Sucrose	(+)	−	(−)	+	+	−	−
Mannitol	+	−	−	+	+	(+)	+
Inositol	−	+	−	−	−		−
Growth in Peptone, 1% with							
0% NaCL	+	+	+	−	+	−	−
7% NaCL	−	−	−	+	−	+	+
11% NaCL	−	−	−	+	−		−

Key: + = 90% or more of strains are positive; (+) = 51–89% of strains are positive; (−) = 10–50% of strains are positive; − = less than 10% of strains are positive; V = variable; K/A = alkaline/acid; K−A/A = alkaline to acid/acid; A/A = acid/acid.

sheen. Their appearance is similar to droplets of mercury. A narrow zone of hemolysis surrounding the colonies may be observed. In gram-stained preparations, the bacterial cells appear as gram-negative, coccoid rods with bipolar staining. The organisms may be identified with the fluorescent antibody staining technique (Plate 4-1, *D*), which permits a rapid, presumptive identification long before colonies develop on culture media.

The gross colony and cellular characteristics are sufficient to allow a laboratory identification, particularly in classic clinical cases. Isolates suspected of being *B. pertussis* should be confirmed by using specific antiserum by either slide agglutination or fluorescent antibody staining techniques. Biochemical tests are not too useful for identification of *B. pertussis* in most instances, but are used to rule out other microorganisms. *B. pertussis* is a strict aerobe and metabolizes carbohydrates oxidatively, never fermentatively. It is catalase-positive, oxidase-positive, nonmotile, and does not produce indole or urease or utilize citrate.

Two other species, *B. parapertussis* and *B. bronchiseptica,* are sometimes recovered from patients with acute respiratory tract infections resembling pertussis. *B. parapertussis* produces a brown pigment on Bordet-Gengou agar, is urease-positive, oxidase-negative, and utilizes citrate. *B. bronchiseptica* is strongly urease-positive and differs from the other two species by being motile. (See Plate 3-3, *J, K,* and *L*) for illustration of the biochemical characteristics of *B. bronchiseptica.*)

BRUCELLA

The genus Brucella which includes the three species pathogenic for man (*B. melitensis* (goats), *B. suis* (pigs) and *B. abortus* (cattle), was named after Sir David Bruce who first described the bacterial agent causative of undulant fever.[15]

Undulant fever in man is a chronic, relapsing febrile illness, characterized by weight loss, lassitude, anorexia, night sweats, and the development of granulomatous inflammation of the reticuloendothelial system, bone, and other tissues. The disease in man tends to become chronic because the organisms are obligate intracellular parasites, and thus they remain partially protected from the cellular and immunologic defenses of the host. Humoral antibodies begin to appear during the second to third week of disease. An agglutination test using the patient's serum and a killed suspension of smooth brucella organisms as antigens is helpful in making a diagnosis especially when attempts to isolate the bacteria are unsuccessful.

Brucellosis in animals is also manifested as a chronic granulomatous disease that may involve virtually any tissue or organ; however, epizootic abortions in cattle, hogs, and goats have the greatest economic impact for farmers. The disease is transmitted to man either by direct contact with infected animals or through ingestion of contaminated dairy products. Because of the mandatory pasteurization of all commercial dairy products and the extensive compulsory vaccination programs for susceptible animals, only 329 human cases of brucellosis were reported in the United States in 1975 by the Center for Disease Control.[21]

Physicians should inform the laboratory when brucellosis is suspected because special selective culture procedures to be discussed in more detail below are required to enhance the recovery of the causative agents. Blood cultures may be positive during the first two weeks of febrile illness. In subacute or chronic cases, bone marrow or tissue biopsy cultures may be required to recover the organisms from their intracellular locations in the reticuloendothelial cells.

Species of Brucella do not grow in a culture medium containing peptone alone; tryptose or trypticase is required, and added enrichments such as liver infusion, calf serum, or yeast hydrolysate may also be needed. Even with enrichments, the organisms grow slowly and cultures should be held at least 3 weeks before discarding as negative. Use of Castaneda bottles containing trypticase soy broth and an agar slat is recommended for blood cultures of patients with suspected brucellosis and these cultures should be subcultured to trypticase agar within 4 to 7 days. Brucella agar, a selective culture medium containing pancreatic digest of casein, peptic digest of animal tissue, yeast autolysate, and sodium bisulfite, is recommended for the recovery of Brucella from specimens that may be contaminated with other species of bacteria. (Crystal violet and antibiotics such as bacitracin are added to inhibit growth of bacteria other than Brucella). This medium is particularly useful for transport of specimens through the mail to a reference laboratory.

Pinpoint colonies, convex with a smooth glistening surface, usually begin to appear within 3 to 7 days on an appropriate culture medium. The organism produces no hemolysis or pigment when growing on media containing blood. Gram-negative coccobacilli without bipolar staining are observed in gram-stained preparations. The bacteria resist counterstaining and it is helpful to have the safranin in contact with the bacteria for at least 2 minutes when performing the Gram's stain.

The important differential characteristics of the genus Brucella are: strictly aerobic growth; catalase and cytochrome-oxidase-positive; nitrates reduced to nitrates; and urease production (at least weakly by most strains, and strongly by *B. suis*).

Definitive identification of Brucella is usually carried out in a state or municipal health department reference laboratory. Table 4-5 lists the characteristics helpful in differentiating the three species of Brucella pathogenic for man. Selective inhibition of bacterial growth by media containing 1:100,000 parts of the dyes thionin or basic fuchsin is most helpful. *B. melitensis* is the most virulent species for man and identification of this organism is an aid in epidemiologic studies.

CAMPYLOBACTER

Campylobacter ("campylo," curved; "bacter," rod) *fetus,* as designated by Sebald and Veron[26] in 1963, is a microaerophilic species of bacteria formerly called *Vibro fetus,* a known cause of contagious abortion in cattle and sheep. The organism has been recovered from the mouth and female genital tract of humans and will also grow in the intestinal tract and gallbladder of both man and animals. The organism has been recovered from blood cultures in humans with septicemia or endocarditis and in dysenteric stools, usually in individuals who have had contact with infected cattle or sheep or following dental extraction.

C. fetus usually does not grow on ordinary blood agar, or MacConkey agar in primary isolation. It requires an enriched medium such as trypticase soy broth. A concentration of carbon of 5 to 10 per cent carbon dioxide is essential for growth.

C. fetus is a small slender, gram-negative curved rod. The organism has a characteristic darting and corkscrew motility from a single polar flagellum at one or both ends of the cell. Colony types on enriched agar include smooth variants, up to 0.4 mm. in diameter, faceted or "cut-glass" variants 1 mm. in diameter, and, less commonly, rough colonies.

The organism is oxidase and catalase-positive and reduces nitrates to nitrites. As shown in Table 4-1, the organism is essentially biochemically inert and does not utilize carbohydrates either fermentatively or oxidatively.

CARDIOBACTERIUM

Cardiobacterium hominis can be recovered from the nose and throat of humans and is a rare cause of endocarditis. The organism will grow on blood agar but requires an increased concentration of CO_2 and an elevated humidity. It does not grow on MacConkey agar but does grow on a KIA agar slant without acidifying the deep of the medium.

Table 4-5. *Characteristics Useful in Differentiating Three Species of Brucella*

Characteristic	Brucella melitensis	Brucella abortus	Brucella suis
5% CO$_2$ required	−	+	−
H$_2$S produced	−	+	+
Urease produced	Weak	Weak	Strong
Growth on dye medium:			
Basic fuchsin, 1:100,000	+	+	−
Thionin, 1:100,000	+	−	+

The biochemical characteristics of *C. hominis* are listed in Table 4-2. In addition to the bacterial species listed in Table 4-2, *C. hominis* may also be confused with the more fastidious Moraxella, Hemophilus, Bordetella, and Eikenella species. *C. hominis* is oxidase-positive and catalase-negative, does not reduce nitrates to nitrites, produces indole weakly, and is urease-negative. These are helpful characteristics in making the species identification.

CHROMOBACTERIUM VIOLACEUM

Chromobacterium violaceum is of little clinical significance in that it causes human infections only extremely rarely. However, this organism has distinctive laboratory features by which it can be easily recognized.

The organism grows well on blood agar, and most strains produce abundant violet pigment (Plate 4-1, *G*). Utilization of carbohydrates may be fermentative or oxidative. Citrate is utilized and nitrates are reduced to nitrites. Casein is strongly hydrolyzed. Other characteristics by which *Chromobacterium violaceum* can be identified are listed in Table 4-4 (See Plate 4-1, *H*).

EIKENELLA

The bacteriology, epidemiology, and clinical significance of *Eikenella corrodens* has been reviewed by Matsen.[20] Attention was first drawn to this gram-negative organism because of its ability to pit the surface of an agar medium. This "corroding bacillus" was initially felt to be an obligate anaerobe, and was designated *Bacteroides corrodens*. Subsequently, it was found that not all strains are true anaerobes, and the microaerophilic variants are now included in the genus Eikenella (named after Eiken[9]). *E. corrodens* is the same organism that is also called HB-1 (Riley, Tatum, and Weaver[25]).

The ability of this organism to grow aerobically is dependent on the presence of hemin in the medium. For this reason it will not grow on MacConkey agar or similar selective media. It grows poorly in broth medium but forms a granular band about 1 cm. below the surface after 3 to 4 days of incubation in a medium such as fluid thioglycollate broth.

In gram-stained smears the cells appear as slender gram-negative rods with parallel sides and rounded ends. Both smooth and rough colonies may be observed, the latter frequently eroding the agar to form pitted depressions.

The biochemical characteristics of *E. corrodens* are listed in Table 4-3. The organism produces cytochrome oxidase and is catalase-negative and nonmotile. Nitrates are reduced to nitrites. Both lysine and ornithine are decarboxylated, important reactions in distinguishing *E. corrodens* from the other bacterial species listed in Table 4-3. Indole and urease are not produced, and carbohydrates are not utilized either fermentatively or oxidatively.

E. corrodens is a normal inhabitant of the

mouth and upper respiratory tract and may also inhabit the gastrointestinal tract. When recovered from clinical sources, it usually is accompanied by other bacteria notably streptococci, *E. coli*, and other facultative anaerobes or obligate anaerobic organisms. *E. corrodens* has been recovered from blood cultures, particularly following tooth extraction, but true septicemia apparently occurs only in compromised hosts. Rare cases of *E. corrodens* endocarditis have been reported. The organism is also frequently recovered from cultures of ulcers, cellulitis and abscesses of the face and neck region. Specimens from such sources should alert the microbiologist to carefully examine culture plates for the characteristic small, pitting colonies of *E. corrodens*.

FRANCISELLA

Francisella tularensis, the causative agent of tularemia, was previously included in the genus Pasteurella.[8] The genus Francisella is named for Edward Francis, an American bacteriologist who first studied the organism. Its strict aerobic growth requirement, the inability to grow on ordinary culture medium without adding enrichments such as cysteine, animal serum, or egg yolk, and its inability to produce catalase are the major features that differentiate *F. tularensis* from members of the genus Pasteurella.

F. tularensis is a tiny gram-negative coccobacillus with bipolar staining. On the cysteine-blood-dextrose agar (of Francis), 1 to 2-mm., clear, droplike colonies are visible within 48 to 72 hours after inoculation. Slight greening of the agar immediately beneath the colony may be observed, particularly after prolonged incubation. These characteristics are usually sufficient to make a presumptive identification.

Confirmatory differential characteristics include the formation of acid but not gas from glucose, maltose, and mannose, and a positive test for H_2S with lead acetate paper when the organism is grown on a medium containing cysteine.

Inoculation of selective medium is best done by aspirating with a syringe or small pipette the suppurative material from a necrotizing lesion or bubo. For optimal recovery of the microorganism, at least 0.5 ml. of inoculum should be placed in the center of a cysteine-glucose blood agar plate, spreading the inoculum evenly over the entire surface with a sterile glass rod.

Tularemia is endemic in a variety of small wild animals such as rabbits and ground squirrels. Humans may become infected by direct contact with diseased animal tissue, or from the bite of a blood-sucking arthropod. *F. tularensis* is capable of penetrating the unbroken skin which makes it necessary to handle laboratory cultures with extreme care. A small pustule develops at the site of inoculation. Also, the organisms invade the regional lymphatics and produce a necrotizing inflammation and enlargement of lymph glands. These lesions are called buboes. A disseminated form of the disease, relatively rare but highly fatal in man, may result from inhalation of infected dust or ingestion of contaminated foodstuff.

HEMOPHILUS

A full discussion of the genus Hemophilus and the more commonly encountered species is presented in Chapter 5. Two species, *Hemophilus aphrophilus* and *Hemophilus vaginalis,* are rarely recovered from clinical specimens and generally are grouped with the unusual bacilli. However, with the use of V agar, to be discussed below, *H. vaginalis* is recovered with increasing frequency from the vaginal canal of symptomatic and asymptomatic females.

H. aphrophilus is a rare cause of endocarditis, septicemia, brain abscess and focal skin infections. One unique characteristic of this organism is the formation of fluffy clumps that cling to the sides of the test tube in broth medium. On chocolate agar, colonies develop within 24 hours if the CO_2 concentration is increased to 5 per cent. Factor X is needed for growth.

The biochemical reactions are listed in Table 4-3. The organism produces an acid deep in agar, is oxidase and catalase-negative and is nonmotile. Nitrate is reduced to nitrite, a characteristic helpful in differentiating it from *H. vaginalis*.

The taxonomic position of *H. vaginalis* is currently uncertain. In 1955, Gardner and Dukes,[10] in a bacteriologic study of several patients with nonspecific vaginitis, first described the tiny, pleomorphic, gram-negative bacillus which they named *Hemophilus vaginalis*. In 1963, Zinnemann and Turner[31] demonstrated that the organism does not require either Factors X or V for growth, and, because of variable gram-staining and cell morphology, they have proposed its reclassification to *Corynebacterium vaginale*. The proposed change was not included in edition 8 of *Bergey's manual* because the cell wall of the organism does not contain DAP and contains 6-deoxytalose in place of arabinose. Although it is generally agreed that this organism does not belong to the genus Hemophilus, further study is needed before making a definite genus assignment.

The laboratory isolation and identification of *H. vaginalis* has been improved through the work of Greenwood and co-workers.[13] Isolation from clinical material is significantly enhanced through the use of V agar, composed of Columbia agar base (BBL), 1 per cent proteose peptone No. 3 (Difco), and 5 per cent whole human blood preserved with citrate phosphate dextrose solution. The proteose peptone No. 3 in particular has been found to be necessary to improve growth, and the unique production of beta hemolysis in the human blood medium makes rapid recognition of these organisms possible in mixed culture. Additional biochemical reactions are recorded in Table 4-3. Greenwood[13] suggests the use of the buffered single-substrate media described in Chapter 3 for carbohydrate utilization studies.

Most gynecologists accept *H. vaginalis* as one of the causes of low-grade vaginitis, particularly in pregnant mothers. Greenwood[13] has found a definite correlation between clinical symptoms and the predominant growth of the organism in culture. In contrast, cultures in which only a few *H. vaginalis* colonies are recovered are most commonly from aymptomatic patients. The organism can be suspected in Gram's stains or in Pap smears of vaginal secretions when the so-called "clue" cells, epithelial cells covered by tiny gram-negative bacilli, are observed (Plate 5-1, *C*).

PASTEURELLA

With the recent reclassification of bacteria, *Pasteurella multocida* is essentially the only species within the genus Pasteurella that is associated with human infections. *P. pneumotropica*, *P. haemolytica*, and *P. ureae* are other species that may be found in healthy and diseased animals. *P. pneumotropica* in particular may cause enzootic disease of mice, rabbits, and other laboratory animals.

P. multocida ("multus," many; "caedo," to kill; that is, killing many animals), the prototype of the genus, cause a variety of endemic diseases in animals: hemorrhagic septicemia in ruminant animals; pneumonia in cattle, sheep, and pigs; and dog-bite infections in dogs, cats, and humans.[16]

P. multocida is part of the normal flora of the upper respiratory tract of dogs and cats and has been recovered from the sputum and bronchial secretions of humans suffering from bronchiectasis and other chronic pulmonary diseases. Most commonly, *P. multocida* recovered from man is associated with suppurative wound infections resulting from the bite or scratch of a dog or cat.[17]

P. multocida grows well on any culture medium containing blood or hemin. Mucoid, smooth, and rough colony variants may be encountered (Plate 4-1, *E*). On blood agar, colonies often have a musty odor, a valuable clue to their identification. Microscopically the bacteria appear as gram-negative, tiny coccobacilli with bipolar staining, simulating closed safety pins.

P. multocida should be suspected when a gram-negative bacillus is recovered from a wound caused by an animal bite that grows on blood agar, not on MacConkey agar, acidifies the deep of KIA agar (ferments glucose), and is oxidase-positive. The glucose fermentation may be weak, producing an intermediate "orange/orange" KIA reaction (Plate 4-1, *F*). Since *P. multocida* ferments sucrose, the acid reaction in the deep of TSI will be more distinct than that produced in KIA. Other characteristics that help confirm the identification of *P. multocida* are indole production, reduction of nitrates to nitrites, and a weak H_2S reaction detected by lead acetate paper (Plate 4-1, *I*).

STREPTOBACILLUS

Rat-bite fever is a rare infection in the United States, caused by *Streptobacillus moniliformis*.[11] The disease is acquired by man from the bite of a rat, or, less commonly from the bite of a mouse, weasel, cat, or squirrel. *S. moniliformis* is also part of the normal flora of laboratory rodents, a potential source of infections for people who work in the laboratory.

S. moniliformis is seldom encountered in a clinical microbiology laboratory because the incidence of disease is so rare and because of the strict growth requirements of the microorganisms. The organism will not grow on ordinary blood agar, MacConkey agar, or KIA. Natural body fluids, such as ascitic fluid, blood, or serum are required for growth. When 10 to 30 per cent ascitic fluid or serum is added to thioglycollate medium the organism grows within 2 to 6 days in the form of characteristic puffball-like colonies.

Microscopically, *S. moniliformis* appears as an extremely pleomorphic, gram-negative bacillus with rounded or pointed ends, arranged in chains and filaments up to 150 μ. in length. Irregular bulbous swellings up to 3 μ. in diameter may give it the appearance of a string of beads.

On soft ascitic fluid or serum agar, incubated in a moisturized candle jar, small discrete colonies with a glistening surface and an irregular, sharply demarcated margin appear within 2 to 3 days. Tiny L-form colonies with a typical "fried egg" appearance may develop.

The characteristics by which *S. moniliformis* can be identified in the laboratory are listed in Table 4-1. Oxidase and catalase reactions are negative and the organism is nonmotile. Nitrates are not produced, lysine decarboxylase activity is absent, and urease is not produced. Glucose, salicin, and maltose are utilized if a basal cystine tryptose agar (CTA) containing 1 per cent carbohydrate and 1 drop of serum per tube is used. Carbohydrate reactions must be held for at least 10 days before discarding as negative.

The laboratory should be notified if *S. moniliformis* infection is suspected so that appropriate culture media can be selected. Cardinal clinical indicators of *S. moniliformis* infection include a patient who develops fever following the bite of a rat (or other rodent) and who develops regional lymphadenitis, upper respiratory symptoms, polyarthritis, and a rash over the palms of the hands and soles of the feet.

VIBRIO

Vibrio cholerae is the etiologic agent of cholera in humans, a potentially severe diarrheal disease that has been the scourge of mankind for centuries.[2] A current worldwide pandemic started in 1961, with local outbreaks reported in at least 24 countries.[1] Biotype El Tor, serotype Ogawa is the specific agent of this pandemic which has been difficult to control because of the widespread travel of newly infected hosts.

The diarrhea of cholera may be very severe. The production of a powerful enterotoxin by the organism in the small intestine affects the mucosa in such a way that there is a profuse outpouring of fluids. The stools appear watery and gray and may contain flecks of mucin, giving the appearance of "rice water." Severe vomiting usually occurs. The disease usually runs its course in 3 to 5 days in less severe infections; however, due to fluid and electrolyte loss, rapidly progressing dehydration, shock, and death may result. The death rate in epidemics may be particularly high if proper therapy for restoration of electrolyte and fluid balance is not available for immediate treatment of all those affected.

The laboratory should be notified if the physician suspects the disease in a symptomatic patient. Although *Vibrio cholerae* is not nutritionally fastidious, growth is suppressed by many of the usual enteric culture media used for the recovery of enteric pathogens from stool specimens. Specimens should be plated promptly because the organism is sensitive to drying, sunlight, an acid pH, and may be rapidly suppressed by the growth of other microorganisms in mixed culture. "Rice water" stools may be cultured, or specimens can be collected with a swab passed through the anal sphincter or though an anal catheter. Organisms can also be recovered from vomitus during the acute stage of the disease.

Thiosulfate citrate bile salts sucrose agar (TCBS) is preferred for the recovery of *Vibrio cholerae.*[1] After 18 to 24 hours of incubation on TCBS, smooth, yellow colonies, 2 to 4 mm. in diameter with an opaque center and transparent periphery, may be noted (Plate 4-2, *A*). Gelatin agar (GA) also serves as a primary plating medium, where transparent colonies surrounded by an opaque halo (as a result of gelatin liquefaction) are observed (Plate 4-2, *B*). A presumptive identification of *Vibrio cholerae* can be made on the basis of these gross colony characteristics.[3]

Vibrio cholerae is cytochrome-oxidase-positive. Additional characteristics helpful in its identification include the so-called "string test." Bacterial colonies are added to a few drops of 0.5 per cent sodium deoxycholate on a glass slide. The mixture becomes extremely viscid with *Vibrio cholerae.* This viscosity can be demonstrated by immersing an inoculating loop into the mixture. The test is positive if a long "string" of the deoxycholate-bacterial mixture forms when the loop is gently raised from the surface of the slide (Plate 4-2, *C* and *D*).

A positive slide agglutination with polyvalent O antiserum is also helpful in identifying *Vibrio cholerae,* particularly in differentiating some closely related strains of *Aeromonas hydrophila* (Plate 4-2, *E*).

Vibrio El Tor is a pathogenic biotype that produces a syndrome of gastroenteritis similar to that caused by *V. cholerae.* Morphologic, staining, and biochemical characteristics cannot be used to differentiate the two and serologic testing is of no aid in that both stains share the same major group O antigens. Vibrio El Tor hemolyzes sheep red cells and agglutinates chicken erythrocytes (Plate 4-2, *F*), characteristics not shared by *V. cholerae.*

Alkaline peptone water is highly recommended as an enrichment medium for the recovery of *Vibrio cholerae.* The high *p*H (8.4) of this medium suppresses the growth of many commensal intestinal bacteria while allowing uninhibited multiplication of *Vibrio cholerae.* Subcultures to TCBS or GA agar within 12 to 18 hours is recommended for optimal recovery since other organisms will begin to grow in the enrichment broth after prolonged incubation.

Watery stools should be examined microscopically in both a direct wet mount and in a Gram's stain preparation. The organisms are motile by a single polar flagellum. This motility can be observed in a direct mount. In a Gram's stain, *Vibrio cholerae* appears as a small gram-negative bacillus, with "c," "s," or commalike forms predominating.

Vibrio cholerae utilizes glucose fermentatively and will produce an acid-deep reaction in KIA. Sucrose is also fermented, producing an acid-deep/acid-slant reaction on TSI. Lysine iron agar will retain an alkaline slant when inoculated with *Vibrio cholerae,* which decarboxylizes lysine. Indole is produced, methyl red is positive, and Voges-Proskauer is negative. Simmons citrate is not utilized and urease is not produced. The biochemical characteristics by which the laboratory identification of *V. cholerae* can be made are shown in Plate 4-2, *G, H* and *I.*

Physicians and clinical microbiologists

must remain alert to the possibility of cholera infections virtually everywhere in the world due to widespread travel of people in and out of endemic areas.

Vibrio parahaemolyticus is a slightly curved gram-negative bacillus resembling *V. cholerae,* and is responsible for outbreaks of gastroenteritis following consumption of contaminated seafood, especially in Japan.[27] The organism grows well on most primary isolation media and may be confused with non-lactose-fermenting members of the Enterobacteriaceae family. However, a positive cytochrome oxidase reaction and the presence of monotrichous polar flagella are important distinguishing characteristics. *V. parahaemolyticus* does not utilize sodium citrate and does not agglutinate in polyvalent group O cholera antiserum, which differentiates it from *V. cholerae.*

YERSINIA

Yersinia pestis, Y. pseudotuberculosis, and *Y. enterocolitica* are three bacterial species formerly included within the genus Pasteurella. The lack of cytochrome oxidase activity by these members of the genus Yersinia is the main biochemical characteristic by which they are differentiated from the pasteurellae. Although the genus Yersinia is presently included within the family Enterobacteriaceae because all species utilize glucose fermentatively and are capable of reducing nitrates to nitrites, discussion will be included in this chapter since their recovery from clinical specimens is relatively rare.

The characteristics useful in differentiating the three species of Yersinia are listed in Table 4-6. All three species grow well on blood agar, MacConkey agar, and other bile-containing media. Lactose is not fermented; therefore, a KIA slant remains alkaline when inoculated with Yersinia species. However, *Y. enterocolitica* produces an acid slant and an acid deep in TSI because it is capable of utilizing sucrose, a feature which can help distinguish it from the other two Yersinia species.

The biochemical characteristics by which Yersinia species can be differentiated from other members of the Enterobacteriaceae are the negative reactions for lysine and ornithine decarboxylase and arginine dihydrolase ("L A O = − − −") and the production of urease in the absence of phenylalanine or lysine deaminase activity (differentiating them from Proteus species which produce both urease and phenylalanine deaminase). Other genus characteristics are: lack of motility at 35° C. (*Y. pseudotuberculosis* and *Y. enterocolitica* are motile at 22° C.); production of acid but not gas from glucose; positive methyl red reaction; production of beta-galactosidase (positive ONPG); and the inability to utilize citrate as the sole source of carbon. The remaining reactions shown in Table 4-6 are variable. They are used to identify species.

The genus name Yersinia honors A. E. J. Yersin, a French bacteriologist who first isolated the causative organism of human plague in 1894. *Yersinia pestis* is the causative organism of plague in man, rats, ground squirrels, and other rodents. The disease is transmitted from rat to rat or from rat to man by fleas which have become infected after feeding on an infected host. Man may also become infected by direct contact with the hide or viscera of diseased animals because the organism has a high propensity to penetrate through small breaks in the skin. For this reason extreme care must be taken while handling *Y. pestis* cultures in the laboratory. This high potential for laboratory infection may be one reason why the biochemical characteristics of *Y. pestis* have not been as extensively studied as those of the other microorganisms.

Plague occurs in three forms: (1) bubonic, in which the lymph nodes draining the infected site become greatly enlarged due to severe suppurative inflammation; (2) pneumonic, primarily a hemorrhagic lobar type of pneumonia, probably resulting from direct inhalation of organisms; and (3) septicemic, a form that can be rapidly fatal with hemorrhage into many organs. The hemor-

Table 4-6.   *Characteristics Useful in Differentiating the Species of Yersinia*

Characteristic	Y. pestis	Y. enterocolitica	Y. pseudotuberculosis
Motility at 22° C.	−	+	+
Hemolysis (blood agar)	−	Alpha	−
Voges-Proskauer	−	+	−
H₂S (lead acetate paper)	+	−	−
Urease	−	+	+
Carbohydrate utilization			
Adonitol	−	+	−
Cellobiose	−	−	+
Inositol	−	+	−
Melibiose	−	−	+
Rhamnose	−	−	+
Sorbose	−	−	+
Sucrose	−	+	−
Xylose	−	−	Delayed

Key: + = 90% or more strains positive; − = 10% or fewer strains positive.

rhage and cyanosis resulting from this last complication were responsible for the name "black death," a recurring scourge of man since prebiblical times. The disease in the United States is currently localized in sylvatic form to rodents in the Southwest, as reported by Reed and associates.[24]

Y. pseudotuberculosis and *Y. enterocolitica* are capable of penetrating the intestinal wall, producing a necrotizing and suppurative infection of the mesenteric lymph glands, a condition called mesenteric adenitis that can clinically simulate appendicitis.[5,19,22,30] Although the disease is most commonly self-limited and localized to the intestinal mesentery, fulminating septicemia may rarely occur and the causative organism can be recovered from blood cultures.

In summary, not every diagnostic bacteriology laboratory can be expected to have on hand the various media required for the differential identification of all the miscellaneous gram-negative bacterial species discussed in this chapter. However, each clinical microbiologist does have the responsibility to remain alert to the potential presence of the organisms discussed here, to know the appropriate methods of specimen collection, and to have available appropriate selective media for the recovery of each species. These items should be specifically listed in the laboratory procedure manual, together with specific instructions on how to properly prepare specimens for transport to a reference laboratory. Instructions for shipping specimens are presented in Chapter 1.

REFERENCES

1. Balows, A.: *Vibrio cholerae.* Check Sample Program, AMB-12. Chicago, American Society of Clinical Pathologists, 1975.
2. Balows, A., Hermann, G. J., and DeWitt, W. E.: The isolation and identification of *Vibrio cholerae*—a review. Health Lab. Sci. *8:*167–175, 1971.
3. Blachman, U.: Distribution of aerobic gram-negative bacilli, Olive View Medical Center. *Personal Communication,* CACMLE Workshop Handout, December, 1977.
4. Block, P. J., et al.: *Actinobacillus actinomycetemcomitans:* report of a case and review of the literature. Am. J. Med. Sci. *166:*387–392, 1973.
5. Braunstein, H., Tucker, E. B., and Gibson, B. C.: Mesenteric lymphadenitis due to *Yersinia enterocolitica.* Am. J. Clin. Pathol. *55:*506–510, 1971.
6. Breed, R. S., Murray, E. G. D., and Smith, N. R.: Bergey's Manual of Determinative Bacteriology. ed. 7. Baltimore, Williams & Wilkins, 1957.

7. Buchanan, R. E., and Gibbons, N. E.: Bergey's Manual of Determinative Bacteriology. ed. 8. Baltimore, Williams & Wilkins, 1974.

8. Eigelsbach, H. T.: *Francisella tularensis. In* Lynette, E. K., Spaulding, E. H., and Truant, J. P. (eds.): Manual of Clinical Microbiology. ed. 2, Chap. 28. Washington, D.C., American Society for Microbiology, 1974.

9. Eiken, M.: Studies on an anaerobic, rod-shaped, Gram-negative microorganism: *Bacteroides corrodens,* n. sp. Acta Pathol. Microbiol. Scand. *43:*404–416, 1958.

10. Gardner, H. S., and Dukes, C. D.: *Haemophilus vaginalis* vaginitis: a newly defined specific infection previously classified "nonspecific vaginitis. Am. J. Obstet. Gynecol. *69:*962–976, 1955.

11. Goldstein, E.: Rat-bite fever. In Hoeprich, P. D. (ed.): Infectious Diseases: A Modern Treatice of Infectious Processes. ed. 2. Hagerstown, Harper & Row, 1977.

12. Graevenitz, A., and Mensch, A. H.: The genus *Aeromonas* in human bacteriology: report of 30 cases and review of the literature. N. Engl. J. Med. *278:*245–249, 1968.

13. Greenwood, J. R., *et al.: Haemophilus vaginalis (Corynebacterium vaginale):* method for isolation and rapid biochemical identification. Health Lab. Sci. *14:*102–106, 1977.

14. Gross, J. E., Gutin, R. S., and Dickhaus, D. W.: Bacterial endocarditis due to *Actinobacillus actinomycetemcomitans.* Am. J. Med. *43:*636–638. 1967.

15. Hausler, W. J., Jr., and Koontz, F. B.: Brucella. *In* Lynette, E. K., Spaulding, E. H., and Truant, J. P. (eds.): Manual of Clinical Microbiology. ed. 2, Chap. 25. Washington, D. C., American Society for Microbiology, 1974.

16. Holloway, W. J., Scott, E. G., and Adams, Y. B.: *Pasturella multocida* infection in man. Report of 21 cases. Am. J. Clin. Pathol. *51:*705–708, 1969.

17. Hubert, W. T., and Rosen, M. N.: *Pasteurella multocida* infection. I. *Pasteurella multocida* infection due to animal bite. Am. J. Public Health, *60:*1103–1108, 1970.

18. Kendrick, P. L.: Transport media for *Bordetella pertussis.* Pub. Health Lab., *27:*85–92, 1969.

19. Knapp, W.: Mesenteric adenitis due to *Pasteurella pseudotuberculosis* in young people. N. Engl. J. Med., *259:*776–778, 1958.

20. Matsen, J. M.: *Eikenella corrodens.* Check Sample Program, AMB-7. Chicago, American Society of Clinical Pathologists, 1974.

21. Morbidity and Mortality Weekly Report: Reported morbidity and mortality in the United States, 1975. Atlanta, Center for Disease Control, 1975.

22. Nilehn, B.: Studies on *Yersinia enterocolitica* —with special reference to bacterial diagnosis and occurrence in human acute enteric disease. Acta Pathol. Microbiol. Scand. [Suppl.], *206:*1–48, 1969.

23. Page, M. I., and King, E. O.: Infections due to *Actinobacillus actinomycetemcomitans:* report of a case and review of the literature. N. Engl. J. Med., *275:*181–188, 1966.

24. Reed, W. P., *et al.:* Bubonic plague in the Southwestern United States. Medicine, *49:*465–486, 1970.

25. Riley, P. S., Tatum, H. W., and Weaver, R. E.: Identity of HB-1 of King and *Eikenella corrodens* (Eiken), Jackson and Goodman. Int. J. Systemic Bacteriol., *23:*75–76, 1973.

26. Sebald, M., and Veron, M.: Teneur en bases de l'ADN et classification des vibrions. Ann. Inst. Pasteur. (Paris), *105:*897–910, 1963.

27. Sukazaki, R., Iwanami, S., and Fukumi, H.: Studies on the enteropathogenic, facultatively halophilic bacteria: *V. parahaemolyticus.* Morphological, cultural and biochemical properties and its taxonomic position. Jap. J. Med. Sci. Biol., *16:*161–168, 1963.

28. Sukazaki, R., Nakaya, R., and Fukumi, H.: Studies on so-called Paracolon C27 (Ferguson). Jap. J. Med. Sci. Biol., *12:*355–363, 1959.

29. Weaver, R. E.: *Actinobacillus actinomycetemcomitans.* Check Sample AMB-18. Chicago, American Society of Clinical Pathologists, 1977.

30. Weber, J., Finlayson, N. B., and Mark, J. B. D.: Mesenteric lymphadenitis and terminal ileitis due to *Yersinia pseudotuberculosis.* N. Engl. J. Med., *283:*172–174, 1970.

31. Zinnemann, K., and Turner, G. C.: The taxonomic position of *"Haemophilus vaginalis" (Corynebacterium vaginale).* J. Pathol. Bacteriol., *83:*213–219, 1963.

5 Hemophilus

As implied by the genus name Hemophilus ("blood loving"), this group of bacteria requires certain factors derived from blood before any of the species will grow on laboratory culture media. Some species require Factor X, a heat-stable, iron protoporphyrin derivative of hemoglobin (heme). Others require nicotinamide adenine dinucleotide (NAD), also known as coenzyme I or Factor V, and some, notably *H. influenzae,* require both growth factors.

Factor X can be derived from a peptic digestion of blood, commercially available as Fildes enrichment, or from heat-disrupted erythrocytes as used in Levinthal's agar and in chocolate agar.

Factor V is heat-labile. It can be derived from extracts of yeasts or potatoes and is produced by certain bacteria such as *Staphylococcus aureus.*

Members of the Hemophilus genus are tiny (0.3 to 1.5 μ.) gram-negative coccobacilli. Gram-stained preparations of infected clinical materials show pleomorphic thread-like filaments, commonly admixed with the coccobacillary forms (Plate 5-1, *A*).

Hemophilus influenzae type B is the strain associated with most human infections.[20] The virulent variants are almost always encapsulated. In specimens containing large numbers of organisms, such as cerebrospinal fluid from cases of suppurative meningitis, type B *H. influenzae* can be rapidly identified by the quellung capsular swelling reaction using type-specific antiserum. Counterimmunoelectrophoresis, discussed in Chapter 7, can also be used to rapidly detect type-specific capsular polysaccharide in biologic fluids infected with *H. influenzae* as well as the direct typing of laboratory isolates.

Species of Hemophilus in general are quite susceptible to chilling and drying. They are apparently strict parasites of man and animals, and are most commonly recovered as commensal inhabitants of the upper respiratory tract. It is estimated that 30 per cent of healthy children carry *H. influenzae* in their oropharyngeal or nasal secretions, and this organism can also be found in the respiratory tract of a relatively high percentage of adults.[10,20]

The mechanism by which certain strains of *H. influenzae* suddenly become virulent and cause rapidly progressive and even life-threatening infections of the epiglottis, larynx, or bronchial tree is not completely known. *Hemophilus influenzae* infections are uncommon in newborns prior to 2 months of age, presumably because of the protection of passively acquired maternal antibodies.[7] *Hemophilus influenzae* meningitis reaches its highest incidence in children between the ages of 2 months and 3 years, a period during which circulating antibodies are at their lowest ebb. In this regard, Whisnant and associates[21] have demonstrated that patients recovering from *Hemophilus influenzae* meningitis have a significantly lower serum antibody response than individuals recovering from acute epiglottitis or pharyngitis. They conclude that the difference in magnitude of the host immunological response in these conditions is under genetic control.

Invasion of the human respiratory tract by colonizing strains of type B *Hemophilus influenzae* may also occur in immunosuppressed hosts receiving chemotherapy.[20] The tissue response in such instances is one of suppurative inflammation, and the exudation may be so thick and the edema so severe that acute airway obstruction may become a life-threatening complication, requiring an emergency tracheostomy.

HUMAN INFECTIONS CAUSED BY HEMOPHILUS SPECIES

The more commonly encountered human infections associated with Hemophilus species, the appropriate specimens to collect for culture, and the predominating clinical manifestations of such infections are listed in Table 5-1. Since special media are required for the isolation of species of Hemophilus, the clinician should always inform the laboratory when he suspects an infection with these bacteria. Because the respiratory carrier rate for Hemophilus is so high (as discussed above), it is not practical for clinical laboratories to screen throat or sputum specimens for this organism. Additional tests are required for identification of isolates and species differentiation, a practice that is time-consuming and costly when in most instances the clinical significance is low. This decision, however, must be made by the laboratory director, who must consider the fact that children with *H. influenzae* respiratory infections are at some risk of developing meningitis.

As seen in Table 5-1, *Hemophilus influenzae* can cause upper and lower respiratory tract infections, otitis media, meningitis, conjunctivitis, and bacterial endocarditis, as well as other diseases.[20] Of the children with *Hemophilus influenzae* meningitis, approximately two thirds had preceding infections of the oropharynx or otitis media and one third had bronchopneumonia caused by type B strains.[7] Therefore, respiratory infections

caused by *H. influenzae* type B should be treated, particularly in children under 3 years of age whose host resistance may not yet be fully functional.

Contagious acute conjunctivitis, known as pinkeye, is classically caused by *H. aegyptius,* although *H. influenzae* may also be implicated, particularly in the wintertime. Localized conjunctivitis epidemics occur among individuals who share towels, handkerchiefs, or other objects that come in direct contact with the skin of the face or the eyes. The diffuse pink color of the sclera and the presence of a serous or purulent discharge is virtually diagnostic of *H. aegyptius* infection.

Hemophilus ducreyi is the causative agent of chancroid, a highly contagious venereal ulcer that may clinically simulate the chancre of syphilis. *Hemophilus ducreyi* ulcers are called "soft chancres" because their margins are not indurated. Chancroid ulcers are also exquisitely painful, another characteristic differentiating them from the syphilitic lesions. The inguinal lymph nodes draining the genital area may enlarge and form swellings known as buboes. These may suppurate, break down, and drain a purulent material from which the organism can be grown in culture.

Gram's stains of the suppurative exudate from the chancroid lesion or from the drainage of buboes may reveal the organism. These appear as minute gram-negative bacilli, often clustered intracellularly within the cytoplasm of polymorphonuclear leukocytes. Extracellular bacteria may form chains of cells in a "boxcar" arrangement (Plate 5-1, *B*).

Some strains of *H. ducreyi* may be difficult to recover in culture. Factor X but not Factor V is required for growth, and most strains can be recovered on chocolate agar after incubation in an atmosphere of increased CO_2 tension (5 to 10%) at 35° C. for 24 to 48 hours. The more fastidious strains can be recovered by inoculating human serum or freshly clotted rabbit blood with the infected secretions and incubating at 35° C. for 24 hours prior to subculture to chocolate agar.

Color Plate
5-1

Plate 5-1
Identification of Hemophilus Species

A.

Photomicrograph of a Gram's stain illustrating tiny, pleomorphic gram-negative bacilli characteristic of Hemophilus species.

B.

Direct mount of purulent exudate illustrating bacilli in a "box-car" arrangement, characteristic of the extracellular forms of *Hemophilus ducreyi.*

C.

Photomicrograph of squamous epithelial cells from the uterine cervix. These so-called "clue cells" are covered with tiny bacillary forms frequently associated with *Hemophilus vaginalis (Corynebacterium vaginale)* infection.

D.

Chocolate agar plates with 48-hour growth of dull gray, somewhat mucoid, opaque colonies of *Hemophilus influenzae* on old (left) and fresh (right) chocolate agar plates. Note improved growth on the fresh plate.

E.

Staphylococcus aureus streak on blood agar illustrating the growth of tiny dewdrop, glistening colonies of *Hemophilus influenzae* in the zone of beta hemolysis. Factor V, produced by the staphylococcus colonies, and Factor X (hemin), from hemolysis of blood, are required for the growth of *Hemophilus influenzae.* This phenomenon is called "satellitism."

F.

Horse blood agar plates made with Casman's base and 10 ug bacitracin susceptibility disks placed in the areas of maximum inoculation. This medium is designed to enrich the growth of Hemophilus species. The bacitracin disk inhibits the growth of most commensal bacteria found in upper respiratory secretions except that of Hemophilus species. Note the growth of tiny beta-hemolytic colonies of *Hemophilus hemolyticus* immediately adjacent to the bacitracin disk in the right plate, while the bacitracin disk in the left plate has inhibited the growth of beta hemolytic streptococci.

G.

Test for X and V growth requirements. *Hemophilus influenzae* requires both X and V factors and thus grows best between the two strips.

H.

Porphyrin test illustrating a positive test in the tube on the left (pink color), compared to a negative control on the right. The tubes contain a substrate composed of delta-aminolevulinic acid hydrochloride. Organisms that do not require Factor X are capable of synthesizing porphobilinogen from the levulinic acid in the substrate, which produces a red color in the aqueous phase when reacted with Ehrlich's reagent.

I.

Supplemented Mueller-Hinton blood agar plate with a 10 ug. ampicillin antibiotic susceptibility disk placed in the zone of inoculation. Notice the wide zone of growth inhibition faintly visible, indicating an ampicillin-sensitive strain of *Hemophilus influenzae.*

J.

Capillary tube beta-lactamase test. The development of a yellow color near the top of the tube in the zone of bacterial inoculation indicates the production of penicilloic acid from destruction of the beta lactam ring of penicillin by beta-lactamase.

K and L.

Filter paper strips impregnated with a penicillin-starch substrate and two drops of iodine reagent. Organisms producing beta-lactamase are capable of reducing the iodine into colorless iodide, resulting in areas of blanching of the iodine-starch complex when colonies are directly applied to the reaction zones. Note positive reactions in the right spots compared to the negative controls on the left.

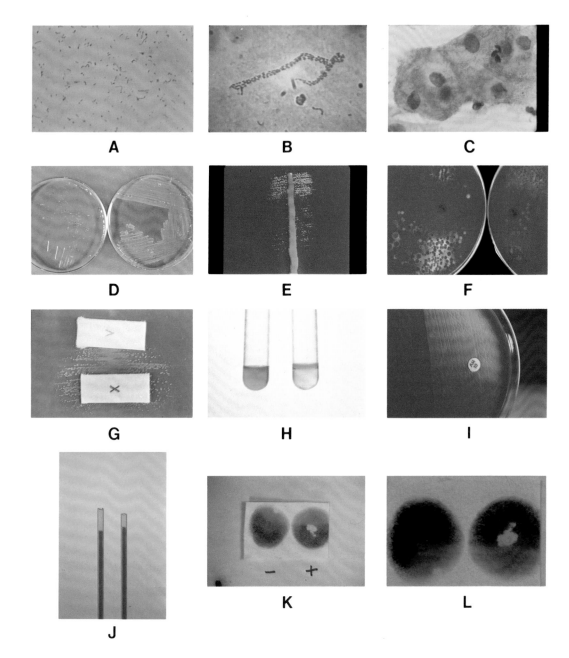

A

B

C

D

E

F

G

H

I

J

K

L

Table 5-1. *Infectious Diseases Associated With Hemophilus Species*

Disease	Hemophilus Species	Specimen for Culture	Clinical Manifestations
Acute pharyngitis Epiglottitis Laryngotracheo-bronchitis	*H. influenzae* (type B)	Throat swab Laryngeal secretions	Reddened mucous membranes, swelling, and patchy, soft, yellow exudate. Croupy cough. Tracheostomy may be necessary in obstructive epiglottiditis or laryngitis due to marked edema. Children are primarily affected; the disease is rare in adults.
Bronchitis Bronchiolitis Pneumonia	*H. influenzae* (type B)	Sputum Translaryngeal aspiration Bronchial washings	Persistent nonproductive cough, wheezing, dyspnea, and often typical asthmatic breathing. Primarily a childhood disease; however, it may be the cause of chronic bronchitis or "senile emphysema" in adults. Bacteremia occurs in one third of patients with pneumonia.
Otitis media	*H. influenzae* (type B > 95%; type A and nontypable strains rare)	Fluid from ruptured tympanic membrane or from middle ear aspiration	Most common in children; rare in adults. Suppurative infection indistinguishable from those caued by *S. aureus, S. penumoniae,* and *S. pyogenes.* Severe pain in ear, feeling of pressure, headache, and fever. On occasion, the discharge is that of serous otitis media, and may be combined with nonbacterial disease.
Meningitis	*H. influenzae* (type B; other types rare)	Cerebrospinal fluid	Signs of meningeal irritation: cervical rigidity, headache, and exaggerated reflexes in older children and adults; in infants, fretful behavior, anorexia, bulging fontanelles, and vomiting.
Conjunctivitis (pinkeye)	*H. aegyptius* *H. influenzae*	Conjunctival swab Cul-de-sac secretions	Intensely red sclera. Scratchy sensation, blurring of vision from exudate. Excess tearing and matting of the lid. Photophobia usually is minimal.
Chancroid ("soft chancre")	*H. ducreyi*	Secretions from genital ulcer Drainage from inguinal lymph node (bubo)	Initial vesicle or papule at site of inoculation progresses to pustulation and ulceration. Ulcer is often extremely painful (in contrast to a syphilitic chancre) with a red overhanging edge and a base covered with a dirty, grayish white exudate. No induration. Inguinal lymph nodes enlarge, suppurate, and drain (bubos). It is transmitted by sexual contact.
Subacute and acute bacterial endocarditis	*H. influenzae* *H. parainfluenzae* *H. aphrophilus**	Blood	Manifestations similar to those of other types of bacterial endocarditis: chills, fever, leukocytosis, anemia, malaise, anorexia, weight loss, and arthralgias in varying degrees of severity
Vaginitis†	*H. (Corynebacterium)‡ vaginalis*		Gray, malodorous vaginal discharge is the chief symptom.

*Classification is in doubt because many variants grow independent of Factor X or increased CO_2 tension.
†Thirty per cent of asymptomatic females are colonized with *H. vaginalis;* therefore, its role in infection is questioned.
‡Taxonomic position is questioned. This organism does not require Factors X or V, excluding it from the genus Hemophilus. Classification within the genus Corynebacterium has been suggested; however, the lack of diaminopimelic acid in their cell wall and the presence of 6-deoxytalose in place of arabinose seem to exclude it from that genus as well.

In 1955, Gardner and Dukes[6] first described the tiny, pleomorphic, gram-negative bacillus, which they named *Hemophilus vaginalis,* as the cause of nonspecific vaginitis in a series of their patients. Symptoms of *H. vaginalis* infection are limited to the production of a vaginal discharge with an offensive odor, thought to represent the protein breakdown products from vaginal squamous epithelial cells invaded by the organism. Segmented neutrophils are not a prominent component of these vaginal secretions, suggesting that the organisms do not invade the deeper tissue. The infection can be suspected by cytologists, who in the examination of "Pap smears" observe large numbers of squamous epithelial cells showing heavy colonization with pleomorphic bacilli, called "clue cells" (Plate 5-1, *C*).

The role of *H. vaginalis* as a cause of vaginitis has been questioned by McCormack and associates,[11] who found that 32.2 per cent of 466 asymptomatic females harbored the organism in their vaginal secretions. This study substantiates a similar previous conclusion arrived at by Dunkelberg and associates.[3]

In 1963, Zinnemann and Turner[22] determined that *H. vaginalis* does not require either Factor X or V for growth and proposed reclassification of the organism as *Corynebacterium vaginale.* This proposed reassignment has not been supported by cell wall analysis, studies of which have disclosed a lack of diaminopimelic acid and the presence of 6-deoxytalose in place of arabinose.[19] This finding is inconsistent with the genus Corynebacterium. *Hemophilus (Corynebacterium) vaginalis* is discussed in more detail with the unusual gram-negative bacilli in Chapter 4.

H. aphrophilus, a relatively uncommon cause of endocarditis and septicemia, is also discussed in detail in Chapter 4. This organism, whose taxonomic status is still uncertain, is morphologically and biochemically more closely related to *Actinobacillus actinomycetemcomitans* than other species of Hemophilus.

THE RECOVERY OF HEMOPHILUS IN CULTURE

Conventional sheep blood agar is not suitable for recovery of any species of Hemophilus that requires Factor V for growth. During the natural process of lysis during storage, sheep blood cells release the enzyme NADase, which inactivates any Factor V contained in the medium. In contrast, blood from rabbits and horses does not release NADase and media containing such blood support the growth of most species of Hemophilus. Unfortunately, neither rabbit nor horse blood supplements are commonly added to the media used for the routine screening of throat or sputum specimens. Because *H. hemolyticus* produces beta-hemolytic colonies indistinguishable from Group A streptococci on these media, additional tests are required to distinguish these two microorganisms.

Primary isolation of species of Hemophilus by culture from clinical specimens is most commonly accomplished by the use of one of four methods:

1. Chocolate agar
2. Staphylococcus streak technique
3. Casman's blood agar and 10-μg. bacitracin disk
4. Levinthal agar or Fildes enrichment agar

Chocolate Agar

Chocolate agar which contains both Factors X and V may be prepared in the laboratory by heating molten blood agar medium to a temperature (about 80° C.) just high enough to lyse the red blood cells to release hematin (Factor X). NADase is also inactivated at this temperature. Since Factor V is heat-labile, care must be taken during preparation that overheating does not occur. Because batch-to-batch consistency is difficult to control, most laboratories currently purchase chocolate agar from commercial suppliers who, in many instances, are better able to standardize the product. Enzymatic

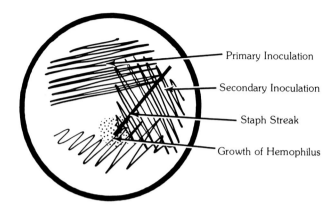

Primary Inoculation

Secondary Inoculation

Staph Streak

Growth of Hemophilus

Fig. 5-1. Use of the staph streak technique for recovery of Hemophilus species from a mixed bacterial culture. The staph streak is made in the zone of secondary inoculation and Hemophilus species may be seen as tiny dew-drop colonies in the hemolytic zone at the terminal portion of the staph streak in the area of lightest inoculation.

methods may be used rather than heat to lyse red blood cells with the addition of specific quantities of yeast extract (Factor V) or defined mixtures of vitamins, cofactors, and other supplements to produce a more consistent growth medium.

The use of chocolate agar has the disadvantage that the hemolytic properties of *H. hemolyticus* cannot be determined, eliminating a valuable means by which this organism can be differentiated from *H. influenzae.* Experience is also required to visually differentiate colonies of Hemophilus from other bacteria growing in mixed culture. Chocolate agar is particularly useful for the recovery of Hemophilus from specimens not commonly contaminated with other species of bacteria, such as cerebrospinal fluid or joint fluids. The appearance of the semiopaque, slightly mucoid, gray-white colonies of *H. influenzae* is illustrated in Plate 5-1, *D.*

Staphylococcus Streak Technique

Many microorganisms, including staphylococci, neisseria, and certain species of yeasts, can synthesize NAD (Factor V). When these organisms are present in mixed cultures, species of Hemophilus requiring Factor V may appear as small "dew-drop" colonies within the zones of NAD production, around colonies of the other bacteria, a phenomenon known as "satellitism."

The production of NAD by certain strains of staphylococci is taken advantage of in the "staph streak" technique. The specimen from which a species of Hemophilus is to be recovered is heavily streaked on the surface of a blood agar plate (sheep blood can be used). Using an inoculating wire, a single narrow streak of a hemolytic staphylococcus known to produce NAD is made through the area where the specimen had been inoculated. After 18 to 24 hours of incubation at 35° C. under CO_2, the tiny, moist, dew-drop colonies of Hemophilus may be observed within the hemolytic zone adjacent to the staphylococcus colonies (Plate 5-1, *E*). It should be pointed out that those species of Hemophilus that also require Factor X grow within the hemolytic zone because the staphylococcus beta hemolysin has lysed the red blood cells releasing hematin. This technique is particularly adaptable to the recovery of *Hemophilus influenzae* from cerebrospinal fluid, blood, or other types of specimens where contamination with other species of bacteria is not a problem.

This procedure can also be used for

screening throat and sputum cultures by making the staph streak through the area of the second isolation streak as shown in Figure 5-1.

Casman's Blood Agar and 10-μg. Bacitracin Disk

Casman's horse blood agar with application of a 10-μg./ml. bacitracin disk in the area of heaviest inoculation provides an excellent method for the selective isolation of species of Hemophilus, particularly from specimens where a heavy growth of mixed bacteria is anticipated (throat cultures, for example). A biplate with sheep blood agar in one half and Casman's horse blood agar in the other half is highly recommended for use in clinical laboratories. Any beta-hemolytic streptococci can be presumptively identified on the sheep blood side while species of Hemophilus can be recognized growing around the bacitracin disk on the Casman's agar side (Plate 5-1, *F*).

Casman's agar base contains peptones, beef extract, dextrose, sodium chloride, and 2.5 per cent agar. Nicotinamide and corn starch are added to enhance the growth of Hemophilus species. The horse blood is rich in Factor X and does not contain NADase. The 10-μg./ml. bacitracin disk inhibits the growth of many gram-positive bacteria commonly encountered in mixed culture, but does not inhibit the growth of most strains of Hemophilus which grow in the zone of inhibition immediately surrounding the disk. Again, refer to Frame *F* in Plate 5-1.

Levinthal Agar and Fildes Enrichment Agar

Levinthal agar and Fildes enrichment agar are commercially available culture media which facilitate the recovery of Hemophilus species. These media are particularly useful in the recovery of Hemophilus species from specimens in which suspicious organisms are seen on Gram's stain. The colonies of *H.*

influenzae and other species of Hemophilus are larger and easier to identify after 24 to 48 hours of incubation on this medium than on chocolate agar. The inclusion of 0.7 μg./ml. of nafcillin in these media make them more selective for the isolation of *H. influenzae*.[10] Also, as described above for use with Casman's blood agar, a 10-μg./ml. bacitracin disk can be used with Levinthal agar and the Fildes enrichment agar to facilitate isolation of species of Hemophilus from mixed bacterial populations.

Fildes enrichment is a peptic digest of sheep blood. The concentrated enrichment can now be obtained commercially and is usually added to an agar base, such as trypticase soy agar or brain-heart infusion agar, in a final concentration of 5 per cent. Since both Factors X and V are contained in Fildes enrichment, there is no necessity to add blood to the medium to isolate Hemophilus species. The use of a clear agar medium has the advantage that the small, colorless, opaque colonies of *Hemophilus influenzae* can be readily recognized by their iridescent appearance when examined by oblique light. The iridescence is characteristic of the encapsulated strains; in contrast, nonencapsulated and avirulent strains of Hemophilus appear small, bluish, transparent, and noniridescent in oblique light.

Levinthal agar is a medium containing brain-heart infusion agar, proteose peptone no. 3 (Difco), defibrinated horse blood (filtered through Whatman filter paper no. 12 and sterilized through a Seitz filter), and 1.5 per cent agar.

Species Identification of Hemophilus

In most clinical laboratories, three characteristics are used to differentiate the species of Hemophilus of clinical importance: (1) hemolytic reactions on horse or rabbit blood; (2) need for Factor V; and (3) the need for Factor X for growth. The differential reactions for differentiating Hemophilus are listed in Table 5-2.

Table 5-2. *Differential Characteristics of Hemophilus Species*

Species	Growth Requirements		Hemolysis on Blood Agar		
	X Factor	V Factor	Sheep	Horse	Human
H. influenzae	+	+	−	−	
H. aegyptius	+	+	−	−	
H. parainfluenzae	−	+	−		
H. hemolyticus	+	+	−	+	
H. parahemolyticus	−	+	+		
H. aphrophilus	+*	−	−		
H. ducreyi	+	−	+/−		
H. vaginalis	−	−	−	−	+

Key: + = 90% or more strains positive; − = 10 per cent or less strains positive.
*Increased CO_2 tension required.

Other differential characteristics, such as the need for increased CO_2 tension, need for serum or other enrichments, production of indole and catalase, and the growth patterns in salt-poor media, are only rarely needed for identification of most isolates from humans but they may be required for differentiation of Hemophilus species associated with infectious disease in animals. All species of Hemophilus except *H. ducreyi* reduce nitrates to nitrites, a helpful characteristic in differentiating Hemophilus from other bacterial species.

As seen in Table 5-2, *H. influenzae* and *H. aegyptius* cannot be differentiated by the characteristics listed. This generally does not cause a problem when the clinical source of the culture is known. *H. aegyptius* is not a common inhabitant of the respiratory secretions and does not cause meningitis. However, differentiation of the two species may be important in determining the cause of conjunctivitis.

Encapsulated strains of *H. influenzae* can be identified by serologic typing; however, nontypable *H. influenzae* can be difficult to differentiate from *H. aegyptius*. In fact, some authorities feel that the two may belong to the same species.[5] The solubility of *H. aegyptius* in 1 per cent sodium desoxycholate (a bile salt) may be used as another characteristic to aid in its differentiation from *H. influenzae*, which is insoluble in bile.

H. parainfluenzae, H. hemolyticus, and *H. parahemolyticus* are not usually differentiated in clinical laboratories because they are not commonly associated with infections. Occasionally symptomatic pharyngitis or bronchitis occurs in the presence of these organisms, often in association with a primary viral infection. Jones and associates[8] have recently reported a case of ampicillin-resistant laryngoepiglottitis caused by *H. paraphrophilus,* an example of how species of Hemophilus other than *H. influenzae* can occasionally cause disease.

Determination of X and V Growth Requirements

Several simplified laboratory procedures are currently in use for the determination of the X and V growth requirements of Hemophilus.

A test using filter paper strips or disks impregnated with either Factor X or V is a commonly used approach to determine X and V growth requirements (Chart 5-1). The organism to be tested is streaked on a medium deficient in Factors X and V such as Mueller-Hinton agar or trypticase soy agar. It is important when picking colonies for inoculation from primary culture plates that none of the chocolate agar or other blood-containing media (containing heme) is transferred to the test plates. Suspending the

Chart 5-1.   *Test for X and V Growth Requirements*

Introduction

Factor X (hemin) and/or Factor V (NAD) are required either singly or in combination to support the growth of various species of Hemophilus on artificial culture media. Filter paper strips or disks impregnated with Factor X or V are commercially available.

Principle

Factors X and V, each being water-soluble, readily diffuse into agar culture media. Filter paper strips or disks impregnated with these factors are placed on the surface of a medium deficient in Factors X and V, such as Mueller-Hinton or trypticase soy agar, which has been inoculated with the test organism. The Factor X and V requirements of the organism can be determined by observing the pattern of colony development around the paper strips (see illustration below).

Media and Reagents

1. Paper strips or disks impregnated with Factor X and Factor V (BBL or Difco)
2. An agar medium deficient in Factors X and V. The Mueller-Hinton agar (the medium used for antimicrobial susceptibility tests) is suitable.
3. Brain-heart infusion broth

Procedure

1. From a pure isolate of the species of Hemophilus to be identified, prepare a light suspension of bacterial cells in brain-heart infusion broth. From this suspension steak inoculate the surface of a Mueller-Hinton agar plate.
2. Place an X and a V strip or disk on the agar surface in the area of inoculation, positioning them approximately 1 cm. apart.
3. Incubate the plate under 5 to 10% CO_2 at 35° C. for 18 to 24 hours.

Interpretation

Visually inspect the agar surface, observing for the presence of visible growth around one or more of the strips (Plate 5-1, *G*). The following patterns indicate the need for Factor X, Factor V, or both:

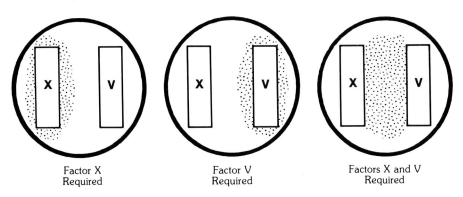

Factor X
Required

Factor V
Required

Factors X and V
Required

Controls

Requires Factor X only: *Hemophilus hemolyticus*
Requires Factor V only: *Hemophilus parainfluenzae*
Requires both Factors X and V: *Hemophilus influenzae*

Bibliography

Parker, R. H., and Hoeprich, P. D.: Disk method for rapid identification of *Hemophilus* sp. Am. J. Clin. Pathol., *37*:319–327, 1962.

organism in a broth deficient in Factors X and V prior to transfer to the test plates is one way to reduce false positive results.

The X and V paper strips or disks are then applied to the surface of the inoculated agar. The plates are incubated under 5 to 10 per cent CO_2 at 35° C. for 18 to 24 hours, and the growth patterns around the strips are observed. Differentiation of the Hemophilus species is then made on the basis of the growth pattern identified below in Chart 5-1 and shown in Plate 5-1, *G*.

The staphylococcus streak method described above can also be used to determine X and V growth requirements. This procedure requires two agar media: (1) a medium deficient in Factors X and V (Mueller-Hinton); and (2) a blood agar medium containing only Factor X, such as chocolatized blood agar that has been held at 60° C. for one hour after the initial 80° C. heat treatment to lyse the red blood cells. Each medium is streaked with a Factor-V-producing strain of *Staphylococcus aureus* after the agar surface had been inoculated with the specimen and incubated at 37° C. in CO_2.

Development of colonies only on the blood agar medium indicates a need for Factor X only. Growth only near the staphylococcus streak on blood agar indicates the need for both Factors X and V, and growth only near the staphylococcus streak on both agar media (with and without blood) indicates the need for Factor V only.

The Porphyrin Test

Kilian[9] has described a simple test for detecting Factor-X-dependent strains of Hemophilus, based on the principle that these strains lack the enzyme porphobilinogen synthase, which converts delta-aminolevulinic acid to porphobilinogen, an initial reaction in the synthesis of heme (Chart 5-2). Thus, the detection of porphobilinogen or porphyrins in the test substrate indicates that the bacterium is capable of endogenous synthesis of heme and does not require an exogenous source of Factor X. This test circumvents some of the problems with the filter strip method; namely, the chance for iron carryover from the primary plate is eliminated and slower-growing strains can be tested.

ANTIMICROBIAL SUSCEPTIBILITY TESTING

Until recently, antimicrobial susceptibility testing of *H. influenzae* was unnecessary because virtually all clinically significant strains were susceptible to penicillin, and notably to ampicillin. Ampicillin treatment failures of patients with *H. influenzae* meningitis were reported as early as 1968[15]; these were attributed to factors other than an inherent resistance of the organism. By 1974, bonified ampicillin-resistant isolates of *H. influenzae* were found.[15,17] By 1975, the Center for Disease Control estimated that the overall incidence of resistance may be as high as 10 per cent.[1] For this reason, it is currently necessary for clinical laboratories to have the capability to perform *H. influenzae* susceptibility testing, at least against ampicillin. Susceptibility testing is usually limited to those *H. influenzae* type B isolates that are recovered from patients with active infection or from individuals who have experienced a treatment failure.

The standard Bauer-Kirby susceptibility test is not applicable to the testing of Hemophilus species because these organisms require Factors X and V not present in the standard Mueller-Hinton agar, and because growth of the organism is too slow for the standard test. Fuchs[5] has outlined a disk agar diffusion method that can be used to assess *H. influenzae* susceptibility to ampicillin, using a modified Bauer-Kirby technique:

1. Prepare a suspension of the organism to be tested in Mueller-Hinton broth, to a turbidity equal to a 0.5 MacFarland standard (10^8 colony-forming units/ml.).

2. Supplement the Mueller-Hinton agar to be used in the test with 5 per cent chocolatized horse or rabbit blood and 1 per cent Isovitalex, BBL (or a comparable supplement such as Fildes enrichment), in a final concentration of 5 per cent.

Chart 5-2. *Porphyrin Test for Differentiation of Hemophilus Species*

Introduction	The porphyrin test is a simple means for detecting those strains of Hemophilus capable of synthesizing heme, a property not shared by strains requiring exogenous Factor X. Interpretation of results can be made rapidly and more dependably than with the filter strip method.
Principle	Delta-aminolevulinic acid is the precursor molecule from which porphobilinogen, porphyrins, and heme are synthesized. Microorganisms possessing the enzyme porphobilinogen synthase can convert delta-aminolevulinic acid into porphobilinogen and do not require heme Factor X for growth. Porphobilinogen can be detected in the test medium by the use of modified Ehrlich's reagent; or, porphyrins can be demonstrated by their fluorescence when observed under ultraviolet light.

Media and Reagents
1. Enzyme substrate:
    A. 2 Mm. Delta-aminolevulinic acid hydrochloride:
       31.8 mg./100 ml. w/v
    B. 0.8 Mm. $MgSO_4$:
       9.62 mg./100 ml. w/v
    C. 0.1 M phosphate buffer, *p*H 6.9
       14.2 g. Na_2HPO_4 in 1 liter of water
       13.61 g. KH_2PO_4 in 1 liter of water

    Add 55.4 ml. of the monohydrogen solution to 44.6 ml. of the dihydrogen solution. Add 31.8 mg. of reagent A, above, and 9.62 mg. of reagent B, above. This buffer solution is stable for at least 6 months when stored at 4° C.
2. Ehrlich's reagent: p-dimethylaminobenzaldehyde, 5 g. in a mixture of 75 mg. amyl alcohol and 25 ml. concentrated HCl
3. Wood's lamp

Procedure
Add a loopful of the test organism to 0.5 ml. of the enzyme substrate. Incubate the mixture at 35° C. for 4 hours (heavy suspension) or 18 to 24 hours (light suspension). At the end of the incubation period, add an equal volume of Ehrlich's reagent and shake the mixture vigorously. Allow substrate and reagent to separate.

Interpretation
The development of a red color in the lower aqueous phase indicates the presence of porphobilinogen and a positive test (organism does not require Factor X). See Plate 5-1, *H.*

Alternately, a red fluorescence in the reagent phase when examined with the aid of a Wood's lamp indicates the presence of porphyrins, and also a positive test.

Controls
Positive control: *Hemophilus parainfluenzae* (can synthesize heme)
Negative control: *Hemophilus influenzae* (cannot synthesize heme)

Bibliography
Kilian, M.: A rapid method for the differentiation of *Haemophilus* strains—the porphyrin test. Acta Pathol. Microbiol. Scand. [B], *82*:835–842, 1974.

3. Proceed with the standard Bauer-Kirby technique (see Chapter 11). A zone of growth inhibition 20 mm. or less in diameter is used to differentiate resistant from susceptible strains. Most susceptible strains have zones of inhibition greater than 25 mm.; resistant strains have zones well below 20 mm. (Plate 5-1, *I*).

Broth and agar dilution methods can also be used, providing that the test medium used has been modified to support adequate growth of Hemophilus species (addition of Fildes enrichment to a final concentration of 5%). The minimal inhibitory concentration (MIC) of susceptible strains is 0.5 μg./ml. or less; resistant strains commonly show MIC's greater than 4 μg./ml.

It is now known that the ampicillin resistance of *H. influenzae,* and other species of bacteria as well, is mediated through the production of beta-lactamase, an extracellular enzyme.[17] Beta-lactamase production is an induced plasmid-linked characteristic,[4] transferrable from ampicillin-resistant strains of *H. influenzae* to previously susceptible strains.[16]

As an alternative to performing antimicrobial susceptibility tests, many laboratories are now using tests to demonstrate the production of beta-lactamase by ampicillin-resistant strains of *H. influenzae.* These tests can be rapidly performed and may be more reliable than the somewhat more difficult to control Bauer-Kirby technique.

Two beta-lactamase test procedures are currently in use in clinical laboratories, as outlined in Chart 5-3. Both of these methods are based on the ability of ampicillin-resistant strains of *H. influenzae* to produce sufficient beta-lactamase to break the beta-lactam ring, with the release of penicilloic acid into the test medium. The capillary tube method, first developed by Rosen[14] to detect penicillin resistance in *Staphylococcus aureus* and later adapted to the testing of *H. influenzae* by Thornsberry and Kirven,[18] is based on detecting the drop in the *p*H and the change to a yellow color in the medium as penicilloic acid is produced (Plate 5-1, *J*). Phenol red is used as the *p*H indicator.

A second method, the iodometric method of Catlin,[2] is based on the ability of the penicilloic acid produced in the test medium to reduce iodine to iodide, resulting in a blanching of the blue iodine-starch complex (Plate 5-1, *K* and *L*).

More recently O'Callaghan and associates[13] have introduced a novel method for detecting beta-lactamase production through the use of nitrocefin, a chromogenic cephalosporin derivative that changes from yellow to red when the beta-lactam ring is broken. The test is easy to perform and is virtually instantaneous in reactivity. In the plate method, when a working solution of nitrocefin is dropped on to a beta-lactamase producing bacteria colony growing on an agar medium, the colony turns an immediate red color. This working solution is made by adding 0.5 ml. of dimethyl sulfoxide to 5 mg. of nitrocefin. Added to this is 9.5 ml. of phosphate buffer *p*H 7.0. This solution can be stored in the dark at 4° C. for up to 14 days. When performing the test in broth (by adding about 0.5 ml. of nitrocefin working solution to a broth culture) an immediate red color reaction is also elicited if beta-lactamase is present, but a 30-minute incubation period is required before the test is considered negative.

The use of the beta-lactamase procedure to detect beta-lactamase-producing bacteria in clinical laboratories is recommended in order to enable detection of clinically significant isolates of *H. influenzae* that produce the enzyme and isolates of other bacterial species, such as *Neisseria gonorrhoeae,* that may also be ampicillin-resistant.

REFERENCES

1. Ampicillin-resistant *Haemophilus influenzae.* CDC Morbidity and Mortality Weekly Report, *24:*205–206, 1975.
2. Catlin, B. W.: Iodometric detection of *Haemophilus influenzae* beta-lactamase: a presumptive test for ampicillin resistance. Antimicrob. Agents Chemother., *7:*265–270, 1975.
3. Dunkelberg, W. E., Jr., *et al.: Haemophilus*

(Text continued on p. 190)

Chart 5-3. *Rapid Tests for Beta-Lactamase Production*

Introduction	Approximately 10% of *Hemophilus influenzae* isolates recovered in clinical laboratories are resistant to ampicillin through the production of the enzyme beta-lactamase. This characteristic is plasmid-linked and can be transferred from resistant to sensitive strains.
Principle	Beta-lactamase is an extracellular enzyme produced by many strains of bacteria that specifically hydrolyze the amide bond in the beta-lactam ring of penicillin analogues, rendering the antibiotic inactive. Penicilloic acid is formed as shown in the following reaction:

Two tests are commonly used for the detection of penicilloic acid in the test medium: (1) the capillary method; and (2) the iodometric method.

The capillary method is dependent upon a color change of phenol red indicater noting a change in *p*H; the iodometric method is dependent upon the ability of penicilloic acid to reduce iodine to iodide, resulting in a decoloration of the blue iodine-starch complex.

Media and Reagents	*Capillary Method*
	Add 2 ml of a 0.5% solution of phenol red (Difco) to 16.6 ml. of sterile distilled water and then add the mixture to a vial containing 20 million units of K penicillin-G (Pfizer). Add 1 M sodium hydroxide drop by drop until the solution turns violet (*p*H approximately 8.5). Either use the test solution immediately or divide into 0.5-ml. portions in screwcap tubes and store at −60° C.

Iodometric Method

1. Iodine reagent:

Iodine	2.03 g.
Potassium iodide	53.2 g.

   Dissolve in 100 ml. of distilled water and store in a brown glass bottle.

2. Starch indicator: add 1.0 g. soluble starch (Merck) to 100 ml. of distilled water. Prepare fresh every 2 days.

3. *p*H Buffer: phosphate buffer, *p*H 5.8, 0.05 M.

KH_2PO_4	6.25 g.
K_2HPO_4	0.696 g.

   Dissolve in 1 liter of distilled water.

Chart 5-3. *Rapid Tests for Beta-Lactamase Production (Continued)*

Media and Reagents (Continued)	4. Penicillin-G solution: prepare in the phosphate buffer a solution of 10,000 units penicillin-G/ml. Dispense in 0.5-ml. volumes.

Procedure

Capillary Method
1. Dip capillary tubes (0.7 to 1.0 mm. outer diameter) into the penicillin-phenol red test solution. Allow capillary action to fill tube to a distance of 1 to 2 cm.
2. Scrape the tip of the capillary tube lightly across several colonies of *H. influenzae* growing on the surface of an agar plate, forming a bacterial plug in the bottom of the tube. Care must be taken that no air is trapped between the test solution and the bacteria.
3. Incubate the capillary tubes containing the test mixture at room temperature in a vertical position, sticking the empty end of the capillary tube into clay in an upright position.
4. Interpretation: if the organism produces beta-lactamase, the solution will turn a bright yellow in 5 to 15 minutes (Plate 5-1, *J*). Beta-lactamase-negative organisms produce no change in the medium or no more than a pink tinge.

Iodometric Method
1. Incubate modified Levinthal agar with a pure culture of the organism to be tested and incubate at 35° C. until good growth is obtained.
2. Suspend a large inoculating loop full of the bacterial growth into the penicillin-buffer mixture to give about 10^9 cells (MacFarland Standard no. 1).
3. After 1 hour of incubation at room temperature, add 2 drops of starch indicator and mix.
4. Add 1 small drop of iodine solution and mix.
5. Interpretation: an initial blue color will rapidly develop due to reaction of the iodine with the starch. Rotate the mixture for up to 1 minute. Persistence of the blue color for more than 10 minutes constitutes a negative test (no beta-lactamase produced). Rapid decoloration of the medium indicates that penicilloic acid was formed, and a positive test (beta-lactamase produced).
6. Alternatively, filter paper strips can be impregnated with an equal mixture of the 1% starch solution and the penicillin-G solution and air dried for future use. At the time the test is performed, a drop of iodine is placed on the filter paper reagent strip, producing an immediate blue color. Using an inoculating loop or an applicator stick, a portion of the colony of Hemophilus to be tested is rubbed in the area of the iodine spot. Blanching of the blue color within 3 to 5 minutes by the inoculum indicates the production of beta-lactamase and a positive test (Plate 5-1, *K* and *L*).

Controls

Positive control: a beta-lactamase-producing strain of *H. influenzae* or *Staphylococcus aureus*
Negative control: a non beta-lactamase-producing bacterial strain

Bibliography

Catlin, B. W.: Iodometric detection of *Haemophilus influenzae* beta-lactamase: rapid presumptive test for ampicillin resistance. Antimicrob. Agents Chemother., 7:265–270, 1975.

Thornsberry, C., and Kirven, I. A.: Ampicillin resistance in *Haemophilus influenzae* as determined by a rapid test for beta-lactamase production. Antimicrob. Agents Chemother., 6:653–654, 1974.

vaginalis among asymptomatic women. Obstet. Gynecol., *20:*629–632, 1962.

4. Elwell, I. E., *et al.:* Plasmid-linked ampicillin resistance in *Haemophilus influenzae* type b. Inf. Immun., *12:*404–410, 1975.

5. Fuchs, P. C.: Ampicillin-Resistant *Hemophilus influenzae* Check Sample MB-80. Chicago, American Society Clinical Pathologists, 1976.

6. Gardner, H. S., and Dukes, C. D.: *Haemophilus vaginalis* vaginitis: a newly defined specific infection previously classified "nonspecific" vaginitis. Am. J. Obstet. Gynecol., *69:*962–976, 1955.

7. Ginsburg, C. M.: Epiglottitis, meningitis and arthritis due to *Haemophilus influenzae* type b presenting almost simultaneously in siblings. J. Pediatr., *85:*492–493, 1975.

8. Jones, R. N., Slepack, J., and Bigelow, J.: Ampicillin-resistant *Haemophilus paraphrophilus* laryngo-epiglottitis. J. Clin. Microbiol., *4:*405–407, 1976.

9. Kilian, M.: A rapid method for the differentiation of Haemophilus strains—the porphyrin test. Acta Pathol. Microbiol. Scand. [B], *82:*835–842, 1974.

10. Liu, C.: Epiglottitis, laryntitis and laryngotracheobronchitis. *In* Hoeprich, P. D. (ed.): Infectious Diseases. Ed. 2. Chap. 25. Hagerstown, Harper & Rowe, 1977.

11. McCormack, W. M. *et al.:* Vaginal colonization with *Corynebacterium vaginale (Haemophilus vaginalis).* J. Infect. Dis., *136:*740–745, 1977.

12. Margolis, C. Z., Collette, R. B., and Grundy, G.: *Haemophilus influenzae* type b: the etiologic agent in epiglottitis. J. Pediatr., *87:*322–323, 1975.

13. O'Callaghan, C. H., *et al.:* Novel method for detection of beta-lactamase by using a chromogenic cephalosporin substrate. Antimicrob. Agents Chemother., *1:*283–288, 1972.

14. Rosen, I. G., Jacobson, J. and Rudderman, R.: Rapid capillary tube method for detecting penicillin resistance in *Staphylococcus aureus.* Appl. Microbiol. *23:*649–650, 1972.

15. Thomas, W. J., *et al.:* Ampicillin resistant *Haemophilus influenzae* meningitis. Lancet, *1:*313, 1974.

16. Thorne, G. M., and Farrar, W. E.: Transfer of ampicillin resistance between strains of *Haemophilus influenzae* type b. J. Infect. Dis., *132:*276–281, 1975.

17. Thornsberry, C., and Kirven, L. A.: Antimicrobial susceptibility of Haemophilus influenzae. Antimicrob. Agents Chemother., *6:*620–624, 1974.

18. ———: Ampicillin resistance in *Haemophilus influenaze* as determined by a rapid test for beta-lactamase production. Antimicrob. Agents Chemother., *6:*653–654, 1974.

19. Vickerstaff, J. M. and Cole, B. C.: Characterization of *Haemophilus vaginalis, Corynebacterium cervis* and related bacteria. Can. J. Microbiol. *15:*587–594, 1969.

20. Weinstein, L.: Haemophilus infections. *In* Harrison, T. R. (ed.): Principles of Internal Medicine. ed. 8, Chap. 142. New York, McGraw-Hill, 1977.

21. Whisnant, J. K., *et al.:* Host factors and antibody response in *Haemophilus influenzae* type b meningitis and epiglottitis. J. Infect. Dis. *133:*448–455, 1976.

22. Zinnemann, K. and Turner, G. C.: The taxonomic position of *Haemophilus vaginalis (corynebacterium vaginale).* J. Pathol. Bacteriol., *85:*213–219, 1963.

6 Packaged Microbial Identification Systems

The introduction of packaged microbial identification systems by several commercial companies in the late 1960's, with which the biochemical testing of bacteria can be easily performed, represents a major milestone in diagnostic bacteriology. Until that time, many of the media formulations and culture techniques employed in diagnostic microbiology laboratories had not changed substantially since the early part of the 20th century.

Robert Koch's introduction of the use of solid media in the early 1880's, although seemingly uneventful, was one of the more important advances in medical science. Solid media made it possible to isolate bacterial colonies in pure culture and to study their physical and biochemical properties *in vitro*.

The marketing of dehydrated culture media in 1914 by Difco Laboratories was another major contribution to diagnostic microbiology. Prior to that time, the need for each laboratory to prepare crude media from raw animal tissues, milk products, or vegetable materials curtailed to varying degrees their ability to provide diagnostic services in microbiology. But with the availability of dehydrated products, it became possible for all laboratories to prepare immediately before use a variety of culture media by merely adding water to a carefully weighed portion of dried powder and sterilizing the mixture. The use of dehydrated media has also provided standardized formulas with batch-to-batch consistency so that the results of different laboratories can be meaningfully compared.

PACKAGED KITS

The concept of packaging a series of differential media, selected to aid in identifying members of a group of bacteria, is certainly a logical development. These kits were initially developed for the identification of members of the Enterobacteriaceae because of their frequency of isolation from clinical specimens and their relatively rapid growth and generally distinct biochemical reactions. At this time, the following six kit systems, which have been on the market long enough for proper evaluation and general acceptance, are commercially available for identification of gram-negative bacilli:

1. **API 20-E:** Analytab Products Inc., Plain View, N.Y.
2. **Enterotube:** Roche Diagnostics, Nutley, N.J.
3. **Entero-Set:** Fisher Diagnostics, Orangeburg, N.Y.
4. **Minitek:** Bioquest (BBL), Division of Becton-Dickinson and Co., Cockeysville, Md.
5. **Micro-ID:** General Diagnostics, Morris Plains, N.J.
6. **r/b Enteric:** Diagnostic Research, Long Island, N.Y.

The compact construction of these kits, requiring little storage space, their ease of use with easy to interpret chemical reactions, their long shelf life, and the standardized quality control provided by the manufacturers make the kits very convenient for use in microbiology laboratories. They are especially useful in low-volume laboratories in aiding in the identification of bacteria that may otherwise require special media to perform conventional tests where quality control is more difficult to maintain.

The use of these kits, with their prescribed set of media, has made it possible to define "biotypes" or biochemical "fingerprints" by which many of the known bacterial species can be further subgrouped. Mathematical formulas have been devised by which biochemical test results are converted into biotype numbers, making it possible to use a computer to assist in the definitive identification of bacteria. Some kit manufacturers have profile directories or registers available which list the biotype numbers of numerous bacterial species. These biotype numbers are based on data obtained from the examination of thousands of biochemical reactions. Following is a description of how these biotype numbers are derived and used in making the identification of a bacterial isolate.

DERIVATION OF BIOTYPE NUMBERS

With the assistance of William R. Dito, Gerald G. Hoffman, and Eugene W. Rypka, Roche Diagnostics developed one of the first numerical coding systems for the identification of the Enterobacteriaceae. This system, called the *Enterobacteriaceae Numerical Coding and Identification System* (ENCISE), developed for use with the Enterotube, will be used to illustrate how numerical codes are derived.

The series of reactions which can be determined with the Enterotube are as follows:

Dextrose
Gas
Lysine decarboxylase
Ornithine decarboxylase
Hydrogen sulfide
Indole

Lactose
Dulcitol
Phenylalanine deaminase
Urea
Citrate

A binary number is based on a system including only the numbers "0" and "1" which can be easily stored in computers. The ENCISE system utilizes binary numbers to represent positive or negative differential characteristics; that is, a positive reaction is designated "1," a negative reaction as "0." Therefore, if the Enterotube characteristics are rearranged, and hypothetical positive ("1") and negative ("0") reactions are assigned for each, an 11-digit binary number can be derived as follows:

Dex	Gas	Lys	Orn	H$_2$S
+	+	+	+	−
1	1	1	1	0
Ind	Lac	Dul	PA	Urea	Cit
−	+	−	−	+	−
0	1	0	0	1	0

Thus, the binary number 11110010010 represents the eleven identification characteristics produced in an Enterotube by the organism being tested.

Reading the Binary Number in the ENCISE Register

All combinations of the 11-digit binary number are presented in sequence in the ENCISE register together with a list of the microorganisms most likely to have the identification characteristics represented by that number. For example, under the binary number 11110010010 in the ENCISE register are listed the following bacterial species:

		VP	ADO	SUC
Enterobacter hafniae	0.8993	V	−	−
Enterobacter aerogenes	0.0889	+	+	+
Klebsiella ozaenae	0.0067	−	+	V
Serratia liquefaciens	0.0051	V	−	+

The decimal numbers represent the frequency with which the bacterial species with binary number 11110010010 can be expected. These frequencies are derived from the data base of Edwards and Ewing, accumulated from the reactions of several hundred Enterobacteriaceae referred to the Center for Disease Control. The most likely identification is *E. hafniae*, with a probability of 0.8993, or 89.9 per cent. However, there is also an 8.9 per cent chance that *E aerogenes* is the correct identification. *K. ozaenae* and *S. liquefaciens* cannot be totally excluded; however, their low frequency (0.67% and 0.51%) makes either choice highly unlikely. The final identification between the four species can be made by determining three additional characteristics as shown in the ENCISE register listing, namely, the acetyl-methyl carbinol (the Voges-Proskauer) test and the ability to utilize adonitol and sucrose. The differential reactions for the four organisms cited above are shown in the right-hand columns of the listing.

Conversion of Binary Numbers to Octal Numbers

The human mind cannot efficiently calculate in binary logic, so most binary codes have been converted into simpler systems called "octals." In order to understand this conversion, picture a series of three light bulbs. By turning different lights "on" and "off," a total of 2^3 or eight combinations is possible, each of which can be represented by one of eight (octal) numbers ranging from 0 to 7. If all lights are "off" ($-$), the combination $- - -$ is equivalent to octal 0. If only the last bulb is turned on ($+$), the combination $- - +$ is equivalent to octal 1. Octal 2 is represented by the binary pattern $- + -$, and octal 3 by the pattern $- + +$.

The octal equivalents of the eight combinations of a 3 digit binary number are as follows:

Binary			Octal
$-$	$-$	$-$	0
$-$	$-$	$+$	1
$-$	$+$	$-$	2
$-$	$+$	$+$	3
$+$	$-$	$-$	4
$+$	$-$	$+$	5
$+$	$+$	$-$	6
$+$	$+$	$+$	7

To illustrate how binary numbers longer than three digits can be converted into their octal equivalents, the binary number cited above can be used:

1 1 1 1 0 0 1 0 0 1

Beginning to the right, since binary numbers are read from right to left, divide the binary number into subsets of three:

1 1 | 1 1 0 | 0 1 0 | 0 1 0

Now it becomes an easy task to convert each 3-digit subset into its octal equivalent using the formula shown above:

1 1 | 1 1 0 | 0 1 0 | 0 1 0

3 | 6 | 2 | 2

The number 3622 is far easier to remember and simpler to enter into a computer than binary number 11110010010. It must be remembered, however, that the number 3622 represents a series of 11 identification characteristics used in the study of an unknown bacterial species. For this reason, these octal derivatives are known as biotype numbers.

It has already been shown that the biotype number 3622 is most likely *Enterobacter hafniae*. In scanning the ENCISE register, *Enterobacter hafniae* is also listed under biotype numbers 3620, 3621, and 3623. What does this mean? Note that only the last digit in the biotype numbers is different. Remember that this last digit represents the last three reactions in the Enterotube, namely, the production of phenylalanine deaminase and urease, and the utilization of citrate. Thus, *Enterobacter hafniae* with biotype number 3620 is negative for these three characteristics; biotype number 3621 is positive for citrate utilization, biotype number

3622 is positive for urease production, and biotype number 3633 is positive for both citrate utilization and urease production.

Biotyping is a valuable aid in recognizing clusters of bacterial isolates. This is needed when conducting epidemiologic investigations or when studying the source of cross infections in hospitals. Analysis of the biotypes of the same bacterial species may also lead to a better understanding of how variance in identification characteristics may be related to known differences in virulence produced by different strains.

REVIEW OF SIX COMMERCIALLY AVAILABLE IDENTIFICATION KITS

It is possible in this discussion to survey only six of the commercial kit systems that have been available to microbiology laboratories long enough to receive general acceptance. A description of the "replicator method" will also be included, because it is a flexible, low-cost alternative to the use of the kits.

A discussion of each kit will follow in alphabetical order, including a brief description of their construction, methods of inoculation, time of incubation, and the appearance of positive and negative reactions.

API 20-E[4,17]

Construction. The API 20-E kit includes a plastic strip with 20 miniaturized cupules containing dehydrated substrates and a plastic incubation chamber with a loosely fitting lid (Plate 6-1, *A* and *B*). Each cupule has a small hole at the top through which the bacterial suspension can be inoculated with a pipette.

Inoculation. The lid is removed from an incubation tray and 5 ml. of tap water is added to provide a humid atmosphere during incubation. An API 20-E strip is removed from the sealed envelope and placed in the incubation tray. A bacterial suspension of the test organism is made by suspending the cells from a well-isolated colony in 5 ml. of sterile 0.85 per cent saline. Using a Pasteur pipette, each cupule tube is filled with the bacterial suspension through the inoculating hole. The three decarboxylase cupules and the urea cupule are then overlaid with sterile mineral oil. This step is quite important to obtain consistent reactions.

Incubation. After the strip is inoculated, the lid is placed on the incubation chamber, and the entire assembly is placed into a 35 to 37° C. incubator for 18 to 24 hours.

Read-out and Interpretation. On the top of the facing page, in order, are the characteristics that can be determined and the visual interpretation of positive and negative reactions (see Plate 6-1, *A* and *B*).

The oxidase test may be performed in either the negative ONPG or H$_2$S cupules, and the nitrate test in the glucose tube. The API 20-E strip can also be used for identifying a number of the nonfermentative gramnegative bacilli and most of the clinically significant anaerobic bacteria as discussed in detail below, although modifications in the technique and an additional 24-hour incubation period may be required.

Also, a 10-test enteric strip is available to microbiologists who do not wish to collect all of the data produced by the API 20-E test strip. A kit of slightly different design is also available for studying the carbohydrate fermentation and assimilation properties of yeasts.

The API 20-E test results can be converted into a 7-digit biotype number from which identification of bacterial cultures can be made with the aid of a profile register or a computer read-out.[10]

Enterotube[13,15,26]

Construction. The Enterotube is a pencil-shaped, self-contained, compartmented plastic tube containing 8 conventional media from which 11 differential characteristics can

API 20-E Characteristics	Visual Reactions Positive	Negative
ONPG	Yellow	Colorless
Arginine dihydrolase	Red-orange	Yellow
Lysine decarboxylase	Red-orange	Yellow
Ornithine decarboxylase	Red-orange	Yellow
Citrate	Dark blue	Light green
Hydrogen sulfide	Blackening	Colorless
Urease	Cherry red	Yellow
Tryptophan deaminase (Add 10% FeCl₃.)	Red-brown	Yellow
Indole	Red ring	Yellow
Voges-Proskauer (Add KOH plus α-naphthol.)	Red	Colorless
Gelatin	Pigment diffusion	No pigment diffusion
Glucose	Yellow	Blue-green
Mannitol	Yellow	Blue-green
Inositol	Yellow	Blue-green
Sorbitol	Yellow	Blue-green
Rhamnose	Yellow	Blue-green
Sucrose	Yellow	Blue-green
Melibiose	Yellow	Blue-green
Amygdalin	Yellow	Blue-green
Arabinose	Yellow	Blue-green

be determined (Plate 6-1; *C* and *D*). An inoculating wire is positioned through the center of all the media chambers and extends out from each end of the tube. One end serves as the inoculating tip; the other as the handle. Both ends are covered with a screwcap when packaged.

Inoculation. A bacterial suspension is not required. Just before use, the screwcaps are removed from each end of the inoculating wire. The pointed end of the wire is touched to the surface of a well-isolated colony on an agar plate. The tube is held firmly in one hand while the handle of the inoculating wire is grasped between the thumb and index finger. With a slow, continuous back-and-forth rotary motion, the inoculating wire is pulled through all of the chambers and completely removed from the tube. The inoculating wire is reinserted into the first four compartments (dextrose, lysine, ornithine, and H₂S/indole) to effect anaerobic conditions. The wire is appropriately scored so that the end can be easily broken off after it has been reinserted. The caps are replaced

on both ends of the tube prior to incubation.

Incubation. The Enterotube is inoculated and placed in a horizontal position into a 35 to 37° C. incubator for 18 to 24 hours and then observed.

Read-Out and Interpretation. Kovac's reagent must be added to the H₂S/indole chamber and 10 per cent ferric chloride to the phenylalanine/dulcitol chamber before reading these reactions. This is most conveniently done by injecting the reagents into the appropriate chambers through the thin plastic back with a syringe and a 27-gauge hypodermic needle.

On the top of the following page, in order, are the characteristics that can be determined with the Enterotube and the visual interpretation of the positive and negative reactions (see Plate 6-1, *C* and *D*).

The Enterotube results are converted into a 4-digit biotype number from which bacterial identifications can be made from a profile register called ENCISE II, or from computer read-outs.

Enterotube	**Visual Reactions**	
Characteristics	*Positive*	*Negative*
Dextrose	Yellow	Red
Gas in dextrose chamber	Bubbles	No bubbles
Lysine decarboxylase	Purple-blue	Yellow
Ornithine decarboxylase	Purple-blue	Yellow
Hydrogen sulfide	Black media	No blackening
Indole (Add Kovac's reagent to the H_2S chamber.)	Red ring	No red ring
Lactose	Yellow	Red
Phenylalanine (Add 10% $FeCl_3$.)	Brown	Light green
Dulcitol (Read in Phenylalanine chamber.)	Yellow	Light green
Urease	Red	Light yellow
Citrate	Deep blue	Light green

Entero-Set 1 and Entero-Set 2[16,27]

Construction. This kit allows a choice of one or two cards. The Entero-Set 1 card is designed for the preliminary screening of bacterial unknowns; the Entero-Set 2 card is designed for a final species identification when needed. Each card is composed of 10 capillary chambers in which reagent-impregnated filter paper strips are sandwiched between a white plastic back and a clear plastic front (Plate 6-1, *E*). An inoculating hole is located at the top of each chamber and a smaller air vent at the opposite end. And hour-glass-shaped bulge is present in the center of each chamber where relatively anaerobic conditions prevail. Any gas formed during incubation of the system can be observed in the center area.

Inoculation. The colony to be tested is transferred from solid media to 5 ml. of brain-heart infusion broth and incubated at 35° C. for 3½ hours. This culture is centrifuged at the end of this time and the sediment resuspended in 1.8 ml. of deionized water. Using a Pasteur pipette, 3 or 4 drops of this suspension are placed in the opening at the top of each chamber, from where the suspension is carried throughout the chamber by capillary action. Care must be taken not to overfill any of the chambers.

Incubation. A disposable plastic incubation chamber that can be humidified by adding a small amount of water is available.

The cards are placed into this chamber and the entire assembly is incubated at 35 to 37° C. The Entero-Set 1 cards can be read within 4 to 8 hours and the Entero-Set 2 cards after 18 to 24 hours of incubation.

Read-Out and Interpretation. Following on the facing page, in order, are the characteristics that can be determined with the Entero-Set 1 and Entero-Set 2 cards, and the visual interpretation of the positive and negative reactions (see Plate 6-1, *E* and *F*).

A profile register based on a binary code, called the Entero-Trak Binary Identification system, is available without cost to each user.

Minitek[5,6,10]

Construction. The fundamental component of this system is a reagent-impregnated filter paper disk to which a broth suspension of the organism to be tested is added. Over 30 different substrate disks are currently available. The following equipment comprises the complete kit:

1. Vials of preformulated inoculating fluid
2. Plastic pipette dispenser
3. Automatic inoculating gun
4. Automatic disk dispenser
5. Plastic incubation plate with 12 wells
6. Humidor incubation chamber
7. Color comparator cards

Entero Sets	**Visual Reactions**	
Characteristics	*Positive*	*Negative*
Entero-Set 1		
Resazurin (growth control)	Pinkish	Blue
Malonate	Blue	Green
Phenylalanine	Dark green	Yellow
(Add 10% FeCl₃.)		
Hydrogen sulfide	Brown-black	Beige
Sucrose	Yellow	Red
ONPG	Yellow	Colorless
Lysine decarboxylase	Gray-purple	Yellow-green
Ornithine decarboxylase	Gray-purple	Yellow-green
Urease	Red	Yellow-orange
Indole	Red	Pale yellow
Entero-Set 2		
Arginine dehydrolase	Gray-purple	Yellow-green
Citrate	Blue	Green
Salicin	Yellow	Red
Adonitol	Yellow	Red
Inositol	Yellow	Red
Sorbitol	Yellow	Red
Arabinose	Yellow	Red
Maltose	Yellow	Red
Trehalose	Yellow	Red
Xylose	Yellow	Red

Inoculation. With an inoculating wire or loop, a small portion of the test colony is transferred from the agar surface to the vial of inoculating fluid. This is incubated at 35° C. sufficiently long to produce a turbidity comparable to a MacFarland No. 5 standard. When this is accomplished, 12 reagent disks are dispensed into the plastic trays, either by using the automatic disk dispenser or manually with sterile forceps (Plate 6-2, *A*). Using the automatic inoculating gun fitted with a pipette from the plastic dispenser, each substrate disk is inoculated with 50 microliters of the bacterial suspension. Each disk is then overlaid with sterile mineral oil by completely filling the remaining portion of the well with the oil.

Incubation. The inoculated plastic trays are fitted with their lids and placed in the humidor chamber. The entire assembly is placed into a 35 to 37° C. incubator for 18 to 24 hours.

Read-Out and Interpretation. The user must select the substrate disks that are applicable to the identification of the organisms to be tested. Listed above are some of the substrates (characteristics) commonly used for the identification of the Enterobacteriaceae and their positive and negative reactions. All color changes can be compared against the color comparator cards (similar to using Komac paint cards). (see Plate 6-2, *B.*)

For the interpretation of biochemical data, the manufacturer has made available the Minitek Minicoder. This is a plastic device, the front face of which has windows forming a matrix, 28 vertical columns × 17 horizontal rows. The columns correspond to particular organisms, the rows to the 17 tests used to differentiate these organisms. A tab that can be moved from positive (+) to negative (−), is available for each biochemical test. Seventeen of these can be inserted at any one time, and from this data all possible bacterial candidates are clearly indicated from the specific profile selected. It must be understood that, although profile numbers

Minitek Characteristics	Visual Reactions	
	Positive	*Negative*
Arginine dihydrolase	Red-orange	Yellow
Citrate	Blue	Green
Esculin	Brown	Off-White
Hydrogen sulfide	Black or gray	White-beige
Indole	Pink to red	Colorless
Lysine decarboxylase	Orange to red	Yellow
Malonate	Blue	Light yellow
Nitrate	Dark rust red	Yellow-white
ONPG	Light yellow	White
Ornithine decarboxylase	Red	Yellow-orange
Phenylalanine	Dark green	Light yellow
Urease	Purple	Light tan
Voges-Proskauer	Pink-purple	Colorless
All carbohydrates	Yellow	Orange red

are generated by the Minicoder, these numbers may not identify a specific biotype unless the user has chosen the same set of characteristics when studying a given group of bacterial isolates.

The Minitek system has also found widespread use in the identification of obligate anaerobes, as discussed below.

The Micro-ID System[11,20]

General Diagnostics* has recently marketed the new Micro-ID system for the rapid differentiation of the Enterobacteriaceae which supersedes their PathoTec system of reagent-impregnated filter paper strips.[11,19]

Construction. The Micro-ID system consists of a molded styrene tray containing 15 reaction chambers and a hinged cover. Each reaction chamber has an opening at the top that serves as the port for inoculation of the test organism suspension (Plate 6-2, *C* and *D*). The first five reaction chambers contain a substrate disk and a detection disk, while the remaining chambers contain only a single combination substrate/detection disk. The disks contain all of the substrate and detection reagents required to perform the biochemical tests included in the strip, with

*General Diagnostics, Division of Warner-Lambert Co., Morris Plains, N.J. 07950.

the exception of the Voges-Proskauer test which requires the addition of 2 to 3 drops of 20 per cent KOH. The surface of the tray is covered with clear, polypropylene tape to contain the organism suspension during incubation while providing complete visibility. The hinged cover can be opened to provide access to the inoculation ports and prevents the loss of moisture when closed. The inside surface of the cover contains a strip of filter paper to absorb any spills resulting from errors in handling.

Inoculation. The inoculum is prepared by selecting several morphologically identical, well-isolated colonies from a primary isolation plate, such as MacConkey agar, EMB agar, SS agar, or Hektoen enteric agar, with an inoculating loop and making a suspension in 3.5 ml. of physiologic saline in a 16 × 100 mm. test tube. The suspension should approximate the turbidity of a McFarland no. 1 standard (as for the Kirby-Bauer susceptibility test).

A Micro-ID unit is removed from the sealed, moisture-proof foil package and placed flat on the workbench. Approximately 0.2 ml. of the organism suspension is pipetted into each inoculation port. The cover is closed and the Micro-ID unit is placed upright in the support rack. It is important to make sure that the organism suspension is in contact with all substrate disks, but the detection disks in the middle

MICRO-ID System Characteristics	Visual Reactions	
	Positive	*Negative*
Voges-Proskauer	Pink to red	Light yellow
Nitrate reduction	Red	Colorless to light pink
Phenylalanine*	Green	Light yellow
H_2S†	Brown to black	White
Indole	Pink to red	Light yellow to orange
Ornithine decarboxylase	Grey to purple	Yellow
Lysine decarboxylase	Grey to purple	Yellow
Malonate	Green to blue	Yellow
Urease	Orange to red	Yellow
Esculin hydrolysis	Brown to black	No color change
ONPG	Light or dark yellow	No color change
Arabinose	Yellow to amber	Purple
Adonitol	Yellow to amber	Purple
Inositol	Yellow to amber	Purple
Sorbitol	Yellow to amber	Purple

*In the test for phenylalanine deaminase, any green color in the organism suspension also indicates a positive test.

†A positive H_2S reaction varies from a thin, dark line at the bottom of the detection disk to the entire disk turning black.

of the first five chambers must not become moistened.

Incubation. Each unit is incubated at 35° C. after inoculation for 4 hours.

Read-Out and Interpretation. After the 4-hour incubation period, place the unit flat on the workbench, open the lid; and add 2 to 3 drops (0.1 ml.) of 20 per cent KOH to the inoculation well of the Voges-Proskauer chamber only. Close the lid and hold the tray upright, making certain that the KOH flows down into the Voges-Proskauer test solution. Rotate the Micro-ID unit clockwise about 90 degrees so that the upper detection disks in the first five chambers become moistened. It may be necessary to gently tap the unit with the snap of the finger to dislodge any suspension trapped under the disks.

All reactions are interpreted immediately, according to the color changes listed above. (see Plate 6-2, *C* and *D*).

The final identification of the organism can be made by comparing the visual reaction results with typical reactions of the Enterobacteriaceae and reading with a differentiation checkerboard that is included with the Micro-ID manual. This identification matrix is constructed so that key recog-

nition tests are used to include or exclude a possible identification based on the user's familiarity with the taxonomy of the Enterobacteriaceae.

The Micro-ID manual also lists possible identification choices for any given set of biochemical reactions based on a 5-digit biotype number. This number is easily computed from the fifteen positive or negative reactions, made simple through the use of the Micro-ID encoding form that is included with the kit.

The Micro-ID system is designed only for the identification of members of the Enterobacteriaceae. Therefore, sufficient preliminary characteristics should be determined to ensure that the organism to be tested belongs to the Enterobacteriaceae. In particular, any organism that is cytochrome-oxidase-positive is not suitable for identification by this system.

The r/b System[9,12,21]

Construction. The r/b system consists of four constricted Beckford tubes that contain sterile media formulated to determine 14 differential characteristics (Plate 6-2, *E* and *F*). Two basic tubes measure 8 parameters which are sufficient for identifying the majority of the Enterobacteriaceae. The expanders, the cit/rham and soranase tubes, are used for further species identification when necessary. The medium above the constriction is poured on a slant to detect aerobic reactions; the medium below the constriction measures anaerobic reactions. All reactions are read visually, with independent color changes occuring in the slant, the deep, and beneath the constriction.

Inoculation. An inoculating needle 4 inches in length is required to reach the constricted portion of the tubes. Inoculation is performed by touching the surface of a well-isolated colony on an agar medium with the tip of the inoculating needle and stabbing the media in the tubes all the way through the constriction into the lower chamber. The slant is then streaked as the needle is removed. The cap is replaced loosely before incubation.

Incubation. All tubes are incubated in an upright position at 35 to 37° C for 18 to 24 hours.

Read-Out and Interpretation. The color changes, resulting from *p*H shifts in indicator dyes or from the formation of colored complexes and precipitates, are ascertained visually (Plate 6-2, *E* and *F*). The manufacturer provides a color chart for comparison with the reactions obtained. Identification of isolates is based on the Edwards and Ewing Schema of arranging the Enterobacteriaceae into five tribes. Below are the characteristics that can be determined by each of the four r/b tubes.

Evaluation of Packaged Identification Systems

The packaged identification systems discussed above have received widespread acceptance in many diagnostic microbiology laboratories for the following reasons:

1. Their accuracy has been proven to be comparable to conventional identification systems. The most extensive studies have been performed at the Center for Disease Control by Smith and associates.[19] Different numbers of cultures were used in these studies, with the same strains tested with each of the products in many instances. In all evaluations, the test cultures were transferred first onto MacConkey agar plates to simulate a primary isolation technique, and, after observing colony morphology, were inoculated into the product under test.

r/b Systems	Visual Reactions	
Characteristics	*Positive*	*Negative*
r/b Basic 1		
Phenylalanine deaminase	Brown slant	Red slant
Lactose	Yellow slant	Red slant
Hydrogen sulfide	Black deep	Red deep
Glucose	Yellow deep	Red deep
Lysine decarboxylase	Blue beneath constriction	Yellow beneath constriction
Gas	Bubbles in deep	No bubbles
r/b Basic 2		
Indole	Pink color	No pink color
Ornithine decarboxylase	Blue beneath constriction	Yellow beneath constriction
r/b Expander 1		
Citrate	Blue slant	Green slant
Rhamnose	Yellow beneath constriction	Green beneath constriction
r/b Expander 2		
DNase	Purple slant	Blue slant
Raffinose	Yellow slant	Blue slant
Sorbitol	Yellow deep	Blue deep
Arabinose	Yellow beneath constriction	Blue beneath constriction

Table 6-1. *Comparison of Per Cent of Agreement of "Kit" Tests With Conventional Tests*

Test or Substrate	Packaged Systems					
	API	Enterotube	Inolex*	Minitek	PathoTec†	r/b
Arabinose	97.0			94.5		
Arginine	98.6			93.0		
Citrate	91.2	79.9		91.0		
Dulcitol		99.0		98.0		
Esculin				89.0	89.4	
Glucose	100.0	100.0		98.5		
Glucose (gas)		86.2				89.0
H₂S	95.6	99.8	98.5	99.5	95.8	99.5
Indole	97.5	99.0	99.0	98.0	96.6	98.5
Inositol	93.8			89.0		
Lactose		99.3		88.0		89.0
Lysine	97.8	99.3	99.5	97.5	99.6	99.0
Malonate			98.0	97.5	95.3	
Mannitol	98.9			97.5		
Melibiose	92.3			94.5		
Nitrate				95.0	98.7	
ONPG				99.0	97.0	
Ornithine	99.2	99.8	99.5	97.0	98.0	98.0
Phenylalanine	99.7	96.1	100.0	98.0	98.5	99.0
Rhamnose	98.6			97.5		
Sorbitol	99.7			98.0		
Sucrose	99.5		94.5	91.0		
Urease	90.4	87.9	99.5	82.5	85.8	
Voges-Proskauer (acetoin)	92.3			96.5	94.3	
Number of cultures used for evaluation	366	414	200	200	471	200

*Inolex is the previous name for the Entero-set kits discussed in this chapter.

†General Diagnostics has replaced its PathoTec identification system with the Micro-ID identification system.

(Courtesy of the Enteric Section, Entereobacteriology Branch, Bacteriology, Bureau of Laboratories, Center for Disease Control, Atlanta, Georgia)

Simultaneously, another technologist identified the culture by conventional methods. In all evaluations, two criteria were used for measuring the performance of a product: (1) a comparison of each test in the product with its conventional counterpart; and (2) the accuracy of identification made using the product. The percent of agreement of individual biochemical characteristics included within two or more of the systems is shown in Table 6-1. In Table 6-2 are listed the per cent of accuracies of some of the more commonly encountered members of the Enterobacteriaceae with the six systems under discussion.

2. Several of the systems have a long shelf life, six months to one year, so that outdat-ing of media, particularly a problem with conventional systems, is minimized.

3. The systems require only a minimum of space during storage and incubation.

4. Some of the systems are as easy or easier to use than conventional systems. Inoculation is simple, reactions are generally clear-cut within 24 hours, and the availability of profile registers and computer programs makes final identification easy and accurate.

Although packaged kits have undoubtedly found a relatively permanent place in diagnostic microbiology laboratories, certain potential disadvantages should be pointed out:

1. *Relatively High Cost.* The packaged

Table 6-2. *Identification Accuracy (by Per Cent) of Six Bacterial Identification Systems*

Bacteria	Identification Systems					
	API	Enterotube	Inolex*	Minitek	PathoTec†	r/b
Arizona	93.1	100	85.7	100	100	100
Citrobacter	91.3	100				94
C. freundii			92.3	100	96	
C. diversus				100	100	
Edwardsiella	100	87.5	90	100	87.5	80
Enterobacter						93
E. hafniae	100	96.7	92.3	100	96.4	89
E. aerogenes	95.5	96.4	94.4	100	85.7	
E. liquefaciens	90.5	90.9		100	(a)	
E. cloacae	100	96.6	100	100	93.1	
E. agglomerans				100	80	
E. coli	92.9	96.4	100	100	93.3	91
Klebsiella	100	100				100
K. pneumoniae			93.3	100	96.9	
K. rhinoscleromatis					80	
K. ozaenae				60	80	
Proteus						100
P. mirabilis	100	100	100	100	96.3	
P. vulgaris	100	100	100	100	100	
P. morganii	100	93.8	100	100		
P. rettgeri	94.7	90.5	100	100	94.4	
Providencia	100	96.8	100	100	100	
Salmonella	100	89.3	100	94.7	93.8	88
Serratia	88.5	100		80	95.9	100
S. marcescens			(c)	(b)		
Shigella	100	100	100	100	100	100
S. sonnei			85.7	100		
Yersinia				100		90
				90		

*Inolex is the previous name for the Entero-set kits discussed in this chapter.
†General Diagnostics has replaced its PathoTec identification system with the Micro-ID identification System.
Key: (a) Results are included with those for Serratia; (b) results are only for *S. rubidaea*; (c) results are included with those for *E. liquefaciens*.
(Courtesy of the Enteric Section, Entereobacteriology Branch, Bacteriology, Bureau of Laboratories, Center for Disease Control, Atlanta, Georgia)

systems are relatively expensive. The cost of the test kit strip or card required for the complete identification of a single bacterial isolate averages $2.00 to $2.50. The cost per test of approximately 10¢ to 15¢ is comparable to that of conventional media. Therefore, packaged kits become cost-effective only if 10 or more differential media are required for the identification of a bacterial isolate. For some laboratories, the use of packaged kits may in fact be less expensive than conventional media when one takes into account the added costs of quality control, outdating, and the ever-present po-

tential for environmental contamination incurred with the use of conventional media.

2. *Personnel Retraining.* The need for personnel to be trained in the use of these systems is becoming less of a problem as the packaged kits find more widespread implementation in clinical laboratories. Some of the differential tests in a given kit system may not be familiar to technologists who have been trained to use conventional systems. Additional time also may be required in gaining experience in the interpretation of reactions that are not clearly positive or negative.

Table 6-3. *Identification Characteristics And Reactions of the Oxi-Ferm Tube*

Compartment	Positive Reaction	Negative Reaction
1. O-F anaerobic dextrose	Yellow	Green
2. Arginine dihydrolase	Purple	Green
3. N$_2$ gas production	Lifting of wax overlay	Wax overlay remains in place.
4. H$_2$S/indole	Blackening of medium	No blackening of medium
	Red color on addition of Kovac's reagent	No red color on addition of Kovac's reagent
5. O-F xylose	Yellow	Green
6. O-F anaerobic dextrose	Yellow	Green
7. Urea	Red-purple	Yellow
8. Simmons Citrate	Deep blue	Green

3. *User Flexibility.* Some of the systems require the user to employ a specified set of differential tests in the identification of a given group of bacteria. Some kits are designed so that all of the tests must be performed. Many microbiologists prefer to use only a minimum number of differential tests in the identification of microorganisms which exhibit highly characteristic colonial and microscopic features in primary isolation media. For example, some lactose-utilizing gram-negative bacilli can be adequately identified with less than 10 differential tests and determining 20 or more characteristics is an unnecessary effort.

In summary, whether or not to use one of the packaged identification systems and which one to select is largely a matter of personal preference. The ease of inoculation, the ability to select only those characteristics to be measured, the manipulation required in adding reagents after incubation, and the availability of interpretive charts or numerical coding devices are the main items that a potential user should consider prior to selecting the system best suited for his needs. If strict attention is paid to the instructions provided by the manufacturer for use of the system, all give essentially the same degree of accuracy and reliability of performance with minor differences in the sensitivity of individual tests.

There is a potential danger that the user may consider the packaged kit as an infallible device for the identification of bacteria. It must be remembered that good diagnostic microbiology does not depend solely on one set of differential characteristics. The biochemical data must be integrated in conjunction with colony characteristics (color, size, texture, odor, hemolytic reactions), Gram's stain reaction, morphologic features, and serologic reactions if available before a final identification can be made with confidence.

USE OF PACKAGED SYSTEMS IN THE IDENTIFICATION OF NONFERMENTATIVE GRAM-NEGATIVE BACILLI

Although the packaged identification systems were initially designed for the identification of bacteria belonging to the family Enterobacteriaceae, documented success has been reported with at least two of these kits in the identification of nonfermentative bacilli as well.

Dowda[1] has recently published a comparative study evaluating the Oxi-Ferm* system and the expanded API 20-E† strip in the identification of commonly encountered nonfermentative and oxidase-positive gram-negative bacilli.

The construction of the Oxi-Ferm tube is similar to that of the Enterotube described above, except for the media contained within the eight different compartments.

*Oxi-Ferm, Roche Diagnostics, Nutley, N.J.
†API 20-E, Analytab Products, Inc., Carle Place, New York.

Nine taxonomic characteristics can be measured, and a summary of the appearance of negative and positive reactions within each compartment is given in Table 6-3.

The manufacturer supplies a color comparator chart and a reaction grid by which the color reactions can be more easily interpreted. The reactions can also be converted to biotype numbers following the method described above, and final interpretations can be made by using *The Computer Coding and Identification System for Oxi-Ferm Tube,* as supplied by the manufacturer.

The API 20-E strip is identical to that used for the identification of the Enterobacteriaceae, and the reactions are interpreted the same way; however, 48 hours of incubation may be required before some reactions turn positive. In the identification of many of the nonfermentative species, the system must be expanded to include the assessment of growth on MacConkey agar, motility in semisolid medium, and the fermentative or oxidative metabolic properties as determined on O-F medium. All reactions can be encoded and identifications can be made by using appropriate sections of the *Analytical Profile Index.*

Dowda[1] found that the Oxi-Ferm system was able to place 75 per cent of 176 isolates into the correct species designation without further testing, while 19.3 per cent required additional testing to make a correct identification. This system failed to identify ten of the isolates tested. The expanded API 20-E identified 86.4 per cent of the isolates studied to the species level and failed to identify 24 strains. Dowda concluded that both systems are valuable tools in the laboratory identification of nonfermentative and oxidase-positive bacilli, however, only after certain precautions are taken to be sure that the organism being identified fits the criteria of organisms identified by the data base of the system being used.

Isenberg and Sampson-Scherer[8] also concluded after a study of 265 cultures of nonfermentative gram-negative bacilli representing 21 species, that the Oxi-Ferm system in addition to selected additional tests is equally as accurate as conventional systems in the identification of this group of bacteria.

The following characteristics of the nonfermentative bacilli make them more difficult to evaluate with packaged systems:

1. Many of the nonfermenting gram-negative bacilli are fastidious and grow only slowly in the packaged media chambers.

2. Many of the metabolites formed by nonfermenting bacteria are only weakly reactive, resulting in delayed or equivocal results. The OxiFerm tube in particular is poorly constructed to provide an inoculum sufficiently heavy to result in detectable biochemical reactions for many of the more fastidious species.

Otto and Pickett[14] have recently reported the use of an oxidative-assimilation (OA) system for the study of nonfermentative bacilli with a compartmented box available from the Bionics Corporation in Carson, California. Although this system shows promise, to date it has been introduced and used in only a few clinical laboratories.

PACKAGED SYSTEMS USED FOR THE IDENTIFICATION OF ANAEROBES

Two packaged systems have been well received by clinical microbiologists for the identification of anaerobes, namely, the API 20-A and the Minitek systems. These two systems have been carefully evaluated in a number of clinical laboratories and reference laboratories, and the results of these evaluations have recently been reviewed by Stargel and associates[22,23]. Both of these systems offer excellent alternative approaches to the conventional biochemical procedures of Dowell and Hawkins,[2] Holdeman and Moore,[7] and Sutter, Vargo, and Finegold,[25] as described in detail in Chapter 10.

The use of each of these systems is similar to that described above for the Enterobacteriaceae. However, the bacterial cells must be suspended in Lombard-Dowell broth, the formula of which is as follows:

Lombard-Dowell Inoculum Broth	
Ingredients	*% Composition*
Trypticase (BBL)	0.50
Yeast extract	0.50
Sodium chloride	0.25
L-tryptophan	0.02
L-cystine	0.04
Hemin	0.001
Vitamin K_1	0.001
Sodium sulfite	0.01
Agar*	0.07

* Agar is present in the Minitek formulation but excluded from the API formulation.

This medium supports good growth of the majority of clinically significant anaerobic bacteria. The relatively low peptone concentration in this medium has the effect of providing less substrate for proteolytic activity of the bacteria, which potentially can result in the formation of alkaline substances that counteract any weak acids that may be formed in carbohydrate test media. The medium also is minimally buffered so that acid *p*H shifts in the carbohydrate tests can be more readily detected.

The inoculated API strips or Minitek plates must be incubated under anaerobic conditions. Moore and associates[12] suggest that the inoculum used in the API system must be quite heavy, approximating the turbidity of a McFarland No. 4 or No. 5 standard; otherwise, readings may be delayed as much as 4 days or more for some of the slower-growing species.

Both of these systems have been evaluated by a number of workers. Starr and associates[24] published the initial evaluation of the API system, subsequently evaluated by Moore and associates.[12] The Minitek system was initially evaluated by Stargel and associates.[23] Hansen and Stewart[6] compared both the API and Minitek systems with the conventional techniques used at the Center for Disease Control. They found both microsystems more rapid, less time-consuming, and easier to work with than conven-

tional methods, when preparation of media, preboiling of media before use, inoculation, and time of incubation were all taken into consideration. In their study of 175 anaerobic strains (158 clinical isolates and 17 reference strains) the overall agreement between the Minitek anaerobe system and conventional methods was 98.9 per cent, compared to a 95.1 per cent agreement found by Stargel *et al.* [23] The percentage correlation of both positive and negative reactions with the API strip ranged from 70.8 to 99.4 per cent. The majority of discrepancies with either microsystem were false negative reactions compared to conventional methodology.

The buffering capacity of the substrate solution, the type of indicator used, age of the inoculum, proportion of inoculum to substrate, and the oxidation-reduction potential of the system used are particularly critical factors in the use of the microsystems. Because of differences in these factors, it is not suprising that differences in results between the API, Minitek, CDC, and VPI[7] methods will exist. It must be emphasized that the results obtained from the use of any of these systems must be derived from the tables formulated for each system. In this way, the results for any given system will be valid and reproducible.

Even though both microsystems provide a battery of 12 to 20 tests, each system must be supplemented with other characteristics if some of the anaerobic species are to be completely and accurately identified. In addition to the evaluation of colonial morphology and microscopic features, the following characteristics may also prove to be of value (these are discussed in more detail in Chap. 10):

1. Catalase
2. Lecithinase
3. Lipase
4. Motility
5. Action on milk
6. Growth on 20 per cent bile
7. Susceptibility to pencillin, kanamycin, rifampin

8. Susceptibility to sodium polyanethol sulfonate (SPS)
9. Detection of metabolic products by gas-liquid chromatography

The use of one of these microsystems, in concert with one or more of the above-listed additional tests, should permit the average clinical laboratory to conveniently and accurately identify most of the clinically significant anaerobic bacteria.

THE REPLICATOR SYSTEM

The replicator system is another method for the identification of bacteria that is being used with frequency in diagnostic laboratories. This method, developed by Fuchs,[3] is easy to use, versatile, and inexpensive, thus overcoming some of the disadvantages of the packaged kits mentioned above. The user can select any number and sequence of identification characteristics to be tested and each determination costs less than a penny in contrast to the cost of 10¢ to 15¢ per unit for tests with commercial and conventional procedures. In addition to the Enterobacteriaceae, other groups of bacteria, such as the gram-positive cocci, can also be studied with this technique. Quality control of this procedure is very easy since stock organisms of known reactivity can be tested along with the unknowns in each determination and not merely with only one sample selected from a batch or lot number, as is the case with other systems.

Construction

The backbone of the replicator system is the Steers replicator, a device used for many years for performing agar dilution antibiotic susceptibility testing (Plate 6-3, *A* and *B*). There are two components: (1) a seed plate, consisting of a square metal block containing 32 or 36 wells into which bacterial suspensions can be placed; and (2) an inoculating head, a metal plate from which 32 or 36 prongs, 0.3 cm in diameter project. The prongs are aligned to exactly fit into the wells of the seed plate. This method utilizes conventional agar differential media which are prepared in standard round or square Petri dishes.

Inoculation

A bacterial suspension of each unknown is prepared by transferring isolated colonies of the organisms to a tube of trypticase soy yeast broth, and incubating at 35° C. for 4 hours. Then 0.5 ml. of each unknown broth culture, together with those of standard control organisms, are placed in the wells of the seed plate. The seed plate is next placed under the inoculating head and the plunger released so that the prongs are immersed into the bacterial suspension. The head is elevated and an agar plate substituted for the seed plate. The inoculating head is again lowered so that the tips of the prongs just touch the agar surface, resulting in multiple circular inocula. This sequence is repeated for as many plates of differential media as the microbiologist wishes to use. Alternatively, a manually operated inoculator can be used, as shown in Plate 6-3, *C*.

Incubation

Most agar plates, except those testing for urease and citrate utilization, are incubated aerobically at 35 to 37° C. in a 5 per cent CO_2 environment for 18 to 24 hours. Motility and gelatin plates are incubated at room temperature. Decarboxylase plates are incubated without an increase in ambient CO_2.

Read-Out and Interpretation

Each agar plate shows visible color changes of the colonies or in the media surrounding the points of inoculation for those organisms giving positive reaction with the differential tests used. The color reactions are essentially the same as those obtained with conventional systems. Some of the more commonly used differential media and typical reaction patterns are shown in

Plate 6-3, *D* through *J*. The microtiter plates shown in Plate 6-3, *K* and *L*, are another approach to organism identification by which a series of biochemical characteristics and MIC antimicrobial susceptibility tests can be simultaneously performed on the same plate. This is a cost-effective approach by which each of these determinations can be performed for about 2¢ per well.

SELECTION OF OPTIMAL CHARACTERISTICS FOR BACTERIAL IDENTIFICATION

One dilemma that has evolved from the widespread implementation of the packaged identification system as well as the increase in the number of individual schematic approaches to bacterial species identification is the lack of standardization in the number, sequence, and types of differential tests used. As mentioned above, the biotype number has evolved as a fingerprint by which microorganisms can be categorized. However, each private entrepreneur and each manufacturer of a commercial system have selected characteristics that are unique to their individual systems.

For example, the biotype number derived for a bacterial species using an Enterotube cannot be compared with the biotype of that same organism when tested with an API 20-E strip or by any other system that uses a different set of characteristics. The existence of so many unrelated biotype numbers has made it difficult to compare the various strains of bacteria that are being recovered in clinical laboratories.

Rypka[18] has developed a rational approach to the selection of the Enterobacteriaceae and other groups of bacteria. This system utilizes a somewhat complex mathematical schema by which the positive and negative reactions of each bacterial species included within a schema are compared with those of each of the other microorganisms in the schema. The final calculations permit the sequencing of the various characteristics in descending order of their ability to iden-

tify the bacteria included within the group, thus allowing selection of only those that will be useful in making a differential identification.

Rypka's method for the selection of optomized test characteristics based on mathematically derived separatory values is illustrated in the paragraphs and tables that follow. Table 6-4 is a prototype of a bacterial identification matrix that is used in many clinical laboratories and can be found in a number of microbiology textbooks. The specific matrix shown in Table 6-4, used as the example for this discussion, is constructed for the species differentiation of seven different gram-positive cocci by assessing 20 physical properties or biochemical characteristics. In this type of chart a variety of notations referring to the reaction patterns obtained are employed. Most commonly, a "+" sign indicates a positive reaction for the character being tested and a "−" sign indicates a negative reaction. Such designations as "v" for variable, "d" for delayed, and "w" for weak reactions are also employed. For the purposes of this discussion, Table 6-4 is constructed so that the number 1 equals a positive reaction 0 equals a negative reaction, and 2 a reaction that is variable or not known.

When using a matrix such as that shown in Table 6-4, the microbiologist must always ask whether is it necessary to determine all of the tests listed in the matrix (20 in this case), or whether fewer characteristics will suffice to identify the organism being tested. If it is possible to use fewer characteristics (certainly desirable from both a cost and time standpoint), how can one select a minimal number of characteristics that will still allow accurate species identification? The answer lies in determination of the S values.

In Table 6-4, note that beneath each vertical row of characteristics there is a double series of numbers designated n_0 and n_1. The number indicates the frequency with which the notation 0 (negative condition) appears for each characteristic being tested. For example, in row number 1, arabinose, the

Table 6-4. Identification Matrix*; Seven Species of Gram-Positive Cocci by 20 Characteristics

Characteristics

Organisms	Arabinose, No. 1	Beta Hemolysis, No. 2	Bile solubility, No. 3	Glycerol, No. 4	Glucose, No. 5	Growth @ 10° C., No. 6	Growth @ 45° C., No. 7	Growth @ 10° C., No. 8	Growth pH 9.6, No. 9	Growth in 40% Bile, No. 10	Growth in 2% NaCl, No. 11	Growth in 6.5% NaCl, No. 12	Hippurate Hydrolysis, No. 13	Lactose, No. 14	Litmus Milk—Clot, No. 15	Maltose, No. 16	Raffinose, No. 17	Survive 60° C for 30 min., No. 18	Trehalose, No. 19	Xylose, No. 20
1. Streptococcus pneumoniae	0	2	1	1	2	2	0	2	0	0	2	0	0	2	2	1	1	0	1	2
2. Streptococcus acidominimus	0	1	0	0	2	0	2	2	0	2	1	0	2	1	2	2	0	0	2	0
3. Streptococcus mitis	0	1	0	0	2	0	2	0	0	0	2	0	0	2	2	1	2	2	2	0
4. Streptococcus thermophilus	2	2	0	0	0	0	1	1	0	2	0	0	0	1	1	0	2	2	0	2
5. Streptococcus, species 3	2	0	0	2	1	2	2	2	2	2	2	2	0	2	2	2	2	2	2	2
6. Streptococcus, species 6	2	2	0	2	2	2	2	2	2	0	2	2	0	1	2	2	2	0	2	2
7. Aerococcus viridans	0	1	2	1	1	1	0	2	2	1	2	2	1	2	0	2	2	0	2	0
$n_0 =$	4	1	5	3	1	3	2	1	4	3	1	4	5	0	1	1	1	4	1	3
$n_1 =$	0	3	1	2	2	1	1	1	0	1	1	0	1	3	1	2	1	0	1	0
$n_0 n_1 = S =$	0	3	5	6	2	3	2	1	0	3	1	0	5	0	1	2	1	0	1	0

*Modified from Rypka, E. W., and Babb, R.: Automatic construction and use of an identification scheme, Med. Res. Eng., 9:9–19, 1970.

Key: 0 = negative; 1 = positive; 2 = variable for the characters indicated.

symbol 0 appears four times; in row number 2, beta hemolysis, the 0 appears only once; and in row number 3, bile solubility, the 0 is present five times, and so on. Similarly, n_1 is determined by totaling the number of 1's in each vertical column. In Table 6-4, there are no 1's in row number 1, three in row number 2, and one in row number 3, and so on. Since the number 2 indicates a variable condition, being either positive or negative, it is not used in the calculation of the S value.

The S value is derived by multiplying the frequency of 0's with the frequency of 1's $(n_0 \times n_1)$, as shown by the bottom row of numbers in Table 6-4. For example, the S value for arabinose is 0; for beta hemolysis, 3; and for bile solubility, 5. It is critical that the meaning of the S value is clearly understood. The S value indicates the number of times that the organisms listed in the matrix can be differentiated from one another when matched as pairs by the specific characteristic being measured. For example, in Table 6-4, the separatory value of characteristic number 2, beta hemolysis, is given as 3. This means that there are three instances in which beta hemolysis can be used to differentiate one of the seven species of bacteria listed from a second species. Thus, *Streptococcus acidominimus*, *Streptococcus mitis*, and *Aerococcus viridans*, all being "positive" for beta hemolysis (1), can each be separated from Streptococcus, species 3, which is non beta-hemolytic (0). Similarly, there are five instances where the bile solubility test can be used to distinguish organism pairs; namely, *Streptococcus pneumoniae* can be differentiated from the remaining species listed except *Aerococcus viridans*, which is variable (2). The reader should examine the remaining tests and review the separatory values listed.

The characteristic that has the highest S value, that is, that has the capability of separating the most organism pairs, is theoretically the best one to test. Therefore, in order to assess the relative value of the characteristics listed in the matrix, they should be rearranged in descending order of their S value. This has been done in Table 6-5. Thus, the characteristic glycerol, originally listed as number 4 in Table 6-4, appears as number 1 in Table 6-5 because of its high S value of 6. Similarly, hippurate hydrolysis with an S value of 5 has been moved from its former position of number 13 to row number 2 in Table 6-5. The remaining characteristics are similarly rearranged in descending order of their S values in Table 6-5.

One observation is immediately obvious when studying Table 6-5. The last six characteristics listed in the matrix have separatory values of 0. This indicates that these characteristics are incapable of separating any of the organisms listed and have no value in the identification schema at all. It is a rule of thumb that any characteristics with an S value of 0 can be excluded from the identification matrix. Thus, of the twenty characteristics originally listed in Table 6-4, only fourteen are of value. Can the matrix be reduced even further?

Figure 6-1 is constructed to illustrate an organism-by-organism comparison of the identification matrix shown in Table 6-5. For example, in Table 6-5, if the first organism listed, *Streptococcus pneumoniae,* is compared with *Streptococcus acidominimus*, the following characteristics can be used to differentiate between the two: number 4, glycerol (1 vs. 0); number 3, bile solubility (1 vs. 0); and number 17, raffinose (1 vs. 0). The numbers of these three characteristics (4, 3, and 17) are placed in the first square in Figure 6-1, which represents the intersect between organism number 1 *(Streptococcus pneumoniae)* and organism number 2 *(Streptococcus acidominimus)*. Similarly, *Streptococcus pneumoniae* (organism number 1) is compared with *Streptococcus mitis* (organism number 3). This pair can be separated by characteristic number 4 (glycerol) and characteristic number 3 (bile solubility), and these numbers appear in the square in Figure 6-1 representing the intersect between organisms number 1 and number 3. Thus, Figure 6-1 is constructed into each of the intersecting squares that will separate

Table 6-5. Identification Matrix: Seven Species of Gram-Positive Cocci by 20 Characteristics
(Arranged in Descending Order of Separate Values)

Characteristics

Organisms	Glycerol, No. 4	Hippurate Hydrolysis, No. 13	Bile Solubility, No. 3	Growth @ 10° C., No. 6	Beta Hemolysis, No. 2	Growth in 40% Bile, No. 10	Maltose, No. 16	Glucose, No. 5	Growth @ 45° C., No. 7	Growth in 2% NaCl, No. 11	Trehalose, No. 19	Raffinose, No. 17	Growth @ 50° C., No. 8	Litmus Milk—Clot, No. 15	Lactose, No. 14	Xylose, No. 20	Arabinose, No. 1	Growth @ pH 9.6, No. 9	Survive 60° C. for 30 min., No. 18	Growth in 6.5% NaCl, No. 12
1. Streptococcus pneumoniae	1	0	1	2	2	0	1	2	0	2	2	1	2	2	2	2	0	0	0	0
2. Streptococcus acidominimus	0	2	0	0	1	2	2	2	2	1	1	0	2	2	1	0	0	0	0	0
3. Streptococcus mitis	0	0	0	0	1	0	1	2	2	2	2	2	0	2	2	0	0	0	2	0
4. Streptococcus thermophilus	0	0	0	2	2	2	0	0	1	0	0	2	1	1	1	2	2	0	2	0
5. Streptococcus, species 3	2	0	0	2	0	2	2	1	2	2	2	2	2	2	2	2	2	2	2	2
6. Streptococcus, species 6	2	0	0	1	2	0	2	2	2	2	2	2	2	2	1	2	2	2	0	2
7. Aerococcus viridans	1	1	2	0	1	1	2	1	0	2	2	2	2	0	2	0	0	2	0	2
$N_0 =$	3	5	5	3	1	3	1	1	2	1	1	1	1	1	0	3	4	4	4	4
$n_1 =$	2	1	1	1	3	1	2	2	1	1	1	1	1	1	3	0	0	0	0	0
$n_0 n_1 = S =$	6	5	5	3	3	3	2	2	2	1	1	1	1	1	0	0	0	0	0	0

Color Plates
6-1 *to* 6-3

Plate 6-1
Packaged Microbial Identification Systems

A number of packaged microbial identification systems are currently available that contain stable reagents and media designed for determining biochemical characteristics.

A.
API 20-E System. Two API strips are illustrated, one uninoculated (top) and one inoculated (bottom) with a test organism producing several positive reactions. Each strip includes 20 small plastic cupulae containing reagents to measure the biochemical characteristics listed on page 195.

B.
API System 50 for identification of Lactobacilli. The 50 systems, in which 50 biochemical characteristics can be determined, are designed for research purposes or special applications.

C.*
Enterotube System. An uninoculated tube is shown at the top and one inoculated with a test organism at the bottom. The Enterotube is a plastic tube that contains eight compartments containing conventional culture media from which

11 differential characteristics can be determined (see p. 196). Each tube has an inoculating wire which is positioned through the centers of the media chambers. One end of the inoculating wire serves as the tip which is touched to the surface of the test colony. The other end is the handle and is used to pull the wire and the inoculum through all of the media-containing compartments.

D.*
Enterotubes inoculated with four different bacterial species, illustrating a variety of biochemical reactions.

E.
Inolex System, consisting of cards with filter paper strips sandwiched between a white plastic back and a clear plastic front (system has been updated and is now called Entero-Set I and Entero-Set II Systems). Within each card are ten capillary chambers in which the reagent impregnated strips are located. After inoculation, a variety of color reactions are observed, representing the identification characteristics listed on page 197.

*Since completion of this chapter, Roche Diagnostics has introduced the New Enterotube II. Although the size and shape of the tube remain unchanged, there are now 12 compartments containing a total of 15 tests. The tests included in Enterotube II are glucose, gas production from glucose, lysine decarboxylase, ornithine decarboxylase, H_2S, indole, adonitol, lactose, arabinose, sorbitol, Voges-Proskauer, dulcitol, phenylalanine deaminase, urea, and citrate.

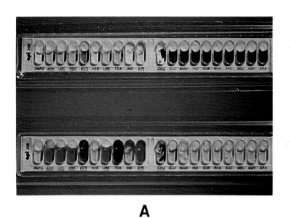

A

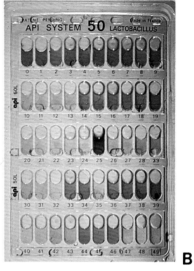

B

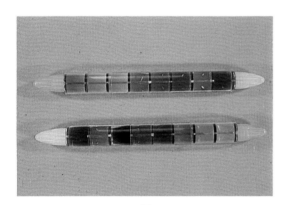

C

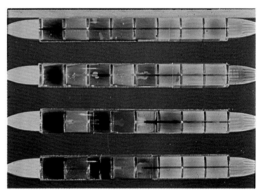

D

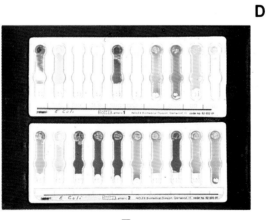

E

Plate 6-2
Packaged Microbial Identification Systems

A.
Minitek system illustrating some of the equipment. The plastic dispenser contains ten cartridges with reagent-impregnated filter paper disks. A large variety of reagent disks are available and each laboratory may make its own selection. The disks selected are dispensed into the plastic reaction tray shown in the left foreground, which are then covered with a standardized suspension of the test organism. The inoculating gun and pipette dispenser are not illustrated here.

B.
Minitek reaction tray with the filter paper disks and suspension of test organism after 18 hours of incubation. A variety of reactions are illustrated and indicated by a color change. Not included in this photograph is the color comparator card which can be used to match negative and positive reactions for each characteristic measured.

C and D.
The Micro-ID System, consisting of a molded styrene plastic tray containing 15 reaction chambers and a hinged cover. Each chamber has an opening at the top that serves as the port of inoculation of a standardized test organism suspension. Reactions are interpreted after 4 hours incubation at 35^0 C., either within the center compartments (first five chambers) or within the suspensions at the bottom compartments in the remaining chambers. The identification characteristics measured by this system are listed on page 199.

E and F.
The r/b system, illustrating a series of constricted Beckford tubes that contain sterile culture media formulated to determine 14 differential characteristics, as listed on page 200. Each tube is inoculated in a similar way as tubes of conventional media are. Reactions beneath the constriction are carried out in an anaerobic environment; those in the slant portion of the upper chamber are exposed to ambient air. Frame *E* illustrates the four different uninoculated tubes; Frame *F* illustrates a variety of reactions after inoculation with a test organism.

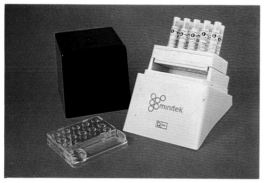

A

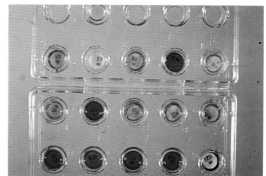

B

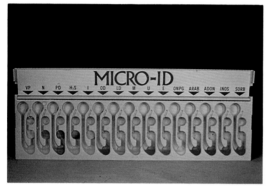

C

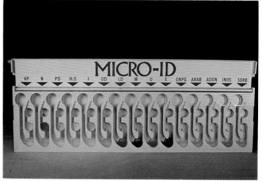

D

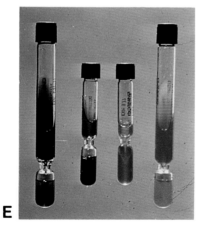

E

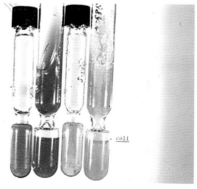

F

Plate 6-3
The Replicator System for Bacterial Identification

The replicator system for the identification of bacteria has proven to be cost effective and accurate. The system requires a multiple-tipped inoculator and a series of agar plates which contain various differential culture media; these are used to simultaneously test a series of unknown bacterial species. The reactions are interpreted similarly to those occuring in conventional media.

A and B.
Photographs of the Steers replicator. A microtube seed plate with 36 wells containing several bacterial suspensions is placed under the inoculating pins (Frame A). The agar plate containing the differential media is next placed under the inoculating head and the pins lowered to touch the surface of agar, thereby producing multiple sites of inoculation. Note that the corner pins in the square inoculating head must be removed when using circular Petri plates.

C.
A hand inocular containing multiple inoculating pins that are spaced to fit into the wells of a microtube tray.

D-J.
A series of replicator agar plates illustrating a variety of biochemical characteristics. The reactions are interpreted by observing various color changes within and surrounding the isolated colonies. Specific reactions shown in the photographs are:

D and E.
MacConkey agar plates with multiple replicator inocula of test organisms. Bacterial species capable of producing acid from lactose fermentation appear red; non-lactose-fermenting bacteria appear colorless.

F.
Plate of OF glucose fermentation medium of Baird-Parker with bromcresol purple as the pH indicator. Bacteria capable of producing acid from glucose change the indicator from purple to yellow. Any carbohydrate can be tested in this manner by adding the appropriate substrate.

G.
Bile esculin test plate. Bacteria capable of hydrolyzing esculin in the presence of 4 per cent bile salts produce a black pigmentation in the medium surrounding the colonies.

H and I.
Citrate utilization test plates. Bacteria capable of utilizing sodium citrate as the sole source of carbon will grow on the medium and produce an alkaline reaction, resulting in a blue color in the medium around the colonies.

J.
Mannitol in trypticase soy agar with phenol red indicator. Bacteria utilizing mannitol produce acid and cause the surrounding agar to turn yellow.

K and L.*
Microtube plates with 96 tiny wells containing a variety of broth culture media, some containing serial dilutions of antibiotics for performance of MIC susceptibility tests and others serving as biochemical substrates. All wells are inoculated with the same organism and the plates placed into a 35^0 C. incubator. After incubation, the reactions are interpreted by observing the presence or absence of growth in the susceptibility wells or the different color reactions similar to those produced in conventional media. Microtube plates are an alternative method for simultaneously testing an organism for a large number of identification characteristics.

*Currently available from Micromedia Systems Inc., Campbell, California, and from Pasco Laboratory, Wheatridge, Colorado, these plates are frozen, packaged microtube systems by which bacteria can be identified and antibiotic susceptibility tests performed using the same plate.

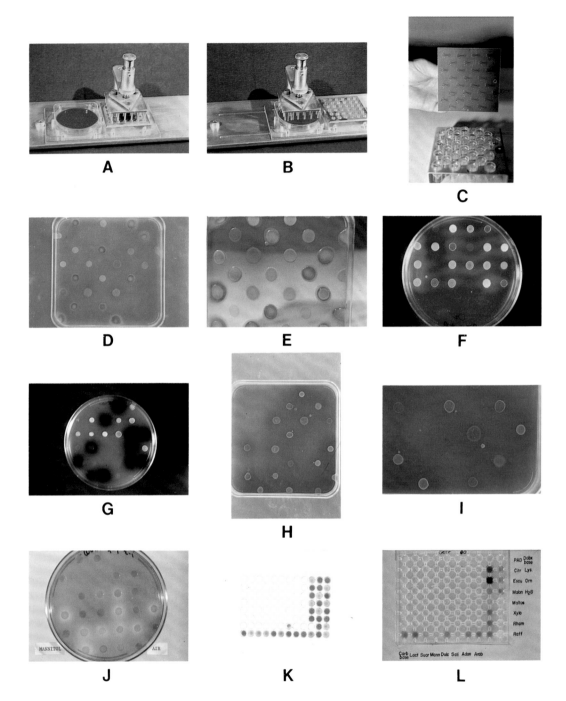

A

B

C

D

E

F

G

H

I

J

MANNITOL   AIR

K

L

PAD Ocbx base
Citr Lys
Escu Orn
Malon H₂S
Maltos
Xylo
Rham
Raff

Carb base Lact Sucr Mann Dulc Sali Adon Arab

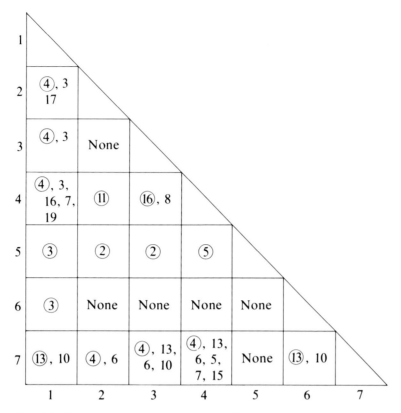

Fig. 6-1. Organism versus organism matrix listing those test numbers that separate the organism pairs represented by the squares of intersect. This matrix is based on pair comparisons of the organisms and tests listed in Table 6-5. For example, tests 4, 3 and 17 separate organism 2 from organism 1. The circled numbers represent essential tests; that is, tests which differentiate between a given organism pair independent of all other tests. See text for details.

each of the organism pairs as they are compared in order.

The following observations can be made from Figure 6-1. Note that characteristic number 4, glycerol, appears in six of the squares. This indicates that glycerol separates six organism pairs, which we already know from the separatory value calculated from the matrix in Table 6-4. Thus, the number of each characteristic will appear in the graph in Figure 6-1, with a frequency identical to its S value. For example, the S value of characteristic number 3, bile solubility, is 5 and the number 3 appears five times in Fig. 6-1.

Note in Figure 6-1 that there are only two

instances where characteristic number 3, bile solubility, separates organism pairs that had not been already distinguished by characteristic number 4, glycerol. Thus, in the square representing the intersect between organism number 1 and organism number 2, characteristic number 4, glycerol, has already distinguished this pair. Characteristics number 3 and number 17, which also appear in this square, add no new information but are confirmatory. However, in the two squares intersecting organism number 1 *(Streptococcus pneumoniae)* with organism number 5 (Streptococcus species 3) and organism number 6 (Streptococcus species 6), the characteristic number 3 appears by itself.

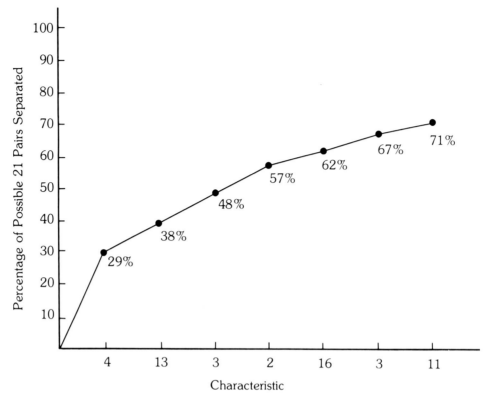

Fig. 6-2. Graph illustrating the optimal percentage rates of separation of the organism pairs listed in Table 6-5 by the sequence of optimized tests represented by the numbers along the abscissa. Thus, the first seven optimized tests included in the matrix in Table 6-5 are capable of separating 71% of the organism pairs listed in the matrix. See text for details.

This indicates that bile solubility (characteristic number 3) is the only test that will distinguish these two organism pairs, and it does it independently of any of the other characteristics in the matrix.

Therefore, Figure 6-1 can be used to further reduce the total number of characteristics to be selected by culling out only those that distinguish between organism pairs independent of any other characteristic. These are designated by circles in Figure 6-1. Note that there are six squares in Figure 6-1 that have the word "none." This indicates that there are six organism pairs in the matrix that cannot be separated by any of the characteristics included in the original set of twenty. This observation illustrates the inadequacy of the identification matrix originally shown in Table 6-4.

Figure 6-2 is a a line graph illustrating the percentage of the organism pairs that can be separated by each of the characteristics listed in the original matrix. With the seven organisms listed in the matrix, there are a total of 21 pair comparisons (shown by the 21 squares in Fig. 6-1) that are possible. Of these twenty-one, six, or 29 per cent, can be separated by determining characteristic number 4 (glycerol) alone. Characteristic number 13, hippurate hydrolysis, can separate an additional two pairs independent of glycerol, bringing the total for these two tests to 38 per cent. Similarly, characteristic number 3 bile solubility, separates an additional two pairs of organisms independent of the first two, bringing the total percentage for the first three tests to 48 per cent. The maximum pair separations possible of the seven organisms listed in the identification matrix shown in Table 6-4 is 71 per cent and

Table 6-6. *Identification Matrix; Seven Species of Gram-Positive Cocci by Seven Optimized Characteristics*

Organisms	Characteristics						
	Glycerol, No. 4	*Hippurate Hydrolysis, No. 13*	*Bile Solubility, No. 3*	*Beta Hemolysis, No. 2*	*Maltose, No. 16*	*Glucose, No. 5*	*Growth in 2% NaCL, No. 11.*
1. *Streptococcus pneumoniae*	1	0	1	2	1	2	2
2. *Streptococcus acidominimis*	0	2	0	1	2	2	1
3. *Streptococcus mitis*	0	0	0	1	1	2	2
4. *Streptococcus thermophilus*	0	0	0	2	0	0	0
5. *Streptococcus, species 3*	2	0	0	0	2	1	2
6. *Streptococcus, species 6*	2	0	0	2	2	2	2
7. *Aerococcus viridans*	1	1	2	1	2	1	2

only seven of the twenty characteristics appear in the matrix are required to reach this maximum (as shown in Fig. 6-2).

Table 6-6 is a reconstruction of Table 6-4, listing only those characteristics which have value in separating the organisms listed independent of other characteristics. The other thirteen characteristics in the matrix have supportive value only or, in the case of the six characteristics with S values 0, have no value at all. This process of selecting only the meaningful differential characteristics is known as optimizing the matrix.

This same process can be applied to any similar identification schema used in the laboratory or published in references. Unfortunately, a large percentage of the identification schema currently in use have not been optimized and often include many characteristics that are of minimal or no value in separating one bacterial species from another. It makes little sense to continue to perform tests that measure suboptimal differential characteristics. Many of the commercially available packaged identification systems currently on the market do not include an optimized selection of test procedures for deriving the biotype numbers. Meaningful comparisons of the bacterial isolates recovered in laboratories all over the world would be possible if they did.

The degree to which bacteria are to be species differentiated varies in different laboratories. In reference laboratories of teaching institutions, the routine determination of a large number of characteristics may be required in order to achieve a desired level of identification. In many diagnostic microbiology laboratories, detailed identification of an isolate may be less important to physicians in respect to patient therapy. Thus, how far to go in the identification of isolates in a clinical laboratory must be decided by the laboratory director or supervisor.

However, the concept of matrix reduction and optimization of test characteristics after deriving the S values as described above can be utilized by microbiologists to aid in the selection of sets of differential tests to use for the identification of various groups of microorganisms that may be encountered.

REFERENCES

1. Dowda, H.: Evaluation of two rapid methods for identification of commonly encountered nonfermenting or oxidase-positive gram-negative rods. J. Clin. Microbiol. 6:605–609, 1977.

2. Dowell, V. R., Jr., and Hawkins, T. M.: Laboratory methods in anaerobic bacteriology, CDC laboratory manual. DHEW Publication No. (CDC) 74-8272, Center for Disease Control, Atlanta, Georgia 30333, 1974.

3. Fuchs, P. C.: The replicator method for identification and biotyping of common bacterial isolates. Lab. Med. *6:*6–11, 1975.

4. Gardner, J. M., Snyder, B. A., and Gröschel, D.: Experiences with the Analytab System for the identification of *Enterobacteriaceae*. Pathol. Microbiol. (Basel), *38:*103–106, 1972.

5. Hansen, S. L., Hardesty, D. R., and Meyers, B. M.: Evaluation of the BBL Minitek system for identification of *Enterobacteriaceae*. Appl. Microbiol. *28:*798–801, 1974.

6. Hansen, S. L., and Stewart, B. J.: Comparison of API and Minitek to Center for Disease Control methods for the biochemical characterization of anaerobes. J. Clin. Microbiol. *4:*227–231, 1976.

7. Holdeman, L. V., and Moore, W. E. C. (eds.): Anaerobic Laboratory Manual. ed. 3. Blacksburg, Va. Virginia Polytechnic Institute and State University, 1975.

8. Isenberg, H. D., and Sampson-Scherer, J.: Clinical laboratory evaluation of a system approach to the recognition of nonfermentative or oxidase-producing gram-negative, rod-shaped bacteria. J. Clin. Microbiol. *5:*336–340, 1977.

9. Isenbert, H. D., et al: R/B expanders: their use in identifying routinely and unusually reacting members of *Enterobacteriaceae*. Appl. Microbiol. *27:*575–583, 1974.

10. Kiehn, T. E., Brennan, K., and Ellner, P. D.: Evaluation of the Minitek system for identification of *Enterobacteriaceae*. Appl. Microbiol. *18:*452–457, 1968.

11. Matsen, J. M., and Sherris, J. C.: Comparative study of the efficiency of seven paper reagent strips and conventional biochemical tests in identifying gram-negative organisms. Appl. Microbiol. *18:*452–457, 1968.

12. Moore, H. B., Sutter, V. L., and Finegold, S. M.: Comparison of three procedures for biochemical testing of anaerobic bacteria. J. Clin. Microbiol. *1:*15–24, 1975.

13. Nord, C. E., Lindberg, A. A., and Dahlback, : Evaluation of five test kits—API, Autotab, Enterotube, PathoTec and R/B—for identification of *Enterobacteriaceae*. Med. Microbiol. Immunol. *159:*211–220, 1974.

14. Otto, L. A., and Pickett, M. J.: Rapid method for identification of gram-negative nonfermentative bacilli. J. Clin. Microbiol. *3:*566–575, 1976.

15. Painter, B. G., and Isenberg, H. D.: Clinical laboratory experiences with the improved Enterotube. Appl. Microbiol. *26:*896–899, 1973.

16. Rhoden, D., et al.: Evaluation of the improved Auxotab 1 System for identifying *Enterobacteriaceae*. Appl. Microbiol., *26:*215–216, 1973.

17. Robertson, E. A., and MacLowery, J. D.: Mathematical analysis of the API Enteric-20 profile register using a computer diagnostic model. Appl. Microbiol., *28:*691–695, 1974.

18. Rypka, E. W., et al.: A model for the identification of bacteria. J. Gen. Microbiol., *46:*407–424, 1967.

19. Smith, P. B.: Performance of six bacterial identification systems. *Personal Communication, 1975.*

20. Smith, P. B.: Rhoden, D. L., and Tromfohrde, K. M.: Evaluation of the PathoTec rapid I-D system for identification of Enterobacteriaceae. J. Clin. Microbiol., *1:*359–362, 1975.

21. Smith, P. B., et al.: Evaluation of the modified R/B system for identification of Enterobacteriaceae. Appl. Microbiol., *22:*928–929, 1971.

22. Stargel, M. D., Lombard, G. L., and Dowell, V. R., Jr.: Alternative approaches to biochemical differentiation of anaerobic bacteria. Am. J. Med. Technol., *44:*709–722, *1978.*

23. Stargel, M. D., *et al.:*Modification of the Minitek miniaturized differentiation system for characterization of anaerobic bacteria. J. Clin. Microbiol., *3:*291–301, 1976.

24. Starr, S. E., Thompson, F. S., and Dowell, V. R., Jr.: Micromethod system for identification of anaerobic bacteria. Appl. Microbiol., *25:*713–717, 1973.

25. Sutter, V. L., Vargo, V. L., and Finegold, S. M.: Wadsworth Anaerobic Bacteriology Manual. ed. 2. Wadsworth Hospital Center, Veterans Administration, Los Angeles, and the Department of Medicine, UCLA School of Medicine, Los Angeles, 1975.

26. Tomfohrde, K. M. *et al.:* Evaluation of the redesigned Enterotube system for identification of *Enterobacteriaceae*. Appl. Microbiol., *25:*304–309, 1973.

27. Washington, J. A., II, Yu, P. K. W., and Martin, W. J.: Evaluation of the Auxotab Enteric system for identification of *Enterobacteriaceae*. Appl. Microbiol., *23:*298–300, 1972.

7 The Gram-Positive Cocci

Excluding the Enterobacteriaceae, the gram-positive cocci are the microorganisms most frequently associated with human infections, and they are commonly recovered from clinical materials in the clinical microbiology laboratory.

Gram-positive bacteria have a number of features that help to distinguish them from gram-negative microorganisms. Of prime importance are the higher peptidoglycan content and lower lipid content of their cells walls. Alcohols and other organic solvents do not penetrate the lipid-poor cell walls of gram-positive cells. This allows gram-positive organisms to retain the crystal violet dye in the Gram's stain and also helps them to resist the germicidal actions of surface-active soaps and detergents.

Differences in cell wall composition help to account for variations in antibiotic susceptibility between gram-positive and gram-negative bacteria. The penicillins and cephalosporins, which exert their antimicrobial effects by inhibiting cell wall synthesis, are more effective against gram-positive bacteria; other antibiotics, such as the aminoglycosides (e.g., gentamicin and kanamycin), that must penetrate the cell wall to produce their inhibitory effect in intracellular metabolic activity, often are less effective against gram-positive than against gram-negative bacteria.

Gram-positive bacteria are usually more resistant to the effects of drying, increased heat, sunlight, and the action of chemicals than are gram-negative bacteria. The gram-positive bacteria are ubiquitous in nature, and their natural habitats include the skin and mucous membrane of man and animals. Some of them can be regularly recovered from the dirt and dust of floors, walls, and a variety of inanimate fomites. Infections in man are most commonly spread by direct contact with infected individuals or from the penetration of skin and mucous membranes by contaminated objects with sharp or pointed surfaces, such as associated with traumatic wounds or surgical procedures.

Staphylococcal infections usually tend to remain localized in the form of an abscess, boil, pustule, furuncle, or carbuncle. The spreading skin inflammation and redness of the throat associated with streptococcal infection are familiar clinical manifestations which essentially all individuals have experienced at one time or another in their lives.

The inflammatory response to infection with gram-positive cocci usually results in an accumulation of pus (composed of live and dead neutrophils and bacterial cells) at the site of infection, commonly called a pyogenic reaction.

Systemic effects of infection with gram-positive cocci are usually due to the effects of toxins produced by the microorganisms. These may be released by living cells as exotoxins. In contrast, gram-negative bacilli more frequently contain lipopolysaccharide endotoxins associated with the bacterial cell wall.[15] These endotoxins are released upon

Chart 7-1. *Catalase*

Introduction	Catalase is an enzyme that decomposes hydrogen peroxide (H_2O_2) into oxygen and water. Chemically, catalase is a hemoprotein, similar in structure to hemoglobin, except that the four iron atoms in the molecule are in the oxidized (Fe^{+++}) rather than the reduced (Fe^{++}) state. Excluding the streptococci, most aerobic and facultatively anaerobic bacteria possess catalase activity. Most anaerobic bacteria decompose H_2O_2 with peroxidase enzymes, similar to catalase except only one ferric ion is contained per molecule.[9]
Principle	Hydrogen peroxide forms as one of the oxidative end products of aerobic carbohydrate metabolism. If allowed to accumulate, hydrogen peroxide is lethal to bacterial cells. Catalase converts hydrogen peroxide into water and oxygen as shown by the following reaction:

$$H_2O_2 \xrightarrow{\text{Catalase}} H_2O + O_2 \text{ (Gas bubbles)}$$

	The catalase test, performed either by the slide or the tube method, is most commonly used to differentiate streptococci (negative) from staphylococci (positive), or in the species differentiation of gram-positive bacilli and the mycobacteria.
Media and Reagents	1. Hydrogen peroxide, 30% stored in a brown bottle under refrigeration 2. An 18- to 24-hour agar plate or slant (preferably without blood since erythrocytes possess catalase activity) containing a pure culture growth of the organism to be tested.
Procedure	*Slide Test:* 1. With an inoculating needle or the tapered tip of an applicator stick, transfer cells from the center of a well-isolated colony to the surface of a glass slide. 2. Add 1 or 2 drops of 3% hydrogen peroxide (prepared by diluting the 30% solution with distilled water). It is recommended that the organism not be added to the reagent (reversing the order), particularly if iron-containing inoculating needles or loops are used, because false positive tests may result.

death and disintegration of the bacteria, and in some cases are responsible for producing vascular collapse and shock in patients with gram-negative sepsis.

APPROACH TO THE LABORATORY IDENTIFICATION OF AEROBIC AND FACULTATIVELY ANAEROBIC COCCI

Schemas used to identify the facultatively anaerobic gram-positive cocci vary considerably from laboratory to laboratory, depending upon the degree to which species identification of isolates is required. Laboratories also differ in the use of names for various groups of gram-positive cocci, and the reasons for this will be discussed.

Aerobic and facultatively anaerobic gram-positive cocci may be recovered from virtually any clinical specimen which contains a mixture of bacteria. These bacteria usually grow quite well on conventional nonselective isolation media, especially on blood agar. Gram-positive cocci are inhibited by selective media such as MacConkey and SS agars which contain bile salts.

Chart 7-1   *Catalase (Continued)*

Procedure *(Continued)*	*Tube or Agar Plate Method:* 1. Dilute the 30% hydrogen peroxide 1:10 with distilled water to give a 3% solution. 2. Add a few drops (about 1.0 ml.) of the 3% hydrogen peroxide reagent directly to the surface of the growth on an agar plate or slant. In the semiquantitative tube test, the agar is poured with a level rather than a slanted surface.
Interpretation	1. *Qualitative Test:* Rapid appearance and sustained production of gas bubbles or effervescence (See Plate 7-1, *G*) is indicative of a positive test. Since some bacteria may possess enzymes other than catalase that can decompose hydrogen peroxide, a few tiny bubbles forming after 20 to 30 seconds is not considered a positive test. 2. *Semiquantitative Test:* Quantitative catalase determinations are performed most commonly in the species identification of mycobacteria (see Chap. 12). The height of gas bubble formation is measured in the tube after the addition of hydrogen peroxide. Bubble formation of 50 mm. or more is considered a strongly positive reaction.
Controls	The H_2O_2 reagent must be tested with positive and negative control organisms each day or immediately before unknown bacteria are tested. Positive control: *Staphylococcus aureus* Negative control: Streptococcus species
Bibliography	Kubica, G. P., et al.: Differential identification of mycobacteria. I. Tests on catalase activity. Am. Rev. Respir. Dis., *94*:400–405, 1966. McLeod, M. B., and Gordon, J.: Catalase production and sensitiveness to hydrogen peroxide amongst bacteria with a scheme of classification based on these principles. J. Pathol. Bacteriol., *26*:326–331, 1923. Taylor, W. I., and Achanzer, D.: Catalase test as an aid to the identification of Enterobacteriaceae. Appl. Microbiol., *24*:58–61, 1972.

When gram-stained, the bacterial cells retain the crystal violet dye and appear as blue (gram-positive) spherules of varying size depending on the species. Variable-staining cells or even gram-negative forms may be observed if stains are made of very young cells (few hours) or of old cultures that contain involutional or dead cells that do not retain the crystal violet after the decolorization step in the Gram's stain.

The number of differential tests employed by various laboratories for the identification of gram-positive cocci is quite varied. In research or large references laboratories, a number of differential characteristics may be evaluated which are not commonly used in clinical laboratories. Examples include testing (1) the ability to grow in media at 10° C. and 45° C.; (2) the ability to grow in the presence of 6.5 per cent sodium chloride; (3) the ability to grow in 40 per cent bile medium; (4) tolerance to heating at 65° C. for 30 minutes; and (5) the ability to hydrolyze a number of compounds such as gelatin, starch, and arginine, among others.

Usually the first practical consideration in

a clinical laboratory is to determine if an isolate is a staphylococcus or a streptococcus. This differentiation can usually be made quickly on the basis of colony appearance, type of hemolysis on blood agar, the arrangement of cells in a gram-stained preparation from a broth culture, and the catalase reaction.

Colonies of staphylococci are commonly large (2 to 3 mm. after 24 hours), convex, opaque, and frequently pigmented (Plate 7-1, *A* and *B*), in contrast to the less than 2 mm. in diameter or pinpoint, translucent to semiopaque colonies of the streptococci (Plate 7-1, *C*). Although staphylococci may be strongly beta-hemolytic on blood agar, the zones of hemolysis are smaller in relation to the size of the colonies than those of hemolytic streptococci (Plate 7-1, *D* and *E*). The staphylococci commonly occur in grapelike clusters (staphylo, "bunch of grapes"; see Plate 7-1, *F*) in contrast to streptococci which occur in pairs or in chains of cells.

Staphylococci are strongly catalase-positive (Plate 7-1, *G*) and the streptococci are catalase-negative. (See chart 7-1).

For clinical purposes, the staphylococci are identified as *S. aureus, S. epidermidis,* and Micrococcus species in most laboratories. *S. aureus* is potentially pathogenic; *S. epidermidis* is only rarely the cause of urinary tract infections or subacute bacterial endocarditis, while the micrococci are usually considered avirulent for man. The definitive identification of these is described in the next column.

In the clinical laboratory the streptococci are initially classified on the basis of their hemolytic reactions on blood agar into the alpha, beta, and gamma strains. The alpha-hemolytic streptococci include a large number of species referred to as the viridans group ("green strep") and *S. pneumoniae.* Although the pneumococci have only recently been classified with the streptococci, the former term *Diplococcus pneumoniae* is descriptive of their morphology and is still used in many laboratories. Some strains of Group D streptococci are also alpha-hemo-

lytic, and one species, *S. faecalis, var liquefaciens,* and ss *zyogenes* may be beta-hemolytic.

Of those producing beta hemolysis only bacteria belonging to Lancefield Group A *(S. pyogenes),* Group B, and Group D are usually specifically identified in most laboratories. Species belonging to Groups C, E, F, and G occasionally cause disease in man; however, the Lancefield serologic grouping technique is required for their identification. In many small clinical laboratories isolates of these microorganisms are referred to a reference laboratory for definitive identification if required.

Lancefield Group D streptococci may be either alpha- or beta-hemolytic; nonhemolytic strains are referred to incorrectly as "gamma"-hemolytic. The Group D streptococci are divided into the enterococcus and the nonenterococcus groups. The nonenterococcus Group D streptococci are generally susceptible to a number of the commonly used antibiotics to which the enterococci are usually resistant. Identification of the streptococci is discussed below.

Identification of Staphylococci

Identification and species differentiation procedures for the staphylococci vary between laboratories but generally poses little problem. The relatively large, raised, dome-shaped opaque colonies with a creamy consistency after 24 hours of incubation on primary medium are easy to recognize (Plate 7-1, *A, B,* and *D*); and the grapelike clusters of gram-positive cocci from a liquid culture are highly characteristic (Plate 7-1, *F*).

On occasion, colonies of staphylococcus and micrococcus can be confused with colonies of some streptococci on agar media. Differentiation between the two can be quickly and easily made by the catalase test, which is rapid and easy to perform (Plate 7-1, *G*). Staphylococci and micrococci decompose hydrogen peroxide (positive catalase test) but streptococci do not. Nonpigmented, white strains of staphylococcus or micrococcus may also be confused with yeasts on primary culture plates and must be differentiated by microscopically examining a gram-stained preparation.

Most clinical laboratories go no further than differentiating *Staphylococcus aureus, Staphylococcus epidermidis,* and Micrococcus species. *Staphylococcus epidermidis* was formerly known as *Staphylococcus albus* because of its white colonies; however, on the basis of mannitol fermentation and coagulase test reactions, white strains may be classified as *S. aureus* despite the lack of yellow pigment.

Upon recognition of a colony of Staphylococcus on a primary isolation culture, the first consideration for the microbiologist is to determine whether or not the microorganism is *Staphylococcus aureus. S. aureus* is a common cause of infectious disease in man, as discussed above. Of the family Micrococcaceae, only *S. aureus* produces coagulase, and the test for coagulase is used in most laboratories as the definitive identifying characteristic of this species.

The ability to produce acid from mannitol (Plate 7-1, *H*) and to reduce tellurite to free tellurium (Plate 7-1, *I*) are two other unique characteristics of *S. aureus* that are used by many microbiologists in making a species identification. Most strains of *S. aureus* also produce deoxyribonuclease (DNase; Plate 7-1, *J*), and some microbiologists use this test in making a species identification; however, DNase is not equivalent to coagulase, and the latter is most widely used in differentiating *S. aureus* from *S. epidermidis.*

The Coagulase Test

Two forms of the coagulase test are commonly used in clinical laboratories, as outlined in Chart 7-2. The slide test is simple to perform and rapid, but has the disadvantage that only "bound" coagulase, or "clumping factor," is detected. Some strains of *S. aureus* produce only "free" coagulase. This gives a false negative reaction in the slide test. Therefore, all negative slide coagulase test reactions must be followed by a tube test to detect both bound and free coagulase.

The slide coagulase test is performed by emulsifying growth from a typical staphlococcus colony in a drop of water on a microscope slide and mixing with a loopful of rabbit plasma. The appearance of white clumps within 5 seconds is indicative of a positive test (Plate 7-1, *K*). The use of citrated plasma is not recommended because false positive reactions may result, as discussed in Chart 7-2.

The tube coagulase test is performed by inoculating 0.5 ml. of a 1:4 dilution of coagulase rabbit plasma with a large loopful of the suspected colony. Coagulase-positive strains usually produce a visible clot within 1 to 4 hours (Plate 7-1, *L*). It is the practice in some laboratories not to read the tube coagulase test until after an overnight incubation. Because the bacteria may also produce fibrinolysins, a clot that forms within 4 hours may be dissolved by the time a 16- or 18-hour reading is taken, giving a false negative interpretation. However, reactions that appear negative in 4 or 6 hours should be incubated overnight and read again at 16 or 18 hours since some strains of *S. aureus* produce coagulase very slowly.

The production of coagulase generally is associated with potential virulence. Most virulent strains of *S. aureus* also produce beta hemolysins; however, this is a variable characteristic and the presence or absence of beta hemolysis on blood agar cannot be used either in the species identification of *S. aureus* or in predicting virulence.

Mannitol Fermentation

S. aureus, in contrast to *S. epidermidis,* can ferment mannitol with the formation of acid. Mannitol salt agar is a highly selective medium for the recovery of pathogenic staphylococci from mixed cultures. This medium takes advantage of the ability of staphylococci to grow in the presence of 7.5 per cent sodium chloride and the ability of *S.*

(Text continues on p. 222)

Chart 7-2.   *Coagulase*

Introduction	Coagulase is a protein enzyme of unknown chemical composition, having a prothrombinlike activity capable of converting fibrinogen into fibrin, which results in the formation of a visible clot in a suitable test system. Coagulase is thought to function *in vivo* by producing a fibrin barrier at the site of staphylococcal infection. *In vivo*, this may serve to localize the organisms into abscesses (carbuncles and furuncles, for example). In the laboratory, the coagulase test is most commonly used to differentiate *Staphylococcus aureus* (coagulase-positive) from the other streptococci and micrococci.
Principle	Coagulase is present in two forms, "free" and "bound," each having different properties which require the use of separate testing procedures:
	1. *Bound Coagulase* (Slide Test): bound coagulase, also known as "clumping factor," is attached to the bacterial cell wall and is not present in culture filtrates. Fibrin strands are formed between the bacterial cells when suspended in plasma (fibrinogen), causing them to clump into visible aggregates when viewed in the slide test. Bound coagulase activity is not inhibited by antibodies formed against free coagulase.
	2. *Free Coagulase* (Tube Test): free coagulase is a thrombinlike substance present in culture filtrates. When a suspension of coagulase-producing bacteria is mixed in equal quantities with a small amount of plasma in a test tube, a visible clot forms as the result of utilizing the plasma coagulation factors in a manner similar to that when thrombin is added.
Media and Reagents	Although human or rabbit plasma obtained from a fresh blood sample may be used, the commercially prepared lyophilized product is recommended because quality control is easier to maintain. Citrated blood products should not be used because citrate-utilizing organisms can release calcium and cause a false positive test. For example, *Streptococcus faecalis* can give a false positive coagulase test in this manner.
	The recommended products are: 1. Difco: Bacto coagulase plasma
	2. BBL: coagulase plasma, rabbit
	Reconstitute only the amount of reagent that will be used within two or three days. Store lyophilized vials in the freezer, reconstituted plasmas in the refrigerator.
Procedure	*Slide Test* (Bound Coagulase):
	1. Place a drop of sterile distilled water or physiologic saline on a glass slide.
	2. Gently emulsify a suspension of the organism to be tested in the drop of water, using an inoculating loop or applicator stick.
	3. Place a drop of reconstituted coagulase plasma immediately adjacent to the drop of the bacterial suspension. Thoroughly mix the two together.
	4. Tilt the slide back and forth, observing the immediate formation of a granular precipitate of white clumps (see Plate 7-1, *K*).

Chart 7-2 Coagulase *(Continued)*

Procedure *(Continued)*	*Tube Test* (Free Coagulase): 1. Aseptically add 0.5 ml. of reconstituted rabbit plasma to the bottom of a sterile test tube. 2. Add 0.5 ml. of an 18- to 24-hour pure broth culture of the organism to be tested (brain-heart infusion or trypticase soy broth). 3. Mix by gentle rotation of the tube, avoiding stirring or shaking of the mixture. 4. Place tube in a 37° C. water bath. Observe for formation of visible clot.
Interpretation	1. *Slide Test:* a positive reaction is usually detected within 15 to 20 seconds by the appearance of a granular precipitate or formation of white clumps (Plate 7-1, *K*). The test is considered negative if clumping is not observed within 2 to 3 minutes. The slide test is considered only presumptive, and all cultures giving negative or delayed positive results should be checked with the tube test because some strains of *Staphylococcus aureus* produce free coagulase that does not react in the slide test. 2. *The Tube Test:* the reaction is considered positive if any degree of clotting is visible within the tube (Plate 7-1, *L*). The test is best observed by tilting the tube. The clot or gel will remain in the bottom of the tube if the test is positive for coagulase. 　　Strongly coagulase-positive bacteria may produce a clot within 1 to 4 hours; therefore, it is recommended that the clot be observed at 30-minute intervals for the first 4 hours of the test. Strong fibrinolysins may also be formed by some *S. aureus* strains and may dissolve the clot soon after it is formed. Therefore, positive tests may be missed if the tube is not observed at frequent intervals. Other *S. aureus* strains may produce only enough coagulase to produce a delayed positive result after 18 to 24 hours of incubation; therefore, all negative tests at 4 hours should be again observed after 18 to 24 hours of incubation.
Controls	The coagulability of the plasma used may be tested by adding 1 drop of 5% calcium chloride to 0.5 ml. of reconstituted rabbit plasma. A clot should form within 10 to 15 seconds. *Staphylococcus aureus,* the coagulase-positive strain, *and Staphylococcus epidermidis,* the coagulase-negative strain, serve as control organisms and each reconstituted vial of plasma should be tested with the control organisms.
Bibliography	Smith, W., and Hale, J. H.: The nature and mode of action of staphyloccus coagulase. Br. J. Exp. Pathol., *25:*101–110, 1944. MacFaddin, J. F.: Biochemical Tests for Identification of Medical Bacteria. pp. 41–52. Baltimore, Williams & Wilkins, 1976. Tager, M.: Current views on the mechanisms of coagulase action in blood clotting. *In* Recent Advances in Staphylococcus Research. Ann. N.Y. Acad Sci., *236:*277–291, 1974.

aureus to ferment mannitol. Colonies of *S. aureus* grow well on the medium and produce a yellow halo in the surrounding agar, indicating the production of acid from mannitol (Plate 7-1, *H* and *I*). The formula for mannitol salt agar is as follows:

Mannitol Salt Agar		
Beef extract	1.00	g.
Peptone	10.00	g.
Sodium chloride	75.00	g.
D-mannitol	10.00	g.
Phenol red	0.025	g.
Agar	15.00	g.
Distilled water to:	1000.0	ml.
pH = 7.4		

Tellurite agar also serves as a selective medium for the recovery of *S. aureus*. The growth of coagulase-negative strains of staphylococci is inhibited by the tellurite in the medium, and *S. aureus* colonies appear black because of the reduction of the tellurite to free tellurium. Occasionally mannitol-positive and/or tellurite-positive strains of staphylococci are found which are coagulase-negative. For this reason the identity of *S. aureus* is based primarily on the coagulase reaction, and the use of either mannitol and/or tellurite tests is used for confirmation only.

S. epidermidis can be differentiated from *Micrococcus* species on the basis of how carbohydrates are utilized. Oxidative-fermentative (O-F) medium, similar to that described in Chapter 3 (Chart 3-1), except that the peptone concentration is 1 per cent rather than 0.2 per cent, can be used to determine if carbohydrates are used fermentatively or oxidatively. *Micrococcus* species produce acid only in the open (oxidative) tube, while *S. epidermidis* produces acid in both open and closed tubes because of its ability to ferment carbohydrates.

Micrococcus species are widely distributed in soil and fresh water. They are commonly found in dust and are frequently recovered in environmental samples taken in infection-control studies. These organisms may be suspected during examination of gram-stained smears because the individual cocci are often larger than staphylococci and tend to form tetrads. In edition 8 of *Bergey's Manual*,[2] only obligately anaerobic cocci in cubical packets are classified in the genus Sarcina and the old term Gaffkya, used previously for certain gram-positive cocci with a tendency to occur in tetrads, has been dropped as a legitimate taxonomic designation.

Identification of Streptococci

The streptococci have an interesting history and as a single genus of bacteria have probably caused more widespread disease and morbidity in man over the centuries than almost any other bacteria, with the possible exception of the tubercle bacillus. As early as 1836, Richard Bright recognized the relationship between scarlet fever, acute glomerulonephritis, and chronic renal failure (Bright's Disease). Pasteur, Koch, and Neisser, working to establish the germ theory of disease, established the streptococci as the cause of puerperal sepsis. The surgeon Frederick Fehleisen recognized that a streptococcus was the etiologic agent of erysipelas, and a later colleague, Alexander Ogston, defined the role of the streptococci in postsurgical wound infections. In 1932, Coburn firmly established the relationship between streptococci and rheumatic fever.

The laboratory study of the streptococci became possible with the introduction of solid culture media toward the end of the 19th century. By the early part of the 20th century, Hugo Schottmuller had demonstrated the hemolytic reaction produced by streptococci on blood agar. Some years later, J. H. Brown, working at the Rockefeller Institute, was the first to describe the different hemolytic reactions (alpha, beta, gamma) of the streptococci (Rockefeller Institute, Medical Research Monograph, Monograph 9, 1919).

In the early 1930's Rebecca Lancefield[8] identified five distinct antigenic groups of streptococci (which she called A, B, C, D, and E), on the basis of serologic differences in the C polysaccharides of the cell wall carbohydrates. Since that time continuing research and study have expanded the number of recognized serologic groups to 18, classified A to H and K to T. In most clinical laboratories only Groups A, B, and D are routinely identified since these groups are responsible for the majority of human infections. The more commonly encountered streptococcal groups, the diseases they cause, and the differential tests used in their identification are shown in Table 7-1.

Although the serologic techniques developed by Lancefield[8] still provide the best laboratory methods for identifying the streptococci and are still used by many reference and research laboratories, most clinical laboratories rely on techniques that demonstrate different morphologic, physiologic, or biochemical characteristics in making a presumptive identification. The initial identification is strongly oriented to assessing the size and appearance of individual colonies on primary plating medium and on the type and amount of hemolysis present on blood agar. However, in order to prevent errors in interpretation based on initial inspection of agar plates, a number of additional simple tests can be employed in establishing a definitive identification.

Hemolytic Properties

The type of hemolysis produced on blood agar is very helpful in the initial identification of streptococci. Technologists must be aware that variations in hemolytic reactions may occur depending upon the species of animal from which the blood was obtained and the type of agar base used for preparing the blood agar. These variations may be particularly noted in Group D streptococci.

For example, Group D enterococci that are usually nonhemolytic on sheep blood agar may be beta-hemolytic when grown on human blood agar, or alpha- or beta-hemolytic or gamma-reacting on horse blood. With experience, each microbiologist becomes accustomed to the hemolytic reactions produced on the type of blood agar being used in a given laboratory. Technolo-

gists, however, may have problems in interpreting cultures if the type of blood agar is changed or if employment is assumed in another laboratory employing a different medium. The increasing use of commercially prepared plating media has resulted in a gradual standardization of blood agar plates used in clinical laboratories. Five per cent sheep blood with trypticase soy agar base, currently the most commonly used primary isolation medium, gives consistent hemolytic reactions for most streptococci.

The agar base employed for blood agar plates should be an infusion product, free of reducing sugars. The lytic action of streptococci on blood is a complex phenomenon and not all of the causes for the inhibition of hemolytic reactions are clearly understood. Reducing sugars, including dextrose, fructose, galactose, and many pentoses, suppress the lytic action of streptococci on the animal erythrocytes in the medium, presumably by lowering the pH. Oxygen-stable streptolysin is inactivated at a low pH; therefore the presence of any reducing sugar in the medium can suppress the expression of beta hemolysis by streptococcal colonies on the surface of agar plates incubated aerobically.

Hydrogen peroxide is another compound that affects the erythrocytes so that they become refractory to hemolysis by the lytic enzymes of the streptococci. The production of peroxidases by streptococci or other bacteria growing in mixed culture varies considerably with the blood agar base being used. This partially explains the varied expressions of hemolytic reactions of bacteria in mixed culture. This variability in hemolytic reactions is one argument advanced by those who advocate the anaerobic incubation of blood agar plates or the use of pour plates in the identification of streptococci. Hydrogen peroxide is not produced in the absence of oxygen. However, one must be cautious in interpreting blood agar plates incubated anaerobically because various other streptococci in addition to Group A appear beta-hemolytic, and some of these

(Text continues on p. 226)

Table 7-1. Some Clinical and Laboratory Aspects of the Streptococci

Lancefield Group	Species	Hemolysis (Sheep Blood)	Normal Human Habitat	Diseases Caused in Man	Laboratory Tests Used in Identification
A	Streptococcus pyogenes	Beta	Pharynx; skin	Primary infections: Acute pharyngitis Erysipelas Wound cellulitis Impetigo Septicemia Postinfection sequelae: Rheumatic fever Glomerulonephritis Rheumatic endocarditis	Lancefield grouping Bacitracin "A" disk Fluorescent staining CIE
B	Streptococcus agalactiae	Beta (Alpha)	Pharynx; vagina; stool Newborn:several sites	Puerperal sepsis Endocarditis Pneumonitis Neonatal infections: Pneumonia Meningitis Septicemia	Lancefield grouping Hippurate hydrolysis CAMP test Bile-esculin (negative) CIE
C	S. equi S. equisimilis S. dysgalactiae	Beta	Pharynx; vagina; skin	Wound infections Puerperal sepsis Cellulitis Endocarditis	Lancefield grouping No growth at 10° C. or 45° C. Hippurate hydrolysis negative Glycerol fermentation +/− Trehalose fermentation negative Sorbitol fermentation negative

Group	Species	Hemolysis	Habitat	Infections	Laboratory identification
D	Enterococcus: S. faecalis, S. faecium, S. durans; Nonenterococcus: S. bovis, S. equinus	Gamma-reactive (Alpha) (Beta)	Large bowel	Urinary tract infection, Pelvic abscesses, Peritonitis, Wound infections, Endocarditis	Lancefield grouping, Bile-esculin hydrolysis, 6.5% NaCl tolerance, CIE
F	S. minutus-angiosus "Strep MG"	Beta	Mouth; teeth; pharynx	Sinusitis, Dental caries, Meningitis, Brain abscesses, Pneumonia	Lancefield grouping, Acid from glucose, maltose salicin, and sucrose, No acid from inulin, xylose, arabinose, and mannitol
G	Streptococcus canis	Beta	Pharynx; vagina; skin	Puerperal infection, Wound infection, Endocarditis	Lancefield grouping, Ammonia from arginine, Inulin fermentation
H	Streptococcus sanguis	Alpha	Mouth; teeth	Dental caries, Endocarditis, Brain abscess, Septicemia	Lancefield grouping*, Inulin fermentation, Production of viscid polysaccharide in 5% sucrose broth
K	Streptococcus salivarius	Alpha	Pharynx; mouth	Endocarditis, Septicemia, Sinusitis, Meningitis	Lancefield grouping*, Acid from glucose, sucrose, maltose, No acid from glycerol, mannitol, sorbitol
None	Streptococcus (Diplococcus) pneumoniae	Alpha	Pharynx; Mouth; Trachea	Lobar pneumonia, Septicemia, Otitis media, Meningitis, Endocarditis	Serotyping, Quellung reaction, Bile solubility, Optochin ("P" disk) susceptibility, CIE

*Lancefield grouping antiserum not commercially available.

may produce a zone of growth inhibition around a 0.04 unit bacitracin disk ("A" disk).

Streptococci produce two hemolysins, streptolysin O, which is antigenic but oxygen-labile (inactivated by oxygen), and streptolysin S, which is nonantigenic but oxygen-stable. Each of these hemolysins produces complete clearing of the blood agar around the colonies (Plate 7-2, *A* and *B*). About 2 per cent of Group A streptococci do not produce streptolysin S and may be missed in aerobically incubated cultures unless provision is made to reduce oxygen tension from at least a portion of the culture medium. It is recommended that several 45-degree angle stabs be made into the medium with the inoculating wire or loop when streaking out the culture to force some of the bacteria beneath the surface where relatively anaerobic conditions prevail. Alternatively, a sterile coverslip can be placed on the agar surface in one portion of the inoculum to prevent contact of the colonies growing under the glass with atmospheric oxygen. Plate 7-2, *C*, illustrates the accentuation of beta hemolysis in the stab areas.

Many microbiologists advocate incubating streptococcus cultures in the candle jar; however, hemolysis may actually be inhibited rather than stimulated due to the increased tendency for streptococci to produce peroxidase in the increased CO_2 atmosphere.[9] By preventing the formation of peroxidases, the anaerobic incubation of blood agar plates or the preparation of pour plates maximizes the expression of the hemolytic reactions of streptococci, as mentioned above. The pour plate technique not only has the advantage of distributing the bacteria beneath the surface of the medium, but also most of the dissolved oxygen is removed from the medium when the base is melted prior to preparation of the pour plates. One disadvantage of pour plates is that it is more difficult to isolate the streptococci from the subsurface colonies in pure culture to use in other tests.

Alpha hemolysis refers to a type of incomplete hemolysis produced by some strains of streptococci in which red blood cells immediately surrounding colonies are partially damaged but not lysed as in beta-hemolytic reactions. There is a characteristic greening of the media due to leakage from the cell of a methemoglobinlike derivation of hemoglobin which undergoes oxidation to biliverdinlike compounds (Plate 7-2, *D*).

Alpha-hemolytic streptococci are frequently referred to as *Streptococcus viridans*, which in fact is not a single species of bacteria but rather a group of streptococci with no serologic specificity. These bacteria are often referred to as "green strep," which are a frequent cause of subacute bacterial endocarditis. These streptococci may also rarely be associated with suppurative otitis media or empyema of the nasal sinuses or pleural cavity. Since the red blood cells in blood agar are not totally destroyed by the alpha-hemolytic reaction, as they are with beta hemolysis, the two can be microscopically differentiated by examining the zones of hemolysis and observing the presence or absence of intact red blood cells. In alpha-hemolytic zones, the red cell membranes are clearly outlined; in beta-hemolytic zones, there are no red cell outlines remaining.

Identification of Group A Streptococci

It is important that Group A beta-hemolytic streptococci be accurately identified in the laboratory because prompt therapy for infected individuals is necessary, not only to allow control of the primary infection (acute pharyngitis, pyoderma, scarlet fever, erysipelas, or cellulitis), but also to prevent potentially serious complications such as rheumatic fever, rheumatic endocarditis and valvulitis, and acute or chronic glomerulonephritis. Streptococci appear as tiny gram-positive cocci in long chains when observed microscopically in Gram's stain preparations (Plate 7-2, *E*).

Bacitracin Susceptibility. Maxted[10] in 1953 found that the growth of Group A streptococci was inhibited by a low concentration (0.02 to 0.04 units) of bacitracin in paper disks on blood agar medium, but most other

Color Plates
7-1 *to* 7-3

Plate 7-1
Identification of Staphylococci

A.
Blood agar plate with white, opaque, smooth colonies characteristic of staphylococci.

B.
Blood agar plates comparing the yellow colonies of *S. aureus* (left) with the white colonies of *S. epidermidis* (right).

C.
Blood agar plate with small, semitranslucent colonies of streptococci, compared to the larger, opaque colonies of staphylococci shown in Frames *A* and *B*.

D.
Blood agar plate illustrating beta hemolysis surrounding colonies of *S. aureus*.

E.
Blood agar plate with beta-hemolytic colonies of *S. aureus*. Note that the staphylococci are not inhibited by the "A" disk (0.04 ug. bacitracin disk) but are inhibited by the 30 ug. neomycin disk. These are helpful differential characteristics for distinguishing beta-hemolytic staphylococci from streptococci.

F.
Photomicrograph of a Gram's stain illustrating the grapelike clusters of blue-staining gram-positive cocci characteristic of staphylococci.

G.
Slide catalase test. Organisms capable of producing catalase release gas bubbles when mixed with a 3 per cent solution of hydrogen peroxide.

H.
Mannitol salt agar. This medium is selective for the growth of staphylococci; the production of a yellow color of the medium is characteristic of *S. aureus*.

I.
Vogel-Johnson agar. This medium is selective for the growth of *S. aureus*. The growth of black colonies (production of free tellurium) surrounded by a yellow halo (mannitol utilization) on this medium is characteristic of *S. aureus*.

J.
Deoxyribonuclease test plate. Organisms producing DNase can be detected by adding dilute hydrochloric acid to the surface of the medium. Destruction of deoxyribonucleic acid by bacteria producing DNase is indicated by a clearing of the medium around the test colonies.

K.
Slide coagulase test. Virulent strains of *S. aureus* agglutinate when emulsified in coagulase plasma (right), compared to coagulase-negative organisms which produce an even suspension (left).

L.
Tube coagulase test. Vitulent strains of *S. aureus* that produce free coagulase result in gel formation of the test plasma after incubation.

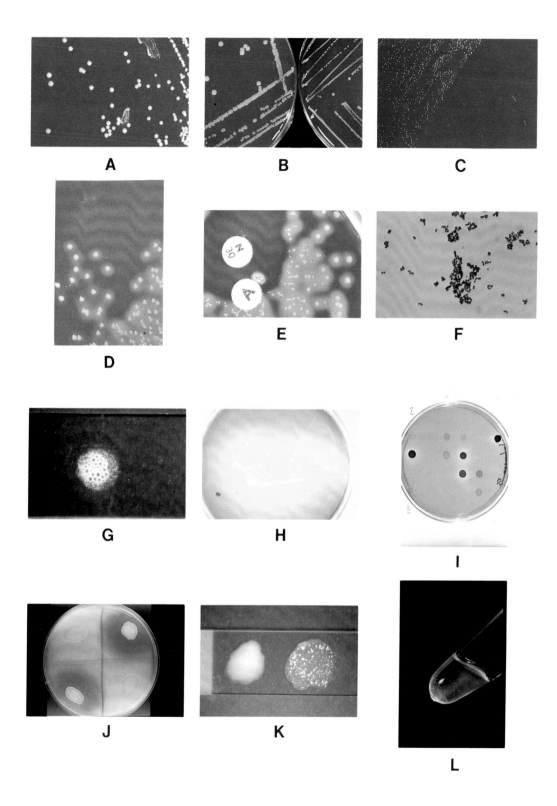

A

B

C

D

E

F

G

H

I

J

K

L

Plate 7-2
Identification of Streptococci

A.
Selective neomycin blood agar plate on which tiny beta-hemolytic colonies of streptococci are growing.

B.
Blood agar plate with a mixed inoculum of group A streptococci and *S. aureus* with a 30 ug. neomycin disk positioned in the area of inoculation. This is a useful method for separating beta-hemolytic streptococci from beta-hemolytic staphylococci when the two are growing in mixed culture. Note that the larger colonies of staphylococci have been inhibited by the neomycin disk while the tiny hemolytic streptococcal colonies are growing in the clear area adjacent to the disk.

C.
A blood agar biplate with colonies of group A beta-hemolytic streptococci. Note the accentuation of the intensity of the hemolytic reaction within the two subsurface stab streak marks. It is recommended that blood agar plates be stabbed obliquely with the inoculating wire in multiple sites in order to detect oxygen labile hemolysins that may be inactivated in surface-growing colonies.

D.
Blood agar plate illustrating tiny colonies of strep-tococci, producing characteristic green, alpha hemolysis.

E.
Photomicrograph of a Gram's stain illustrating long chains of gram-positive streptococci.

F.
Blood agar plate on which are growing colonies of beta-hemolytic streptococci. The inhibition of growth around the "A" disk presumptively identifies this as a Group A strain.

G and H.
Blood agar plates with colonies of Group B beta-hemolytic streptococci. The growth of colonies up to the "A" disk tends to exclude these organisms from Group A. The lack of inhibition of growth by the neomycin disk also excludes these bacteria from the staphylococcus or micrococcus groups.

I.
Hippurate hydrolysis test. Group B streptococci are capable of hydrolyzing sodium hippurate, resulting in a cloudy suspension within the test substrate (left), compared to the clear appearance of the negative control (right).

J and K.
The CAMP test for identification of Lancefield Group B streptococci is based on production of a substance called the CAMP factor that has the property of enhancing the beta hemolysis of certain strains of staphylococci. Note the arrow-shaped zone of hemolysis at the junction of the staphylococcus streak and the inoculum line of the Group B streptococcus. This is considered a positive test and diagnostic of Group B streptococcus.

L.
Growth on bile esculin agar. Lancefield Group D streptococci can be differentiated from other streptococci by the ability to grow on bile esculin medium and producing a dark brown discolora-tion of the medium secondary to esculin hydrolysis.

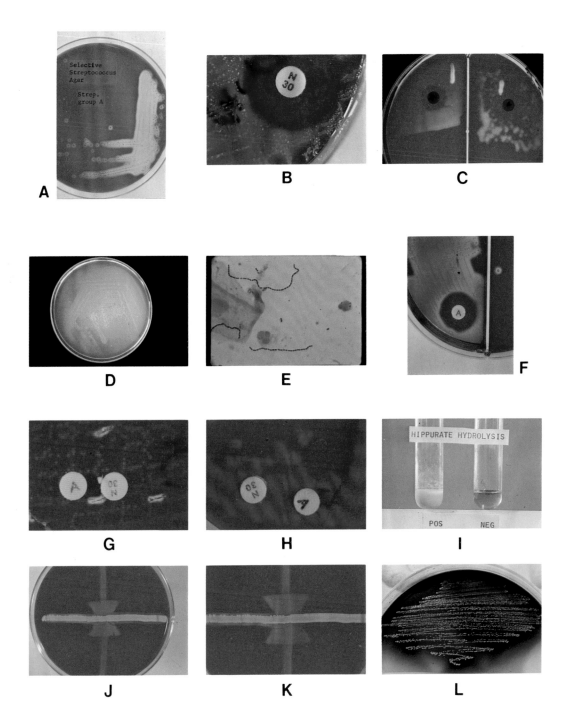

A

B

C

D

E

F

G

H

I

J

K

L

Plate 7-3
Identification of *Streptococcus pneumoniae*

A and B.
Photomicrographs of Gram's stains of a sputum specimen containing lancet-shaped, gram-positive diplococci characteristic of *Streptococcus pneumoniae*. Note the presence of capsules around the pairs of pneumococci, a strong indication of virulence.

C and D.
Blood agar plates with relatively large, glistening, semitranslucent colonies of *S. pneumoniae*. The characteristic alpha hemolysis produced by *S. pneumoniae* is well illustrated in Frame *C*.

E.
Blood agar plate on which are growing alpha-hemolytic colonies of *S. pneumoniae*. These colonies represent the rough colony type. Note the inhibition of growth around the optochin ("P") disk.

F.
Tube bile solubility test. Both tubes contain a suspension of pneumococci. A few drops of 10 per cent sodium deoxycholate have been added to the tube on the right. The clearing of the suspension compared to the control tube indicates solubility of the organism in bile, which is characteristic of *S. pneumoniae*.

G.
Plate bile solubility test. Blood agar plate with colonies of *S. pneumoniae*. Near the center of the photograph, two or three drops of 2 per cent sodium deoxycholate were added and the plate was placed in a 35⁰ C. incubator for 30 minutes. Note dissolution of the bile-soluble colonies, leaving only a residual of the background hemolytic zones.

H and I.
Optochin (ethylhydroxycupreine hydrochloride) susceptibility test. Both of these photographs are of blood agar plates with colonies of *S. pneumoniae*. The colonies on the plate in Frame *H* are mucoid; those on the plate in Frame *I* are non-mucoid but more clearly alpha hemolytic. In both instances, colonial growth is inhibited by the optochin ("P") disk placed in the area of inoculation. This characteristic provides an acceptable identification of *S. pneumoniae*.

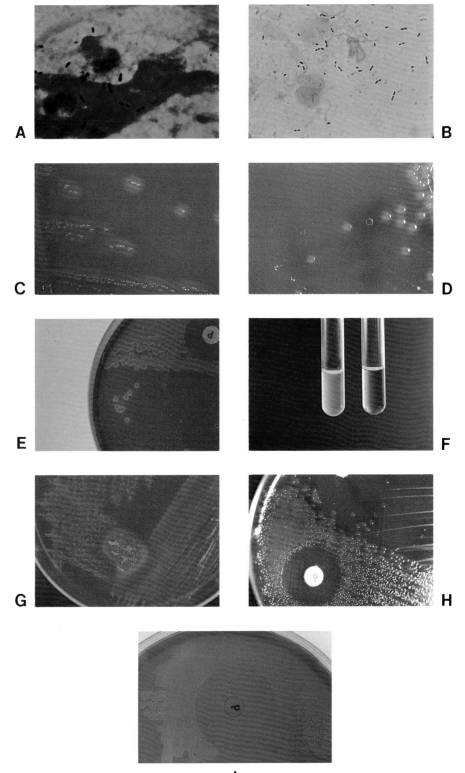

streptococci were not inhibited. Use of a low-concentration bacitracin, or "A" disk is the method most commonly employed in clinical laboratories for the presumptive identification of Group A streptococci. Plate 7-2, *F,* illustrates a positive "A" disk reaction exhibited by a strain of Lancefield Group A *Streptococcus pyogenes.* Although use of "A" disks is quite practical for presumptive identification of Group A streptococci, an estimated 5 to 15 per cent of bacitracin-susceptible streptococci recovered from clinical sources may belong to groups other than Group A. For example, 6 per cent of Group B and 7.5 per cent of Groups C and G beta-hemolytic streptococci are bacitracin-sensitive.[6] This relatively high rate of false positive results can be reduced by carefully evaluating the type of hemolysis produced by isolates. About 7.5 per cent of alpha-hemolytic streptococci are also bacitracin-susceptible.[6] Further differentiation can be made by determining additional characteristics, to be discussed below. Most physicians accept this percentage of false positive results and treat patients symptomatically.

Since many strains of bacitracin-sensitive streptococci of groups other than Group A exhibit zones of inhibition of 10 mm. or less, it has been suggested that an identification as Group A streptococcus be made only if the zone of growth inhibition exceeds 10 mm. This zone size criterion is not universally accepted. Microbiologists must take care not to use the 10-microgram bacitracin disks as used for antimicrobial susceptibility testing as an aid for presumptive identification of streptococci because the concentration of the drug is too high (10 μg. versus 0.04 μg.).

Also, 2 to 7 per cent of A disk tests may be false negative.[6,12] This is potentially a more serious problem than false positive results since Group A streptococcal infection can go unrecognized and the patient may not receive proper treatment. It is common practice in many laboratories, particularly those in physicians offices, to place the "A" disk directly on the primary isolation plate.

It should be emphasized that the bacitracin disk procedure has been designed for use only with pure cultures and not for identification of streptococci growing in mixed culture. The false negative rate may be particularly high if the direct technique is the only procedure used for the screening and identification of Group A streptococci. The production of peroxidases by the other bacteria growing in mixed culture may inhibit the production of hemolysins as discussed above. The overall rate of inaccurate results for the presumptive identification of beta-hemolytic streptococci by the direct technique has been reported as high as 30 per cent.[12]

The examination of throat swabs by the direct fluorescent antibody technique is used by many laboratories for the rapid identification of Group A streptococci.[11] Virtually 100 per cent correlation with standard procedures can be obtained within 6 hours. Throat swabs are placed directly into Todd-Hewitt broth and incubated for 3 to 5 hours. The bacteria are then concentrated by centrifugation, and smears are prepared from the sediment for fluorescent staining. The potential errors as described for the primary culture plate technique with a bacitracin disk applied directly are avoided by use of the fluorescent antibody technique.

Identification of Lancefield Group B Streptococci

Many clinical laboratories are being called on to identify Group B streptococci because of their etiologic role in puerperal sepsis and newborn infections. In some clinical practices, vaginal cultures for the recovery of Group B streptococci are obtained routinely during the third trimester. The organism can be recovered even more frequently in stool swabs than in vaginal or cervical cultures. Because of the relatively high rate of recovery of Group B streptococci from healthy mothers and newborns, the practice of routine culturing during pregnancy has been largely abandoned.

Lancefield Group B streptococci should

Chart 7-3. *Hippurate Hydrolysis Test*

Introduction   Group B streptococci contain the enzyme hippuricase which can hydrolyze hippuric acid. Other beta-hemolytic streptococci lack this enzyme.

Principle   The production of hippuricase, a hydrolytic enzyme by Group B streptococci, results in the hydrolysis of sodium hippurate with the formation of sodium benzoate and glycine, as shown by the following reaction:

Hippuric Acid   Sodium   Glycine
                                 Benzoate

The hydrolysis of hippuric acid can be detected by one of two methods:

1. *Test for Benzoate Using 7% Ferric Chloride:* protein, hippurate, and benzoate are all precipitated by ferric chloride; however, protein and hippurate are more readily soluble than benzoate in an excess of ferric chloride. Thus, the persistence of a precipitate in the hippurate culture broth after adding excess ferric chloride indicates the presence of benzoate and a positive test for hippurate hydrolysis.

2. *Test for Glycine Using Ninhydrin Reagent:* Ninhydrin is a strong oxidizing agent that deaminates alpha amino groups with release of NH_3 and CO_2. The released ammonia reacts with residual ninhydrin to form a purple color. Ninhydrin can be used to detect glycine, the only alpha amino compound formed in the hippurate test, indicating hydrolysis. The test is sensitive and rapid, with interpretations often possible after 2 to 3 hours of incubation.

Media and Reagents   *Benzoate Test:*

1. Sodium hippurate medium:

Heart-infusion broth	25 g.
Sodium hippurate	10 g.
Distilled water	1000 ml.

2. Ferric chloride reagent:

$FeCl_3$ 6 H_2O	12 g.
2% aqueous HCl	100 ml.

Glycine Test:

1. Sodium hippurate reagent:

Sodium hippurate	1 g.
Distilled water	100 ml.

2. Ninhydrin reagent:

Ninhydrin	3.5 g.
1:1 acetone/butanol	100 ml.

Procedure   *Benzoate Test:*

1. Inoculate a tube of sodium hippurate medium with the organism to be tested. Incubate at 35° C. for 20 hours or longer.

2. Centrifuge the medium to pack the cells and pipette 0.8 ml. of the clear supernate into a clean test tube.

3. Add 0.2 ml. of the ferric chloride reagent to the test tube and mix well.

Chart 7-3. *Hippurate Hydrolysis Test (Continued)*

Procedure *(Continued)*	**Glycine Test:** 1. Inoculate a large loop of a beta-hemolytic streptococcus culture suspension into a 0.4-ml. aliquot of the 1% sodium hippurate reagent in a small test tube. 2. Incubate the tube for 2 hours at 35° C. 3. Add 0.2 ml. of the ninhydrin reagent and mix well.
Interpretation	1. *Benzoate Test:* a heavy precipitate will form upon addition of the ferric chloride reagent (Plate 7-2, *I*). If the solution clears within 10 minutes, the reaction is nonspecific due to interaction with unhydrolyzed hippurate or the protein in the medium. If the precipitate persists beyond 10 minutes, sodium benzoate has been formed, indicating hydrolysis and a positive test. 2. *Glycine Test:* the appearance of a deep purple color after addition of the ninhydrin reagent indicates the presence of glycine in the mixture and a positive test for hippurate hydrolysis.
Controls	Positive control: Group B streptococcus Negative control: Group D streptococcus
Bibliography	Ayers, S. H., and Rupp, P.: Differentiation of hemolytic streptococci from human and bovine sources by the hydrolysis of sodium hippurate. J. Infect. Dis., *30*:388–399, 1922. Facklam, R. R., *et al.*: Presumptive identification of group A, B and D streptococci. Appl. Microbiol. 27:107–113, 1974. Hwang, M. N., and Ederer, G. M.: Rapid hippurate hydrolysis method for presumptive identification of group-B streptococci. J. Clin. Microbiol., 1:114–115, 1975.

be suspected when a beta-hemolytic streptococcus is isolated from newborns or from vaginal cultures, particularly if the "A" disk test shows no zone of growth inhibition (Plate 7-2, *G* and *H*). The great majority of Group B streptococci are beta-hemolytic, although an occasional pathogenic non-hemolytic strain is encountered. Only 6 per cent of Group B streptococci are inhibited by the "A" disk, and many of these produce so-called narrow zones (less than 10 mm.) of hemolysis. The sodium hippurate hydrolysis and CAMP tests are available to allow confirmation of the presumptive identification of streptococcal isolates on the basis of susceptibility to bacitracin ("A" disk) and their hemolytic reactions.

Hydrolysis of Sodium Hippurate. Group B streptococci *(S. agalactiae)* have the capability to hydrolyze sodium hippurate but the majority of other beta-hemolytic streptococci do not. The principle and the laboratory technique for performing the hippurate hydrolysis test are given in Chart 7-3. Bacteria possessing the enzyme hippuricase are capable of hydrolyzing hippuric acid to benzoic acid and glycine. The hydrolysis reaction can be detected either by using ferric chloride to detect benzoic acid (Plate 7-2, *I*), or by ninhydrin reagent to detect glycine as described by Hwang and Ederer.[7] The ninhydrin-glycine test allows detection of hippurate hydrolysis within 4 hours as compared to the 24 hours required for the ferric chloride-benzoate test. Because the ninhydrin-glycine reaction is more sensitive, additional tests may be in order to exclude beta-hemolytic enterococci since some

Chart 7-4.   *The CAMP Test*

Introduction	The laboratory identification of Group B hemolytic streptococci has been simplified by implementation of the CAMP test which is easy to perform and well within the capabilities of small laboratories. The CAMP phenomenon was first reported in 1944 by Christie, Atkins, and Munch-Peterson, whose contribution is acknowledged in the acronym "CAMP."
Principle	The hemolytic activity of staphylococcal beta lysin on erythrocytes is enhanced by an extracellular factor produced by Group B streptococci, called the CAMP factor. Therefore, wherever the two reactants overlap in a sheep blood agar plate, an accentuation of the beta-hemolytic reaction will be noted.
Media and Reagents	Sheep or bovine blood must be used. Washed sheep blood resuspended in physiologic saline produces the most reproducible results. The test is performed in a standard 100-mm. Petri dish containing sheep blood agar.
Procedure	The CAMP test is performed by making a single streak of the streptococcus (to be identified) perpendicular to a strain of *Staphylococcus aureus* that is known to produce beta lysin. The two streak lines must not touch one another (see diagram A, below). The inoculated plates must be incubated in an ambient atmosphere. Although incubation in a candle jar may accelerate the reaction, more false positive Group A streptococci will be noted. The plates should not be incubated in an anaerobic environment because many Group A streptococci are positive in the absence of oxygen.

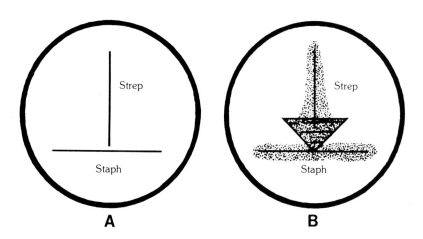

Interpretation	As illustrated in diagram B above, the zone of increased lysis assumes the shape of an arrow-head at the junction of the two streak lines (Plate 7-2, *J* and *K*). Any bacitracin-negative, CAMP-positive, bile-esculin-negative streptococcus can be reported as "Group-B streptococcus, presumptive by CAMP." Presumptive identifications can also be made of those Group D streptococci that are bacitracin-positive and/or non beta-hemolytic.

Chart 7-4.   *The CAMP Test* (Continued)

Controls	Positive control: Group B streptococcus *(S. agalactiae)* Negative control: Group A streptococcus (CAMP-negative) or Group D streptococcus
Bibliography	Christie, R., Atkins, N. E., and Munch-Peterson, E. A.: A note on a lytic phenomenon shown by group B streptococci. Aust. J. Exp. Biol. Med. Sci., *23*:197–200, 1944. Facklam, R. R.: A review of the microbiological techniques for the isolation and identification of streptococci. CRC Crit. Rev. Clin. Lab. Sci., *6*:287, 1976. ———: Isolation and Identification of Streptococci: Part III. DHEW Public Health Service Publication, Center for Disease Control, Atlanta, Georgia, 30333.

strains will react positively with ninhydrin. Group D streptococci, including the enterococci, are bile-esculin-positive, but Group B streptococci are bile esculin-negative. Therefore, beta-hemolytic streptococci that hydrolyze hippurate but are bile-esculin-negative can be presumptively identified as Group B streptococci.

The CAMP Test. Although the CAMP phenomenon was described in 1944 by Christie, Atkins, and Munch-Peterson ("CAMP"),[3] the test has only recently been used for the identification of Group B streptococci. The principle and the procedure for performing the test are outlined in Chart 7-4.

The CAMP factor is an extracellular substance produced by Group B streptococci that enhances the lysis of red blood cells by staphylococcal beta lysin. The test is performed by streaking a blood agar plate with a beta-lysin-producing strain of staphylococcus perpendicular to the streak line of the beta streptococcus to be tested. A zone of enhanced hemolysis, detected by the formation of an arrowhead-shaped zone of clearing at the junction of the two streak lines (see illustration in Chart 7-4 and the photograph in Plate 7-2, *J* and *K*) indicates a positive identification of Group B streptococcus. It is important that the plates not be incubated anaerobically because some strains of Group A beta streptococci produce a positive CAMP reaction if oxygen is absent.

Identification of Group D Streptococci

Lancefield Group D streptococci are divided into two groups: (1) enterococci and (2) nonenterococci.

It is important that this differentiation be made because enterococci are less susceptible to penicillin than are other streptococci. In some cases of severe infection, combined therapy with two or more antimicrobial agents must be used. Group D streptococci most commonly cause urinary tract infections in humans. However, they also are recovered with some frequency from wound infections, peritoneal and deep pelvic abscesses, and, most important, cases of subacute endocarditis and septicemia.

S. faecalis and *S. faecium* are the species of enterococcus most commonly isolated in laboratories which identify enterococci. However, many laboratories do not attempt to differentiate them into species, but simply report "enterococcus." Similarly, *S. bovis* and *S. equinus* may not be differentiated but reported as "Group-D streptococci, not enterococci." Laboratories are urged to use procedures which allow streptococci to be differentiated into species. Group D strepto-

Chart 7-5. *Bile-Esculin Test*

Introduction

The bile-esculin test is based on the ability of certain bacteria, notably the Group D streptococci, to hydrolyze esculin in the presence of 1% to 4% bile. Esculin chemically is a coumarin derivative (6-B-glucoside-7-hydroxy-coumarin), structurally belonging to the class of compounds known as glycosides. Glycosides by definition have two moieties connected by an oxygen (glycoside bond) bridge. Each moiety may be a monosaccharide such as glucose; or, in the case of esculin, the second moiety may be a noncarbohydrate, called an aglycone. Esculin is composed of glucose and the aglycone 7-hydroxycoumarin, as shown in the reaction below.

Principle

Bacteria capable of growing in bile and also hydrolyzing esculin produce glucose and the aglycone esculetin (7,7 dihydrooxycoumarin) in an appropriate medium. Esculetin reacts with an iron salt to form a dark brown or black complex, resulting in a diffuse blackening of the bile-esculin medium, which contains ferric citrate as the source of ferric ions.

Esculin $\xrightarrow{H_2O}$ Glucose + Esculetin $\xrightarrow{Fe^{+++}}$ Black Complex

The exact chemical formula of the phenolic iron complex formed with esculetin is not known. Some bile-esculin formulations also include sodium azide to inhibit the growth of gram-negative organisms, making the medium selective for streptococci. Culture media without sodium azide usually contain 4% bile, making them inhibitory to gram-negative organisms but more selective for Group D streptococci.

Media and Reagents

Bile-esculin medium is usually prepared as slants in tubes. Bile-esculin agar plates have also been used successfully, particularly with the replicator method of bacterial identification discussed in Chapter 6. The formulation is:

Peptone	5.0 g.
Beef extract	3.0 g.
Oxgall (bile)	40.0 g.
Esculin	1.0 g.
Ferric citrate	0.5 g.
Agar	15.0 g.
Distilled water	1000 ml.

$$pH = 7.0$$

Commercial Products

Pfizer: selective enterococcus medium (SEM; contains sodium azide)
Difco: bile-esculin medium (BEM; horse serum added)
    Modified bile-esculin medium (MBEM; no horse serum added)
BBL: bile-esculin medium (BEM), selective enterococcus agar (contains sodium azide)
General Diagnostics: bile-esculin strip (PathoTec)

Chart 7-5. *Bile-Esculin Test (Continued)*

Interpretation	Esculetin, being water-soluble, diffuses into the agar medium. A positive test is read by observing diffuse blackening of the slant, or, in the case of the agar test, a brown or black halo develops around the growing colonies. (see Plate 7-2, *L*). With the PathoTec strip, blackening is observed in the reaction zone with organisms hydrolyzing esculin.
Controls	Positive control: Group D streptococcus Negative control: streptococcus, not Group D
Bibliography	Facklam, R. R.: Recognition of group D streptococcal species of human origin by biochemical and physiological tests. Appl. Microbiol., *23*:1132–1139, 1972. Swan, A.: The use of bile-aesculin medium and of Maxted's technique of Lancefield grouping in the identification of enterococci, group D strepto-cocci. J. Clin. Pathol., 7:160–163, 1954. Weatherall, C., and Dible, J. H.: Aesculin fermentation and hemolysis by enterococci. J. Pathol. Bacteriol., *32*:413–417, 1929.

cocci may show beta or alpha hemolysis, but more commonly are nonhemolytic on blood agar. The same strain of Group D strepto-coccus may vary in its hemolytic properties depending upon the type of animal blood used. Group D streptococci can readily be differentiated from virtually all other strep-tococci by their ability to hydrolyze esculin in the presence of 4 per cent bile. Less than 2 per cent of the viridans group of streptococci are bile-esculin-positive.

The principle and the procedure for the bile-esculin test are shown in Chart 7-5. The test depends on the ability of the organism to grow in the presence of 4 per cent bile, and its ability to hydrolyze esculin which results in a dark brown color in bile-esculin medium (Plate 7-2, *L*). Some bile-esculin culture media contain only 1 per cent bile and are less selective for Group D streptococci.[9] For example, a significantly higher percentage of the viridans group of streptococci produce positive reactions in bile-esculin media that contain lower concentrations of bile.

It is important to understand that certain enterococcus-selective media, such as SF, KF, and BAGG (see Chart 7-6) cannot be used as a substitute for the bile-esculin test in the differential identification of Group D streptococci. The commercial product, PSE (Pfizer selective enterococcus medium), does not substitute for the test described in Chart 7-5 because it has a lower concentra-tion of bile. These selective media are designed primarily to recover streptococci, particularly but not exclusively Group D streptococci, from specimens in which high concentrations of other bacterial species may be present (e.g., feces or sewage sam-ples).

These media generally contain sodium azide as an inhibitor to the growth of most nonstreptococcal bacteria, one or more car-bohydrates, and an indicator to detect acid production. Growth and acid production in these media are highly indicative of the presence of Group D streptococci; however, bile-esculin and other differential tests should be performed to confirm this identifi-cation.

Once a streptococcus is identified as be-longing to Lancefield Group D (positive bile-esculin reaction), it is necessary to de-termine whether or not it is an enterococcus or one of the nonenterococcus species. En-terococci will grow and produce acid in heart-infusion broth containing 6.5 per cent sodium chloride; nonenterococci are inhib-

Chart 7-6.   *Enterococcus-Selective Media*

Introduction

A number of selective culture media for enterococci, including *Streptococcus faecalis* broth (SF), Kenner fecal streptococcal broth (KF), and buffered-azide-glucose-glycerol broth (BAGG), have been developed to selectively recover enterococci from water, food, urine, or fecal material. All coliforms and gram-negative bacilli are inhibited; Group D streptococci grow in these media and produce a visible color change due to the production of acid from the glucose present.

Principle

Sodium azide is the active growth inhibitor contained in all of these media. Most aerobic and facultatively anaerobic bacteria obtain their energy through respiratory metabolism, usually requiring respiratory enzymes called cytochromes. Sodium azide acts to inhibit the transfer of hydrogen through the cytochrome system by tying up the iron in the cytochrome molecule in the ferric state, thus preventing the final electron transfer to molecular oxygen. Streptococci, which lack the cytochrome enzymes, are capable of growing in enterococcus-selective media containing sodium azide and form acid from fermentation of the glucose in the medium which produces a visible yellow color. The BAGG medium includes glycerol which serves to accelerate acid production, often producing a change in the *p*H indicator within 12 to 18 hours.

Media and Reagents

Streptococcus faecalis broth has the following formulation:

Tryptone	20.0	g.
Dextrose	5.0	g.
Monopotassium phosphate	1.5	g.
Dipotassium phosphate	4.0	g.
Sodium azide	0.5	g.
Sodium chloride	5.0	g.
Bromcresol purple	0.032	g.
Distilled water	1000.0	ml.

$$pH = 6.9$$

Kenner fecal streptococcal broth differs from SF broth in that the *p*H is buffered at 7.3 and two carbohydrates, maltose and lactose, are included instead of glucose to better detect the fermentative ability of the enterococci. BAGG broth differs by having glycerol to enhance glucose utilization and a reduction in the concentration of bromcresol purple to 0.015 g., which increases its ability to indicate acid production in the medium.

Procedure

The tube of selective medium can be inoculated directly with the specimen swab, or with a suspension of bacteria taken from a mixed culture. The tubes are placed in a 37° C. incubator for 18 to 24 hours before interpretation.

Interpretation

The development of turbidity in the broth is highly suggestive of the growth of Group D streptococci. If the medium also has converted to a yellow color, indicating acid production from the carbohydrate present, a presumptive identification of enterococci can be made. Use of additional tests may be required to rule out unusual strains of fermentative Group D nonenterococci, and for definitive species identification of isolates.

Chart 7–6. *Enterococcus-Selective Media* (Continued)

Controls	Positive control: *Streptococcus faecalis* (enterococcus)
	Negative control: *Streptococcus bovis* (nonenterococcus)
Bibliography	Facklam, R. R.: Recognition of Group D streptococcal species of human origin by biochemical and physiological tests. Appl. Microbiol., *23*:1131–1139, 1972.
	Hajna, A. A., and Perry, C. A.: Comparative study of presumptive and confirmative media for bacteria of the coliform group and for fecal streptococci. Am. J. Public Health, *33*:550–556, 1943.
	Kenner, B. A., Clark, H. F., and Kabler, P. W.: Fecal streptococci. I. Cultivation and enumeration of streptococci in surface waters. Appl. Microbiol., *9*:15–20, 1961.

ited from growing in this concentration of salt. Thus, with the bile-esculin and salt tolerance tests, three presumptive identifications are possible: (1) strains that are positive in both tests are enterococci; (2) strains that are negative in both tests belong to the viridans group of streptococci; and (3) strains that give a positive bile-esculin test and no growth in 6.5 per cent sodium chloride broth medium are presumptively termed nonenterococcal Group D streptococci.

A summary of the laboratory characteristics by which the clinically significant species of streptococci may be identified is given in Table 7-2. Although the tests and procedures described above for the determination of these characteristics are sufficient to make a presumptive identification of the clinically significant streptococci, in some instances it may be desirable to make a final determination by serologic grouping techniques. In many laboratories these serologic tests are performed routinely since the techniques are not difficult and quality antisera are commercially available.

The procedures for the extraction and serologic identification of streptococci are described in Chart 7-7. The Rantz and Randall autoclave extraction technique is

Table 7-2. *Laboratory Identification of Commonly Encountered Streptococci*

Streptococcus Group	Hemolysis	Bacitracin Sensitivity	Hippurate Hydrolysis	Bile-Esculin Hydrolysis	6.5% NaCl Tolerance
A	Beta	+	–	–	–
B	Beta	– *	+	–	V
Non Group A, B, or D	Beta	– *	–	–	– *
Group D enterococcus	Gamma-reactive	–	V	+	+
S. faecalis	(Alpha)				
S. faecium	(Beta)				
S. durans					
Group D nonenterococcus	Gamma-reactive	–	–	+	–
S. bovis	(Alpha)				
Viridans (Non Group D)	Alpha	V	– *	–	–
	(Gamma)				

Key: + = 90% or more strains positive; – = 10% or less of strains positive; V = variable
*Rare strain positive.

Chart 7-7.   *Extraction and Serological Identification of Streptococci*

Introduction	Streptococci are arranged serologically into Lancefield Groups A, B, C, D, *etc.*, based on the antigenic characteristics of group-specific carbohydrate C substances within their cell walls.

Principle Serologic grouping of the streptococci is accomplished by performing precipitin tests. The antigenic determinants must first be extracted from the streptococcal cells before reacting them with absorbed antisera of known specificity. Extraction can be carried out using hot acid (Lancefield), hot formamide (Fuller), heat (autoclave method of Rantz and Randall), nitrous acid (El Kholy), *Staphylococcus albus* enzyme (Maxted), *S. albus* lysozyme (Watson), or Pronase B enzyme (Ederer), as discussed by Facklam.[5] Only the Lancefield and Rantz-Randall procedures will be described here. The precipitin test is most commonly performed with capillary tubes, in which antiserum is layered over the antigen extract (Lancefield), or vice versa (CDC). The CDC technique is described here because it works better than the Lancefield technique for antisera of low potency.

Media and Reagents
1. A 16- to 24-hours growth of the streptococcus to be tested in 30 ml. of Todd-Hewitt broth
2. Metacresol purple, 0.04%
3. HCl, 0.2 N made up in 0.85% NaCl
4. NaOH, 0.2 N (made up in distilled water)
5. Vaccine capillary tubes, 1.2 to 1.5 mm. outside diameter, both ends open, lightly fire-polished (Kimble borosilicate glass)
6. Group-specific antisera
7. Wooden block racks, 12 inches in length, with a groove containing Plasticine to hold capillary tubes

Procedure   *Lancefield Hot-Acid Extraction*
1. Pack the streptococcal cells grown overnight in 30 ml. of Todd-Hewitt broth by centrifugation.
2. Discard the supernatant fluid; save the cells.
3. Add 1 drop of 0.04% metacresol purple and about 0.3 ml. of 0.2 N HCl to the sedimented cells. Mix well and transfer to a Kahn tube. If the suspension is not a definite pink (pH 2.0 to 2.4), add another drop or so of the 0.2 N HCl.
4. Place in a boiling water bath for 10 minutes, shaking the tube several times.
5. Remove from water bath and pack the cells by centrifugation.
6. Decant supernatant into clean Kahn tube; discard the sediment.
7. Neutralize the extract by adding 0.2 N NaOH drop by drop until it is slightly purple (pH 7.4 to 7.8). A deep purple indicates that the pH is too high. Adjust back to light purple with 0.2 N HCl because too high a pH may cause nonspecific cross-reactions. Try to avoid having to readjust.
8. Clarify by centrifugation and decant supernatant fluid into a small screw-cap vial. This can be stored at 4° C.
9. React the extract with grouping antisera.

Chart 7-7. *Extraction and Serological Identification of Streptococci (Continued)*

Procedure *(Continued)*	**Rantz-Randall Autoclave Extraction** 1. Pack the streptococcal cells grown overnight in 30 ml. of Todd-Hewitt broth by centrifugation. 2. Discard the supernatant fluid; save the cells. 3. Add 0.5 ml. of 0.85% NaCl solution. Shake to suspend the cells. 4. Autoclave the tube and cells for 15 minutes at 121° C. 5. Centrifuge to sediment the cellular debris. 6. Decant the supernatant fluid into a clean sterile container; discard sediment. 7. React the extract with grouping antisera. **CDC Capillary Precipitin Test** 1. Dip capillary tube into serum until a column about 1 cm. long has been drawn in by capillary action. 2. Wipe off tube with facial tissue, taking care to hold tube so that air does not enter the end. 3. Dip the tube into streptococcal extract until an amount equal to the serum column is drawn up. If an air bubble separates the serum and the extract, discard tube and repeat. 4. Wipe tube carefully. Fingerprints, serum, or extract on the outside of the tube may simulate or obscure a positive reaction. 5. Plunge the lower end of the tube into plasticine or clay until a small plug fills the opening. Do not let the reactants mix. 6. Invert tube and insert gently into a plasticine-filled groove of a rack.
Interpretation	After 5 to 10 minutes, examine the capillary tubes with a bright light against a dark background. A white cloud or ring at the center of the column represents a positive test. A strong reaction appears within 5 minutes; a weaker reaction develops more slowly. Since after 30 minutes the reaction may fade or a false positive may appear, examine the capillary tubes at frequent intervals between 10 and 30 minutes.
Controls	Streptococci of the groups to be tested, having a known reactivity
Bibliography	Facklam, R. R.: Isolation and Identification of Streptococci. Part II: Extraction and Serological Identification. Atlanta, Center for Disease Control Publication, 1977.

relatively simple to perform and sufficiently accurate for the identification of most clinical isolates. The original Lancefield hot-acid extraction technique is most effective and still serves as the standard reference method, although somewhat complex and time-consuming. Other extraction techniques utilizing hot formamide, nitrous acid, *Staphylococcus albus* enzyme or lysozyme, and Pronase B enzyme have been reviewed by Facklam.[5] The effectiveness of all extraction techniques depends on the quality of the antisera used in the precipitin tests. Test strains of streptococci of known reactivity should be available for quality control testing of each new lot of antisera used.

Chart 7-8. *Bile Solubility*

Introduction   Bile salts, specifically sodium deoxycholate and sodium taurocholate, have the capability to selectively lyse *Streptococcus (Diplococcus) pneumoniae* when reagent is added to actively growing bacterial cells in an artificial culture medium. *S. pneumoniae* produces a self-lysing enzyme that accounts for the central depression or umbilication characteristic of older colonies on agar media. The addition of bile salts is thought to accelerate this process, augmenting the lytic reaction associated with the lowering of surface tension between the medium and the bacterial cell membrane.

Principle   The bile solubility test can be performed either with a broth culture of the organism to be tested or directly on colonies isolated on a solid medium. The turbidity of a broth suspension visibly clears upon addition of bile salts if the organism is soluble; on solid medium, bile-soluble colonies disappear when overlaid with the reagent. Since sodium deoxycholate may precipitate at a *p*H of 6.5 or less, the broth culture medium used must be buffered at *p*H 7.0 if a false negative insoluble reaction is to be prevented.

Media and Reagents   1. A pure culture of the test organism incubated at 35° C. for 18 to 24 hours
    A. Todd-Hewett broth (or equivalent)
    B. Sheep blood agar plate, 5%
2. Sodium deoxycholate (preferred), 10% solution, or 10% sodium taurocholate (BBL or Difco)
3. Phenol red *p*H indicator
4. Sodium hydroxide (carbonate free), 10 N (40%). Prepare a 0.1 N working solution by diluting stock reagent 1:100.

Procedure   **Broth Test**
1. Transfer approximately 0.5 ml. of an 18- to 24-hour broth culture to each of two clean test tubes. Alternatively, a saline suspension of organisms taken from an 18- to 24-hour growth on blood or nutrient agar may be made. Add 0.1 N sodium hydroxide to adjust the *p*H to 7.0 if required.
2. Add 1 drop of phenol red *p*H indicator to each of the two test tubes.
3. Add 0.5 ml. of 10% sodium deoxycholate to one of the two test tubes (marked "test").
4. Add 0.5 ml. of sterile normal saline to the second test tube (marked "control").
5. Gently agitate both test tubes and place them in a 35° C. incubator or water bath.

Agar Plate Test
1. To a well-isolated colony of the test organism on a 5% sheep blood agar plate, add a drop of 2% sodium deoxycholate (dilute 10% reagent 1:5 with distilled water).
2. Without inverting the plate, place in a 35° C. incubator for 30 minutes.

Chart 7-8 *Bile Solubility (Continued)*

Interpretation	1. *Broth Test:* bile-soluble (positive reaction), visible clearing of the turbid culture within 3 hours. The saline control tube should remain turbid (Plate 7-3, *F*). 2. *Agar Plate Test:* a bile-soluble (positive test) colony disappears, leaving a partially hemolyzed area where the colony had been. Colonies insoluble in bile remain intact (Plate 7-3, *G*).
Controls	Positive (bile-soluble) control: *Streptococcus (Diplococcus) pneumoniae* Negative (bile-insoluble) control: an alpha-hemolytic streptococcus
Bibliography	Anderson, A. B., and Hart, P. D.: The lysis of pneumococci by sodium deoxycholate. Lancet, 2:359–360, 1934. Blazevic, D. J., and Ederer, G. M.: Principles of Biochemical Tests in Diagnositic Microbiology. Pp. 7–11. New York, John Wiley & Sons, 1975. MacFaddin, J. F.: Biochemical Tests for Identification of Medical Bacteria. Pp. 9–15. Baltimore, Williams & Wilkins, 1976.

Identification of the Pneumococci

The pneumococci are now included with the genus Streptococcus in edition 8 of *Bergey's Manual*,[2] reflecting basic similarities between pneumococci and streptococci, and conforming with the terminology used in Europe for a number of years.

Pneumococci are normal inhabitants of the upper respiratory tract and for that reason their recovery from sputum or lower respiratory secretions may be difficult to interpret. Generally the recovery of a predominance of pneumococci from the sputum of susceptible hosts, particularly with radiologic evidence of classic lobar pneumonia, is considered significant but cannnot be judged diagnostic.

The inability to recover pneumococci from upper respiratory secretions or sputum samples in up to 50 per cent of patients with known pneumococcal pneumonia is currently a major diagnostic problem.[1] Inadequate sputum samples, overgrowth of pharyngeal organisms antagonistic to the growth of pneumococci, delay in specimen transport, and use of improper culture procedures are common reasons for the poor recovery rate. Mouse inoculation with sputum samples can increase the rate of recovery of pneumococci by 47 per cent[13]; however, the technique is cumbersome and relatively expensive. Transtracheal aspiration techniques have doubled the rate of recovery of pneumococci from respiratory secretions in one reported study.[4] The addition of 5 µg./ml. of gentamicin to sheep blood agar plates also significantly improves the laboratory recovery of pneumococci from sputum samples by inhibiting the growth of antagonistic bacteria.[4]

Pneumococci have a propensity to cause infections in the lungs, meninges, endocardium, and certain other sites, particularly in hosts with compromised resistance. Alcoholics and older people with debilitating disease are especially susceptible. Prior to the advent of antibiotics, the pneumococcus was known as the "old man's friend"; however, its exquisite susceptibility to the penicillin derivatives makes the rapid cure of most cases of pneumococcus infections possible, particularly if the diagnosis is made early.

Examination of gram-stained smears of sputum, joint fluid, cerebrospinal fluid, and certain other body secretions for the presence of pneumococci can be very helpful. In stains of biologic specimens, the pneumococci appear as gram-positive diplococci, with the pairs of cocci tapered in the shape of a lancet (Plate 7-3, *A* & *B*). In heavy infections, the presence of pneumococci may be confirmed and serotyped by use of the Neufeld quellung reaction. Approximately two thirds of human cases of pneumococcal pneumonia are caused by serotypes 1 to 10.[1] The presence of prominent capsules surrounding the pneumococci is suggestive of virulence.

The quellung reaction is performed by mixing approximately equal quantities of the specimen, that is, sputum, cerebrospinal fluid sediment, *etc.*, with type-specific pneumococcal antiserum. Suitable antisera are available from commercial laboratories in the United States and a pneumococcal "omniserum," reacting with all 82 known pneumococcal types, is available through the Staten Seruminstitut in Copenhagen, Denmark. After adding the antiserum, the mixture is examined microscopically using a $100 \times$ (oil immersion) objective. If the quellung reaction is positive, the capsules of the pneumococci will appear quite prominent compared to the same specimen mixed with saline as a control. The increased prominence of the capsule or "swelling" is apparently due to an alteration of its refractive index from reacting with the antiserum.

Counterimmunoelectrophoresis (CIE), to be described in the next section of this chapter, is also a rapid and useful tool in the identification of soluble capsular polysaccharides in biologic secretions. CIE may also be more sensitive than the quellung capsular swelling technique in the typing of isolates of pneumococci and is very useful in detecting pneumococcal polysaccharides in cerebrospinal fluid or in sputum samples.

More commonly the initial identification of pneumococci is made by observing the appearance of colonies on blood agar after 18 to 24 hours of incubation. Virulent strains with abundant capsular polysaccharide produce moist, mucoid, transparent colonies that tend to run together (see Plate 7-3, *C* & *D*). Poorly encapsulated strains of pneumococci produce small, round, translucent colonies that are initially convex, but with time develop a central depression because of autolysis (see Plate 7-3, *E*). A 2- to 3-mm. zone of alpha hemolysis virtually always surrounds the colonies.

Bile salts, sodium deoxycholate and sodium taurocholate, selectively lyse colonies of *S. pneumoniae* while other streptococci are refractory to these agents. The test may be performed either in a test tube or by adding the bile salt solution directly to the colony, as described in Chart 7-8.

Optochin (ethylhydroxycupreine hydrochloride), a quinine derivative, also has a detergentlike action and causes selective lysis of pneumococci. The optochin test is most conveniently performed by using an optochin-impregnated paper disk, similar to that used for antimicrobial susceptibility tests. Characteristically, a growth-inhibition zone with a size 15 mm. or more in diameter is obtained with *S. pneumoniae*. Other alpha-hemolytic streptococci have zones less than 10 mm. in diameter (Plate 7-3, *H* and *I*). The principle of the optochin susceptibility test and the laboratory procedure are shown in Chart 7-9.

DETECTION OF BACTERIAL ANTIGENS AND ANTIBODIES IN BODY FLUIDS

Microscopic examination of gram-stained preparations and culture techniques still serve as the foundation for the laboratory diagnosis of infectious disease. However, in some acute infections these techniques may be misleading or too slow to provide the physician with sufficient guidelines to insti-

Chart 7-9.   *Optochin Susceptibility Test*

Introduction	Ethylhydroxycupreine hydrochloride (Optochin), a quinine derivative, selectively inhibits the growth of *S. pneumoniae* in very low concentrations (5 μg./ml. or less). Optochin may inhibit other alpha-hemolytic streptococci, but only at higher concentrations.
Principle	Optochin is water-soluble and diffuses readily into agar medium. Therefore, filter paper disks impregnated with optochin can be used directly on the surface of agar plates in the optochin test. Optochin has a detergentlike action. Pneumococcal cells exposed to the Optochin which has diffused into the medium surrounding the disk are lysed due to changes in surface tension and a zone of growth inhibition is produced.

Media and Reagents
1. Well-isolated colonies of the organism to be tested on an agar medium
2. A 5% sheep blood agar plate
3. Optochin disk (5 μg.):
   A. Difco: Bacto-differentiation disks, Optochin
   B. BBL: Taxo "P" disks
   Disks should be stored at 4° C. when not in use.

Procedure
1. Using an inoculating loop or wire, select a portion of a well-isolated colony of the organism to be tested and streak an area 3 cm. in diameter on a 5% sheep blood agar plate.
2. Place an optochin disk in the center of the streaked area. With sterile forceps, gently press the disk so that it adheres to the agar surface.
3. Invert the plate and incubate at 35° C. for 18 to 24 hours. (*Note:* use of a CO_2 incubator should be avoided in that zone sizes may be altered due to the acid *p*H shift on the surface of the medium.)

Interpretation
The growth of *S. pneumoniae* is characteristically inhibited by the Optochin disk (see Plate 7-3, *H* and *I*). Usually the zone of inhibition will be 15 mm. or greater; any organism showing a zone of 10 mm. or less should be tested for bile solubility. There are occasional bile-insoluble strains of alpha-hemolytic streptococci that may show a narrow zone of growth inhibition around the Optochin disk.

Controls
Positive control: *Streptococcus (Diplococcus) pneumoniae*
Negative control: *Streptococcus faecalis*

Bibliography
Blazevic, D. J., and Ederer, G. M.: Principles of Biochemical Tests in Diagnostic Microbiology. Pp. 87–89. New York, John Wiley & Sons, 1975.
Bowen, M. K., *et al.:* The Optochin sensitivity test: a reliable method for identification of pneumococci. J. Lab. Med., 49:641–642, 1957.
Lund, E.: Diagnosis of pneumococci by the optochin and bile tests. Acta Pathol. Microbiol. Scand., 47:308–315, 1959.

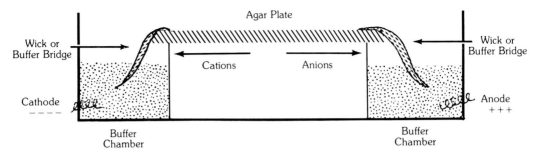

Fig. 7-1. Electrophoresis apparatus. Diagram of CIE apparatus showing agar plate, buffer chambers, electrolyte wicks, and power supply leads. (From Earl Edwards, U.S. Naval Research Laboratory, San Diego, California)

tute specific therapy. Countercurrent immunoelectrophoresis (CIE) has emerged as a simple and accurate technique by which a number of infectious diseases can be quickly diagnosed on the basis of soluble antigens or type-specific antibodies detected in body fluids of patients.

Developed initially in 1970 for detection of hepatitis B antigen, CIE was soon found to be suitable for detecting the antigens of *S. pneumoniae* in the serum of patients with pneumonia and *Neisseria meningitidis* antigen in patients with fulminant meningococcemia.[14] CIE has been particularly useful in the immediate diagnosis of pneumococcal, meningococcal, and hemophilus type b meningitis by allowing direct detection of polysaccharide antigens in the spinal fluid. Some success has also been reported in the detection of antigens of staphylococci, pseudomonas, and klebsiellae. Potential use of CIE for detecting antigens of nonbacterial agents such as *Cryptococcus neoformans* and antibodies to parasitic and viral agents is currently under investigation.

CIE is also a convenient technique for identifying bacterial type-specific antibodies in culture extracts, instead of using the more conventional Lancefield precipitin technique.

Basic Principles of Counterimmunoelectrophoresis (CIE)

The basic principles of electrophoresis and immunodiffusion are combined in CIE. In a suitably buffered medium of proper ionic strength and *p*H, proteins (including antigens and antibodies) become negatively charged and migrate toward the anode in an electrical field. The apparatus used for CIE consists of an agar plate that is supported above two buffer chambers, as shown in Figure 7-1.

A filter paper or cloth wick connects the agar surface with the buffer solution in each chamber. Electrical connections are made from each buffer chamber, one attached to the positive pole (anode) of a constant voltage power supply, the other attached to the negative pole (cathode).

Agarose, 1 per cent, is used for the diffusion medium, and sodium barbitol, *p*H 8.2 to 8.6, ionic strength 0.015, is the buffer solution added to the chambers. The agarose plates are prepared with barbitol buffer of ionic strength 0.075. with a series of small wells punched out of the agar as shown in Figure 7-2. These wells are spaced 5 mm. apart from center to center.

One set of wells receives the antigen (body fluid), the other receives the type-specific antibodies (antisera). The agar plate is placed in the apparatus so that the antigen wells are toward the cathode side and the antibody wells are on the anode side. At a buffer *p*H of 8.2 or 8.6, both the antigens and the antibodies become negatively charged; however, the antibodies carry much less negative charge than the antigens. When a current is applied to the electrophoresis system, the negatively charged antigens migrate toward the anode. The antibodies, being far less negatively charged and

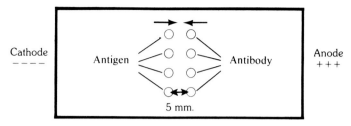

Endosmotic Ion Flow

Fig. 7-2. Diagram showing migration of antigen and antibody in CIE agarose plate.

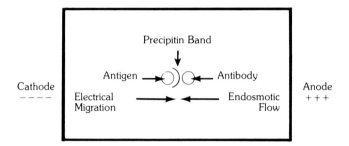

Fig. 7-3. Diagram showing the formation of precipitin bands.

almost neutral in electrical potential, migrate virtually not at all toward the anode and consequently are swept toward the cathode by the counter endosmotic flow of the buffer ions. At some point between the wells, the migrating antigens and antibodies meet. If they are antigenically homologous, a distinct precipitin band will form (see Fig. 7-3).

Suitable antisera are available from the following sources:

H. influenzae (types A–F):	Hyland Laboratories, Costa Mesa, Cal.
Pneumococcal omni serum and pools A–I:	Statens Seruminstitut, Copenhagen, Denmark
N. meningitidis, types A–D, x, Y, and Z:	Burroughs-Wellcome, Research Triangle Park, N.C.
Streptococcal Groups A–G:	Burroughs-Wellcome, Research Triangle Park, N.C.

Electrical power units and disposable agarose plates are available from Hyland Laboratories, although almost any electrophore-

sis unit can be used. A fluorescent illuminator is helpful for interpreting the plates. Purified agarose is available from the Sigma Chemical Company, Saint Louis, Missouri.

This CIE technique is rapid, sensitive, and quite suitable for detecting bacterial antigens in spinal fluid and serum, especially in the diagnosis of pneumococcal, meningococcal, and *H. influenzae* infections. It can also be used for Lancefield grouping of streptococci, using either autoclaved or acid extracts of the bacterial cells.

REFERENCES

1. Barrett-Conner, E.: The nonvalue of sputum culture in the diagnosis of pneumococcal pneumonia. Am. Rev. Respir. Dis., *103:*845–848, 1971.
2. Buchanan, R. E., and Gibbons, N. E.: Bergey's Manual of Determinative Bacteriology. ed. 8. Baltimore, Williams & Wilkins, 1974.
3. Christie, R., Atkins, N. E., and Munch-Petersen, E.: A note on a lytic phenomenon shown by group-B streptococci. Aust. J. Exp. Biol., *22:*197–200, 1944.

4. Dilworth, J. A., *et al.:* Methods to improve detection of pneumococci in respiratory secretions. J. Clin. Microbiol., *2:*453–455, 1975.

5. Facklam, R. R.: Isolation and Identification of Streptococci. Part II: Extraction and Serological Identification. Atlanta, Center for Disease Control, 1977.

6. Facklam, R. R., *et al.:* Presumptive identification of groups A, B and D streptococci. Appl. Microbiol., *27:*107–113, 1974.

7. Hwang, M., and Ederer, G. M.: Rapid hippurate hydrolysis method for presumptive identification of group-B streptococci. J. Clin. Microbiol., *1:*114–115, 1975.

8. Lancefield, R. C.: A serological identification of human and other groups of hemolytic streptococci. J. Exp. Med., *57:*571–595, 1933.

9. MacFaddin, J. F.: Biochemical Tests for Identification of Medical Bacteria. Baltimore, Williams & Wilkins, 1976.

10. Maxted, W. R.: The use of bacitracin for identifying group-A haemolytic streptococci. J. Clin. Pathol., *6:*224–226, 1953.

11. Moody, M. D., *et al.:* Fluorescent antibody identification of group-A streptococci from throat swabs. Am. J. Public Health, *53:*1083–1092, 1963.

12. Murray, P. D., *et al.:* Bacitracin differentiation for presumptive identification of Group-A b-hemolytic streptococci. Comparison of primary and purified plate testing. J. Pediatr., *89:*576–579, 1976.

13. Rathgun, K. H., and Govani, I.: Mouse inoculation as a means of identifying pneumococci in sputum. Johns Hopkins Med. J., *120:*46–48, 1967.

14. Rytel, M. W.: Counterimmunoelectrophoresis in diagnosis of infectious disease. Hosp. Pract., *10:*75–82, 1975.

15. Shands, J. M., Jr.: The physical structure of bacterial lipopolysaccharides. *In* Weinbaum, G., Gadis, S., and Ajl, S. J. (eds.): Microbial Toxins. Vol. 4, pp. 127–144. New York, Academic Press, 1971.

8 Neisseria

The Neisseria are gram-negative cocci that in stained smear preparations characteristically occur in pairs with adjacent sides flattened, simulating kidney beans (Plate 8-1, *A*). There are morphologic and other similarities between the genera Neisseria, Moraxella, and Acinetobacter, all of which are taxonomically included in the family Neisseriaceae. For this reason, observation of gram-negative, bean-shaped diplococci in gram-stained preparations of clinical material cannot be used as the sole criterion for the identification of Neisseria.

The Neisseria include a number of species, most of which are relatively fastidious, requiring enriched media such as chocolate agar to provide a rich source of iron. There is species variation in specific requirements for amino acids, purines, pyrimidines, and vitamins, which can be used as the basis for auxotyping, an identification tool useful in epidemiologic studies.[3,18] A relatively high humidity and an ambient CO_2 in the range of 5 to 10 per cent are required by most species for growth. Some Neisseria are especially susceptible to drying and grow only within narrow temperature (35°C. to 37°C.) and *p*H (7.2 to 7.6) ranges. All Neisseria produce catalase and cytochrome oxidase, important characteristics which aid in differentiating them from morphologically similar bacteria. Some species produce xanthophyll pigments, and colonies of these may appear yellow or brown.

Neisseria catarrhalis is now called *Branhamella catarrhalis* as suggested by Catlin,[4] because this organism has differences in growth requirements and in biochemical, physiologic, and genetic properties which exclude it from the genus Neisseria. Only *N. gonorrhoeae* and *N. meningitidis* frequently cause significant disease in man. Although other species of Neisseria are frequently encountered in the laboratory as part of the "normal flora," these are rarely pathogenic in man and will not be discussed here.

NEISSERIA GONORRHOEAE

Gonorrhea, primarily an infectious disease of the urogenital tract, has reached an epidemic proportion in modern society. An estimated one to five million new cases occur annually in the United States alone and the prevalence rate may be even higher in other parts of the world.

Humans serve as the only natural host for *N. gonorrhoeae,* and the disease is transmitted almost exclusively by sexual contact; therefore, the disease has no social or geographic boundaries. Its current incidence is particularly high among sexually active teenagers and young adults.

Although the disease may involve only the mucous membranes of the uterine cervix or urethra, salpingitis and bartholinitis may develop in females or epididymitis and periurethral abscess may develop in males through local spread of the bacteria.

Other mucous membranes, including the anal canal, oropharynx, and conjunctiva, may also be infected with *N. gonorrhoeae.* Less commonly, disseminated gonococcal infections may occur, manifesting as arthritis, cutaneous lesions, and septicemia. Endocarditis, meningitis, and hepatitis caused by *N. gonorrhoeae* are rare.

Gonococcal infections are not always symptomatic, and both males and females may serve as carriers. This makes eradication of the disease virtually impossible. Therefore, it is necessary to culture healthy and asymptomatic individuals as well as subjects included within high-incidence groups if epidemiologic control measures are to be effective. In males, the scant urethral secretions may be insufficient for culture and prostatic massage may be indicated. Culture of the urine may allow isolation of *N. gonorrhoeae* when urethral cultures are negative.[6] The first morning sample of urine should be collected, preferably after a 12-hour overnight water fast. The initial 10 to 20 ml. of the sample should be examined, and mucin strands in particular selected for Gram's stain and culture since the microorganisms tend to become trapped within these fibers.

The diagnosis of clinical gonorrhea requires close cooperation between the physician and the laboratory. Particular attention must be paid to the following:

1. Correct specimen collection by the physician
2. Selection of the appropriate transport container and prompt delivery to the laboratory
3. Selection of the appropriate culture medium
4. Recovery and correct identification of the organism in the laboratory

Specimen Collection

Because the gonococcus is so susceptible to drying, specimens for culture must be placed immediately into a transport container or transferred directly to appropriate culture media. The use of disinfectants or lubricants must be avoided when preparing the patient because even trace amounts of certain chemicals can be injurious to the organisms that may be present in small quantities in some infections.

In females, the endocervix is the optimal site for recovery of *N. gonorrhoeae*, and samples should be taken by direct vision through a speculum because the adjacent vaginal mucosa harbors other microorganisms that may inhibit the growth of *N. gonorrhoeae* in mixed culture. Any excess cervical mucin should be removed with gauze or cotton held in a ring forceps. The swab used to collect the specimen should be inserted into the endocervical canal, moved gently from side to side, and left in the canal for 10 to 30 seconds to allow the swab to absorb the secretions. Because cotton fibers may contain fatty acids inhibitory to *N. gonorrhoeae*, Dacron or calcium alginate swabs are recommended for the collection of these specimens, particularly if such swabs are to be placed in a transport medium.

The diagnosis of gonorrhea can be made in approximately 80 per cent of infected females by endocervical cultures alone. However, the rate of recovery of *N. gonorrhoeae* from females, especially asymptomatic patients, is increased even more if the urethra and anal canals are also cultured.

In males with suspected gonorrhea, the best specimen to collect is usually urethral exudate. In homosexuals, however, rectal and oropharyngeal samples should also be taken.

Samples may be collected from the urethra using either a platinum inoculating loop about 2 mm. in diameter or a thin calcium alginate urethral swab, inserting either of these for a short distance into the urethral orifice. Urethral specimens should not be collected until at least one hour has passed after the patient has urinated.

Rectal specimens are somewhat more difficult to obtain without contaminating the swab with feces. The swab should first be moistened with sterile water and inserted into the anal canal just beyond the anal sphincter. Allow sufficient time (10 to 30 seconds) for material to adsorb into the fibers of the swab before removing it.

Direct smears for Gram's stain should be prepared at the time each of these specimens is collected. These smears can allow an immediate presumptive identification of suspicious bacterial cells and may also serve to

Color Plate
8-1

Plate 8-1
Identification of Neisseria Species

A.
Photomicrograph of a Gram's stain of a purulent urethral exudate revealing a cluster of tiny, gram-negative, intracellular diplococci in a neutrophil, characteristic of *Neisseria gonorrhoeae*.

B.
Two chocolate agar plates, one uninoculated, the other illustrating dull, gray-white, opaque colonies of *N. gonorrhoeae* on chocolate agar plate incubated in CO_2 (left), compared to no growth on the plate incubated in ambient air, illustrating CO_2 dependence of this organism.

C.
Chocolate agar with multiple colonies of *N. gonorrhoeae* in pure culture. These colonies are not distinctive, and a cytochrome oxidase test, Gram's stain and additional identification characteristics must be determined before a species identification can be established.

D, E and F.
Gonococcal colony types. Frame *D* illustrates a Type 2 colony having a relatively small size and a dark, defined edge. Type 1 colonies are approximately the same size as Type 2 colonies but generally are more transparent in appearance and lack the dark edge. Both Type 1 and Type 2 are known to be infective for man. Type 3 colonies are larger than Type 1 and Type 2 (Frame *E*) colonies and generally have a dark appearance, although light variants are seen. Type 4 colonies are slightly smaller than Type 3 colonies but larger than either Type 1 or Type 2 colonies. Type 4 colonies are usually colorless or have a slight suggestion of pale yellow (Frame *F*). Type 3 and Type 4 colonies are noninfective and are thought to be cultural variants of Type 1 and Type 2 colonies. The identification of colonial types has not found practical applications in most diagnostic laboratories, but have been found to be useful in applied research.

G.
Gram's stain of a smear made from *N. gonorrhoeae* from a 48-hour growth on chocolate agar, illustrating small, gram-negative cocci arranged in pairs.

H.
Cytochrome oxidase test strips illustrating the blue color of a positive reaction in the zone where growth from a bacterial colony had been applied. An uninoculated control strip is next to the positive strip.

I.
Carbohydrate utilization reactions of *N. gonorrhoeae* in CTA semisolid agar, revealing acid production (yellow color) in only the tube containing glucose.

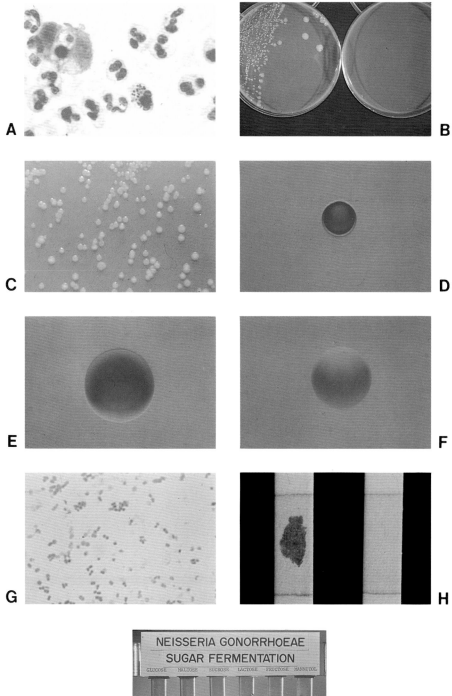

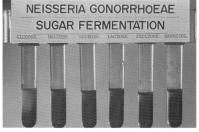

indicate whether negative cultures should be repeated. Specific antimicrobial therapy for gonorrhea may be instituted in males on the basis of classic symptoms and the presence of typical intracellular gram-negative diplococci in direct smears. However, identification of the gonococcus in gram-stained smears from females is less reliable because of "look-alike" bacteria belonging to the Moraxella and Acinetobacter groups, as discussed above. All presumptive Gram's stain identifications should be confirmed by culture and appropriate differential tests.

Specimen Delivery to the Laboratory and Use of Transport Media

Ideally, all specimens collected for the recovery of *N. gonorrhoeae* should be inoculated immediately to appropriate culture media and placed in a CO_2 incubator at 35° C. However, because most patients with suspected gonorrhea are seen in physicians' offices or clinics where the equipment and trained personnel necessary to carry out proper culture techniques may be lacking, a variety of transport systems have been designed for the recovery of *N. gonorrhoeae*.

The Transgrow bottle is currently one of the more popular transport systems used for the recovery of *N. gonorrhoeae* from clinical specimens in the United States.[14] This system consists of a screwcap bottle containing an agar slant of modified Thayer Martin medium (MTM) and 10 per cent CO_2.[12] The original Thayer-Martin medium,[21] composed primarily of chocolate agar to which was added ristocetin and polymyxin B, was designed to selectively grow *N. gonorrhoeae* (or other neisseria) from clinical specimens containing a mixed bacterial flora. Following removal of ristocetin from the market, vancomycin, colistin, and nystatin were found to work as effectively and are currently used in the modified Thayer-Martin medium (MTM).

In the most recent modification of the Thayer-Martin medium,[12] the agar concentration has been increased from 1 per cent to 2 per cent to reduce the swarming effect of

species of Proteus if present in specimens, and the glucose content has been increased from 0.1 per cent to 0.25 per cent to enhance the growth of *N. gonorrhoeae*. Trimethoprim lactate has also been advocated to reduce the growth and swarming of species of Proteus; however, this antibiotic has been found to inhibit the growth of occasional strains of *N. gonorrhoeae* as well.

The procedure listed below should be followed when inoculating the Transgrow bottle:

1. Allow the medium to come to room temperature before use.

2. Remove the screwcap of the bottle only when ready to inoculate the medium. During inoculation, keep the bottle in the upright position at all times to reduce the loss of CO_2 (heavier than air).

3. Inoculate the surface of the agar quickly by rolling the swab in a Z-shaped motion starting at the bottom of the bottle.

4. Immediately replace the lid and tighten it securely.

5. For optimal recovery of *N. gonorrhoeae*, it is recommended that the Transgrow bottle be incubated at 35° C. for 12 to 24 hours before shipment to a reference laboratory if prompt delivery is not possible.

When receiving Transgrow bottles, reference laboratories are urged to loosen the lid and place the bottle immediately into a CO_2 incubator for an additional 24 to 48 hours of incubation before discarding the culture as negative. These bottles can cause problems for reference laboratories if they have not been incubated before shipment, and it cannot always be guaranteed that the bottle was not tipped or only loosely capped, allowing CO_2 to escape during transport.

JEMBEC is a recently marketed commercial transport system that overcomes some of the disadvantages of the Transgrow bottle for the recovery of *N. gonorrhoeae*.[13,19] Thin, flat, rectangular plastic plates are used instead of the thick glass bottles, making observation of colonies and subculturing from these plates to secondary media less difficult. These plates contain modified Thayer-Martin (MTM) medium as described above and are packaged in plastic bags. They can be stored in the refrigerator for many weeks. After inoculating the medium, the CO_2 generating tablet is placed in a special well in the plastic plate. The plate is returned to the bag, which is tightly sealed. The humidity within the bag decomposes the tablet into gaseous CO_2. This provides the proper atmosphere during transport. The JEMBEC system is slightly more expensive than Transgrow bottles, but may result in an improved recovery of *N. gonorrhoeae* in situations in which the use of the latter cannot be properly controlled.*

Selection of Primary Isolation Medium

A variety of culture media have been recommended for the primary recovery of *N. gonorrhoeae* from clinical specimens. MTM medium is described above, and because of its selective nature it is most useful for the recovery of *N. gonorrhoeae* from specimens contaminated with other species of bacteria that may suppress the growth of the gonococcus. For clinical materials such as synovial fluid or cerebrospinal fluid, chocolate agar supplemented with yeast extract and without antibiotics is an excellent primary recovery medium. Because the antibiotics contained in MTM medium have been found to inhibit the growth of certain strains of *N. gonorrhoeae*, noninhibitory chocolate agar without antibiotics should be set up in parallel, particularly with patients whose cultures have been negative in the face of strong clinical evidence of gonorrhea.

A number of other culture media have recently been developed that are said to be equal or superior to MTM medium for the isolation of *N. gonorrhoeae*. Many of these use IsoVitaleX enrichment (BBL), which contains a variety of vitamins, amino acids, purines, and other supplements. Martin,

Billings, Hackney, and Thayer[15] reported a 13 per cent increase in the recovery of *N. gonorrhoeae* by supplementing MTM medium with this enrichment. Jones and Talley[8] have reported good results with a simplified Texas Tech University (TTU) medium that has fewer supplements than supplied by IsoVitaleX. In addition to supporting good growth of *N. gonorrhoeae*, TTU medium has the advantage that it is less expensive than MTM medium, a distinct cost saving when large batches of media are used in research studies involving nutritional, genetic, and physiologic experiments. Payne and Finkelstein[17] advocate the substitution of Imferon, an iron-dextran complex, for the ferric nitrate in the IsoVitaleX formula. In their studies, the addition of Imferon to GC primary recovery media led to enhanced growth of both gonococci and meningococci, and a three- to six-fold increase in the diameter of their colonies was observed.

Laboratory Identification of *Neisseria gonorrhoeae*

For practical purposes, the diagnosis of gonorrhea can be confirmed in males or females with classical clinical symptoms if the following criteria are met:

---

1. The recovery of opaque, convex, gray-white, glistening colonies on MTM medium (or its equivalent) after 24 to 48 hours of incubation at 35° C. under increased CO_2 tension (Plate 8-1, *B* and *C*).
   (*Note:* four different types of gonococcal colonies have been described.[2,9] Colony types 1 and 2, recovered in primary cultures from clinical materials, are pileated in appearance and infective for man; types 3 and 4 are nonpileated mutants observed *in vitro* after multiple subcultures, and are not virulent (Plate 8-1, *D* through *F*). In order to identify these variants, it is necessary to examine colonies with a dissecting microscope and special reflected light. The ability to differentiate these colonial types has found little clinical application in most clinical laboratories as yet, but has been of some use in genetic transformation studies.[11])

---

*The JEMBEC system is available from the Ames Co., Division of Miles Laboratories, Inc., Elkhart, Indiana 46514, and the ANA/MED Laboratories, Inc., Maryland Heights, Missouri 68043.

Table 8-1. *Differentiation of N. gonorrhoeae, N. meningitidis, and N. lactamica With Carbohydrate-Utilization Tests*

Species	Carbohydrate-Utilization Tests			
	Glucose	Maltose	Sucrose	Lactose
N. gonorrhoeae	+	−	−	−
N. meningitidis	+	+	−	−
N. lactamicus	+	+	−	+

2. Demonstration that the characteristic colonies on MTM medium are composed of typical gram-negative diplococci by examining a gram-stained smear (Plate 8-1, *G*)
3. The colonies of gram-negative cocci on MTM medium are cytochrome-oxidase positive (Plate 8-1, *H*).

N. gonorrhoeae and *N. meningitidis* cannot be differentiated from each other by the above criteria. Although infections involving these two species can generally be differentiated on clinical grounds, this is not always the case. *N. meningitidis* is occasionally recovered from the genitourinary tract and *N. gonorrhoeae* can be recovered from the oropharynx and from cerebrospinal fluid in rare cases of gonococcal meningitis. Definitive identification of isolates of Neisseria from these sources should therefore always be made, in addition to isolates recovered from specimens of joint effusions, cutaneous pustules, or from blood cultures in suspected septicemia.

Clinical microbiology laboratories should have the capability of differentiating Neisseria species, particularly *N. gonorrhoeae* and *N. meningitidis*. *N. lactamicus* should also be identified although usually it is avirulent. This species can occasionally be recovered from cerebrospinal fluid and confused with *N. meningitidis* because of biochemical similarities, particularly in carbohydrate-utilization studies. The differential reactions are listed in Table 8-1 (Plate 8-1, *I*).

Cystine trypticase agar (CTA) is the basal medium most commonly used to determine carbohydrate utilization by the Neisseria. This medium supports the growth of most strains of *N. gonorrhoeae* and *N. meningitidis* that are recovered from clinical specimens. The details of the CTA carbohydrate-utilization test are presented in Chart 8-1.

Difficulties may be encountered in interpreting the CTA carbohydrate tests if certain details in the test procedure are not followed as outlined in Chart 8-1. Following are precautions that should be taken:

1. Tubes with a size greater than 13 × 100 mm. should not be used. A larger surface area may result in false negative reactions.
2. A 24-hour subculture of the organism to be tested should be used. False negative results may occur if organisms are tested directly from primary culture since carbohydrate utilization may be demonstrated only with subculture of some isolates.
3. Make sure that only the top 5 mm. of the medium is inoculated with a large number of organisms. Most species of Neisseria are obligate aerobes and grow only on the surface where there is exposure to ambient air. The semisolid consistency of the medium tends to restrict the acids formed to a narrow band just below the surface.
4. It is important that the lids of the tubes be tightly closed after inoculation, particularly if they are incubated in a CO_2 atmosphere. If CO_2 enters the tubes, the carbonic acid formed may be sufficient to drop the pH below 6.8. This will give a false positive reaction.
5. Allow sufficient time for incubation, at least 72 hours, before interpreting reactions as negative. Reactions may be accelerated by incubating the tubes in a 35° C. water bath or heating block.
6. If irregular or equivocal results occur, subculture the test organism to be sure that contamination with an unwanted species has not occurred.

Chart 8-1. *Carbohydrate Utilization by Neisseria Species, Cystine Trypticase Agar (CTA) Method*

Introduction

Cystine trypticase agar (CTA) supports the growth of many fastidious organisms, including species of Neisseria. Since the medium contains no meat or plant extracts and is bacteriologically tested by the manufacturers for the absence of fermentable carbohydrates, CTA is particularly well suited for use in carbohydrate-utilization studies.

Principle

CTA must be autoclaved prior to the addition of carbohydrates because some carbohydrates are degraded by heat. Various carbohydrates can be added to the CTA base to a final concentration of 1%, using stock solutions that have been sterilized by passing them through a Millipore filter. Phenol red indicator is included in the medium and will develop a yellow color in the presence of acid, at a *p*H of 6.8 or less.

Media and Reagents

CTA Basal Medium:

L-Cystine	0.500 g.
Peptone	20.000 g.
Sodium chloride	5.000 g.
Sodium sulfite	0.500 g.
Agar	2.500 g.
Phenol red	0.017 g.
Distilled water to:	1000 ml.

*p*H = 7.3

Dispense 2.7 ml. of medium into 13 × 100-mm. screwcap tubes while still warm after autoclaving. Store at 4° C. until time of use.

Carbohydrate Solutions:
Prepare 10% stock solutions of the various carbohydrates to be tested in distilled water and sterilize by filtration. Dispense in small quantities (1 to 2 ml.) into separate tubes and store at −20° C. until time of use.

Heavy Bacterial Suspension:
Inoculate a plate of chocolate agar supplemented with yeast extract but free of antibiotics with the isolate to be tested and incubate for 18 to 24 hours. Harvest the entire growth on the plate in approximately 0.5 ml. of distilled water.

Procedure

1. Prewarm as many of the tubes containing CTA medium as the number of carbohydrates to be tested. To each tube add 0.3 ml. of the stock carbohydrate solutions.
2. With the tip of a Dacron or calcium alginate swab, sample the heavy bacterial suspension prepared just prior to the performance of the test.
3. Using the swab, inoculate only the upper 5 mm. of the medium with the bacterial suspension. Use a separate swab to inoculate each carbohydrate to be tested.
4. Fasten the caps of the tubes tightly and place in a 35° C. incubator, preferably one without CO_2, and incubate for at least 72 hours before discarding as negative.

Chart 8-1. *Carbohydrate Utilization by Neisseria Species, Cystine Trypticase Agar (CTA) Method (Continued)*

Interpretation	The development of a yellow band in the upper portion of the medium indicates production of acid and is interpreted as a positive test for utilization of the carbohydrate present (Plate 8-1, *I*). Although reactions may occur as early as 24 hours after inoculation, some reactions are delayed and negative results should not be interpreted before 72 hours of incubation. *N. gonorrhoeae* utilizes only glucose but *N. meningitidis* utilizes glucose and maltose with the production of acid.
Controls	Positive glucose: *N. gonorrhoeae* Positive maltose: *N. meningitidis* Positive lactose: *N. lactamicus* Negative control: *Branhamella catarrhalis*
Bibliography	Thayer, J. D.: *Neisseria gonorrhoeae* (gonococcus). *In* Lennette, E. H., Spaulding, E. H., and Truant, J. P. (eds.): Manual of Clinical Microbiology. Washington, D.C., American Society for Microbiology, 1974.

Brown[1] has described a rapid procedure to determine the carbohydrate utilization of Neisseria species as an alternative to the CTA test. This procedure utilizes a heavy bacterial suspension and buffered salt solution and gives results in only 4 hours. The procedure is performed as follows:

Reagents

1. Buffered Salt Solution (BSS):

K_2HPO_4	.04 g.
KH_2PO_4	.01 g.
KCl	.80 g.
Phenol red, 1% aqueous	.20 ml.
Distilled water to:	100 ml.

2. Carbohydrate Solutions:

   Prepare 20 per cent stock solutions of the various carbohydrates to be tested, using sterile distilled water. Disperse in small quantities (1 to 2 ml. per tube) and store at $-20°$ C.

3. Heavy Bacterial Suspension:

   Grow a pure isolate of the organism to be tested on chocolate agar. After 18 to 24 hours of incubation, harvest the entire surface growth into 0.3 ml. of BSS.

Procedure

1. To tubes containing 0.1 ml. of BSS, add one drop of each carbohydrate solution to be tested and one drop of the heavy bacterial suspension. Mix well.
2. Place all tubes in a 35° C. water bath or heating block.
3. Read for acidification (yellow color) after exactly 4 hours of incubation.

Positive and negative control strains of bacteria should be used to test each new lot of media and newly prepared carbohydrate solutions.

Slifkin and Pouchet[19] have suggested another approach to test isolates of Neisseria to see how they use carbohydrates. They report an uncomplicated sensitive test that traps the CO_2 released from the utilization of carbohydrate with barium hydroxide, producing a visible white precipitate of barium carbonate. A radiometric system is also commercially available that measures the amount of $^{14}CO_2$ metabolized by isolates of Neisseria from ^{14}C-labeled carbohydrate test vials.

Another concern in the laboratory evaluation of gonococcal disease is the recent emergence of penicillinase-producing strains. As of 1976, twelve strains of penicillinase-producing gonococci had been isolated in the United States,[16] eleven linked to individuals traveling in the Far East, where the penicillin-resistant strains were first reported. Although the problem of penicillin resistance by gonococci is currently not widespread in the United States, laboratories should be alerted to the fact that susceptibility tests may be required in certain clinical situations.

Penicillinase-producing strains of *N. gonorrhoeae* can be detected by using a 10-unit penicillin susceptibility disk placed on a chocolate agar plate which has been inoculated by the method used for the Kirby-Bauer procedure. Zones of inhibition less than 20 mm. in diameter suggest a penicillinase-producing strain.[10] Otherwise, the Bauer-Kirby disk diffusion susceptibility test should not be used for gonococci since the procedure has not been standardized for these organisms. The direct test for detection of beta-lactamase was discussed in Chapter 5.

NEISSERIA MENINGITIDIS

The incidence of meningococcal disease in the United States continues to decline, so that the average clinical laboratory may encounter relatively few cases each year. In 1974 only 1273 cases were reported in civilians,[7] a 7 per cent decline from 1973. The majority of cases were found in patients in the age groups of less than 1 year and 1 to 4 years. The development of bactericidal and/or other protective antibodies following asymptomatic infections may serve to protect older individuals.

N. meningitidis may be classified antigenically into four groups, A, B, C, and D. Groups A and C meningococci are encapsulated and produce mucoid colonies on agar media; group B meningococci are not encapsulated and must be identified serologically with agglutination rather than quellung reactions. Meningococci that do not react serologically with the A, B, C, and D subgroups are given other designations, such as Groups X, Y, and Z. In endemic meningococcal disease, Groups B (45 per cent), C (32 per cent), and Y (18 per cent) are the most common isolates.[15a]

Laboratory personnel should be aware that a significant percentage of meningococcal strains are resistant to sulfonamides. Jacobson[7] reports that in 1974, 23 per cent of all isolates referred to the Center for Disease Control were resistant to sulfonamides, with the following group distribution: group A, 33 per cent; group B, 4.2 per cent; group C, 68 per cent; group Y, 1.4 per cent.

Meningococcal meningitis is potentially a serious infection because it may be a manifestation of generalized disease. A petechial rash may develop in meningococcemia and organisms can be recovered from the skin lesions. Disseminated intravascular coagulation (DIC), a process in which diffuse clotting occurs within small arterioles and capillaries, is a common and dangerous complication of meningococcemia, resulting in consumption of circulating blood-clotting factors and potential fatal hemorrhage. Hemorrhage into the adrenal glands, known as the Waterhouse-Friderichsen syndrome, resulting in shock and death, can be the result of meningococcal septicemia.

Gram-negative diplococci may be seen in Gram's stains of centrifuged cerebrospinal fluid sediment, and, when accompanied by numerous polymorphonuclear leukocytes, a presumptive diagnosis of meningococcal meningitis can be made. If the infection is heavy and many organisms are present, the diagnosis may be further confirmed by demonstrating a positive quellung reaction, using type-specific antisera. The detection of capsular polysaccharide from the cerebrospinal fluid by the counterimmunoelectrophoresis technique described in Chapter 7 can also provide a rapid presumptive diagnosis. Culture confirmation, however, should always be sought.

N. meningitidis is extremely sensitive to chilling and dehydration, and specimens should be inoculated to prewarmed chocolate agar or MTM medium as soon after collection as possible. Specimens should not be refrigerated if there is to be a delay in processing; rather, an appropriate transport medium as described above should be employed.

The procedures for the recovery, culturing, and identification of *N. gonorrhoeae* as discussed above also apply to *N. meningitidis*. Selective antibiotic-containing media should be used in conjunction with nonselective chocolate agar because some strains of *N. meningitidis* are inhibited by some of the antibiotics incorporated in these media. The colony characteristics of *N. meningitidis* cannot always be distinguished from *N. gonorrhoeae*. However, observation of highly mucoid colonies is suggestive of *N. meningitidis*. Definitive identification in most instances requires the use of carbohydrate-utilization tests, following the procedures described above.

Fluorescent antibody-staining techniques are now being used for the identification of *N. gonorrhoeae* and *N. meningitidis*. However, direct fluorescent methods, similar to those currently being used for the identification of Lancefield Group A streptococci in throat cultures, and using commercially available reagents, are generally considered unsatisfactory for use in diagnostic microbiology laboratories. The fluorescein-conjugated globulins are not sufficiently specific for either *N. gonorrhoeae* or *N. meningitidis* and cross-react with organisms belonging to other species and genera of Neisseria. The direct method is currently used in only a few specialized laboratories where personnel are experienced with the technique and reagents are prepared and used under strict quality control.

An indirect fluorescent method[22] designed to detect gonococcal antibodies in the sera of females with uncomplicated disease has been introduced, particularly for the screening of subjects with subclinical infections or who may be acting as carriers. This approach has found only limited application at this time.

REFERENCES

1. Brown, J. W.: Modification of the rapid test for *Neisseria gonorrhoeae*. Appl. Microbiol., *27:*1027–1030, 1974.
2. Brown, J. W., and Kraus, S. J.: Gonococcal colony types. J.A.M.A., *228:*862–863, 1974.
3. Carifo, K., and Catlin, B. W.: *Neisseria gonorrhoeae* auxotyping: differentiation of clinical isolates based on growth response on chemically defined media. Appl. Microbiol., *26:*223–230, 1973.
4. Catlin, B. W.: Transfer of the organism named *Neisseria catarrhalis* to *Branhamella gen. nov.* Int. J. System Bacteriol., *20:*155–159, 1970.
5. Cherry, W. B.: Immunofluorescent techniques. *In* Lennette, E. H., Spaulding, E. H., and Truant, J. P. (eds.): Manual of Clinical Microbiology. ed. 2. Chap. 4. Washington, D.C., American Society for Microbiology, 1974.
6. Girsh, I., Karsh, H., and Koneman, E. W.: Asymptomatic chronic gonorrhea in a male patient. Rocky Mt. Med. J., *73:*36–40, 1976.
7. Jacobson, J. A.: Trends in meningococcal disease, 1974. J. Infect. Dis., *132:*480–484, 1975.
8. Jones, R. T., and Talley, R. S.: Simplified complete medium for the growth of *Neisseria gonorrhoeae*. J. Clin. Microbiol., *5:*9–14, 1977.
9. Juni, E., and Heyn, G. A.: Simple method for distinguishing gonococcal colony types. J. Clin. Microbiol., *6:*511–517, 1977.
10. Kellogg, D. S., Jr., Holmes, K. K., and Hill, G. A.: Laboratory Diagnosis of Gonorrhea. Cumitech 4. Washington, D.C., American Society for Microbiology, 1976.
11. Kellogg, D. S., Jr., *et al.*: Neisseria gonorrhoeae I. Virulence genetically linked to colonial variation. J. Bacteriol., *85:*1274–1279, 1963.
12. Martin, J. E., Jr., Armstrong, J. H., and Smith, P. B.: New system for cultivation of *Neisseria gonorrhoeae*. Appl. Microbiol., *27:*802–805, 1974.
13. Martin, J. E., Jr., and Jackson, R. L.: A biological environment chamber for the culture of *Neisseria gonorrhoeae*. J. Am. Vener. Dis. Assoc., *2:*28–30, 1975.
14. Martin, J. E., Jr., and Lester, A.: Transgrow, a medium for transport and growth of *Neisseria gonorrhoeae* and *Neisseria meningitidis*. HSMHA Health Rep., *86:*30–33, 1971.
15. Martin, J. E., Jr., *et al.*: Primary isolation of *Neisseria gonorrhoeae* with a new commer-

cial medium. Public Health Rep., *82:*361–363, 1967.

15. Meningococcal Disease Surveilance Group, Center for Disease Control: Analysis of endemic meningococcal disease by serogroup and evaluation of chemoprophylaxis. J. Infect. Dis., *134:*201-204, 1976.

16. Morbidity and Mortality Weekly Report, Follow-up on penicillinase-producing *Neisseria gonorrhoeae*. Center for Disease Control, *25:*307, 1976.

17. Payne, S. M., and Finkelstein, R. A.: Imferon agar: improved medium for isolation of pathogenic neisseria. J. Clin. Microbiol., *6:*293–297, 1977.

18. Short, H. B., *et al.:* Rapid method for auxotyping multiple strains of *Neisseria gonorrhoeae*. J. Clin. Microbiol., *6:*244–248, 1977.

19. Slifkin, M., and Pouchet, G. R.: Rapid carbohydrate fermentation test for confirmation of the pathogenic neisseria using Ba(OH)$_2$ indicator. J. Clin. Microbiol., *5:*15–19, 1977.

20. Symington, D. A.: Improved transport system for *Neisseria gonorrhoeae* in clinical specimens. J. Clin. Microbiol., *2:*498–503, 1975.

21. Thayer, J. D., and Martin, J. E., Jr.: Improved medium selective for cultivation of *Neisseria gonorrhoeae* and *Neisseria meningitidis*. Public Health Rep., *81:*559–562, 1966.

22. Welch, B. G., and O'Reilly, R. J.: An indirect fluorescent technique for study of uncomplicated gonorrhea. I: Methodology. J. Infect. Dis., *127:*69–76, 1973.

9 The Aerobic Gram-Positive Bacilli

Aerobic gram-positive bacilli are commonly encountered in the clinical microbiology laboratory. The majority of these bacilli are ubiquitous in nature, inhabiting soil, water, and the skin and mucous membranes of various animals, including humans. The virulence of these organisms varies from *Bacillus anthracis,* one of the most highly pathogenic microorganisms known to mankind, to others that are common laboratory contaminants, capable of producing disease only in individuals with compromised host resistance due to underlying disorders.

It is essential for the clinical microbiologist to be able to isolate and learn to recognize the various aerobic gram-positive bacilli and presumptively identify those species that are of particular medical significance. The finding of gram-positive bacilli on direct microscopic examination of a smear made from a clinical specimen or from the broth of a turbid blood culture bottle allows the microbiologist to make only a presumptive identification to the genus level. Such a microorganism could be a member of one of the genera listed below.

Obligate Anaerobes
  Actinomyces
  Arachnia
  Bifidobacterium
  Clostridium
  Eubacterium
  Propionibacterium
Facultatively Anaerobic or Aerobic Bacteria
  Bacillus
  Corynebacterium

  Erysipelothrix
  Lactobacillus
  Listeria
  Nocardia
  Streptomyces

Certain members of this group may demonstrate morphologic characteristics referred to as "diphtheroid," that is, gram-positive bacilli with a distinctive variable morphology as will be described below.

Diphtheroid Gram-Positive Bacilli

The diphtheroid bacterial cells are of varying shapes and sizes, ranging from coccoid to definite rod forms. They often stain unevenly in the Gram's stain procedure. "Chinese figures" or "picket fence" arrangements of the cells are frequently observed in smears, presumably due to "snapping" after the cells divide (Plate 9-1, *A*).[2] Members of the genera Bacillus, Clostridium, and Lactobacillus rarely exhibit diphtheroidal morphology. Some members of the genus Streptococcus, not a bacillus, may appear elongated and pleomorphic in some smears, but usually do not form the characteristic "picket fence" arrangement of the cells. The remaining organisms listed above frequently demonstrate the diphtheroid appearance in Gram's stains. Bifid or branching forms may also be seen with some species.[9] The outline on page 256 can aid in differentiating the gram-positive bacilli that have a diphtheroid appearance.

I. Spore-forming rods, catalase usually produced*   —Bacillus
II. Non-spore-forming rods, not acid-fast; facultatively anaerobic
   A. Catalase-positive
      1. Motile at 25° C. (motility weak or absent at 35° C.)
         a. Transluent colony, beta-hemolytic (13% nonhemolytic[24])   —Listeria
         b. Opaque colony, no hemolysis or alpha hemolysis   —Corynebacterium,
      2. Nonmotile   —Corynebacterium
                                      (majority of species)
   B. Catalase-negative
      1. Major amounts of lactic acid produced, grow in media with a low   —Lactobacillus
         pH (e.g., LBS agar (BBL))
      2. Not as above, "brush growth" in gelatin stab   —Erysipelothrix
III. Non-spore-forming; branched rods or filaments; obligate aerobes
      1. Acid-fast[†]   —Nocardia
      2. Not acid-fast   —Aerobic actinomycetes
                                      other than Nocardia

* A positive catalase test rules out aerotolerant species of Clostridium such as *C. tertium* (see Chap. 10), which are defined as catalase-negative.
† On occasion, acid-fast species of Mycobacterium are encountered which are weakly gram-positive. Branching is sometimes present in mycobacteria, but is not usually a prominent feature.

Subdivision of Gram-Positive Bacilli

The gram-positive bacilli are further subdivided on the basis of their oxygen tolerance. The species listed on page 255 as obligate anaerobes are included in Chapter 10 and will not be discussed further here. The aerobic nocardia and streptomyces are discussed along with the filamentous fungi in Chapter 13. Only the facultatively anaerobic bacteria belonging to the genera Corynebacterium, Erysipelothrix, Listeria, Lactobacillus, and Bacillus will be discussed in this chapter.

The differentiation of this group of facultatively anaerobic organisms to a genus level can be accomplished by observing several morphologic and biochemical characteristics, which are outlined in Table 9-1. Also, the determination of metabolic products by gas-liquid chromatography, quite useful in the identification and classification of the anaerobic bacteria (Chap. 10), appears promising in the differentiation of these facultatively anaerobic gram-positive bacilli as well. For example, the genera Lactobacillus and Listeria produce large quantities of lactic acid, whereas certain species of Corynebacterium produce acetic, formic, and propionic acids and only small amounts of lactic and succinic acids.[25]

The major pathogens in this group of organisms and the diseases they commonly cause are as follows:

Species	Disease
Corynebacterium diphtheriae	Diphtheria
Listeria monocytogenes	Listeriosis
Erysipelothrix rhusiopathiae	Erysipeloid
Bacillus anthracis	Anthrax
Bacillus cereus	Food poisoning

CORYNEBACTERIUM

Classification

The genus Corynebacterium includes a heterogenous group of bacteria of uncertain family affiliation. The type species is *Corynebacterium diphtheriae*. Characteristics this genus have in common include: (1) gram-positive pleomorphic bacilli; (2) no endospore formation; (3) "snapping" when cells divide, which results in "chinese letters"; (4) aerobic or facultatively anaerobic; (5) catalase-positive; and (6) cytochrome-oxidase-negative.

The genus Corynebacterium is currently divided into three major groups in the edition 8 of *Bergey's Manual*[5]:

Group I. Human and animal parasites and pathogens
Group II. Plant-pathogenic Corynebacteria
Group III. Nonpathogenic Corynebacteria

Table 9-1. *Key Differential Characteristics of Aerobic and Facultatively Anaerobic, Gram-Positive Non-spore-Forming Rods**†

Organism	Cellular Morphology	Beta Hemolysis	Motility	H₂S/TSI	Esculin	Glucose	Mannitol	Salicin
Corynebacterium species	Medium size, diphtheroidal	− (+)	− (+)	−	−	+	+ (−)	− (+)
Listeria monocytogenes	Short, thin, coccobacillary to diphtheroidal	+	+ 25°C	−	+	+	−	+
Erysipelothrix rhusiopathiae	Same as above, but may form long filaments	−	−	+	−	+	−	−
Lactobacillus species	Long, slender to short coccobacilli; chain formation common	−	−	−	−	+	+ (−)	+ (−)

*Based on data and methods of Weaver, R. E., Tatum, H. W., and Hollis, D. G.: The Identification of Unusual Pathogenic Gram-Negative Bacteria. Atlanta, Center for Disease Control, 1974.
†Other bacteria that can be confused with those included in this table are *Propionibacterium, Actinomycetes,* and *Arachnia* species, because some of these organisms will grow in 5 to 10 per cent CO_2 or in air. Their identification is discussed in Chapter 10. The identification of obligate aerobes which produce long, branching filaments such as *Nocardia, Streptomyces,* and *Actinomadura* species is discussed in Chapter 13.
Key: + = 90% or more strains positive; (+) = 51–89% strains positive; (−) = 10–50% strains positive; − = less than 10% strains positive.

The anaerobic diphtheroids *C. acnes, C. avidum,* and *C. granulosum* in edition 7 of *Bergey's Manual* are now excluded from the genus Corynebacterium in edition 8. These bacteria are anaerobic, produce propionic acid as a major end product of fermentation, and possess a cell wall composition different from that of the type species. These species are now included in the genus Propionibacterium. The species of Corynebacterium most frequently encountered in the clinical laboratory include:

C. diphtheriae
C. ulcerans
C. equi
C. haemolyticum
C. aquaticum
C. hofmannii (pseudodiphtheriticum)
C. xerosis
C. pyogenes
C. renale
C. ovis (pseudotuberculosis)
C. bovis

Habitat

The corynebacteria are widely distributed in nature and are commonly found in the soil and water and reside on the skin and mucous membranes of man and other animals. Except for *Corynebacterium diphtheriae,* coryneform bacilli are usually regarded as contaminants when recovered in the clinical laboratory. On the other hand, the repeated isolation of Corynebacterium species from blood, cerebrospinal fluid, and other body fluids that are normally sterile is suggestive that the organism may be the cause of an infectious process. *C. diphtheriae* infection usually involves the upper respiratory tract but has also been recovered from wounds, the skin of infected individuals, and from healthy carriers. This species is not found in animals.

Diseases Caused by Species of Corynebacterium

Diphtheria is an acute, contagious, febrile illness caused by *Corynebacterium diphtheriae.* The disease is characterized by a combination of local inflammation and pseudomembrane formation of the oropharynx (see Plate 9-1, *B*) and damage to the heart and peripheral nerves caused by the action of a potent exotoxin. In this country, the disease occurs primarily in the southern United States, both in endemic and epidemic forms.[20] An association between the full immunization of the population and a decrease in the incidence of diphtheria has been noted. The annual incidence of the disease in the United States declined from 200 cases per 100,000 population in 1920 to 0.1 cases per 100,000 population in 1976, at which time there was only a total of 128 cases reported.[7]

C. diphtheriae is spread primarily by convalescent and healthy carriers via the respiratory route. During infection, *C. diphtheriae* grows in the nasopharynx or elsewhere in the upper respiratory tract, producing an exotoxin that causes necrosis and superficial inflammation of the mucosa. A grayish "pseudomembrane" is formed, which is an exudate composed of neutrophils, necrotic epithelial cells, erythrocytes, and numerous bacteria embedded in a meshwork of fibrin. The organisms do not invade the submucosal tissue. The exotoxin produced locally in the throat is absorbed through the mucosa and is carried in the circulation to distant organs. The major sites of action of the toxin are the heart and peripheral nervous system, although other organs and tissues may be affected as well.

Diphtheria carries a 10 to 30 per cent mortality rate, most commonly secondary to congestive heart failure and cardiac arrhythmias caused by the toxic myocarditis. Obstruction of the airway is a severe complication, especially when the pseudomembrane involves the larynx or trachea.

Diphtheria toxin is produced only by strains of *C. diphtheriae* that have been infected by a specific bacteriophage called beta-phage. Nontoxigenic strains of *C. diphtheriae* are commonly isolated from carriers, particularly during a diphtheria outbreak. Such strains may cause pharyngitis but do not produce the systemic manifestations of diphtheria. Therefore, the definitive laboratory identification of *C. diphtheriae* includes animal testing for toxigenic effects of the strain isolated and the morphologic and biochemical criteria listed in Figure 9-1.

Other conditions that must be differentiated from diphtheria include streptococcal pharyngitis, adenovirus infection, infectious mononucleosis, and Vincent's angina. The physician usually must make a presumptive diagnosis of diphtheria from clinical criteria and immediately initiate therapy without waiting for laboratory confirmation.

C. ulcerans can also elaborate diphtherial toxin, and this species has been recovered from the throat of individuals with a diphtherialike disease. Some strains of *C. ulcerans* elaborate a second toxic substance that is not an exotoxin.

C. haemolyticum, an organism morphologically similar to *C. diphtheriae,* has also been recovered from patients with symptomatic pharyngitis; however, it does not produce toxin. Therefore, it must be distinguished from *C. diphtheriae* and *C. ulcerans.*[15]

Other species of Corynebacterium associated with humans cause disease only under rare circumstances.[15] *C. hofmannii,* a normal inhabitant of the pharynx, has been recovered from the blood of patients with subacute bacterial endocarditis; *C. xerosis,* commonly encountered in conjunctival sacs, has been recovered from patients with prosthetic-valve endocarditis; and *C. ovis, C. pyogenes, C. equi,* and *C. bovis* have all been incriminated in occasional human infections, particularly in compromised hosts or in patients with endocarditis complicating chronic valvular heart disease or prosthetic heart valve replacement.[17]

Selection, Collection, and Transport of Clinical Specimens for Culture

When diphtheria is suspected, clinically, the swab should be rubbed vigorously over any inflammatory lesion to obtain suitable material for laboratory examination. Nasopharyngeal specimens, not cultures of the anterior nares, should be submitted for culture in addition to the throat swab. If the swab cannot be transported immediately to the laboratory, the specimen should be inoculated directly on to Loeffler's serum medium or tellurite medium as described below. If the personnel working in a given laboratory are not experienced in the recov-

Color Plates
9-1 *to* 9-2

Plate 9-1
Identification of Corynebacterium Species

A.

Gram's stain of a bacterial colony revealing gram-positive bacilli clustered in "Chinese letter" or "picket fence" arrangements. This is the so-called "diphtheroidal" appearance characteristic of Corynebacterium species.

B.

Photographs of the oral cavity illustrating acute reddening of the mucous membrane (acute pharyngitis) with focal fibrinous membrane formation characteristic of diphtheria. In the top photograph the membrane is seen on either side of the tonsillar pillars extending onto the surface of the posterior portion of the tongue. In the lower left photograph, the membrane is clearly seen in the left tonsillar fossa; in the lower right photograph, in the right tonsillar fossa.

C.

Methylene blue stain prepared from colonies growing on Loeffler's medium, illustrating the diphtheroidal arrangement of bacilli with dark metachromatic granules characteristic of *Corynebacterium diphtheriae*.

D.

Blood agar plates on which are growing tiny colonies of *C. diphtheriae*. This strain is producing an alpha hemolysin resulting in a colonial appearance similar to streptococci. Gram's stain must be performed to avoid a potential misidentification based on visual examination of the primary isolation plate.

E, F and G.

Tinsdale tellurite agar illustrating several black colonies surrounded by a brown halo, the characteristic appearance of *C. diphtheriae* when recovered on this medium. The black colonies with the sharp, discrete borders are *Staphylococcus aureus*. Distinction between the two can be made by the brown "halo" characteristic of *C. diphtheriae* as well as by Gram's stain. Some Proteus species will also reduce tellurite and produce a brown halo on Tinsdale agar, but can be differentiated from *C. diphtheriae* by a Gram's stain and biochemical characteristics.

H.

Photograph of a shaved section of the skin of the side of a rabbit at the site of *in vivo* toxigenicity tests for *C. diphtheriae*. In the left upper field of the photograph two distinct areas of necrosis are seen at the sites where a small portion of the test culture had been inoculated prior to the administration of diphtheria antitoxin. In the squares immediately below these two necrotic areas, notice the tiny, nonsuppurative inoculation sites of the cultures after administration of antitoxin, which serve as controls. The two cultures inoculated at the sites that subsequently became necrotic are therefore confirmed as containing toxigenic strains of *C. diphtheriae*.

I.

Elek *in vitro* toxigenicity test for *C. diphtheriae*. Three strains of *C. diphtheriae* have been streaked perpendicularly to a horizontal strip containing diluted diphtheria antitoxin. Note the white precipitin lines extending at 45^0 angles from the two bacterial streak lines on the right, while no such lines are observed adjacent to the first streak. The two strains producing the precipitin lines are thereby confirmed as toxin producing *C. diphtheriae*.

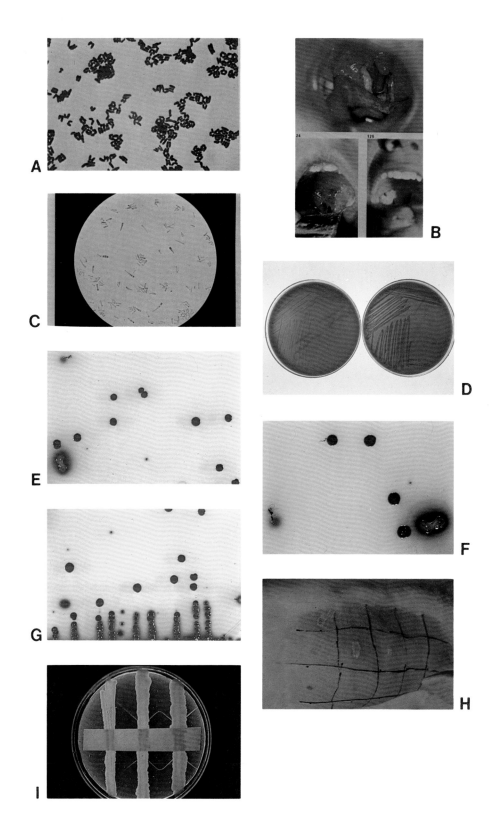

Plate 9-2 (Part I)
Identification of *Listeria monocytogenes*

A.
Gram's stain preparation of a Bacillus Species showing gram-positive bacilli similar to those of *Listeria monocytogenes*.

B.
Blood agar plate with tiny, pinpoint colonies surrounded by zones of beta hemolysis. This appearance is highly suggestive of beta-hemolytic streptococci. If colonies similar to these are recovered in pure culture from spinal fluid, vaginal or chorionic membranes or cultures from septic newborns, the possibility of *L. monocytogenes* must be strongly considered. Notice in the photograph that the zones of beta hemolysis appear less intense than those of most beta-hemolytic streptococci, almost having a ground-glass appearance. A Gram's stain should always be performed when colonies and hemolysis of this type are observed.

C.
Semisolid agar motility test medium illustrating the characteristic "umbrella" type of motility exhibited by *L. monocytogenes* when incubated at 22° C.

D.
KIA agar slant exhibiting an alk/alk reaction with production of a small amount of H_2S in the deep of the medium adjacent to the inoculum stab line. This characteristic is produced by *Erysipelothrix rhusiopathiae*, which, in primary culture may resemble *L. monocytogenes* (the differential characteristics are presented in Table 9-1).

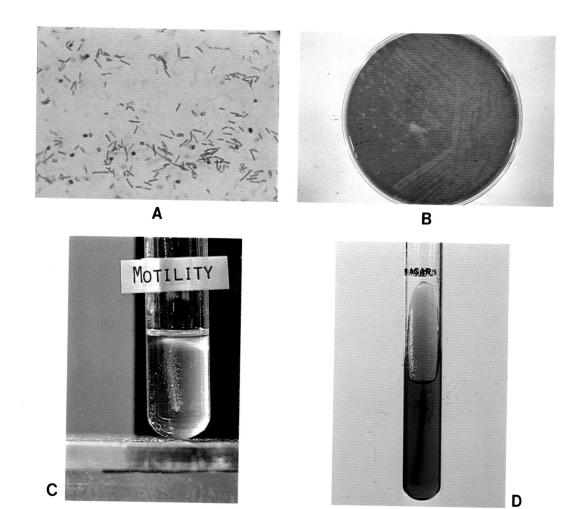

A

B

MOTILITY

C

D

Plate 9-2 (Part II)
Identification of Bacillus Species

E.
Gram's stain preparation illustrating over-decolorized gram-positive bacilli containing subterminal endospores, a microscopic picture characteristic of members of the genus Bacillus. This appearance is also suggestive of species of Clostridium. Distinction between the two can be made on the basis that Bacillus species are aerobic or facultatively anaerobic in contrast to Clostridia species, which are obligate anaerobes.

F.
Nutrient agar plate on which are growing gray-white, opaque colonies with irregular margins, from which comma-shaped outgrowths are observed (so-called "Medusa-head" formations). This type of colony is characteristic of *Bacillus anthracis*, one of the most virulent organisms that can cause human disease.

G.
Blood agar plate with spreading, green-gray, hemolytic colonies of Bacillus species. Note the absence of the irregular margins of the colonies and the hemolytic zones, two characteristics helpful in differentiating Bacillus species from *B. anthracis*.

H.
Photomicrograph of a Gram's stain preparation of *B. anthracis* illustrating the characteristic large, gram-positive bacilli with ovoid, subterminal endospores that do not bulge the bacterial cells. The ends of the individual cells have a typical squared-off appearance.

I.
Heart infusion blood agar plate with a colony of *B. anthracis*. Note the lack of a hemolytic reaction and the irregular outline of the colony margin.

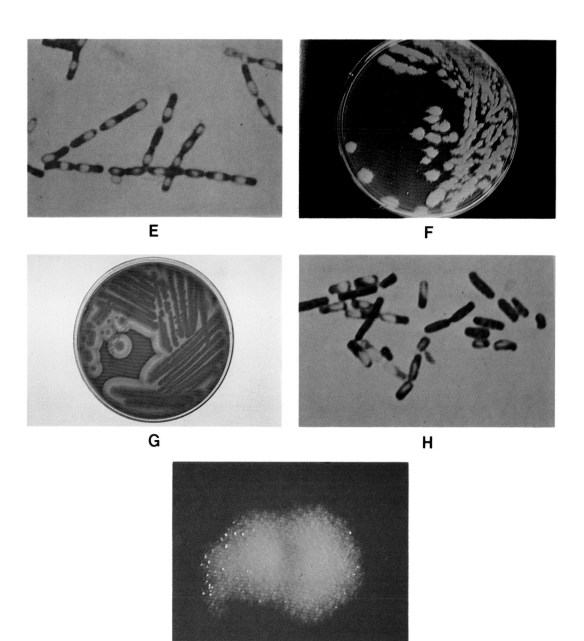

E

F

G

H

I

Chart 9-1. *Loeffler's Methylene Blue Stain*

Introduction Methylene blue is a simple stain that is particularly useful in the identification of species of Corynebacterium.

Principle The metachromatic granules of *C. diphtheriae* readily take up methylene blue dye and appear deep blue. Although some authors have stated that the cytoplasmic granule formation characteristic of *C. diphtheriae* is rarely seen with saprophytic species of Corynebacterium this criterion is unreliable and cannot be used for definitive identification without further studies.

Media and Reagents Methylene blue (80% dye content)   0.3 g.
Ethyl alcohol (95%)   30.0 ml.
Distilled water   100.0 ml.

Procedure 1. Heat-fix the smear.
2. Flood the surface of the smear with the methylene blue staining solution for 1 minute.
3. Wash the slide with water and blot dry.
   Historically it was necessary to add alkali to the above solution before use. However, methylene blue dyes prepared in recent years do not require this additional step because acid impurities found in older stains have been removed.

Interpretation The corynebacteria are pleomorphic bacilli that range in size from 0.5 to 1 μ. in width and from 2 to 6 μ in length, and appear as straight, curved, or club-shaped rods. Characteristic for the microorganism are the metachromatic granules which take up the methylene blue stain and appear dark blue (Plate 9-1, C). Although this finding is characteristic of the corynebacteria, species of Propionibacterium, some of the actinomycetes, pleomorphic strains of streptococci, and other bacteria may also morphologically resemble the corynebacteria and must be differentiated by other cultural and biochemical characteristics.

ery and identification of *C. diphtheriae,* the specimens should be sent to a reference laboratory. Swabs for bacteriologic examination should be shipped dry in packets or tubes containing desiccants such as silica gel.[18]

Processing of Throat and Nasopharyngeal Specimens

Figure 9-1 is an overview of one approach that can be used in the processing of clinical specimens in which the presence of *C. diphtheriae* is suspected. The laboratory may be asked to process specimens for *C. diphtheriae* from acutely ill patients, from patients who are recovering from diphtheria, from healthy carriers, or, rarely, from patients with suspected skin or wound diphtheria.

Smears should be prepared from the clinical specimen, one stained by the Gram's method and the other by Loeffler's methylene blue stain (Albert's stain). Loeffler's methylene blue stain is described in Chart 9-1. Direct examination of smears is valuable because early presumptive information may aid the clinician long before organisms have grown in culture.

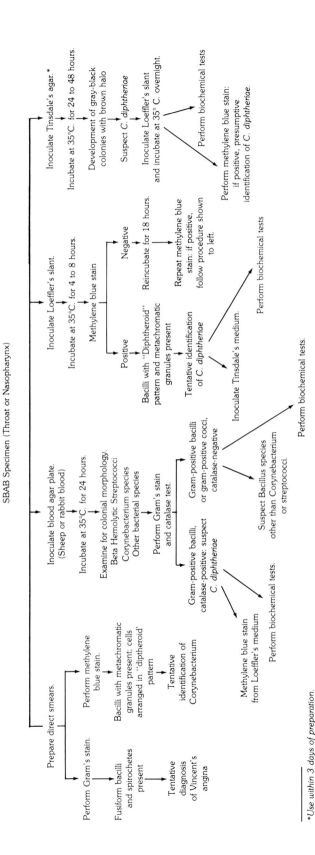

Fig. 9-1. Schema for specimen processing in the laboratory diagnosis of diphtheria.

*Use within 3 days of preparation.

The direct Gram's stain of a throat swab in a suspected case of diphtheria is valuable in differentiating Vincent's angina, a condition that may clinically resemble diphtheria, from diphtheria. The presence of large numbers of fusiform gram-negative bacteria and spirochetes are the characteristics that are helpful in making a presumptive diagnosis of Vincent's angina. The corynebacteria, on the other hand, have a diphtheroid appearance.

The bacterial cells of the corynebacteria can best be visualized in the direct smear stained with Loeffler's methylene blue stain. The characteristic "V", "L," and "Y" arrangements of the cells may be seen, but most important, one should search for the typical deep-blue-staining metachromatic granules of the corynebacteria (Plate 9-1, *C*). Unfortunately, the morphologic characteristics of *C. diphtheriae* in the methylene blue smear preparation are not distinctive and may be identical to other corynebacteria comprising the commensal flora of the throat and nasopharynx. Therefore, the results of direct microscopic examination can only be reported as "gram-positive pleomorphic bacilli present, morphologically resembling *C. diphtheriae*", and the physician must determine if this finding is clinically relevant. In most instances, additional cultural, biochemical, and toxigenicity studies must be performed before a definitive diagnosis can be made.

The Selection and Use of Primary Isolation Media: Cultural Characteristics of *C. diphtheriae*

The primary recovery of *C. diphtheriae* from clinical specimens requires the use of both selective and nonselective culture media. As illustrated in Figure 9-1, the following media are recommended:

1. Blood agar plate (The medium employed for this routine cultivation of beta-hemolytic streptococci is adequate.)
2. A slant of Loeffler's serum medium
3. Medium containing potassium tellurite (Tinsdale agar and/or cystine-tellurite blood agar).

Blood Agar Plate. All throat and nasopharyngeal specimens should be inoculated to blood agar even if diphtheria is clinically suspected. It is important that these specimens be examined for the presence of Group A beta-hemolytic streptococci because the patient may have streptococcal pharyngitis or a mixed streptococcal-diphtherial infection. Some strains of *C. diphtheriae* are hemolytic (Plate 9-1, *D*); however, differences in Gram's stain morphology readily distinguish these species from the hemolytic streptococci. Blood agar also serves as a valuable back-up to recover those strains of *C. diphtheriae* that may not grow on tellurite-containing media.

Loeffler's Serum Medium. Loeffler's serum medium may be used both for the direct recovery of species of Corynebacterium from clinical specimens or for the subculture of colonies suspicious for *C. diphtheriae* on tellurite media. Growth on Loeffler's medium is used to enhance the granule formation in methylene blue stains to demonstrate the characteristic cellular morphology of *C. diphtheriae*. However, in contrast to Tinsdale's medium, the colonies on Loeffler's medium show no characteristic differential features to distinguish the corynebacteria from other aerobic gram-positive bacilli. Primary colonies isolated on Loeffler's medium are usually transferred to tellurite medium or subjected to other biochemical tests or toxigenicity studies. The preparation and use of Loeffler's medium is reviewed in Chart 9-2.[3]

Tellurite Medium. Media containing potassium tellurite should be used for the primary recovery of *C. diphtheriae* from clinical materials in cases in which the diagnosis of diphtheria is clinically suspected. Potassium tellurite inhibits the growth of most of the normal flora of the upper respiratory tract, allowing *C. diphtheriae* and other saprophytic corynebacteria to grow. All corynebacteria produce grayish-black colonies on tellurite media after 24 to 48 hours of incubation at 35° C.

(Text continues on p. 264)

Chart 9-2. *Loeffler's Serum Medium*

Introduction	Loeffler's serum medium is used primarily for the recovery of *C. diphtheriae* from clinical specimens. Because of its serum content, the medium may also be used more generally to determine the proteolytic activity of various microorganisms.
Principle	The characteristic staining reaction of *C. diphtheriae* is based on the microscopic examination of stained smears prepared from the growth of the organism on this medium. The medium is also helpful in the determination of pigment production.

Media and Reagents

Formula:

Beef serum	70.0 g./liter
Infusion dextrose broth (dry powder)	2.5 g./liter
Egg (whole, dried)	7.5 g./liter

Final *p*H = 7.6

Preparation:

To rehydrate the medium, dissolve 80 g. of Loeffler's medium (BBL) in one liter of distilled water and warm to 42 to 45° C. The powder should be gradually added while gently rotating the flask to minimize mixing air into the suspension. The medium should be dispensed in tubes and coagulated-sterilized in the autoclave as follows:

1. When the suspension is uniform, dispense in tubes.
2. Arrange the tubes in a slant position not more than four deep with several layers of newspaper or paper towels below and above the tubes to prevent rapid coagulation.
3. Tightly close the autoclave, turn on the steam, and allow pressure to remain at 10 pounds/square in. for 20 minutes.
4. During this time allow no air or steam to escape.
5. Adjust the steam-inlet valve and open the air-escape valve so as to maintain a pressure of 10 ppi. Abrupt changes in pressure may cause the medium to bubble.
6. Close the outlet valve when all air has been replaced by steam and allow the pressure to reach 15 ppi and hold there for 15 minutes.
7. Allow the autoclave to cool slowly. When properly prepared the slants are smooth and gray-white. The slants should be incubated before inoculation for 24 hours at 35° C. as a sterility check.

Procedure	When *C. diphtheriae* is suspected, inoculate the Loeffler's medium as soon as possible after collection of the specimen. Examine the slants for growth in 8 to 24 hours after incubation. Prepare smears and stain with methylene blue.
Interpretation	See Chart 9-4 for interpretation of these stained smears.
Comment	Since Loeffler's medium is difficult to prepare, acquisition of commercially prepared sealed tubes is recommended.
Bibliography	Buck, T.: A modified Loeffler's medium for cultivating *Corynebacteria diphtheriae*. J. Lab. & Clin. Med., *34*:582–583, 1949. BBL Manual of Products and Laboratory Procedures. ed. 5, pp. 118–119. Cockeysville, Md., BBL, Division of Becton, Dickinson and Co., 1973.

Chart 9-3. *Tinsdale's Agar (as Modified by Moore and Parsons)*

Introduction	Tinsdale's medium supports the growth of all species of Corynebacterium while inhibiting the growth of most of the normal flora of the upper respiratory tract. Moore and Parsons modified the original Tinsdale formula, simplifying the composition but retaining the specificity for the recovery of *C. diphtheriae* and *C. ulcerans.*
Principle	The potassium tellurite is deposited within the colonies of Corynebacterium turning them black. Tinsdale medium is cystine-sodium thiosulfate tellurite, which is specifically helpful in the identification of *C. diphtheriae*, colonies of which are surrounded by a brown halo (see Plate 9-1, *E, F,* and *G*).
Media and Reagents	*Formula* (Modified Tinsdale Base):

Thiotone peptone   20.0 g.
Sodium chloride   5.0 g.
Agar   14.0 g.
L-Cystine   0.24 g.
Sodium thiosulfate   0.24 g.
Distilled water   1 liter

Preparation:
Suspend 39 g. of powder in 1 liter of distilled water and heat with agitation. Boil for 1 minute. Autoclave for 15 minutes at 121° C.
Cool the modified base to 50° C. and add:
Sterile bovine serum   100 ml.
Tellurite solution, 1%   30 ml.
Pour 15 to 20 ml. of this medium into Petri plates and allow to harden.

The Tinsdale agar base medium without serum and tellurite is stable indefinitely if stored in closed screwcapped tubes or bottles. However, the medium is ordinarily stable for only two or three days when stored in the refrigerator following the addition of serum and tellurite.

Procedure	Streak the plate so as to obtain well-isolated colonies. It is recommended that the agar should be stabbed at intervals since browning of the medium can be detected early in the stab areas.
Interpretation	A brown halo around the colony is regarded as presumptive evidence of the presence of *C. diphtheriae.* This can sometimes be seen after 10 to 12 hours of incubation, although 48 hours may be required for the appearance of typical dark brown halos. The only related species other than *C. diphtheriae* that produces this halo is *C. ulcerans.* Other bacteria, such as coagulase-positive staphylococci, grow well on this medium but do not have a brown halo. Bacteria, such as species of Proteus which produce a heavy diffuse blackening of the medium, can be distinguished by their Gram's stain reaction and biochemical characteristics.
Bibliography	Moore, M. S., and Parsons, E. I.: A study of modified Tinsdale's medium for the primary isolation of *Corynebacterium diphtheriae.* J. Infect. Dis., *102*:88–93, 1958.
	Tinsdale, G. F. W.: A new medium for the isolation and identification of *C. diphtheriae* based on production of hydrogen sulfide. J. Pathol. Bacteriol., *59*:61–66, 1947.

Chart 9-4. *Cystine Tellurite (CT) Blood Agar*

Introduction

CT blood agar is a medium used for the primary isolation of *Corynebacterium diphtheriae*. It has advantages over Tinsdale's medium in being easier to prepare and having a longer shelf life.

Principle

The potassium tellurite in this medium serves to inhibit growth of many of the bacterial species comprising the normal flora of the upper respiratory tract, including most species of Streptococcus and Staphylococcus. *C. diphtheriae* grows well, producing grayish or black colonies after 24 to 48 hours of incubation.

Media and Reagents

Formula:

Heart infusion agar, 2% solution	100 ml.
Potassium tellurite, 0.3% solution	15 ml.
L-Cystine	5 mg.
Sheep blood	5 ml.

Preparation:

Melt the sterile heart infusion agar solution and cool to 45 to 50° C. Carefully maintain this temperature while performing subsequent steps. Aseptically add the sterile potassium tellurite solution (previously sterilized by autoclaving) and the sheep blood. Thoroughly mix. Add the cystine powder, mix well, and pour the medium into sterile Petri dishes. Shake the flask frequently while pouring into plates because the cystine does not go into solution entirely.

Procedure

Streak the plate with the clinical material to be cultured to obtain well-isolated colonies.

Interpretation

C. diphtheriae develops gray or black colonies after 24 to 48 hours of incubation. On occasion, colonies of staphylococci not inhibited by the medium will also appear black; however, these can be distinguished by their characteristic Gram's stain morphology.

Three biotypes of *C. diphtheriae* exist and these can be distinguished on CT blood agar: (1) biotype *gravis* colonies are flat, large, dark gray with radial striations, have an irregular edge, and are dry in appearance; (2) biotype *mitis* colonies are small, black, shiny, convex with an entire edge, and have a moist appearance; (3) biotype *intermedius* colonies are small, flat, and have a raised, black center.

Bibliography

Frobisher, M. Jr.: Cystine-tellurite agar for *C. diphtheriae*. J. Infect. Dis., *60*:99–105, 1937.

Tinsdale's medium and cystine-tellurite (CT) blood agar are the two tellurite-containing media most commonly used in clinical laboratories.[14,21] If possible, both should be used. CT medium is simple to prepare and has a long shelf life (1 month), in contrast to the more involved preparation of Tinsdale's medium which must be used within two or three days. If it is impractical to use both media, the modified Tinsdale's medium should be selected because *C. diphtheriae* can be more readily differentiated from the saprophytic corynebacteria because of the development of a brown halo around the black *C. diphtheriae* colonies (Plate 9-1, *E, F,* and *G*). The preparation and use of Tinsdale's medium is presented in Chart 9-3, and the preparation and use of CT blood agar is presented in Chart 9-4.

If an organism is suspected of being *C. diphtheriae* or *C. ulcerans* based on characteristic colony morphology on tellurite medium and/or the cellular morphology observed in methylene-blue-stained smears, a series of biochemical characteristics must be performed to confirm this tentative identification.

Biochemical Characteristics of the Corynebacteria

The media and methods recommended for the biochemical characterization of the clinically important species of Corynebacterium have been described by Weaver,[27] as summarized in Table 9-2. The colonial morphology of *C. diphtheriae* on Tinsdale's medium, as described above, is the most important characteristic in differentiating it from other species of corynebacteria. Additional characteristics, including the production of urease, reduction of nitrate, liquefaction of gelatin, and fermentation reactions for various carbohydrates as outlined in Table 9-2, help in identifying the various species within the genus Corynebacterium.

Tests for Toxigenicity. Individuals may carry nontoxigenic strains of *C. diphtheriae* in their throat or nasopharynx; therefore, for epidemiologic and clinical reasons, the definitive laboratory identification of pathogenic strains must include tests for toxigenicity. *In vivo* and *in vitro* methods are available, and one or the other may be performed as follows:

In vivo Toxigenicity Tests for C. Diphtheriae (Fraser and Weld[13])
1. From a pure culture of an isolate grown on Loeffler's serum medium, inoculate a 10-ml.

tube of brain-heart infusion broth (pH 7.8 to 8.0) and incubate at 35° C. for 48 hours.
2. Clip the hair close from the back and sides of a white rabbit or light-skinned guinea pig. After the hair is removed, mark the skin off into 2-cm. square areas with a marking pencil (see Plate 9-1, *H*).
3. A syringe graduated to 0.1 ml. and fitted with an 0.5-inch, 24-gauge needle is filled with 1 to 2 ml. of the broth culture to be tested.
4. Inject the rabbit or guinea pig with 0.2 ml. of each broth to be tested intracutaneously. The square immediately below each injection site will be used for control injections. Refrigerate the syringes containing the remaining portions of broth suspensions.
5. Five hours after the test injections, administer 500 units of diphtheria antitoxin, intraperitoneally into the guinea pig if a guinea pig was used or into the marginal ear vein if a rabbit was used. Wait 30 minutes.
6. With the refrigerated syringe, inject 0.2 ml. of broth culture intracutaneously into the square immediately beneath the corresponding test site.
7. Preliminary readings can be made in 24 hours; final readings can be made at 48 hours.
8. If the isolate being tested is a toxigenic strain of *C. diphtheriae,* a necrotic area, usually about 5 to 10 mm. in diameter, will appear at the site of the first injection, while the corresponding control site will show only a pinkish nodule without evidence of necrosis (see Plate 9-1, *H*). Any culture that does not elicit the response does not contain a toxigenic strain of *C. diphtheriae.*

More recently, Hermann and Bickham[19] have described a modification of the *in vivo* test. A pair of guinea pigs are used for this test. One of the animals is injected intraperitoneally with 250 units of diphtheria antitoxin. After waiting two hours, this animal is injected subcutaneously or intraperitoneally with 4 ml. of the broth culture containing the test organism. The second guinea pig is injected directly with 4 ml. of the broth culture without prior administration of the diphtheria antitoxin.

If the isolate is a toxigenic strain of *C. diphtheriae,* the unprotected guinea pig will die within 24 to 96 hours, while the animal that has received the antitoxin will remain healthy. If both animals die, the organism is

Table 9-2. Some Key Characteristics of Corynebacterium Species Encountered in Clinical Specimens*

Species	Tinsdale's Agar-Halo	Hemolysis	Catalase	Motility	Urea	Nitrate	Gelatin	Glucose†	Maltose†	Lactose†	Sucrose†
C. diphtheriae	+	Variable‡	+	−	−	+	−	+	+	−	−
C. ulcerans	+	Variable	+	−	+	−	−37°C. +25°C.	+	+	−	Variable
C. equi	−	−	+	−	Variable	Variable	−	−	−	−	−
C. haemolyticum	−	+	−	−	−	−	−	+	+	+	Variable
C. aquaticum	−	−	+	+	−	−	Variable	+	+	−	+
C. hofmannii (pseudodiphtheriticum)	−	−	+	−	+	+	−	−	−	−	−
C. xerosis	−	−	+	−	−	+	−	+	+	−	+
C. pyogenes	−	+	−	−	−	−	+	+	+	+	+
C. renale	−	−	+	−	+	−	−	+	−	−	−
C. ovis (pseudotuberculosis)	−	Variable	+	−	+	+	−	+	+	Variable	Variable
C. bovis	−	−	+	−	−	−	−	+	+	−	−

*Data from Rogosa et al.[22] and Weaver, Tatum, and Hollis.[27]

†Carbohydrate-fermentation tests performed in peptone-meat extract broth fermentation base.[27] (Exception: the O-F Base is used for C. equi and C. hofmannii).

‡Of the three biotypes of C. diphtheriae, only mitis-type strains are usually hemolytic.

Key: + = 90% strains positive; − = less than 10% strains positive.

probably not *C. diphtheriae.* If neither animal succumbs, the organism is most likely a nontoxigenic strain of *C. diphtheriae* or one of the nonpathogenic species of Corynebacterium.

Modified Elek[10a] *in vitro* Test for *C. diphtheriae*

1. Add 2 ml. of sterile rabbit serum and 1 ml. of 0.3 per cent potassium tellurite to 10 ml. of KL virulence agar (Difco) that has been warmed to 55° C. in a water bath.
2. Pipette 10 ml. of medium into a Petri dish and rotate 20 times to mix.
3. Before the medium solidifies, place a 1 x 8-cm. filter paper strip that has been saturated with diphtheria antitoxin (diluted to contain 100 units of antitoxin/ml.) across the diameter of the plate. This strip should sink through the medium and rest on the bottom of the plate. Sterile forceps can be use to submerge the strip.
4. Allow the medium to solidify and then place the plate in the incubator with the lid ajar to allow the surface moisture to evaporate.
5. Inoculate the plate within two hours after drying by streaking a 24-hour culture of a toxin-producing strain of *C. diphtheriae* at right angles to the antitoxin strip. Similarly streak a negative control strain.
6. Streak the unknown cultures to be tested parallel to the streaks of the control cultures.
7. Incubate the plate for 24 to 48 hours at 35° C. Examine for white lines of precipitation that extend out from the line of bacterial growth, forming an angle of about 45 degrees (see Plate 9-1, *I*). These white precipitin lines form where the toxin from pathogenic strains of *C. diphtheriae* combines, with the antitoxin from the paper strip, indentifying those strains that produce the toxin.

LISTERIA

The family affiliation of the genus Listeria is presently uncertain. The species listed in edition 8 of *Bergey's Manual*[5] are:

Listeria monocytogenes
Listeria denitrificans
Listeria grayi
Listeria murrayi

Listeria monocytogenes is the clinically important species most commonly encountered in clinical laboratories.

Habitat

Listeria monocytogenes is found in a wide variety of habitats, including the normal microbiota of healthy ferrets, chinchillas, ruminants, and humans, and in environmental sources such as sewage, silage, soil, fertilizers, and decaying vegetation. Seeliger and Welshimer[23] report that this bacterium has been isolated from over 50 species of warm-blooded and cold-blooded animals.

Incidence of Diseases Caused by Listeria and Their Clinical Manifestations

L. monocytogenes is encountered in the clinical laboratory more frequently than the other pathogenic aerobic gram-positive bacilli listed in Table 9-1. Yet, less than 200 cases of human listeriosis are reported in the United States each year.[8] It is likely that more cases occur but either go unrecognized or are unreported.

Listeriosis in man occurs sporadically and may present in a variety of clinical forms: conjunctivitis (oculoglandular), cervicoglandular (often associated with pharyngitis,) pneumonic, often with generalized symptoms simulating typhoid fever, or cutaneous. The more commonly encountered manifestations are meningoencephalitis, genital infection with habitual abortion, or perinatal infant septicemia.

Infections in newborn infants have occurred in epidemics, one of the largest of which was reported in East Germany involving over 200 cases in the same year.[10] Neonates acquire their infection *in utero* from the mother, resulting in stillbirth, septicemia, or meningitis.[6] The organism may also cause meningitis in newborn infants, a disease similar to that which occurs in animals.[1] In older children and adults, listeriosis is seen primarily in patients with compromised host resistance, particularly in individuals with advanced malignancy, after renal transplantation, or with out-of-control diabetes mellitus. Septicemia and meningitis are the most common clinical manifestations of listeriosis in these compromised patients.

Bacteriology of Listeriosis

Of 125 isolates of *L. monocytogenes* from 99 patients referred to the Center for Disease Control, 47 were isolated from blood cultures and 52 from the cerebrospinal fluid.[8] Other anatomical sites from which the organism has been recovered include the uterus and the vaginal canal, as well as the vaginal lochia and placenta. Listeriosis can involve the formation of multiple, focal, acute inflammatory lesions or abscesses throughout the viscera in disseminated cases.

Humans may acquire the disease after contact with infected dogs, through the ingestion of contaminated milk or infected meat, or after handling infected newborn calves (mainly livestock producers and veterinarians). Therefore, the disease is generally considered one of the zoonoses. Although livestock and poultry constitute a prime reservoir for the organism, many recent cases in the United States have occurred in residents of urban areas where there have been no known animal contacts.[6] The reported mortality rate for patients with listeriosis ranges between 42 and 50 per cent.[6] Strains of *L. monocytogenes* vary in their susceptibility to various antimicrobial agents; however, ampicillin and tetracycline are reported to be the drugs of choice.[16]

Collection, Transport, and Processing of Specimens for Culture

Cerebrospinal fluid, blood, or other materials for culture are collected, transported, and processed as outlined in Chapter 1. Because *L. monocytogenes* may be difficult to isolate from certain clinical specimens, particularly from tissues removed at surgery or at autopsy, a cold "enrichment" technique has been recommended.[16] This involves exposing tissues or other contaminated material to cold temperature (4° C.) for a period of from several days to a few weeks and taking cultures at frequent intervals until recovery has been accomplished.

Microscopic Examination of Smear Preparations

L. monocytogenes is a non-spore-forming, short, gram-positive bacillus with cells varying from 0.4 to 0.5 × 1.0 to 2.0 μ., which is somewhat smaller than other species of aerobic gram-positive bacilli. The cells are coccobacillary; on occasion, diplobacilli occurring in short chains may be observed (Plate 9-2, *A*). In gram-stained smears of cerebrospinal fluid sediments, the organism may occur intra- or extracellularly, at times occurring in pairs, in which case the organisms can be mistaken for pneumococci. If the Gram's stain is overdecolorized, the bacterial cells of *L. monocytogenes* may appear gram-negative and can be confused with Hemophilus. In other smear preparations, the organisms may assume the pleomorphic, palisade forms of diphtheroids.

Primary Isolation and Cultural Characteristics

On sheep blood agar incubated at 35° C. for 24 hours, the growth is generally light. Growth may also be obtained on blood agar plates incubated in 5 to 10 per cent CO_2 or anaerobically. The colonies are small, translucent, and gray, and most strains produce a narrow zone of beta hemolysis around the colonies (Plate 9-2, *B*). On occasion this beta hemolysis may be confused with that produced by beta-hemolytic streptococci, and a Gram's stain should always be performed when this type of colony is recovered in cultures of spinal fluid, blood, or vaginal secretions. *L. monocytogenes* is never alpha-hemolytic and does not form a white pigment, characteristics helpful in differentiating it from other species of gram-positive bacilli. Additional characteristics by which *L. monocytogenes* may be identified are as follows:

Catalase-positive
Optimal motility at 25° C.
Growth at 4° C.
Narrow zone of beta hemolysis on blood agar
Fermentation of glucose, trehalose, and salicin
Hydrolysis of esculin
H_2S-negative

The results of these reactions are compared with those of other gram-positive bacilli in Table 9-1.

The catalase test is performed by dropping 3 per cent hydrogen peroxide on colonies isolated on brain-heart infusion agar. *L. monocytogenes* produces catalase whereas streptococci and lactobacilli do not.

Motility can be determined either by the hanging-drop technique or in semisolid motility medium. When examined in the hanging-drop or wet-mount preparations, the bacterial cells of *L. monocytogenes* that have been grown in a 6-hour broth culture at 25° C. exhibit a "tumbling" or "head-over-heels" motility. The use of the phase contrast microscope aids in the microscopic examination of these preparations.

In semisolid agar, the motility of *L. monocytogenes* should be determined at room temperature. An umbrellalike zone of growth approximately 2 to 5 mm. below the surface of the medium is characteristic (see Plate 9-2, *C*). Motility at 35° C. incubation is either absent or extremely sluggish.

Test for Pathogenicity

The ocular test of Anton is one of the tests for pathogenicity performed in reference laboratories. The test is performed by instilling a drop of a 24-hour broth culture of the organism into the conjunctival sac of a young rabbit or guinea pig. The conjunctival sac of the opposite eye serves as an uninoculated control. *L. monocytogenes* produces a severe purulent conjunctivitis within 24 to 36 hours.

ERYSIPELOTHRIX

As with the genus Listeria, the family affiliation for the genus Erysipelothrix is uncertain in edition 8 of *Bergey's Manual.*[5] *Erysipelothrix rhusiopathiae*, the only species of this genus, is rarely encountered in most clinical laboratories.

Habitat

E. rhusiopathiae is widely distributed in nature, and the organism has been isolated from soil, food, and water, presumably contaminated by infected animals. Various animal hosts have been found for this organism, including several species of fish, shellfish, and birds, and it has been isolated from the gastrointestinal tracts of healthy swine.

Diseases Caused by *E. rhusiopathiae*

E. rhusiopathiae, a pathogen seen primarily in veterinary medicine, causes infectious diseases in swine, turkeys and other birds, mice, rabbits, fish, and crustaceans. Swine erysipelas, an inflammatory disease of the skin and joints, is of major economic importance in the United States.

In humans, the organism causes a cutaneous inflammatory disease, usually of the hands and fingers, called erysipeloid. Erysipeloid is largely an occupational disease of persons who handle meat, poultry, fish, or crustaceans. The organism is thought to enter the skin through minor abrasions, leading to raised, erythematous areas of inflammation of the hands and fingers. The lesions are painful and tend to spread peripherally while the central areas fade. The organism is able to survive for long periods of time outside the animal body as a soil contaminant and is not killed by salting, smoking, or pickling procedures.

In rare instances, infection in humans may be serious. A total of 30 cases of human endocarditis and septicemia has been reported since 1912, occurring in patients ranging in age from 10 to 69 years.[4] The majority of the human cases involved males whose occupations predisposed them to infection with this organism. Fifteen of the thirty individuals reported in this series died of the infection.

Collection and Processing of Clinical Specimens for Culture

In patients with clinical erysipeloid, it is best to obtain a biopsy through the full thickness of the infected skin at the advancing margin of the lesion. The skin surface should be first cleansed and disinfected with alcohol and/or iodine prior to the biopsy procedure. Blood specimens should be obtained in suspected cases of endocarditis or septicemia.

Selective media are not required for the isolation of the organism from skin or tissue aspirates, providing the skin surface was properly decontaminated during collection. The organism grows well on routinely used blood agar media. Weaver[27] has recommended that cutaneous biopsy specimens should be placed in an infusion broth containing 1 per cent glucose and incubated aerobically under 5 to 10 per cent CO_2 at 35° C. The broth is then subcultured to a routine blood agar plate at 24-hour intervals.

Identification of *E. rhusiopathiae*

Both smooth and rough colonies develop on blood agar. The smooth colonies are smaller, measuring 0.5 to 1.0 mm. in diameter, and are convex, circular, and transparent. The larger rough colonies show a mat surface with a fimbriated edge. Greenish discoloration of the blood medium adjacent to the colonies may be seen after prolonged incubation.

Cells from smooth colonies typically appear as short, slender, straight or slightly curved gram-positive bacilli, measuring 0.2 to 0.4 × 1.0 to 2.5 μ. There is also a tendency for the cells to form long filaments (4 to 15 μ. in length).

E. rhusiopathiae is nonmotile, does not produce catalase, and produces either alpha hemolysis or no hemolysis on blood agar, which helps to distinguish it from Listeria. The important biochemical characteristics

for the identification of *E. rhusiopathiae* are listed in Table 9-1.

The ability of this organism to produce H_2S in Kligler iron agar or triple sugar iron agar is a helpful identification feature from the other species of gram-positive bacilli (Plate 9-2, *D*). The fermentation reactions for *E. rhusiopathiae* should be determined in fermentation base medium. In addition, another helpful characteristic in the identification of this organism is the "test tube brush" pattern of growth exhibited in gelatin stab cultures.

LACTOBACILLUS

A discussion of the genus Lactobacillus is included here because these organisms comprise an important component of the human indigenous flora. Lactobacilli are commonly encountered in the clinical laboratory as commensals or as isolates of little clinical significance. Rare cases of endocarditis have been reported.[12]

The genus Lactobacillus consists of nonsporulating, gram-positive bacilli which are classified in the family Lactobacillaceae. The genus is defined in part by the metabolic products produced, and the majority of species are homofermentative; that is, they form lactic acid from glucose as the major fermentation product. Heterofermentative species may be encountered which produce about 50 per cent lactic acid and varying amounts of CO_2, acetic acid, and ethanol from glucose.

Habitat

Lactobacilli are widely distributed in nature and are ubiquitous in humans. They inhabit the mouth, gastrointestinal tract, vaginal canal, and other sites. A number of older textbooks used the term "Döderlein's bacillus" for a variety of human vaginal strains of Lactobacillus. It is now recognized that the Döderlein's bacilli include *L. acidophilus, L. casei, L. fermenti, L. cellobiosus,* and *Leuconostoc mesenteroides.*[24] It is of no clinical significance to identify these organisms.

Diseases Caused by Species of Lactobacillus

A limited number of species of Lactobacillus have been incriminated in serious human infections. *L. plantarum* and *L. casei* have been recovered from patients with endocarditis, and other species have been incriminated in patients with meningitis. The isolation of various lactobacilli from human clinical specimens has been reviewed by Finegold.[12]

Laboratory Identification of Lactobacilli

The lactobacilli are non-spore-forming, rod-shaped bacteria, varying from long and slender forms to short coccobacilli, at times producing short chains. Pleomorphic forms are at times encountered with some tendency to form palisades. Most species are nonmotile.

The lactobacilli are generally grown on blood agar and chocolate agar media. Good growth is also obtained on Rogosa's selective tomato juice agar medium[22] (LBS medium, BBL) which has an acid *p*H. Additional characteristics for differentiating the lactobacilli from other species of gram-positive bacilli are shown in Table 9-1. The negative catalase reaction, the production of major quantities of lactic acid (as determined by gas-liquid chromatography), and the lack of lateral outgrowth from the stab line on a gelatin tube are the most helpful differentiating features.

BACILLUS

The genus Bacillus, classified within the family Bacillaceae in edition 8 of *Bergey's Manual*,[5] is composed of several species of aerobic or facultatively anaerobic gram-positive bacilli that produce endospores (Plate 9-2, *E*). The organism grows well on blood agar, producing large, spreading, gray-white colonies with irregular margins. Many species are beta-hemolytic, a helpful characteristic in the differentiation of Bacillus species from *Bacillus anthracis,* which is not hemolytic (Plate 9-2, *F* and *G*). Catalase is produced by most species.

Most Bacillus species encountered in the clinical laboratory are saprophytic contaminants or members of the normal flora. Although rarely encountered in the United States,[7] *Bacillus anthracis* is the most important member of this genus, causing anthrax in humans and animals. *B. cereus,* another species of importance to man, has been associated with outbreaks of human food poisoning.[26]

Habitat

Bacillus species are ubiquitous in nature, inhabiting soil, water, and airborne dust. Some species are included in the normal intestinal microbiota of humans and other animals.

Diseases Caused by Species of Bacillus

Anthrax is primarily a disease of herbivorous animals and can be transmitted to humans by direct contact with certain animal products, principally wool and hair. Anthrax spores can remain infectious for more than 30 years. This is an important factor to consider in the epidemiology and control of this disease.

Anthrax is usually encountered in humans as an occupational disease of veterinarians, agricultural workers, and various individuals who handle animals and animal products. Approximately 90 per cent of human cases reported in recent years occurred in mill workers handling imported goat hair.

About 90 per cent of human cases of anthrax are cutaneous infections, beginning 1 to 5 days after contact with the infected materials as a small, pruritic, nonpainful papule at the site of inoculation. The papule then develops into a hemorrhagic vesicle, which ultimately ruptures, leading to a very slow healing ulcer that is covered with a black eschar surrounded by edema. The infections may spread to involve the lymphatics, and regional adenopathy may develop. In rare instances, cutaneous infections may develop into septicemia.

Inhalation anthrax, a severe hemorrhagic

mediastinal adenitis resulting from inhalation of anthrax spores, is virtually 100 per cent fatal. Meningitis may also complicate both cutaneous and inhalation forms of the disease.

Penicillin is usually the drug of choice in the treatment of anthrax; tetracycline is an acceptable alternative. A vaccine is now available for use in humans, but has been recommended only for laboratory workers who are involved in species identification and for employees of mills handling goat hair.[11] An effective vaccine for use in animals has probably been the major factor in reducing the incidence of this disease.[11]

Food poisoning from *Bacillus cereus* has been recognized as a disease of increasing frequency in recent years. Ten outbreaks of *B. cereus* gastroenteritis were reported to the Center for Disease Control between 1966 and 1975, involving a total of 133 persons.[26] The disease, characterized by vomiting, abdominal cramps, and diarrhea, typically occurred within six hours following ingestion of contaminated rice. The disease, simulating staphylococcal food poisoning, usually lasts for only 8 to 12 hours. *B. cereus* may be found in uncooked rice, although spores can survive boiling and will germinate if the boiled rice is not refrigerated, a common practice in many Chinese restaurants. The diagnosis of *B. cereus* food poisoning can be confirmed by the isolation of 10^5 or more organisms per gram of suspected food sample.

Specimen Collection and Processing for Culture

When anthrax is suspected, the state public health laboratory and the Center for Disease Control should be notified immediately. Specimens that may be collected include material from cutaneous lesions and blood or any other material that potentially may be infected. Laboratory safety is of utmost importance when working with any material thought to contain *B. anthracis*.[11] All specimens and cultures should be processed and examined with great care in a biological safety cabinet.[11] Every precaution should be taken to avoid the production of aerosols of the infected material. Laboratory personnel should wear protective coats or gowns, masks, and surgical gloves when processing the samples. This safety apparel should be autoclaved before it is reused or discarded. When the work is finished, all surfaces in the biologic safety cabinet and laboratory workbenches must be decontaminated with 5 per cent hypochloride or 5 per cent phenol and all instruments used for processing the specimen must be autoclaved. Persons working directly with spore suspensions, contaminated animal tissues, or hair must be properly immunized.[11]

In cutaneous anthrax infections, specimens to collect include swab samples of the serous fluid of vesicles or of material beneath the edge of the black eschar. With inhalation anthrax, a sputum sample and blood cultures should be obtained. Gastrointestinal anthrax is a third major form of the disease and gastric aspirates, feces, or food may be cultured. A blood culture should also be obtained.

Cultural Characteristics

B. anthracis species are large, gram-positive bacilli in Gram's stains, measuring 1 to 1.3 × 3 to 10 μ. and the individual cells have square or concave ends (Plate 9-2, *H*). Ovoid, subterminal spores that do not cause any significant swelling of the cells may be observed. On Gram's stain, they are seen as unstained areas within the cytoplasm. Individual spores can also be completely separated from the vegetative cells. Endospores may be seen in direct smears prepared from animal or human tissue; however, they are best demonstrated after the organisms have grown in artificial media. Capsules, however, do not form in artificial culture media, but are found only in smears prepared from infected tissues.

B. anthracis grows well on ordinary blood agar within 18 to 24 hours at 35° C. Typically the colonies are flat, and irregular, measure 4 to 5 mm. in diameter, and have a slightly

Table 9-3. *Some Key Characteristics for Differentiation of Bacillus anthracis and Other Species of Bacillus**

Characteristic	B. anthracis	B.cereus and Other Species of Bacillus
Hemolysis (sheep blood agar)	−	+
Motility	−	+ (usually)
Gelatin hydrolysis (7 days)	−	+
Salicin fermentation	−	+
Growth on PEA medium†	−	+

*Data from Feeley, J. C., and Brachman, P. S.: *Bacillus anthracis. In* Lenette, E. H., Spaulding, E. H., and Truant, J. P.(eds.): Manual of Clinical Microbiology. Chap 15. Washington, D. C., American Society for Microbiology, 1974.
†PEA is prepared by the addition of 0.3 per cent phenylethyl alcohol to heart infusion agar (Difco).

undulate margin when grown on heart infusion blood agar (Plate 9-2, *I*). The organism is not hemolytic on sheep blood agar, a helpful feature in differentiating *B. anthracis* from alpha or beta-hemolytic isolates of Bacillus species (Plate 9-2, *G*). Under the dissecting microscope, numerous comma-shaped outgrowths consisting of long filamentous chains of bacilli may be seen (so-called "Medusa-head" appearance).

The biochemical characteristics which aid in differentiating *B. anthracis* from *B. cereus* and other species of Bacillus are shown in Table 9-3. Except for the identification of *B. anthracis*, it is not clinically relevant for most laboratories to identify members of the genus Bacillus. However, any species of Bacillus that is not hemolytic on blood agar and that has the microscopic Gram's stain characteristics suggestive of *B. anthracis* should be immediately submitted to the state public health laboratory or the Center for Disease Control for final confirmation.

REFERENCES

1. Albritton, W. L., Wiggins, G. L., and Feeley, J. C.: Neonatal listeriosis: distribution of serotypes in relation to age at onset of disease. J. Pediatr., *88:*481–483, 1976.
2. Barksdale, L.: *Corynebacterium diphtheriae* and its relatives. Bacteriol. Rev., *34:*378–422, 1970.
3. BBL Manual of Products and Laboratory Procedures. ed. 5, pp. 118–119. Cockeysville, Md. BBL, Division of Becton, Dickinson & Co., 1973.
4. Borchardt, K. A., et al.: *Erysipelothrix rhusiopathiae* endocarditis. West. J. Med. *125:*149–151, 1977.
5. Buchanan, R. E., and Gibbons, N. E., (eds.): Bergey's Manual of Determinative Bacteriology. ed. 8. Baltimore, Williams & Wilkins, 1974.
6. Busch, L. A.: Human listeriosis in the United States, 1967–1969. J. Infect. Dis., *123:*328–332, 1971.
7. Center for Disease Control: Reported Morbidity and Mortality in the United States, 1976. Annual Summary, Vol. 25, no. 53, August, 1977.
8. Center for Disease Control: Zoonoses Surveillance, Listeriosis. Annual Summary 1971, issued August, 1972.
9. Dowell, V. R., Jr., Stargel, M., and Allen, S. D.: *Propionibacterium acnes.* Microbiology Check Sample MB-85. Chicago, American Society of Clinical Pathologists, 1976.
10. Editorial: *Listeria monocytogenes* and encephalitis. Arch. Intern. Med. *138:*198–199, 1978.
10a.Elek, S. D.: The plate virulence test for diphtheria. J. Clin. Pathol., *2:*250–258, 1949.
11. Feeley, J. C , and Brachman, P. S.: *Bacillus anthracis. In* Lennette, E. H., Spaulding, E. H., and Truant, J. P. (eds.): Manual of Clinical Microbiology, Chap. 15. Washington, D.C. American Society for Microbiology, 1974.
12. Finegold, S. M.: Anaerobic Bacteria in Human Disease. New York, Academic Press, 1977.
13. Fraser, D. T., and Weld, C. B.: The intracutaneous "virulence test" for *Corynebacterium diphtheriae.* Trans. Roy. Soc. Can. Sect. V, *20:*343–345, 1926.
14. Frobisher, M.: Cystine-tellurite agar for *C., diphtheriae.* J. Infect. Dis., *60:*99–105, 1937.

15. Goldstein, H., and Hoeprich, P. D.: Diphtheria. In Hoeprich, P. D.(ed.): Infectious Diseases. ed. 2, Chap. 24, Hagerstown, Md. Harper & Row, 1977.

16. Gray, N. L., and Killinger, A. H.: *Listeria monocytogenes* and listeric infections. Bacteriol. Rev., *30:*309–381, 1966.

17. Hande, K. R., *et al.:* Sepsis with a new species of *Corynebacterium.* Ann. Intern. Med. *85:*423–426, 1976.

18. Hermann, G. J.: Corynebacterium. *In* Bodily, H. L., Updyke, E. L. and Mason, J. O. (eds.): Diagnostic Procedures for Bacterial, Mycotic and Parasitic Infections. ed. 5, Chap. 5, New York, American Public Health Association, 1970.

19. Hermann, G. J., and Bickham, S. T.: Corynebacterium. *In* Lennette, E. H., Spaulding, E. H., and Truant, J. P. (eds.): Manual of Clinical Microbiology, Ed. 2, Chap. 12. Washington, D. C., American Society for Microbiology, 1974.

20. Marcuse, E. K., and Grand, N. G.: Epidemiology of diphtheria in San Antonio Texas, 1970. J.A.M.A. *224:*305–310, 1973.

21. Moore, M., and Parsons, E. I.: A study of a modified Tinsdale's medium for the primary isolation of *Corynebacterium diphtheriae.* J. Infect. Dis., *102:*88–93, 1958.

22. Rogosa, M., Mitchell, J. A., and Weiseman, R. F.: A selective medium for the isolation and enumeration of oral and fecal lactobacilli. J. Bacteriol, *62:*132–133, 1951.

23. Seeliger, H. P. R., and Welshiner, H. J.: Genus *Listeria. In* Buchanan, R. E., and Gibbons, N. E. (eds.): Bergey's Manual of Determinative Bacteriology. ed. 8. Baltimore, Williams & Wilkins, 1974.

24. Sonnenwirth, A. C.: Gram-positive bacilli. *In* Frankel, S., Reitman, S., and Sonnenwirth, A. C.: Gradwohl's Clinical Laboratory Methods and Diagnosis. ed. 7, Vol. 2, Chap. 65. St. Louis, C. V. Mosby, 1970.

25. Tasman, A., and Branwyk, A. C.: Experiments on metabolism with diphtheria bacillus. J. Infect. Dis., *63:*10–20, 1938.

26. Terranova, W., and Blake, P. A.: *Bacillus cereus* food poisoning. N. Engl. J. Med., *298:*143–144, 1978.

27. Weaver, R. E., Tatum, H. W., and Hollis, D. G.: The Identification of Unusual Pathogenic Gram-Negative Bacteria. pp. 1–12. Atlanta, Center for Disease Control, 1974.

10 The Anaerobic Bacteria

For the purposes of this discussion, anaerobic bacteria are defined as those bacteria that fail to multiply on the surface of freshly prepared solid media in the presence of air.[37] These organisms are widespread in soil, marshes, lake and river sediments, the oceans, sewage, foods, and in animals. In humans, anaerobic bacteria normally are prevalent in the oral cavity around the teeth, in the gastrointestinal tract, especially in the colon where they outnumber coliforms by at least 1000:1,[16] in the orifices of the genitourinary tract, and on the skin.[14,37] Most of these anaerobic habitats have a low oxygen tension and reduced oxidation-reduction potential (E_h) resulting from the metabolic activity of microorganisms which consume oxygen through respiration.[7] If there is no replacement oxygen, the microenvironment stays anaerobic.

Based on their ability to form spores and on the morphologic characteristics observed in gram-stained preparations, the anaerobic bacteria are broadly classified as listed below.

Classification of the Genera of Anaerobic Bacteria

I. Spores Formed
  A. Gram-positive bacilli
    Clostridium
II. Spores Not Formed
  A. Gram-positive bacilli
    1. Actinomyces
    2. Arachnia
    3. Bifidobacterium
    4. Eubacterium
    5. Lachnospira
    6. Lactobacillus
    7. Propionibacterium
  B. Gram-positive cocci
    1. Peptococcus
    2. Peptostreptococcus
    3. Ruminococcus
    4. Sarcina
    5. Streptococcus
  C. Gram-negative bacilli
    (curved and spiral forms)
    1. Bacteroides
    2. Borrelia
    3. Butyrivibrio
    4. Campylobacter
    5. Fusobacterium
    6. Leptotrichia
    7. Selenomonas
    8. Succinivibrio
    9. Treponema
  D. Gram-negative cocci
    1. Acidaminococcus
    2. Veillonella

---

*Modified from Dowell, V. R., Jr.: Methods for isolation of anaerobes in the clinical laboratory. Am. J. Med. Technol., 41:32–40, 1975.

Anaerobic infections in man and animals can involve virtually every organ when conditions are suitable. Some of the more commonly involved sites are shown in Fig. 10-1. Based on other reports in the literature,[14,38] the relative incidence of anaerobes in infections is listed in Table 10-1.

Most deep-seated abscesses and necrotizing lesions involving anaerobes are polymicrobial, including aerobic or facultatively anaerobic bacteria such as coliforms or en-

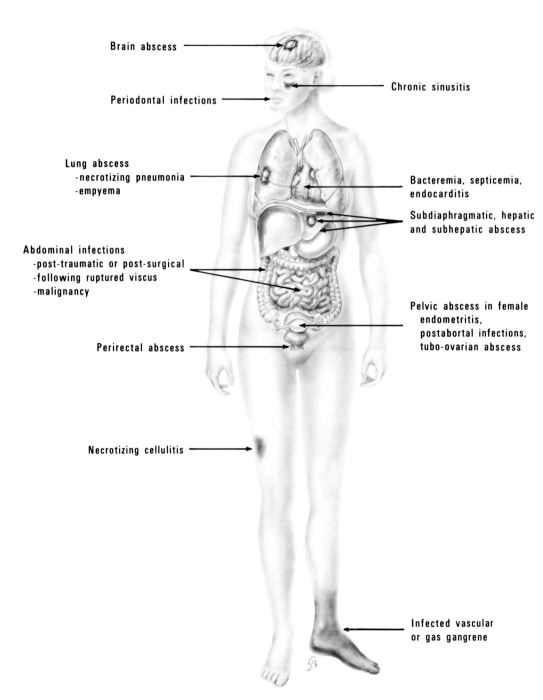

Brain abscess

Chronic sinusitis

Periodontal infections

Lung abscess
-necrotizing pneumonia
-empyema

Bacteremia, septicemia,
endocarditis

Subdiaphragmatic, hepatic
and subhepatic abscess

Abdominal infections
-post-traumatic or post-surgical
-following ruptured viscus
-malignancy

Pelvic abscess in female
endometritis,
postabortal infections,
tubo-ovarian abscess

Perirectal abscess

Necrotizing cellulitis

Infected vascular
or gas gangrene

Fig. 10-1. Common locations of infections involving anaerobic bacteria.

terococci. These aerobic species, acting in concert with trauma, vascular stasis, or tissue necrosis, lower tissue oxygen tension and the oxidation-reduction potential, and provide favorable conditions for obligate anaerobes to multiply. Historically, infections and diseases involving anaerobes from exogenous sources have been best known (see the list of anaerobic infections of exogenous origin, below).

Anaerobic Infections of Exogenous Origin

Botulism or wound botulism
Clostridium perfringens gastroenteritis
Myonecrosis ("gas gangrene")
Tetanus
Crepitant cellulitis
Benign superficial infections
Infections following animal or human bites
Septic abortion

Within the past decade, however, endogenous anaerobic infections have become far more common. There are two probable explanations. One explanation is that the laboratory recovery of anaerobic bacteria has improved so that endogenous infections are no longer misdiagnosed or overlooked as they were in the past. Secondly, a larger proportion of the patient population is receiving immunosuppressive drugs for malignancy and other disorders, resulting in compromised host resistance. Primary anaerobic infections easily become established in areas of tissue damage, and bacteremia, metastatic spread of bacteria with the formation of distant abscesses, and a progressive chain of events resulting in a fatal outcome may occur. The more common endogenous anaerobic infections are listed above.

Anaerobic Infections of Endogenous Origin

Abscess of any organ
Actinomycosis
Complications of appendicitis or cholecystitis
Crepitant and noncrepitant cellulitis
Clostridial myonecrosis
Periodontal infection
Endocarditis
Meningitis, usually following brain abscess
Osteomyelitis
Otitis media
Peritonitis
Thoracic empyema
Septic arthritis
Sinusitis
Tetanus

It is essential to isolate and identify anaerobic bacteria because: (1) these infections are associated with high morbidity and mortality; and (2) the treatment of the infection varies with the bacterial species involved. Antibiotic therapy for certain anaerobic infections is different from that employed for many infections caused by aerobic or facultatively anaerobic bacteria. Prompt surgical intervention, including debridement of necrotic tissue or amputation of a limb, may be of extreme importance, particularly in cases of clostridial gas gangrene or in loculated abscesses where antibiotics may be ineffective until the exudate is drained.

Prior to the mid 1960's, clostridial infections predominated;[17] during the 1970's, 85 per cent of anaerobes isolated from properly selected clinical specimens are accounted for by Bacteroides, Fusobacterium, Peptostreptococcus, and Peptococcus species and the gram-positive, non-spore-forming bacilli. This current trend is summarized in the data

Table 10-1. *Incidence of Anaerobes in Infections*

Infection Type	Incidence (%)
Bacteremia	10–20
Brain abscess	60–89
Dental infections, chronic sinusitis	50
Pleuropulmonary infection	30–90
Intra-abdominal/pelvic sepsis	60–100
Urinary tract infection	1

Table 10-2. *Distribution of Anaerobes Isolated from Human Clinical Materials at Four Medical Centers**

Anaerobes	TM[73] (%)	MA[70] (%)	MA[71] (%)	MIN[73] (%)	IU[73] (%)	IU[74] (%)	IU[75] (%)
Bacteroides species	35	44	37	42	35	39	40
Fusobacterium species	10	8	7	1	3	4	2
Peptococcus species	15	15	17	13	21	18	18
Peptostreptococcus species	11	15	8	3	3	7	5
Other cocci	7	X	3	1	2	1	2
Gram-positive non-spore-forming bacilli	17	X	20	25	25	20	11
Clostridium species	4	13	8	11	11	11	12

*TM,[73] Temple University, 1973[42]; MA,[70,71] Mayo Clinic, 1970 and 1971[31]; MIN,[73] University of Minnesota, 1973[38]; IU,[73-75] Indiana University, 1973–1975.[1]

collected from four medical centers (Table 10-2).

The most common disease-producing, gram-negative, non-spore-forming bacilli in humans are the *Bacteroides fragilis* and *Bacteroides melaninogenicus* groups and *Fusobacterium nucleatum*. The percentage distribution of 313 isolates of the *B. fragilis* group isolated from materials submitted to the laboratory at the Indiana University Medical Center in 1975 is as follows: *B. fragilis*, 65 per cent; *B. thetaiotaomicron*, 17 per cent; *B. vulgatus*, 5 per cent; *B. distasonis*, 4 per cent; and *B. ovatus*, 1 per cent.[1] *B. fragilis* is particularly important because it may be isolated from a variety of infections and because of its resistance to the action of the penicillin analogues, tetracyclines, and aminoglycosides.[14,30]

Penicillin is the antibiotic of choice for clinical infections caused by most other anaerobic bacteria, except for occasional infections with species of Fusobacterium and *B. melaninogenicus*. Clostridia are recovered from anaerobic clinical infections far less frequently, but can be responsible for life-threatening complications. *C. perfringens* is commonly associated with gas gangrene (myonecrosis). This organism is an inhabitant of the large bowel and may contaminate the perianal skin and other body sites. The more common species of Clostridium submitted to the Center for Disease Control from clinical sources include *C. perfringens, C. ramosum, C. septicum, C. sordellii, C. bifermentans, C. butyr-icum, C. innocuum, C. sporogenes, C. suberminale,* and *C. tertium.*[12] Many of these species are commonly encountered as contaminants; however, *C. septicum* in particular may be clinically significant, and has been reported to have a high association with bacteremia in patients with neoplastic disease.

At least five species of the gram-positive, non-spore-forming bacilli can cause actinomycosis in humans[19]: *Actinomyces israelii, A. naeslundii, A. viscosus, A. odontolyticus,* and *Arachnia proprionica.* Although encountered in relatively low frequency in routine specimens, these species are well-documented pathogens. *Bifidobacterium eriksonii,* a common isolate from pulmonary anaerobic infections, is the only documented pathogenic species of this genus.[14] *Propionibacterium acnes,* usually a contaminant in clinical specimens, has been recovered from cases of endocarditis and other diseases, frequently associated with implanted prosthetic devices.

The anaerobic cocci most commonly encountered in clinical specimens include *Peptococcus asaccharolyticus, Peptococcus prevotii, Peptostreptococcus anaerobius,* and *Streptococcus intermedius.*

ISOLATION OF ANAEROBIC BACTERIA

The steps involved in the laboratory diagnosis of anaerobic bacterial infections are similar to those described in Chapter 1. It is particularly important that attention be paid to the proper selection, collection, and transport of clinical specimens for the recovery of anaerobic bacteria. The processing of specimens, selection of media, inoculation and incubation methods, and the inspection of positive cultures are laboratory procedures that must be carefully quality controlled.[11] Failure to perform any one step correctly may lead to erroneous results, thus potentially supplying misinformation to the physician.[11]

Since Chapter 1 covers each of these steps in detail, only a few comments will be included here as they pertain specifically to the anaerobic bacteria.

Selection of Specimens for Culture

With few exceptions, all material collected from sites not harboring an indigenous flora, such as body fluids other than urine, exudates from deep abscesses, transtracheal aspirates or direct lung aspirates, and tissue biopsies, should be cultured for anaerobic bacteria. However, since anaerobes normally inhabit the skin and mucous membranes as part of the normal indigenous flora, the following specimens are unacceptable for anaerobic culture because the results cannot be interpreted:

1. Throat or nasopharyngeal swabs
2. Gingival swabs
3. Sputum or bronchoscopic specimens
4. Gastric contents, small bowel contents, feces, rectal swabs, colocutaneous fistulae, colostomy stomata
5. Surfaces of decubitus ulcers, swab samples of encrusted walls of abscesses, mucosal linings, and eschars
6. Material adjacent to skin or mucous membranes other than the above which have not been properly decontaminated
7. Voided urine
8. Vaginal or cervical swabs

Collection and Transport of Specimens

When collecting cultures from mucous membranes or the skin, stringent precautions must be taken to properly decontami-

nate the surface. A surgical soap scrub, followed by application of 70 per cent ethyl or isopropyl alcohol and tincture of iodine, and then removal of the iodine with alcohol, is the method recommended.

A needle and syringe should be used whenever possible for the collection of specimens for anaerobic culture. Collection of swab specimens should be discouraged because they dry out and also because they expose anaerobes, if present, to ambient oxygen. Once collected, particular precautions should be taken to protect specimens from oxygen exposure and to promptly deliver them to the laboratory.

Blood culture techniques should always allow for the recovery of obligate anaerobes as well as aerobes, facultative anaerobes, and microaerophiles. The culture medium should be nutritionally adequate to support the growth of fastidious strains, and inclusion of a reducing agent such as cysteine and an anticoagulant such as sodium polyanethol sulfonate helps to improve recovery of anaerobes. Most commercial blood culture media are placed in evacuated bottles relatively free of oxygen and with added CO_2.

Because some anaerobic bacteria may grow slowly in blood culture media and do not produce visible cloudiness, blind subcultures should be performed after 48 hours of incubation and gram-stained smears made as a routine. Negative cultures should be held for 5 to 7 days or as long as 10 days before reporting as negative.

Direct Examination of Clinical Materials

Gross examination of specimens is particularly valuable in alerting one to the possible presence of anaerobes. A foul odor, purulent appearance of fluid specimens, and the presence of necrotic tissue and gas or sulfur granules are all valuable clues.[14]

The importance of microscopic examination of clinical specimens has been emphasized by several authors,[12,27,40] and the information derived may give immediate presumptive evidence that anaerobes are present. The type of background and cellular characteristics of the smear should be

Table 10-3. *Representative Media for Recovery of Anaerobes from Clinical Specimens*

Medium	Purpose
Blood agar (BAP)	Nonselective plating medium for general use
Chocolate agar (CA)	Used primarily for recovery of species of Hemophilus and Neisseria; however, it may be required for recovery of some fastidious species of anaerobes as well.
MacConkey agar (MAC)	Recovery of aerobic and facultatively anaerobic gram-negative bacilli
Phenylethyl alcohol blood agar (PEA)[12]	Recovery of facultatively anaerobic gram-positive cocci, and for gram-positive and gram-negative obligate anaerobes
Kanamycin-vancomycin blood agar (KV)	Selective recovery of anaerobic gram-negative bacilli
Thioglycollate medium (BBL-0135C) enriched with hemin and vitamin K_1 (THIO)	An enrichment broth that can be used to supplement plating media, particularly if the specimen is scanty
Modified Thayer-Martin medium (MTM)	Used as a supplement to anaerobic media when *Neisseria gonorrhoeae* or *Neisseria meningitidis* is suspected in the specimen

observed, and the Gram's stain reaction, the size, shape and arrangement of bacteria, and the relative number of organisms present should be recorded. The presence of spores, their shape and position in the bacterial cell, and other distinctive morphologic features such as branching, filaments with spherical bodies, pointed ends, and granular forms should be noted. Although the Gram's stain is ordinarily satisfactory for determining cellular characteristics, the Giemsa and Wright's stains may occasionally reveal valuable additional information. The direct microscopic morphology in smears is illustrated in the photomicrographs included in Plates 10-1, 10-3, and 10-4.

Selection and Use of Media

The media employed for the recovery of anaerobes from specimens should include nonselective, selective, and enrichment types, as illustrated in Table 10-3. Other media may also be included or substituted for those listed in Table 10-3. For example, chopped meat glucose medium is commonly used instead of thioglycollate broth; eosin-methylene blue (EMB) may be selected over MacConkey agar; and colistin-nalidixic acid (CNA) agar may be used instead of phenylethyl alcohol (PEA) agar. However, the results obtained with these alternative media may not be entirely comparable and must be interpreted accordingly.

The anaerobe blood agar currently used at the Center for Disease Control and at Indiana University is recommended. It contains 5 per cent defibrinated rabbit or sheep blood added to trypticase soy agar (BBL), with L-cystine, yeast extract (Difco), vitamin K_1, and hemin added. The formula is:

Trypticase soy agar (BBL)	15.0 g.
Phytone (BBL)	5.0 g.
Sodium chloride	5.0 g.
Agar	20.0 g.
Yeast extract (Difco)	5.0 g.
Hemin	5.0 mg.
Vitamin K_1 (3-phytylmenadione)	10.0 mg.
L-cystine	400.0 mg.
Demineralized water	1000.0 ml.
Blood (sheep or rabbit), defibrinated	50.0 ml.

Plates of this medium are commercially available from Carr-Scarborough Microbiologicals, Stone Mountain, Georgia, and Nolan Laboratories, Tucker, Georgia. The commercially prepared anaerobe blood agar plates can be stored in the refrigerator within cellophane bags for up to six weeks.

Prior to use, the plates are held overnight in an anaerobic jar or an anaerobic glove box in an atmosphere of 85 per cent N_2, 10 per cent H_2, and 5 per cent CO_2. An added benefit of the CDC anaerobe blood agar described above is that the added L-cystine in the medium permits the growth of certain thiol-dependent or sulfur-containing amino-acid-requiring bacteria such as *Fusobacte-*

rium necrophorum[2,29] and fastidious strepto-cocci that have been isolated from patients with endocarditis. This medium also supports excellent growth of the strict anaerobes *Clostridium novyi* type B, and *Clostridium haemolyticum*.[2]

The phenylethyl alcohol blood agar is prepared by supplementing the anaerobe blood agar described above with 0.25 per cent phenylethyl alcohol. Similarly, the kanamycin-vancomycin blood agar is prepared by adding 100 μg. of kanamycin and 7.5 μg. of vancomycin per ml. of the blood agar medium.

The enriched thioglycollate medium (BBL-135C with hemin and vitamin K_1 supplement) is primarily recommended as a back-up to the plating media.[12] This medium is particularly helpful for cultivating slow-growing species of Actinomyces and Arachnia.[11] Chopped meat glucose broth is useful for isolation of Clostridum species by the spore selection technique and as a holding medium for anaerobic cultures in general. Prereduced anaerobically sterilized (PRAS) media in roll tubes are recommended by the Virginia Polytechnic Institute (VPI) anaerobe laboratory[27] for isolation of anaerobes. These media are now available from Carr-Scarborough Microbiologicals, Stone Mountain, Georgia, and Scott Laboratories, Fisherville, Rhode Island.

ANAEROBIC SYSTEMS FOR THE CULTIVATION OF ANAEROBIC BACTERIA

Comparative studies have shown that the following systems are satisfactory for the cultivation of anaerobic bacteria commonly associated with human disease.[28a,35]

1. Jar techniques
   Evacuation-replacement
   GasPak method
2. Anaerobic glove box techniques
3. Roll tube and roll-streak tube with prereduced anaerobically sterilized media.

The following general principles must be followed for optimal results:

1. Proper collection and transport of the clinical specimen
2. Processing of specimens with minimal exposure to atmospheric oxygen
3. Use of fresh or prereduced media
4. Proper use of an anaerobic system with inclusion of an active catalyst to allow removal of oxygen (from jar or glove box systems)

Anaerobic Jar Techniques

The different jars in current use for the cultivation of anaerobic bacteria include the Brewer, Baird-Tatlock, GasPak, McIntosh-Fildes, and Torbal jars. The GasPak jar, illustrated in Fig. 10-2, is the system most commonly used in clinical laboratories in the United States. The basic principle of these jars is the same; namely, the removal of oxygen from the chamber by reaction with hydrogen added to the system in the presence of a catalyst. Oxygen is reduced to water as follows:

$$2H_2 + O_2 \longrightarrow 2H_2O$$

The use of an active catalyst in each system is important. The older Brewer jar technique used a palladium catalyst in the lid of the jar (modified McIntosh-Fildes jar), which had to be heated with an electric current to be fully active. The GasPak jar uses a "cold" catalyst, composed of palladium-coated alumina pellets, which does not require heating. The cold catalyst is not only more convenient, but also has no explosion hazard. The cold palladium catalyst can be inactivated in the jar by the production of hydrogen sulfide or other volatile metabolic products of the bacteria. It is recommended that the catalyst pellets be replaced with new or rejuvenated pellets each time the jar is used.[12] The pellets can be rejuvenated or restored to full activity by heating them in a dry-heat oven at 160 to 170° C. for 2 hours. After heating, the pellets are stored at room temperature in a clean, dry container or in a desiccator until the time of use.

Anaerobic conditions can be produced in jar systems with either the disposable Gas-Pak hydrogen-carbon dioxide generator (BioQuest) or by the evacuation-replacement procedure. The evacuation-replace-

Fig. 10-2. The GasPak (BBL; Division of Becton and Dickinson and Co., Cockeysville, Md. 21020) anaerobic system. The jar contains inoculated plates, broth tubes, a GasPak hydrogen and carbon dioxide generator envelope, a disposable methylene blue indicator strip and a catalyst basket in the lid.

ment procedure, in which the air in the jar is flushed and replaced with a mixture of 85 per cent N_2, 10 per cent H_2, and 5 per cent CO_2, is more economical than GasPak generators and allows anaerobic conditions to be established more rapidly. Any airtight container can be used, including a GasPak jar with a vented lid, a Brewer jar, Torbal jar, or even a modified pressure cooker.

Whaley and Gorman[46] have recently described an inexpensive device for evacuating and gassing anaerobic systems. This device can be used with an in-house vacuum, thereby eliminating the need for a vacuum pump when the evacuation-replacement procedure is used. Air is evacuated from the jar by drawing a vacuum of 20 to 24 inches of mercury. This procedure is repeated three times. The jar is then filled with N_2 after the first two evacuations and the final replace-

ment is made with the 85 per cent N_2, 10 per cent H_2, and 5 per cent CO_2 gas mixture.

The disposable GasPak H_2 and CO_2 generators are used by opening the generator envelope and placing it into the anaerobe jar to be used. Approximately 10 ml. of water is added to allow the generation of hydrogen and carbon dioxide, and the lid is tightly sealed. If the lid is not warm to the touch within a few minutes after it is sealed, or if condensation does not appear on the inner surface of the glass within 30 minutes, the jar should be opened and the generator envelope discarded. A defective gasket in the lid that allows escape of gas or inactivated catalyst pellets are the two most common causes of failure of this system.

Anaerobic conditions should also be monitored when using either of the two jar techniques by the inclusion of an oxidation-reduction indicator. Methylene blue strips are currently available commercially (BBL). Alternatively, a 13 × 100-mm. test tube containing a few milliliters of methylene blue-$NaHCO_3$-glucose mixture[12] can be placed in the jar. Methylene blue is blue when oxidized, white when reduced. The color changes at $+11.0$ mv. Thus, if anaerobic conditions are achieved, the methylene blue indicator solution will gradually turn colorless and will remain that way if there are no leaks which allow additional oxygen to enter the system. If the solution turns blue after being colorless, this indicates a failure to properly establish anaerobic conditions within the jar and the culture results may not be valid.

Use of the Anaerobic Glove Box

An anaerobic glove box is a self-contained anaerobic system which allows the microbiologist to process specimens and perform most bacteriologic techniques for isolation and identification of anaerobic bacteria without exposure to air. Glove boxes suitable for cultivation of anaerobes can be constructed from various materials, including steel, acrylic plastic, or fiberglass (Fig. 10-3). The flexible vinyl plastic anaerobic

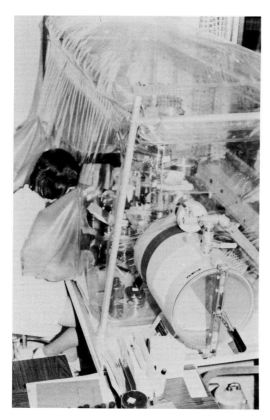

Fig. 10-3. The anaerobic glove box (Coy Laboratory Products, Inc., Ann Arbor, Michigan 48106) anaerobic system. Materials are passed in and out of the large flexible plastic chamber through the entry lock in the foreground. Anaerobic conditions are maintained by the constant recirculation of the atmosphere within the plastic chamber (85% N_2, 10% H_2, 5% CO_2) through palladium catalyst. Cultures are incubated either within a separate incubator inside the glove box or by maintaining the entire chamber at 35°C. through use of heated catalyst boxes.

Fig. 10-4. Uninoculated (left) and inoculated (right) brain heart infusion agar roll tubes following incubation. The agar in the tube at right was streaked by slowly drawing the edge of the inoculating loop against the agar from bottom to top while the tube revolved. Tubes are gassed with CO_2 during inoculation and subculture.

chamber developed by Freter and coworkers[3] at the University of Michigan has enjoyed wide popularity, and a modification of this design is available in varying sizes from the Coy Manufacturing Co., located in Ann Arbor, Michigan.

An anaerobic glove box, if properly constructed, is economical to operate because it permits the use of conventional plating media and the cost for gases for operation of the system is minimal. Once set up, the major expense is for the nitrogen and the nitrogen-hydrogen-carbon dioxide gas mixture used to replace the air in the entry lock when materials are passed into the glove box chamber.

The Roll-Streak System

The roll-streak system, developed by Moore and colleagues[27] at the Anaerobe Laboratory, Virginia Polytechnic Institute

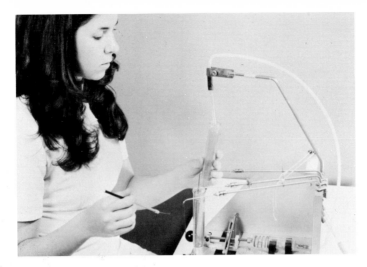

Fig. 10-5. Use of the Virginia Polytechnic Institute (VPI) anaerobic culture system (Bellco, Inc., Vineland, N.J. 08360). Included in the system are foot-treadle-operated cannulas that deliver oxygen-free carbon dioxide into the roll tubes during inoculation and subculture, a streaker that turns the tubes during inoculation, and a semiautomatic inoculator.

Fig. 10-6. Close-up view of roll tube showing at least two colony types on the agar surface.

and State University, is a modification of the roll tube technique developed by Hungate and associates for culturing anaerobic bacteria from the rumen of cows and other herbivorous animals. Equipment for the Virginia Polytechnic Institute anaerobic culture system is available commercially from Bellco Glass Corp, Vineland, New Jersey.

The roll-streak system utilizes prereduced, anaerobically sterilized (PRAS) media which are prepared in tubes with rubber stoppers (Fig. 10-4). After autoclaving, the tubes of agar media are cooled in a rolling machine which results in a thin coating of the inner surfaces of the tubes with solidified medium. Both the roll-streak tubes and the PRAS liquid media require the addition of a reducing agent, such as L-cysteine-hydrochloride, which is added just before autoclaving to help maintain a low oxidation-reduction potential within the system. All inoculating and subculturing of the PRAS solid and liquid media are performed under a stream of oxygen-free carbon dioxide which minimizes exposure to air and helps to maintain a reduced oxidation-reduction potential in the media before and after growth of the obligate anaerobes (Fig. 10-5).

This system has certain advantages and disadvantages. The major advantages are: (1) inspection and subculture can be done at any time without undue exposure of anaerobic bacteria to atmospheric oxygen Fig. (10-6); and (2) only minimal bench and incubator space is required since each tube has its own self-contained anaerobic environment. Some of the disadvantages include the need for the special equipment, a period of training for personnel to become adept at using the system, and the relatively high cost of the roll tube media.

Use of the Anaerobic Holding Jar

A modification of the Martin holding jar procedure[31] is a convenient and inexpensive adjunct to the jar and glove box anaerobic systems, which allows primary plating, inspection of cultures, and subculture of colonies at the bench with only minimal exposure of anaerobic bacteria to atmospheric oxygen[2]. The holding jar assembly is illustrated in Fig. 10-7, and its use is briefly described as follows:

---

1. Three holding jars are used, the first to hold uninoculated media, the second for plates on which are growing colonies to be subcultured, and the third to receive freshly inoculated plates of media.
2. Commercially prepared agar plates or agar media freshly prepared in the laboratory can be used. These may be held in a refrigerator for up to six weeks if each individual plate is placed in a cellophane bag.
3. The plates to be used on any given day should first be placed in an anaerobic glove box or a GasPak jar for 18 to 24 hours prior to use in order to reduce the media.
4. As needed, the reduced media are placed in the first holding jar and continuously flushed with a gentle stream of nitrogen gas.
5. The plates of reduced media are surface inoculated, one at a time, in ambient air and immediately placed in the third holding chamber which is also flushed with nitrogen gas. The second holding jar is used to hold any plates removed from the GasPak jar that require subculture.
6. After the jar holding the newly inoculated plates is filled, the plates can be transferred to a conventional anaerobic system such as a GasPak jar or into an anaerobic glove box for incubation at 35° C.

Inexpensive commerical-grade N_2 can be used in the holding jar system. Open the small needle valve on the gas manifold (Fig. 10-7) and set the gas-tank regulator to 4 lbs./in.² for 20 to 30 seconds to rapidly purge the jar of air. Then turn the regulator pressure down to about ½ to 1 lb./in.² and regulate the flow to each jar at 50 to 100 ml./min., using the small needle valve on the manifold. This is equivalent to a flow rate of 1 to 2 bubbles/sec. when the rubber tubing in the jar is placed just beneath the surface in a beaker of water. Alternatively, CO_2 passed through a tube of heated copper catalyst (Sargent furnace) can be used in the holding jars instead of N_2.[2,31]

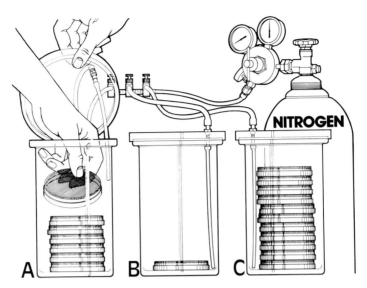

Fig.10-7. Illustration of the anaerobic holding jar system. The flow rate of nitrogen to each jar is regulated by the needle valves on the manifold (three gang valve, available where aquarium supplies are sold). Jars A, B and C contain uninoculated plates, plates with colonies to be subcultured, and freshly inoculated plates, respectively.

INCUBATION OF CULTURES

In most instances, 35 to 37° C. is the temperature most satisfactory for the primary isolation of anaerobic bacteria from clinical specimens. Plates inoculated at the bench and placed in anaerobic jars should be incubated for at least 48 hours and preferably for 72 to 96 hours before the jars are opened for the colonies to fully develop; some anaerobes, such as certain species of Actinomyces, Arachnia, and Eubacterium, grow rather slowly, and colonies may not be detected if jars are opened sooner. Also, if the jar is opened too soon, some of the slow-growing organisms may be killed due to oxygen exposure. In emergency situations, duplicate sets of plating media can be incubated in two different jars, one set incubated for 18 to 24 hours and the other for 3 to 5 days. This procedure allows rapid isolation of fast-growing anaerobes in the 18- to 24-hour jar and the later recovery of slow growers in the jars left for delayed incubation. If clostridial myonecrosis is clinically suspected, plates can be inspected as early as 6 to 12 hours after inoculation.

Prolonged exposure of freshly inoculated plates or those that have been previously incubated to the ambient air must be avoided. Certain anaerobes commonly encountered in clinical specimens, such as *Peptostreptococcus anaerobius,* may either fail to grow or may exhibit a prolonged growth lag when freshly inoculated plates are held in ambient air for as short a time as 2 hours. Thus, if a holding jar procedure is not used, inoculated plates must be immediately placed in an anaerobic system (Gas-Pak jar or anaerobic glove box) to allow for effective cultivation of these anaerobes.

Enriched thioglycollate and chopped meat glucose media should also be incubated after inoculation in an anaerobic system to allow maximum recovery of anaerobes. PRAS media in rubber-stoppered tubes can also be used. It is no longer necessary to boil the tubes of enriched thioglycollate or chopped meat glucose broth if they are prepared in tight-fitting screwcap tubes and gassed in a glove box after auotclaving. Unless growth is visually apparent, broth cultures should be held a minimum of 2 weeks before discarding as negative.

INSPECTION AND SUBCULTURE OF COLONIES

After incubation, anaerobic and CO_2 plates should be examined with a hand lens and a dissecting microscope. If anaerobe jars are used, a holding jar system should be employed at the time of colony examination and subculture to minimize exposure of oxygen-sensitive isolates to air. Of course, the anaerobic glove box and the VPI roll streak systems both allow inspection and subculture of colonies in the absence of air.

Use of a stereoscopic dissecting microscope during the examination of colonies is extremely useful because a number of anaerobes have distinctive colony features.[12] The dissecting microscope is also a valuable aid during the subculture of colonies to obtain pure culture isolates.

During the inspection of colonies, any action on the medium, such as hemolysis of blood agar or clearing of egg yolk agar, as well as the size and distinctive features of the colonies should be recorded.[12,13] A number of characteristic colonies of anaerobes are illustrated in Plates 10-1, 10-3, and 10-4. When recording colony characteristics, the following should be noted: the age of the culture and the name of the medium, the diameter in millimeters of each colony in addition to its color, surface features (glistening, dull), density (opaque, translucent), consistency (butyrous, viscid, membranous, brittle), and other descriptive features (see Fig. 1-30).

Gram-stained smears of colonies from the anaerobic and CO_2-incubated plates should also be examined. Do not assume on the basis of colony and microscopic features only that colonies on plates which have been incubated in an anaerobic system are obligate anaerobes. Although the morphology and colony characteristics of certain anaerobes are distinctive, it is often impossible to distinguish some facultative anaerobes from obligate anaerobes without aerotolerance tests, even when the CO_2-incubated plates show no growth.

The number of different colony types on the anaerobe plates should be determined

and a semiquantitative estimate of the number of each type should be recorded (light, moderate, or heavy growth). Using a needle or a sterile Pasteur capillary pipette, transfer each different colony to another anaerobe blood agar plate to obtain a pure culture of each. If colonies are well separated on the primary isolation plate, inoculate a tube of enrichment broth, such as enriched thioglycollate or chopped meat glucose medium, to provide a source of inoculum for differential tests or for aerotolerance studies to be described below.

In general, a tube of enriched thioglycollate medium is recommended for the study of non-spore-forming anaerobes. The chopped meat glucose medium is more suitable for cultivation of clostridia that are to be tested with various differential media or when clostridial toxins are to be demonstrated from broth cultures.

After incubation, gram-stain the enriched thioglycollate and chopped meat glucose subcultures. If the organisms appear to be in pure culture, they can be used to inoculate appropriate differential media for identification of isolates.

Examine enriched thioglycollate and chopped meat glucose cultures that were incoulated with the original specimen along with all primary isolation plates. If no growth is evident on the primary anaerobic plates, or if the colonies isolated fail to account for all the morphologic types found in the direct gram-stained smear of the specimen, subculture each broth medium to anaerobe blood agar plates for anaerobic incubation and also to blood agar plates for aerobic CO_2 incubation. These subculture plates should then be examined as described above.

PERFORMANCE OF AEROTOLERANCE TESTS

Each colony type from the anaerobic isolation plate is subcultured to an aerobic-CO_2 (5% CO_2, or candle jar) and anaerobic blood agar plate for overnight incubation. It is expedient to inoculate quadrants or sixths of one anaerobe blood agar and one plain

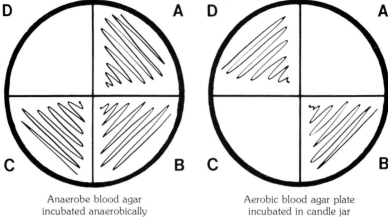

Anaerobe blood agar
incubated anaerobically

Aerobic blood agar plate
incubated in candle jar

Fig. 10-8. Illustration of quadrant plating technique used for aerotolerance testing of four anaerobe isolates. The left plate has been incubated in an anaerobe jar for 18 to 24 hours, while the right-hand plate was incubated in a candle jar. Isolates A and C are obligate anaerobes. Isolate B is facultatively anaerobic. Isolate D is either a microaerophile or obligate aerobe and should be further tested for its ability to grow in ambient air compared with the environment containing increased CO_2 (candle jar).

aerobic blood agar plate for testing the aerotolerance of 4 to 6 colonies from a primary isolation plate (as shown in Fig. 10-8).

PRELIMINARY REPORTING OF RESULTS

Organisms that are shown to be obligate anaerobes should immediately be reported to the clinician together with the results of Gram's stain observations and colonial morphology. However, it is not justified to report the "confirmed" presence of an obligate anaerobe until aerotolerance studies have been completed.

Unfortunately, a period of 3 days or longer often is required for these studies to be completed. Clinicians should be made aware that this lengthy time cannot be avoided with some slow-growing anaerobes (e.g., some species of Actinomyces, Arachnia, and Propionibacterium). Fortunately, the colonial and microscopic morphology of certain anaerobic bacteria is often so distinctive that *preliminary* or *presumptive* reports of these isolates can be made prior to aero-

tolerance studies. Examples include *Clostridium perfringens*, members of the *Bacteroides fragilis* group, *B. melaninogenicus*, and others.

IDENTIFICATION OF ANAEROBIC BACTERIA

Classification of Anaerobes

In edition 7 of *Bergey's Manual,*[6] various morphologic features were used as primary characteristics for the classification of anaerobic bacteria. This system, however, was soon found to be inadequate and it became evident that additional tests were needed to accurately characterize the anaerobes. In addition to the classification in edition 8 of *Bergey's Manual,*[8] other systems of nomenclature were published during the early 1960's.[5,33,34,47] This created a chaotic situation. Not only was there a plethora of different genus and species names, but there was also considerable disagreement among various workers regarding criteria for differentiating species.

In 1968, Dowell and Hawkins compiled

differential tables from data tabulated from the characterization of reference strains and numerous isolates received at the CDC anaerobe laboratory from state health department and federal laboratories throughout the United States. In contrast to the identification schema published in edition 7 of *Bergey's Manual*,[6] the Dowell-Hawkins tables were based on data generated from batteries of biochemical and cultural tests performed on all anaerobe isolates received by this reference laboratory during the 1960's.

For example, the majority of the saccharolytic bacteroides isolates were placed in five groups: *Bacteroides fragilis, B. incommunis, B. oralis, B. variabilis,* and *B. terebrans.* It was emphasized at that time that the species names and group designations in the CDC tables were tentative, pending the adoption of acceptable nomenclature by taxonomists.[10]

Soon thereafter, Moore and associates at the Virginia Polytechnic Institute reported the use of metabolic product analysis by gas-liquid chromatography (GLC) and described various other differential tests for the characterization of anaerobes based on reactions in prereduced anaerobically sterilized (PRAS) media. They also proposed various changes in the classification of non-spore-forming anaerobes, many of which were subsequently included in edition 8 of *Bergey's Manual.*[8] The saccharolytic bacteroides listed in the 1970 VPI anaerobe manual are given below, and the key characteristics which were proposed for differentiating the five subspecies of *B. fragilis* are presented in Table 10-4.

Saccharolytic Bacteroides Listed in the VPI Anaerobe Laboratory Manual, 1970*

Bacteroides fragilis
  ss *fragilis*
  ss *distasonis*
  ss *ovatus*
  ss *thetaiotaomicron*
  ss *vulgatus*
Bacteroides oralis
  ss *oralis*
  ss *elongatus*
Bacteroides trichoides
Bacteroides biacutus

*Adapted from Anaerobe Laboratory Manual, Virginia Polytechnic Institute: Outline of Clinical Methods in Anaerobic Bacteriology, Blacksburg, Va., VPI, 1970.

During 1972, the CDC anaerobe laboratory began using a new battery of tests for characterizing anaerobes which allowed for differentiating *B. fragilis* into subspecies. Species of previously identified saccharolytic bacteroides were reexamined using these tests and the results are shown in Table 10-5.

The major differences between the new and old data were that approximately 10 per cent of the strains previously identified as *B. fragilis* were found to be *B. fragilis* ss *fragilis,* and that other strains initially called *B. oralis* were identified as *B. fragilis. B. fragilis* ss *ovatus* was conspicuously absent from the clinical isolates submitted to the CDC for identification during the period from April 1971 to March 1972 (Table 10-6.)

To date (January 1978), *B. fragilis* ss *ovatus* has not been isolated from a properly collected clinical specimen of a normally sterile area of the body in any samples re-

Table 10-4. *Key Characteristics for Subspeciating Bacteroides fragilis*

Subspecies	Indole	Mannitol	Trehalose	Rhamnose
fragilis	−	−	−	−
distasonis	−	−	+	Variable
ovatus	+	+	+	+
thetaiotaomicron	+	−	Variable	+
vulgatus	−	−	−	+

*Adapted from Anaerobe Laboratory Manual, Virginia Polytechnic Institute: Outline of Clinical Methods in Anaerobic Bacteriology, Blacksburg, Va., VPI, 1970.

Table 10-5. *Nomenclature Changes Resulting from Reexamination of 50 Saccharolytic Bacteroides, CDC, 1972**

Number of Strains	Former Name	New Name	Number of Strains
35	B. fragilis	B. fragilis ss fragilis	32
		ss distasonis	3
2	B. incommunis	B. fragilis ss vulgatus	2
6	B. variabilis	B. fragilis ss thetaiotaomicron	6
2	B. terebrans	B. trichoides†	2
5	B. oralis	B. oralis	1
		B. fragilis ss fragilis	4

*Modified from Dowell, V. R., Jr., *et al*: Biotyping of anaerobic bacteria associated with human disease. *In* Isenberg, H. D., and Balows, A. (eds.): Biotyping in the Clinical Microbiology Laboratory. Springfield, Charles C Thomas, 1978.
†Subsequently identified as *Clostridium ramosum*.[25]

ferred to the CDC. However, this subspecies appears to reside in large numbers in the human intestinal tract as part of the normal microbiota.[16,26]

Clarification of the taxonomy of the saccharolytic bacteroides groups has now paved the way for meaningful clinical studies to assess the role of these organisms in disease processes. Thus, *B. fragilis* ss *fragilis* is by far the most common anaerobic, non-spore-forming, gram-negative bacillus isolated from clinical specimens, although it may be outnumbered by other subspecies of *B. fragilis* in the gastrointestinal tract. This suggests a fundamental pathogenic difference between subspecies *fragilis* and other subspecies of *B. fragilis*.

The organism listed as *Bacteroides trichoides* in Table 10-5 (as well as those formerly called *Bacteroides terebrans*, *Catanabacterium filamentosum*, and *Eubacterium filamentosum*) is now classified as *Clostridium ramosum*.[25] Some strains of this species are easily confused with species of Bacteroides because of their tendency to stain gram-negative and their failure to produce spores on media routinely used for culturing anaerobes.[12,26] In 1976, the subspecies of *B. fragilis* were reinstated as *B. fragilis*, *B. ovatus*, *B. distasonis*, *B. thetaiotaomicron*, and *B. vulgatus*, since the former subspecies of *Bacteroides fragilis* have been shown to be genetically distinct.[9]

There has been recent progress, not only in improving the current taxonomic status of saccharolytic bacteroides, but also in the classification of the various other anaerobes, much to the credit of workers at the VPI anaerobe laboratory.[27] A list of changes in the nomenclature of certain anaerobes encountered in human specimens is given in Table 10-7. A more exhaustive list of synonyms and nomenclature changes has been published by Finegold.[14]

Table 10-6. *Distribution of B. fragilis Subspecies* in the Clinical Isolates of Saccharolytic Bacteroides Identified by the CDC from April 1971 to March 1972*

Subspecies	Number
B. fragilis ss fragilis	49
ss thetaiotaomicron	10
ss distasonis	6
ss vulgatus	5
ss ovatus	0

*Modified from Dowell, V. R., Jr., *et al*: Biotyping of anaerobic bacteria associated with human disease. *In* Isenberg, H. D., and Balows, A. (eds.): Biotyping in the Clinical Microbiology Laboratory. Springfield, Charles C Thomas, 1978.
†Subspecies identified as described in 1970 VPI Manual. Anaerobe Laboratory, Virginia Polytechnic Institute: Outline of Clinical Methods in Anaerobic Bacteriology, Blacksburg, Va., VPI, 1970.

Table 10-7.   *Changes in the Nomenclature of Anaerobic Bacteria**

Current Name	Former Name(s)
Arachnia propionica	*Actinomyces propionicus*
Bacteroides fragilis	*Bacteroides fragilis* ss *fragilis*
	Bacteroides fragilis
Bacteroides distasonis	*Bacteroides fragilis* ss *distasonis*
	Bacteroides fragilis
Bacteroides vulgatus	*Bacteroides fragilis* ss *vulgatus*
	Bacteroides incommunis
Bacteroides thetaiotaomicron	*Bacteroides fragilis* ss *thetaiotaomicron*
	Bacteroides variabilis
Bacteroides ovatus	*Bacteroides fragilis* ss *ovatus*
	Bacteroides ovatus
Bacteroides melaninogenicus	*Bacteroides melaninogenicus*
ss *melaninogenicus*	*Bacteroides melaninogenicus*
ss *asaccharolyticus*	*Bacteroides melaninogenicus*
ss *intermedius*	*Bacteroides melaninogenicus*
Bacteroides oralis	
ss *oralis*	*Bacteroides oralis*
ss *elongatus*	*Bacteroides oralis*
Bacteroides ochraceus	*Bacteroides oralis* ss *elongatus*
Bacteroides pneumosintes	*Dialister pneumosintes*
Bifidobacterium eriksonii	*Actinomyces eriksonii*
Clostridium cadaveris	*Clostridium capitovale*
Clostridium clostridiiformis	*Bacteroides clostridiiformis* ss *girans*
	Fusobacterium girans
	Bacteroides clostridiiformis ss *clostridiiformis*
Clostridium limosum	*Clostridium* species group P$_1$
Clostridium ramosum	*Catenabacterium filamentosum*
	Bacteroides trichoides
	(*B. terebrans*)
Clostridium symbiosum	*Fusobacterium symbiosum*
Eubacterium alactolyticum	*Ramibacterium alactolyticum*
	Ramibacterium pleuriticum
Eubacterium lentum	*Corynebacterium diphtheroides* (CDC Manual), *Bifidobacterium cornutum*, *Corynebacterium* group 3
Fusobacterium necrophorum	*Sphaerophorus necrophorus*
Fusobacterium nucleatum	*Fusobacterium fusiforme*
Fusobacterium mortiferum	*Fusobacterium ridiculosum*
	Sphaerophorus ridiculosum
Lactobacillus catenaforme	*Catenabacterium catenaforme*
Propionibacterium acnes	*Corynebacterium acnes*
	Corynebacterium parvum
	Corynebacterium anaerobium Prévot (some strains)
	Corynebacterium liquefaciens
Propionibacterium freudenreichii ss *freudenreichii*	*Propionibacterium freudenreichii*
Propionibacterium freudenreichii ss *shermanii*	*Propionibacterium shermanii*
Propionibacterium granulosum	*Corynebacterium granulosum*
Peptococcus asaccharolyticus	Peptostreptococcus CDC group 1
Peptococcus magnus	Peptostreptococcus CDC group 2
Peptococcus prevotii	Peptostreptococcus CDC group 2
Peptostreptococcus anaerobius	Peptostreptococcus CDC group 3
Streptococcus constellatus	*Peptococcus constellatus*
Streptococcus intermedius	*Peptostreptococcus intermedius*
	Streptococcus MG
Streptococcus morbillorum	*Peptococcus morbillorum*

*Modified and updated from Dowell, V. R., Jr., and Hawkins, T. M.: Laboratory Methods in Anaerobic Bacteriology. CDC Laboratory Manual. DHEW Publication No. (CDC) 74-8272, Atlanta, Ga., Center for Disease Control, 1974.

Table 10-8.   *Media and Incubation Conditions for Presumptive Identification of Anaerobic, Non-spore-forming Gram-Negative Bacilli***

Media	Incubation Conditions (35°C)
Anaerobe blood agar	Anaerobic system
Blood agar (¼ plate)	CO_2 incubator
Blood agar (¼ plate)	Air
Enriched thioglycollate medium	Anaerobic system
Presumpto plate	Anaerobic system
Anaerobe blood agar plus 2-U. penicillin, 15-μg. rifampin, and 1000 μg. kanamycin disks	Anaerobic system

*From Dowell, V. R., Jr., and Lombard, G. L.: Presumptive Identification of Anaerobic Nonsporeforming Gram-Negative Bacilli. DHEW Publication, Atlanta, Ga., Center for Disease Control, p. 355, 1977.

Presumptive Identification of Anaerobic Bacteria

Reference laboratories, such as those of the CDC anaerobe section, commonly employ rather large batteries of tests in the characterization of anaerobe isolates referred to them for identification or confirmation. These are listed below (see the list, Characterization of Anaerobic Bacteria). The data derived from the characterization of cultures with a large number of tests provide a valuable base for compiling tables of differential characteristics such as those published by the CDC, VPI, and Wadsworth Anaerobe Laboratories. However, in most clinical diagnostic laboratories it is not practical or economically feasible to use such a large number of differential media and biochemical determinations to identify isolates from clinical specimens.

Characterization of Anaerobic Bacteria

Relation to O_2
Colonies
Gram reaction
Microscopic features
Motility
Growth in liquid media
Biochemical tests
Metabolic products (GLC)
Antibiotic susceptibility
Serologic tests
Toxicity, toxin neutralization, pathogenicity in animals

Fortunately, certain characteristics (listed under the heading Some Cardinal Identifying Characteristics of Anaerobes, below) are especially useful in the identification of anaerobes. These characteristics form the basis of a practical approach for the identification of anaerobe isolates.[13]

Some Cardinal Identifying Characteristics of Anaerobes

Relation to O_2
Colonial characteristics
Pigment
Hemolysis
Pitting of medium
Gram reaction
Morphology
Spores
Motility
Flagella
Growth in thioglycollate broth, catalase, lecithinase, lipase, indole, esculin, gelatin; fermentation of key carbohydrates, e.g., glucose, mannitol, mannose; growth in presence of bile, penicillin, rifampin, kanamycin; toxins produced; metabolic products

Only a minimal number of media and biochemical tests are required as listed in Table 10-8. Use of such a procedure for presumptive identification of isolates in a clinical laboratory is feasible because only a limited number of species of Bacteroides and Fusobacterium are commonly isolated from properly collected clinical specimens associated with disease in humans. Species

Table 10-9. *Bacteroides and Fusobacterium Species from Clinical Materials Identified in the CDC Anaerobe Section (1975)*

Species or Group	Number
Bacteroides	
B. fragilis	603
B. thetaiotaomicron	107
CD group F1	54
B. vulgatus	53
CDC group F2	39
B. distasonis	24
B. melaninogenicus ss asaccharolyticus	15
B. melaninogenicus ss intermedius	14
Fusobacterium	
F. nucleatum	114
F. necrophorum	74
F. mortiferum	58
F. varium	15

of Bacteroides and Fusobacterium frequently submitted to the CDC Anaerobe Reference Laboratory from 1966 to 1975 are listed in Table 10-9. Notice that *B. fragilis* (603 isolates) was by far the most commonly submitted species during that period. This agrees with data collected from the four medical centers listed in Table 10-2.

However, this was not the case in 1977 since a larger number of public health and clinical laboratories are now able to identify *B. fragilis* and other commonly encountered anaerobes. Consequently, fewer of these species are now referred to the CDC laboratory than in the past.[13]

The Presumptive Identification of Bacteroides and Fusobacterium Species

Bacteroides or Fusobacterium can be suspected in purulent clinical materials if gram-negative bacilli are observed in gram-stained preparations. The cells of Bacteroides are pale, irregularly staining, and pleomorphic

Table 10-10. *Media and Characteristics of Cultures Which Can Be Identified in the Procedure for Presumptive Identification of Bacteroides and Fusobacterium Species*

Media	Characteristics
Blood agar	Relation to O_2, colonial characteristics, hemolysis, pigment, fluorescence with ultraviolet light (Wood's lamp), pitting of agar, cellular morphology, gram reaction, spores, motility (wet mount); inhibition by penicillin, rifampin, or kanamycin.
Enriched thioglycollate medium	Rapidity of growth, appearance of growth, gas production, odor, cellular morphology
Presumpto plate	
LD agar	Indole, growth on LD medium, catalase*
LD esculin agar	Esculin hydrolysis, H_2S, catalase
LD egg yolk agar	Lipase, lecithinase, proteolysis
LD bile agar	Growth in presence of 20% bile (2% oxgall), insoluble precipitate under and immediately surrounding growth

*The catalase test can be performed by adding 3% hydrogen peroxide to the growth on LD agar but reactions after addition of H_2O_2 to catalase-positive cultures are more vigorous on LD esculin agar.

with irregular swellings. Gram-negative bacilli with pointed ends are more suggestive of Fusobacterium (See Plate 10-1).

Illustrations of the colonial types of the more commonly encountered species of Bacteroides and Fusobacterium are shown in Plate 10-1. The black colonies of *B. melaninogenicus* and the opalescent colonies of *F. nucleatum* are distinctive.

Although Gram's stain morphology and colonial characteristics may be helpful in the identification of species of Bacteroides and Fusobacterium, metabolic and chemical characteristics are required for a more definitive identification. The media and characteristics commonly used for identification of these bacteria are listed in Table 10-10.

The Presumpto Plate.[13] The presumpto plate listed in Table 10-10 is a 4-quadrant Petri dish containing the following media: Lombard-Dowell (LD) agar, LD esculin agar, LD egg yolk agar, and LD bile agar. The preparation of these four media, including a description of the formulas and procedures, is as follows:

Media

1. *LD Agar*

Trypticase (BBL)	5.0 g.
Yeast extract (Difco)	5.0 g.
Sodium chloride	2.5 g.
Sodium sulfite	0.1 g.
L-Tryptophan	0.2 g.
Vitamin K₁ (3-phytylmenadione)	0.01 g.
Agar	20.0 g.
Distilled water	1000.0 ml.
L-cystine	0.4 g.
Hemin	0.01 g.

Dissolve L-cystine and hemin in 5 ml. of 1 N sodium hydroxide before adding to the medium. The vitamin K_1 is added from a 1 per cent stock solution prepared in absolute ethanol. Autoclave at 121° C. for 15 minutes. Final pH of medium should be 7.4 ± 0.1.

2. *LD Esculin Agar*

Trypticase (BBL)	5.0 g.
Yeast extract (Difco)	5.0 g.
Sodium chloride	2.5 g.
L-tryptophan	0.2 g.
Vitamin K₁ (1 ml. of a 1% solution in ethanol)	0.01 g.
L-cystine	0.4 g.

Hemin	0.01 g.
Esculin	1.0 g.
Ferric citrate	0.5 g.
Agar	20.0 g.
Demineralized H₂O	1000.0 ml.

The hemin and L-cystine are dissolved in 5 ml. of 1 N sodium hydroxide before adding to the other ingredients. Autoclave at 121° C. for 15 minutes, Final pH of medium should be 7.4 ± 0.1.

3. *LD Egg Yolk Agar*

LD agar supplemented with glucose, 2 g., Na_2HPO_4, 5 g., and 5 per cent $MgSO_4$, 0.2 ml. per 1000 ml. After the base is autoclaved at 121° C. for 15 minutes and cooled to 55 to 60° C., sterile egg yolk suspension (Difco), 100 ml. per liter, is added, and the medium is dispensed into the quadrant plates.

4. *LD Bile Agar.* The LD bile agar is prepared by supplementing LD agar with 20 g. of oxgall (Difco) and 1 g. of glucose per liter.

Reagents

Paradimethylaminocinnamaldehyde Reagent for Detection of Indole

Dissolve 1 g. of Paradimethylaminocinnamaldehyde and dilute to 100 ml. with dilute hydrochloric acid (10 ml. of concentrated HCl plus 90 ml. of distilled water). Store in a refrigerator at 4° C.

Inoculation and Reading of the Presumpto Plate. The procedure for the inoculation of the media in the presumpto plates, the method of incubation, and the use of selective antimicrobial susceptibility disks are as follows:

Inoculation of Media. As soon as the anaerobe has been isolated on a solid medium, either a turbid cell suspension (McFarland No. 1) in LD broth prepared from isolated colonies or a 24- to 48-hour enriched thioglycollate medium subculture from an isolated colony can be used for inoculating the media for the presumptive identification of the isolate.

1. Place one or two drops of cell suspension or broth culture on each quadrant of the presumpto plate and streak three fourths of the medium with the capillary pipette.

2. Place a sterile, blank, ¼ inch in diameter paper disk on the LD agar near the outer periphery of the quadrant. This disk is used in the test for indole after incubation of the plates.

3. Evenly inoculate the surface of an anaerobe blood agar plate with a sterile swab which has been dipped in the cell suspension or culture.

4. Place the antibiotic disks (penicillin, 2 units, rifampin, 15 μg., kanamycin, 1 mg.)* on the blood agar with sterile forceps. Evenly space the disks so that overlapping zones of inhibition will not be a problem.

Incubation. Incubate the Presumpto Plates and the anaerobe blood agar with antibiotic disks in an anaerobic system such as an anaerobic glove box or an anaerobic jar (e.g., GasPak jar) at 35° C. for 48 hrs.

Observation and Interpretation of Results

1. *LD Agar:*

A. Note and record the degree of growth on LD agar (light, moderate, heavy).

B. Test for indole by adding 2 drops of paradimethylaminocinnamaldehyde reagent to the paper disk on the medium. Observe for the development of a blue or bluish-green color in the disk within 30 seconds which indicates a positive reaction for indole. Development of another color (pink, red, violet) or no color is negative for indole.

2. *LD Egg Yolk Agar:*

A. Formation of a zone of insoluble precipitate in the medium surrounding the bacterial colonies is positive for lecithinase production. This is best seen with transmitted light.

B. The presence of an iridescent sheen ("pearly layer") on the surface of colonies and on the medium immediately surrounding the bacterial growth (best demonstrated with reflected light) is indicative of lipase production.

C. Proteolysis. Clearing of the medium in the vicinity of the bacterial growth indicates proteolysis as exhibited by certain proteolytic clostridia.

3. *LD Esculin Agar:*

A. A positive test for esculin hydrolysis is indicated by the development of a reddish-brown to dark brown color in the esculin agar surrounding the bacterial growth after exposure of the presumpto plate cultures to air for at least 5 minutes. Further evidence for esculin hydrolysis can be obtained by examining the esculin agar quadrant under a Wood's lamp. Esculin agar exhibits a bright blue fluorescence under the ultraviolet light which is not present after the esculin is hydrolyzed.

B. H_2S production. Blackening of the bacterial colonies on the esculin agar is indicative of H_2S production. The blackening dissipates very rapidly after exposure to air. Therefore the bacterial growth should be observed for blackening under anaerobic conditions (anaerobic glove box) or immediately after opening anaerobic jars in air.

C. Catalase. To test for hydrogen peroxide degradation as an indication of catalase, expose the plates to air for at least 30 minutes and then flood the esculin agar quadrant with a few drops of fresh, 3 per cent hydrogen peroxide. Sustained bubbling after addition of the H_2O_2 is interpreted as a positive reaction for catalase. In some cases rapid bubbling may not be evident until after 30 seconds to a minute.

4. *LD Bile Agar:*

A. Compare the degree of bacterial growth on the LD bile agar with that on the plain LD agar and record as L (growth less than on the LD agar control) or E (growth equal to or greater than on the LD agar control).

B. Using transmitted light, look for the presence or absence of an insoluble white precipitate underneath and/or immediately surrounding the bacterial growth. If in doubt, inspect under a stereomicroscope using transmitted light.

5. *Anaerobe Blood Agar—Inhibition by Antibiotics:*

Observe for zones of inhibition around the antibiotic disks and record as follows:

A. Penicillin, 2-unit disk: sensitive "S" if zone of growth inhibition is 12 mm. or greater in diameter; and resistant "R" if the zone is less than 12 mm.

B. Rifampin, 15-μg. disk: sensitive "S" if zone of growth inhibition is 15 mm. or larger; and resistant "R" if the zone is less than 15 mm.

---

*Disks available from BioQuest, Cockeysville, Md.

Table 10-11. *Characteristics of Commonly Isolated Bacteroides Species (B. fragilis Group)*

Characteristic	B. distasonis	B. fragilis	B. thetaiotaomicron	B. vulgatus
Anaerobe Blood Agar:				
Relation to O$_2$	OA	OA	OA	OA
Colonies	Convex, semi-opaque, entire edge	Convex, mottled surface, entire edge	Convex, opaque, entire edge	Convex, semi-opaque, entire edge
Hemolysis, sheep blood	−	−	−	−
Hemolysis, rabbit blood	−	−	−	−
Black pigment	−	−	−	−
Red fluorescence	−	−	−	−
Pitting of agar	−	−	−	−
Cellular morphology	Small rods, variable in length	Small rods, variable in length	Small rods, variable in length	Small rods, variable in length
Gram reaction	−	−	−	−
Spores	−	−	−	−
Motility	−	−	−	−
Penicillin (2-U. disk)	R	R	R	R
Rifampin (15 μg. disk)	S	S	S	S
Kanamycin (1-mg disk)	R	R	R	R
Enriched Thioglycollate:				
Rapidity of growth	Moderate to rapid	Moderate to rapid	Moderate to rapid	Moderate to rapid
Appearance	Flocculent	Flocculent	Flocculent	Flocculent
Gas	− +	− +	− +	− +
Odor	Butyrous	Butyrous	Butyrous	Butyrous
Cellular morphology	Medium rods, vacuolated	Medium rods, vacuoated	Medium rods, vacuolated	Medium rods, vacuolated
Presumpto Plate:				
Growth on LD agar	Moderate	Moderate	Moderate	Moderate
Indole	−	−	+	−
Lecithinase	−	−	−	−
Lipase	−	−	−	−
Proteolysis	−	−	−	−
Esculin hydrolysis	+	+	+	− +
H$_2$S	−	−	−	−
Catalase	+	+	V	− +
20% Bile agar, growth	E	E	E	E
Precipitate in 20% bile agar	−	+	−	−

Key: OA = obligate anaerobe; + = positive reaction in >90% of strains tested; − = negative reaction in >90% of strains tested; − + = usually negative but may exhibit a weak or delayed positive reaction; R = resistant; S = sensitive; E = equal or greater than growth on LD agar control; I = less than growth on LD agar control; V = variable reaction.[13]

C. Kanamycin, 1000-μg. disk: Record as sensitive "S" if zone of growth inhibition is 12 mm. or greater; and resistant "R" if zone is less than 12 mm.

Summaries of the characteristics for the presumptive identification of species of Bacteroides and Fusobacterium are found in Tables 10-11, 10-12, and 10-13, including the reactions on the presumpto plate.[13] Selected reactions on the presumpto plate are shown in Plate 10-2.

Bacteroides and Fusobacterium species may also be characterized by detecting the metabolic products in PYG broth cultures with analysis by a gas chromatograph. This procedure is described in detail below and the metabolic products for these organisms are listed in Table 10-18.

Table 10-12. *Characteristics of Commonly Isolated Bacteroides (Miscellaneous Species and Groups)*

Characteristic	B. corrodens	CDC F1	CDC F2	B. melaninogenicus ss asaccharolyticus	B. melaninogenicus ss intermedius
Anaerobe Blood Agar:					
Relation to O₂	OA	OA	OA	OA	OA
Colonies	Pinpoint, convex, irregular edge	Pinpoint, convex, entire edge	Pinpoint, convex, entire edge	Small to medium, convex, entire edge	Small to medium, convex, entire edge
Hemolysis, sheep blood	–	–	–	+	+
Hemolysis, rabbit blood	–	–	–	+	+
Black pigment	–	–	–	+	+
Red fluorescence	–	–	–	+	+
Pitting of agar	+	–	–	+	–
Cellular morphology	Small, slim rods, variable in length	Small, slim rods, variable in length	Small, slim rods, variable in length	Tiny coccoid rods	Tiny coccoid rods
Gram reaction	–	–	–	–	–
Spores	–	–	–	–	–
Motility	–	–	–	–	–
Penicillin (2-U. disk)	S	S	S	S	S
Rifampin (15-μg. disk)	S	S	S	S	S
Kanamycin (1-mg disk)	S	S	S	R	R
Enriched Thioglycollate:					
Rapidity of growth	Slow	Slow	Slow	Moderate	Moderate
Appearance	Cloudy	Cloudy	Cloudy	Cloudy	Cloudy
Gas	–	–	–	–	–
Odor	–	–	–	Acrid	Acrid
Cellular morphology	Small slim rods	Small slim rods	Small slim rods	Coccoid rods with some pleomorphic forms	Coccoid rods with some pleomorphic forms
Presumpto Plate:					
Growth on LD agar	Light	Light	Light	Moderate	Moderate
Indole	–	–	–	+	+
Lecithinase	–	–	–	–	–
Lipase	–	–	–	–	–
Proteolysis	–	–	–	–	–⁺
Esculin hydrolysis	–	–	–	–	–
H₂S	–	–	–	–	–
Catalase	–	–	–	–	–
20% Bile agar, growth	I	I	I	E or I	I
Precipitate in 20% bile agar	–	–	–	–	–

Key: OA = obligate anaerobe; + = positive reaction in >90% of strains tested; – = negative reaction in >90% of strains tested; –⁺ = usually negative but may exhibit a weak or delayed positive reaction; R = resistant; S = sensitive; E = equal or greater than growth on LD agar control; I = less than growth on LD agar control; V = variable reaction.[13]

The Presumptive Identification of the Anaerobic Non-Spore-Forming Gram-Positive Bacilli

Included in this group of anaerobes are members of the genera Actinomyces, Arachnia, Bifidobacterium, Eubacterium, Lactobacillus, and Propionibacterium. The characteristics by which this group of anaerobes can be presumptively identified are listed in Table 10-14. The Presumpto Plate is helpful in determining some of the charac-

Table 10-13.   *Characteristics of Commonly Isolated Fusobacterium Species*

Characteristic	F. mortiferum	F. necrophorum	F. nucleatum	F. varium
Anaerobe Blood Agar:				
Relation to O_2	OA	OA	OA	OA
Colonies	"Fried egg"	Raised with opaque centers	Opolescent appearance	"Fried egg"
Hemolysis of sheep blood	$-$ $^+$	$-$	$-$ (greening)	$-$ $^+$
Hemolysis of rabbit blood	$-$	$+$	$-$	$-$
Black pigment	$-$	$-$	$-$	$-$
Red fluorescence	$-$	$-$	$-$	$-$
Pitting of agar	$-$	$-$	$-$	$-$
Cellular morphology	Highly pleomorphic bacilli, filaments, large bodies, *etc.*	Highly variable in length and width	Slim rods, even diameter, with or without pointed ends	Small rods, variable in length, rounded ends
Gram reaction	$-$	$-$	$-$	$-$
Spores	$-$	$-$	$-$	$-$
Motility	$-$	$-$	$-$	$-$
Penicillin (2-U. disk)	R or S	S	S	R or S
Rifampin (15-µg. disk)	R	S	S	R
Kanamycin (1-mg. disk)	S	S	S	S
Enriched Thioglycollate:				
Rapidity of growth	Moderate to rapid	Slow to moderate	Slow to moderate	Moderate to rapid
Appearance	Flocculent	Granular or fluffy	Granular or fluffy	Diffuse turbid

teristics listed in Table 10-14 for this group of anaerobes as well.

The microscopic morphology and colonial appearance of *Actinomyces israelii* and *Eubacterium lentum* are shown in Plate 10-3.

Presumptive Identification of Commonly Isolated Anaerobic Spore-Forming Gram-Positive Bacilli

The anaerobic gram-positive spore-forming bacilli by definition are all members of the genus Clostridium. The key characteristics by which the Clostridium species most commonly associated with disease in humans can be presumptively identified in the laboratory are listed in Table 10-15.

Most of the procedures for characterization of the clostridia as outlined in Table 10-15 have been described above. Some of the key reactions for the identification of *Clostridium perfringens* are illustrated in Plate 10-4. The double zone of hemolysis, production of lecithinase on egg yolk agar, and

stormy fermentation in litmus milk agar (proteolysis) are characteristic of this microorganism.

The catalase test is helpful in differentiating Clostridium species from Bacillus species (positive) if aerotolerance tests are equivocal. Information on additional tests and procedures for identifying Clostridium species in clinical specimens can be found in the references cited.[12,27,40]

Presumptive Identification of Anaerobic Cocci

Anaerobic cocci include members of the genera Acidaminococcus, Peptococcus, Peptostreptococcus, Ruminococcus, Sarcina, Streptococcus, and Veillonella. Although representatives of these groups may be associated with the normal microbiota of humans and other animals, only Peptococcus, Peptostreptococcus, Streptococcus, and Veillonella are encountered with any frequency from properly selected and handled

Table 10-13. *Characteristics of Commonly Isolated Fusobacterium Species* (Continued)

Characteristic	F. mortiferum	F. necrophorum	F. nucleatum	F. varium
Enriched Thioglycollate: (Continued)				
Gas	+	+	−	+
Odor	Butyrous	Acrid	Acrid	Butyrous
Cellular morphology	Highly pleomorphic	Pelomorphic rods and filaments	Slim rods and filaments	Rods variable in length
Presumpto Plate:				
Growth on LD agar	Moderate	Moderate	Light	Moderate
Indole	−	+	+	+ −
Lecithinase	−	−	−	−
Lipase	−	+	−	−
Proteolysis	−	−	−	−
Esculin hydrolysis	+	−	−	−
H_2S	+	+	−	+
Catalase	−	−	−	−
20% Bile agar, growth	E	I	I	E
Precipitate in 20% bile agar	−	−	−	−

Key: OA = obligate anaerobe; + = positive reaction in >90% of strains tested; − = negative reaction in >90% of strains tested; − + = usually negative but may exhibit a weak or delayed positive reaction; + − = usually positive but may exhibit a negative reaction; R = resistant; S = sensitive; E = equal or greater than growth on LD agar control; I = less than growth on LD agar control; V = variable reaction; "Fried egg" = raised opaque center with translucent entire edge.[13]

clinical specimens. The identifying characteristics used for this group of organisms are included in Table 10-16.

Peptostreptococcus anaerobius can be readily distinguished from the other anaerobic cocci by the sodium polyanetholsulfonate (SPS; Grobax, Roche Diagnostics, Nutley, N.J.) disk test.[17] Heavily inoculate a CDC-anaerobe blood agar plate with the unknown organism to be tested. Apply a ¼ in. sterile blank disk (Difco) to which has been added 20 µl. of 5% SPS and incubate the plate 48 h at 35°C. Anaerobic gram-positive cocci showing a 12 mm. or greater zone of inhibition can be presumptively identified as *P. anaerobius*.[32]

DETERMINATION OF METABOLIC PRODUCTS BY GAS-LIQUID CHROMATOGRAPHY

In addition to the characteristics listed in Tables 10-11 through 10-13, the species of anaerobic bacteria can be further identified on the basis of the metabolic products produced in peptone-yeast extract-glucose (PYG) broth cultures. These products include volatile short-chain fatty acids, nonvolatile organic acids, and alcohols. Practical inexpensive procedures using gas-liquid chromatography (GLC) are now available which permit the rapid determination of these products in the clinical microbiology laboratory.

Gas chromatographs are now relatively inexpensive, safe, simple to operate, and reliable, and are commercially available from various scientific instrument manufacturing companies. Some of the chromatographs currently available and their specifications are listed in Table 10-17. In general, thermal conductivity detectors are more commonly used; however, hydrogen flame ionization detectors can also be used effectively. Recorders should have a full scale

(Text continues on p. 304)

Table 10-14. *Some Key Differential Characteristics of*

Species	Relation to Oxygen	Rapidity of Growth	Colonies on Blood Agar	Red Pigment on Blood Agar	Appearance in Enriched Thioglycollate Broth
Actinomyces bovis	M or OA	Moderate	Smooth	−	Diffuse
A. israelii	M or OA	Slow	Rough	−	Granular or diffuse
A. naeslundii	F	Moderate	Smooth	−	Diffuse
A. odontolyticus	M or OA	Moderate	Smooth	+	Diffuse
A. viscosus	F	Rapid	Smooth	−	Diffuse
Arachnia propionica	M or OA	Slow	Rough	−	Granular or diffuse
Bifidobacterium eriksonii	OA	Rapid	Smooth	−	Diffuse
Eubacterium alactolyticum	OA	Slow	Smooth	−	Diffuse
E. lentum	OA	Moderate	Smooth	−	Diffuse
E. limosum	OA	Rapid	Smooth	−	Diffuse
Lactobacillus catenaforme	OA	Rapid	Smooth	−	Diffuse (granular)
Propionibacterium avidum	F	Rapid	Smooth	−	Diffuse
P. acnes	OA^F	Moderate	Smooth	−	Diffuse (granular)
P. granulosum	F	Rapid	Smooth	−	Diffuse

Key: + = positive reaction for 90%–100% of strains tested; − = negative reaction for 90%–100% of strains tested; superscript indicates the reaction shown with 11–25% of strains tested; V = variable reaction; ( ) = variable; F = facultatively anaerobic; M = microaerophilic; OA = obligately anaerobic; A = acetic acid; B =

Table 10-15. *Some Key Characteristics of Clostridium Species*

Species	Aerotolerant	Double Zone Hemolysis	Terminal Spores	Motility	Lecithinase (EYA)	Lipase (EYA)	Proteolytic (Milk)
C. bifermentans	−	−	−	+	+	−	+
C. botulinum	−	−	−	+	−$^+$	+	V
C. butyricum	−	−	−	+	−	−	−
C. difficile	−	−	−	+	−	−	−
C. innocuum	−	−	+	−	−	−	−
C. limosum	−	−	−	+	+	−	+
C. novyi type A	−	−	−	+	+	−	−
C. perfringens	−	+	−	−	+	−	+
C. ramosum	−	−	+	−	−	−	−
C. septicum	−	−	−	+	−	−	+
C. sordellii	−	−	−	+	+	−	+
C. sporogenes	−	−	−	+	−$^+$	+	+
C. subterminale	−	−	−	+	+	−	(+)
C. tetani	−	−	+	+	−	−	−
C. tertium	+	−	+	+	−	−	−

*For additional information on definitive identification of these species and other clostridia which may be encountered in clinical specimens, see Dowell and Hawkins,[12] Holdeman and Moore,[27] and Sutter, Vargo, and Finegold.[40]

Key: + = positive reaction for 90%–100% of strains tested; − = negative reaction for 90%–100% of strains tested; superscript

Anaerobic, Gram-Positive Non-Spore-Forming Bacilli

Cellular Morphology in Enriched Thioglycollate Broth	Catalase Production	Esculin Hydrolysis	Indole Production	Glucose Fermented	Metabolic Products in PYG Broth, 48 Hours, 35° C.
Diphtheroidal	−	+	−	+	A,L,S
Branching filaments or diphtheroidal	−	+ −	−	+	A,L,S
Diphtheroidal, branching	−	+ −	−	+	A,L,S
Diphtheroidal, branching	−	V	−	+	A,L,S
Diptheroidal, branching	+	+	−	+	A,L,S
Branching filaments or diphtheroidal	−	− +	−	+	A,P
Thin rods, bifid ends, bulbous ends	−	+	−	+	A,L
Thin rods, V-forms, "cross stich" arrangements	−	−	−	+	A,B,C
Short coccoidal rods, diphtheroidal	−	−	−	−	A
Plump rods, bulbous and bifid forms	−	V	−	+	A,B,(IB,IC)
Short rods in chains or singly	−	+	−	+	A,L
Diphtheroidal	+	+	−	+	A,P
Diphtheroidal	+	−	+ −	+	A,P
Diphtheroidal	+	−	−	+	A,P

butyric acid; C = caproic acid; L = lactic acid; P = propionic acid; S = succinic acid; IB = isobutyric acid; IC = isocaproic acid. From Dowell, V. R.: Clinical Veterinary Anaerobic Bacteriology, DHEW Publication, Atlanta, Ga., Center for Disease Control, pp. 1–25, 1977.

Associated with Disease in Humans

Indole Production	Esculin Hydrolysis	Glucose Fermented	Lactose Fermented	Mannitol Fermented	Urease	Volatile Metabolic Products (GLC) in PYG, 48 Hours, 35° C.
+	V	+	−	−	−	A,IC,(P),(IB),(B),(IV)
−	V	+	−	−	−	A,(P),(IB),B,IV,(V),(IC)
−	+	+	+	−	−	A,B
−	+	+	−	+	−	A,IB,B,IV,IC
−	−	+	+	+	−	A,B
−	−	−	−	−	−	A
−	V	+	−	−	−	A,P,B
−	V	+	+	−	−	A,B,(P)
−	+	+	+	+ −	−	A
−	+	+	+	−	−	A,B
+	−	+	−	−	+	A,IC,(P),(IB),(IV)
−	+	+	−	−	−	A,P,IB,B,IV,V,IC
−	− +	−	−	−	−	A,IB,B,IV,(P)
V	−	−	−	−	−	A,(P),B
−	+	+	+	+	−	A,B

indicates the reaction shown with 11%–25% of strains tested; V = variable reaction; ( ) = variable; A = acetic acid; P = propionic acid; IB = isobutyric acid; IV = isovaleric acid; V = valeric acid; IC = isocaproic acid.

Table 10-16. *Some Key Characteristics of Anaerobic Cocci Commonly Isolated From Clinical Specimens*

Species	Relation to Oxygen	Enriched Thioglycollate Broth, Microscopic Chains* Produced	*Gram Reaction	"Catalase" Production	Indole Production	Esculin Hydrolysis	Glucose Fermented	Acid Metabolic Products In PYG, 48 Hours, 35° C.	Other
Peptococcus asaccharolyticus	OA	–	+	–	+	–	–	A,B	
P. magnus	OA	–	+	–	–	–	–	A	
P. prevotii	OA	–	+	–	–	–	–	A,(P),B	
P. saccharolyticus	OA	–	+	+	–	–	+	A	
Peptostreptococcus anaerobius	OA	+	+	–	–	–	+	A,(P),IB,B,IV,IC	Inhibited by SPS
Streptococcus intermedius	OA or F	+	+	–	–	+	+	(A),L	
Veillonella alcalescens	OA	–	–	+	–	–	–	A,P	
V. parvula	OA	–	–	–	–	–	–	A,P	

*Definite chains of ten or more cells.

Key: SPS = sodium polyanethol sulfonate; OA = obligately anaerobic; F = facultatively anaerobic; ( ) = variable; + = positive reaction in 90%–100% of strains tested; – = negative reaction in 90%–100% of strains tested; A = acetic acid; B = butyric acid; P = propionic acid; IB = isobutyric acid; IC = isocaproic acid; IV = isovaleric acid; L = lactic acid. From Dowell, V. R.: Clinical Veterinary Anaerobic Bacteriology, DHEW Publication, Atlanta, Ga., Center for Disease Control, pp. 1–25, 1977.

Table 10-17. Specifications for Two Commercial Gas-Liquid Chromatographs

Chromatograph	Type of Detector	Type of Column	Column Material	Column Temperature	Injection Port Temperature	Carrier Gas	Flow Rate	Attenuation
Beckman Model GC2A	Thermal conductivity	6 foot × ¼ inch stainless steel	Resoflex LAC-1-R-296 standard concentration (P), Burrell Corp., Pittsburgh, Pa.	149° C.	198° C.	Helium	120 ml./min.	×1
Fisher Series 2400	Thermal conductivity	6 foot × ¼ inch stainless steel	Volatile acids: 15% SP-1220/1% H_3PO_4 on 100/120 Chromasorb WAW, Supelco Inc., Bellefonte, Pa. Nonvolatile acids: 10% SP 1000/1% H_3PO_4 on 100/120 Chromasorb WAW	145° C.	210° C.	Helium	100 ml./min.	×2

response of 1 second or better and it must be a 1-millivolt full response. The most commonly used carrier gas is helium.

Column packing materials that have been satisfactory for determining metabolic products include:

---

1. 20% LAC-1-296 Resoflex (Burrell Corporation; Pittsburg, Pa.)
2. 15% CPE 2225 on 45/60 mesh Chromasorb (CAPCO, Sunnyvale, Cal.)
3. 15% SP-1220/1% H_3PO_4 on 100/120 Chromasorb W AW (Supelco Inc., Bellefonte, Pa.)
4. 10% SP-1000/1% H_3PO_4 on 100/120 Chromasorb W AW (Supelco Inc., Bellefonte, Pa.)

Equipment and procedures for determining metabolic products by GLC are described in more detail by Dowell and Hawkins,[12] Holdeman and Moore,[27] and Sutter, Vargo, and Finegold.[40]

Procedure for Identification of Volatile Fatty Acids

This procedure is used for the identification of short-chain volatile fatty acids that are soluble in ether. The acids detected with this procedure include acetic, propionic, isobutyric, butyric, isovaleric, valeric, isocaproic, and caporic acids. Pyruvic, lactic, and succinic acids, however, are not detected by this procedure.

These nonvolatile acids are identified after preparation of methylated derivatives. The procedure is as follows:

1. Inoculate 7- to 8-ml. tubes of prereduced peptone-yeast extract-glucose (PYG) broth with a few drops (0.05–0.1 ml.) of an actively growing culture.
2. Incubate under anaerobic conditions for 48 hours or until adequate growth is obtained.
3. Transfer 2.0 ml. of the culture to a clean, 13 × 100-mm. screw-cap tube.
4. Acidify the culture to pH 2.0 or below by adding 0.2 ml. of 50 per cent (V/V) aqueous H_2SO_4.
5. Add 1 ml. of ethyl ether, tighten the cap, and mix by gently inverting the tube about 20 times.

6. Centrifuge briefly in a clinical centrifuge (1500–2000 r.p.m.) to break the ether-culture emulsion.
7. Place the ether-culture mixture in a −20° C. freezer (or lower) or in an alcohol-dry ice bath until the aqueous portion (bottom) is frozen.
8. Rapidly pour off the ether layer into a clean screwcap tube.
9. If desired, add 1 or 2 anhydrous $CaCl_2$ pellets to the ether extract to allow removal of residual water.
10. Inject 14 microliters of the extract into the column of a gas chromatograph.
11. Identify volatile acids by comparing elution times of products in extracts with those of a known acid mixture (volatile fatty acid standard) chromatographed under the same conditions on the same day. A representative standard tracing is shown in Fig. 10-9, and an example of the GLC results from an unknown anaerobe *(Fusobacterium mortiferum)* is shown in Fig. 10-10.

Procedure for the Analysis of Nonvolatile Acids

1. Transfer 1 ml. of the original PYG culture to a clean 13 × 100-mm. screwcap tube.
2. Add 0.4 ml. of H_2SO_4 (V/V) and 2 ml. of methanol. Place the tube in a 55° C. water bath overnight.
3. Add 1 ml. of distilled water and 0.5 ml. of chloroform and centrifuge briefly to break any emulsion in the chloroform layer (chloroform will be in the bottom of the tube).
4. Fill a syringe with the chloroform extract after placing the tip of the needle beneath the aqueous layer.
5. Wipe off the outside of the needle with a clean tissue and inject 14 microliters.
6. Analysis of chloroform extracts is performed using the same column packing materials and conditions as the volatile acids.
7. Identify nonvolatile or methylated acids by comparing elution times of products in extracts with those of known acids chromatographed on the same day.

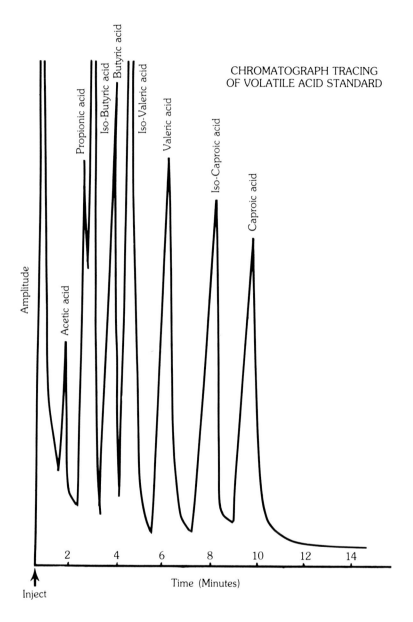

CHROMATOGRAPH TRACING
OF VOLATILE ACID STANDARD

Fig. 10-9. A typical volatile acid standard chromatogram. The time elapsed between the injection of an ether extract of the standard solution and the peak for each acid (retention time) is used to identify the acids. Note, for example, that the retention time for acetic acid is 1.8 minutes and for valeric acid is 6 minutes (Instrument used: Dohrmann Anabac, Clinical Analysis Products Co., Sunnyvale, Ca. 94086; Detector: thermal conductivity; Column packing: 15%, SP-1220/1% H_3PO_4 on 100/120 Chromasorb W/ AW from Supelco Inc., Bellefonte, Pa. 16823).

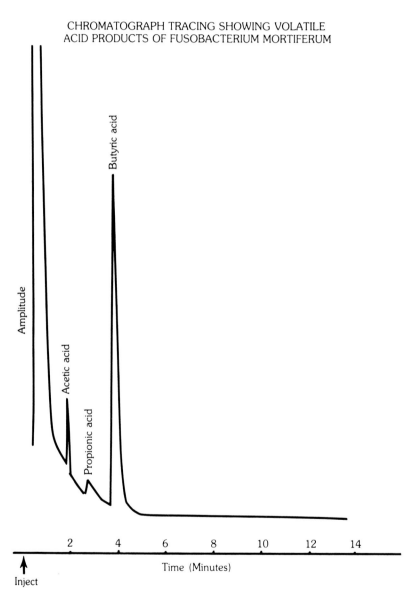

Fig. 10-10. Volatile acid chromatogram of a 48-hour peptone-yeast extract-glucose broth culture of *Fusobacterium mortiferum*. The retention times of the products in the broth culture are compared with those of the standard tracing (Fig. 10-9.) to identify the unknown acids. The peak heights indicate major amounts of butyric and acetic acids but only a minor amount of propionate for this culture. The same instrument and operating conditions were used as for the tracing in Fig. 10-9.

Color Plates
10-1 *to* 10-4

Plate 10-1

Identification of Anaerobic Bacteria
Bacteroides and Fusobacterium Species

A.

Bacteroides fragilis. Gram's stain of cells in 48-hour thioglycollate broth culture.

B.

Bacteroides fragilis. Colonies in anaerobe blood agar after 48 hours incubation at 35⁰ C.

C.

Bacteroides melaninogenicus. Gram's stain of cells from a 48-hour colony on blood agar.

D.

Bacteroides melaninogenicus. Black colonies on blood agar after 5 days incubation at 35⁰ C. Note hemolysis.

E.

Fusobacterium nucleatum. Gram's stain of cells from a 48-hour colony on anaerobe blood agar. Note long gram-negative bacilli with pointed ends.

F.

Fusobacterium nucleatum. Characteristic colonies on anaerobe blood agar after 48 hours incubation at 35⁰ C., illustrating the opalescent effect.

G.

Fusobacterium necrophorum. Gram's stain of cells from a 48-hour colony in anaerobe blood agar. Note pleomorphism.

H.

Fusobacterium necrophorum. Colonies on anaerobe blood agar after 48 hours at 35⁰ C.

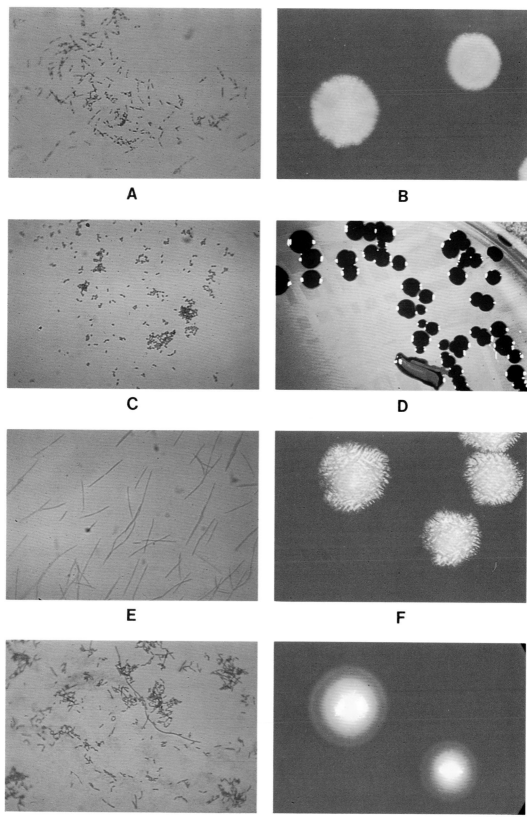

A

B

C

D

E

F

G

H

Plate 10-2
Identification of Anaerobic Bacteria
Presumptive Identification of Bacteroides and Fusobacteria
(Presumpto Plates)

A.

Photograph of a presumpto quadrant plate and a plate of anaerobe blood agar used for presumptive identification of Bacteroides and Fusobacterium species. After inoculation with an active broth culture or cell suspension of the isolate to be identified, antibiotic disks (penicillin, 2 units; rifampin, 15 ug. kanamycin, 1000 ug.) are placed on the blood agar medium and a blank filter paper disk is placed on the LD agar portion of the quadrant plate for use in detection of indole production. The presumpto quadrant plate contains the following media: LD agar, LD esculin agar, LD egg yolk agar, and LD bile agar (see p. 294).

B.

Bacteroides fragilis. Growth in quadrant plate after 48-hours incubation at 35⁰ C. On the first quadrant (top, far right), LD agar shows moderate growth. A drop of paradimethylaminocinnamaldehyde reagent will be added to the paper disk to test for the presence of indole. The LD esculin agar to the left of the LD agar is diffusely dark due to the hydrolysis of the esculin.* The LD egg-yolk agar to the right of the esculin agar shows good growth but no lecithinase, lipase, or proteolytic activity. There is abundant growth in the LD bile agar (bottom, far right) and a characteristic precipitate in the medium is typically exhibited by *B. fragilis.*

C.

Bacteroides fragilis. Bubbling after addition of 3 per cent hydrogen peroxide to the LD esculin agar quadrant, which indicates catalase activity.

D.

Bacteroides fragilis. A zone of growth inhibition is seen around the 15 ug. rifampin disk, but no

growth inhibition is seen around the 2-unit penicillin disk or the 1000 ug. kanamycin disk. This pattern is characteristic of *B. fragilis.*

E.

Bacteroides thetaiotaomicron. Reactions on the presumpto quadrant plate. The first quadrant (top, far right) shows weak, positive indole as indicated by the pale blue color of the disk on LD agar after addition of paradimethylaminocinnamaldehyde reagent. Black (amber) appearance of esculin agar (left of first quadrant) indicates esculin hydrolysis. Adequate growth but no lecithinase, lipase or proteolysis on LD egg yolk agar (bottom, far left). Good growth on LD bile agar but no precipitate as exhibited by *B. fragilis* (bottom, far right).

F.

Fusobacterium necrophorum. Reactions on presumpto quadrant plate after 48 hours incubation at 35⁰ C. The first quadrant (bottom, far right) shows strong indole reaction as evidenced by the dark blue color of the paper disk on LD agar after the addition of paradimethylaminocinnamaldehyde reagent. Growth is inhibited on LD bile agar (top, right). Although not visible in this photograph because of the lighting arrangement, there is good growth in LD egg-yolk agar (top, far left) and characteristic lipase activity as evidenced by an iridescent sheen "pearly layer" on the surface of the colonies and in the medium immediately surrounding the bacterial growth. This is best demonstrated with reflected light. On LD esculin agar (bottom, left), the *F. necrophorum* shows good growth and H₂S production, as evidenced by the black appearance of the colonies, but no darkening of the medium to suggest esculin hydrolysis.

---

*Note: In these photographs, the LD esculin agar appears black upon hydrolysis of esculin due to the black background. In transmitted light, the agar appears deep amber in color when esculin is hydrolyzed.

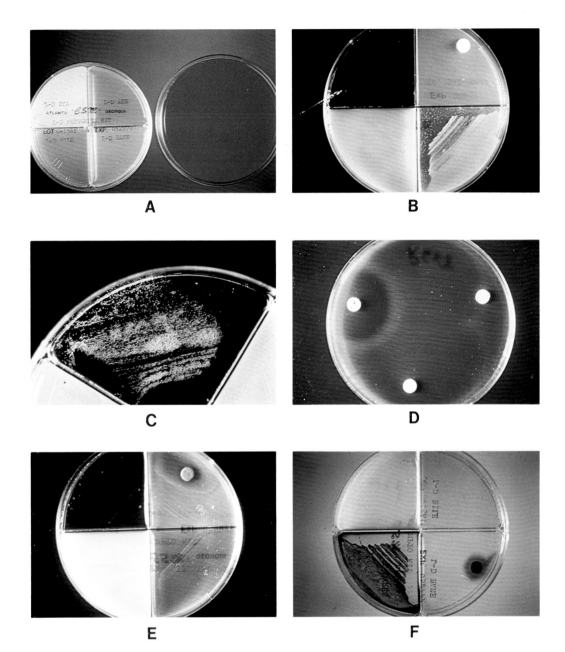

A

B

C

D

E

F

A.
Gram-stained direct smear of a purulent exudate from an intraabdominal abscess showing segmented neutrophils, gram-positive cocci in pairs and short chains, and tiny gram-negative coccoid rods. Anaerobic infections usually contain a mixed bacterial flora.

B.
Gram-stained direct smear of a purulent exudate showing large numbers of gram-positive cocci in chains and gram-negative bacilli. This picture is suggestive of mixed infection with anaerobic cocci and Bacteroides species.

C.
Actinomyces israelii. Gram-stained preparation of growth from a colony on blood agar. Note branching of cells.

D.
Actinomyces israelii. Characteristic "molar tooth" colonies produced on brain heart infusion agar after 7 days of anaerobic incubation at 35⁰ C.

E.
Eubacterium alactolyticum. Gram's stain of cells from growth in enriched thioglycollate broth after 48 hours incubation at 35⁰ C.

F.
Eubacterium alactolyticum. Forty-eight-hour colonies on anaerobic blood agar.

G.
Gram's stain of cells from enriched thioglycollate broth culture which are suggestive of a Peptococcus species.

H.
Colonies of an anaerobic Streptococcus species on blood agar. Note hemolysis.

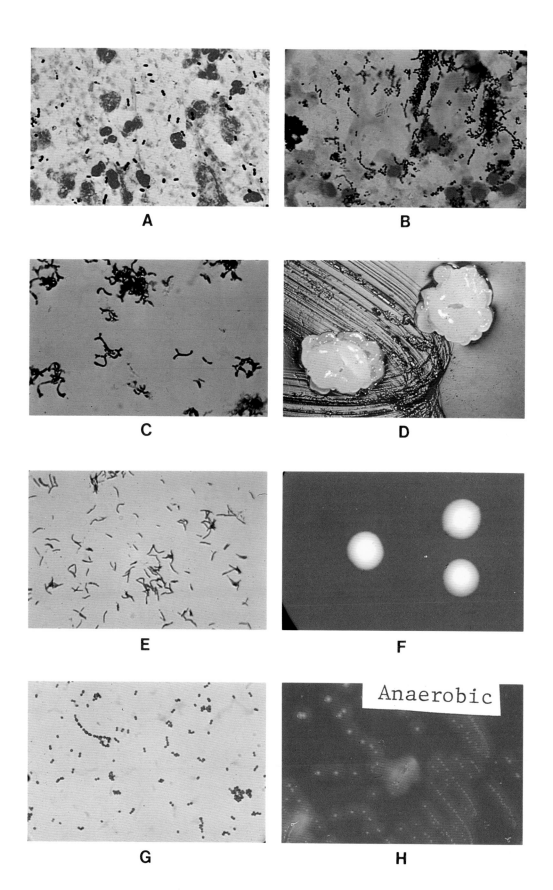

A

B

C

D

E

F

G

H

Anaerobic

Plate 10-4 (Part I)
Identification of Anaerobic Bacteria
Clostridia

A.

Clostridium perfringens. Gram's stain of cells from a 24-hour colony on blood agar. Note absence of spores and of some cells that tend to stain red (gram-negative).

B.

Clostridium perfringens. Gram's stain of cells from a 24-hour thioglycollate broth culture. Note absence of spores and of a few filamentous forms.

C.

Clostridium perfringens. Typical appearance of *C. perfringens* on blood agar after a 24-hour incubation at 35⁰ C. Note the double zone of hemolysis. The inner zone of complete hemolysis is due to theta toxin and the outer zone of incomplete hemolysis to alpha toxin (lecithinase activity).

D.

Clostridium perfringens. Colonies of *C. perfringens* on modified McClung egg-yolk agar. The precipitate surrounding the colonies is indicative of lecithinase activity of alpha toxin produced by the organism.

E.

Clostridium perfringens. Photograph to illustrate the use of the Nagler test for presumptive identification of *C. perfringens*. Prior to inoculation with the organism, a drop of *C. perfringens*-type antitoxin was spread over one-half of the egg yolk medium (right side). Note the lack of lecithinase activity on this side, compared to the clearing of the agar on the left side, where antitoxin is not present. This represents a positive Nagler reaction. *C. bifermentans* and *C. sordellii* produce positive Nagler reactions; however, these two species can be differentiated by a few simple tests.

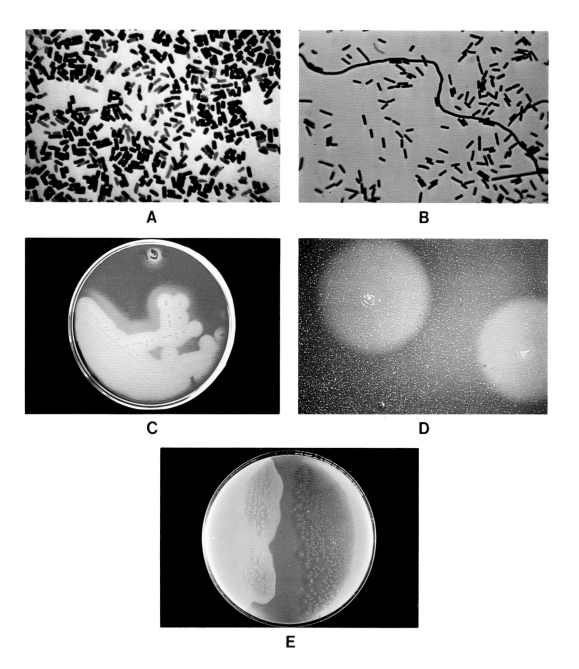

Plate 10-4 (Part II)
Identification of Anaerobic Bacteria
Clostridia

F.

Reactions exhibited by clostridia in milk medium are useful in the identification of the micro-organism. This is a photograph of reactions in litmus milk. The tubes on either side show coagulation and gas production, which is frequently called "stormy fermentation." The center tube illustrates coagulation and digestion of the milk by a proteolytic Clostridium species.

G.

Clostridium tetani. Gram's stain of cells from a chopped meat glucose broth culture. Some of the cells have round, terminal spores, which are characteristic of *C. tetani.*

H.

Clostridium tetani. Colonies of *C. tetani* on stiff blood agar (4 per cent agar) which is used to inhibit the swarming of the microorganism so that it can be isolated from other bacteria present in mixed cultures.

I.

Lipase production on egg-yolk agar. A few clostridia, such as *C. botulinum, C. sporogenes,* and *C. novyi,* Type A, exhibit lipase activity on egg-yolk agar, as shown in this photograph. Note the irridescent "pearly layer" on the surface of the colonies extending onto the medium immediately surrounding them.

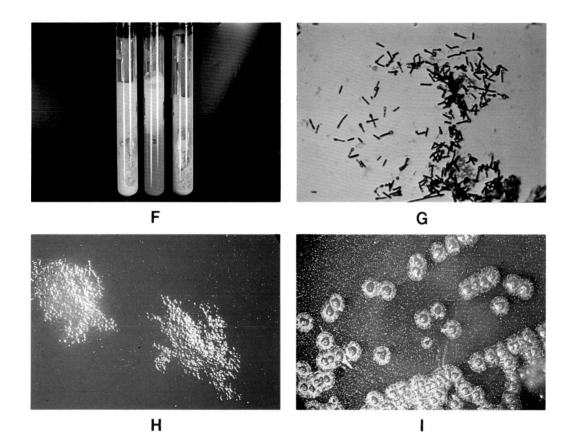

F

G

H

I

Table 10-18. *Metabolic Products of Common Bacteroides and Fusobacterium Species in Peptone Yeast Extract Glucose Broth Cultures After 48 Hours at 35° C.*

Species or Group	A	P	IB	B	IV	V	IC	C	L	S
Bacteroides corrodens	+	−	−	−	−	−	−	−	−	−
B. distasonis	+	+⁻	−	−	V	−	−	−	+⁻	+
B. fragilis	+	+⁻	V	−	V	−	−	−	+	+
Bacteroides F-1	+	−	−	−	−	−	−	−	−	V
Bacteroides F-2	+	−	−	−	−	−	−	−	−	V
B. melaninogenicus ss asaccharolyticus*	+	+	+	+	+	−	−	−	−	−
B. melaninogenicus ss intermedius	+	−	+⁻	−	+	−	−	−	−	+
B. thetaiotaomicron	+	+⁻	−⁺	−	V	−	−	−	+	+
B. vulgatus	+	+	−⁺	−	−⁺	−	−	−	+	+
F. mortiferum	+	+⁻	−	+	−	−	−	−	−	−
F. necrophorum	+	+	−	+	−	−	−	−	−	−
F. nucleatum	+	+	−	+	−	−	−	−	−	−
F. varium	+	+	−	+	−	−	−	−	−	−

*Currently classified as *Bacteroides asaccharolyticus* by Finegold and Barns, who have proposed that the saccharolytic and asaccharolytic strains presently classified with *Bacteroides melaninogenicus* be reclassified in two species, *Bacteroides melaninogenicus* and *Bacteroides asaccharolyticus*.

Key: A = acetic; P = propionic; IB = isobutyric; B = butyric; IV = isovaleric; V = valeric; IC = isocaproic; C = caproic; L = lactic; S = succinic acid, + = major peak; − = no major peak; +⁻ = usually a major peak but may be negative; −⁺ = usually negative but may be present; V = variable.[13]

8. After testing approximately 20 methylated samples, recondition the packing material by injecting 14 microliters of methanol into the gas chromatograph column.

GLC Controls

1. Standard solutions containing 1 mEq./100 ml. of each volatile acid and nonvolatile acid should be examined each time unknowns are tested. The volatile acid standard should contain at least the following acids: acetic, propionic, isobutyric, butyric, isovaleric, valeric, isocaproic, and caproic acids. The nonvolatile acid standard should contain at least pyruvic, lactic, and succinic acids.

2. A tube of uninoculated medium should be examined in the same manner since various lots of PYG broth may contain significant quantities of these acids.

Lombard and associates *(unpublished data)* have found that several other liquid media can be used instead of PYG broth for the performance of GLC. This observation takes advantage of the fact that certain anaerobes grow much better in media other than PYG. These media include enriched thioglycollate broth with dextrose (BBL-135C), chopped meat glucose and Lombard-Dowell-glucose broth (Carr-Scarborough Microbiologicals, Inc., Forest Park, Ga.), and modified Schaedler broth (BBL). When grown in these other broth media, however, results for a given organism may differ from those attained in PYG. Thus, caution should be exercised in interpreting products from different media when the identification tables have been prepared from a PYG data base. A further note of caution is warranted in relation to the second paragraph under the heading GLC Controls, above. The amount of acetic, lactic, and other acids present in certain liquid media such as chopped meat glucose broth may make it difficult or at times impossible to determine if the acetic or lactic acid peak was produced by the unknown isolate or was only that of the uninoculated medium.

The metabolic products of common species of Bacteroides and Fusobacterium are listed in Table 10-18.

MICROMETHODS FOR IDENTIFICATION OF ANAEROBES

The packaged "microsystems" for the identification of various species of bacteria were discussed in detail in Chapter 6. Of these systems, only two have received acceptance by clinical microbiologists for the biochemical characterization of the anaerobes, namely, the API 20-A system (Analytab Products Inc., Plainview, N.Y.) and the Minitek system (Bioquest, BBL, Baltimore, Md.). A discussion of the specific applications of these two systems for the identification of anaerobes is beyond the scope of this text; however, these systems have been carefully evaluated in a number of clinical laboratories, as fully described in the references cited.[21,22,23]

ANTIMICROBIAL SUSCEPTIBILITY TESTING OF ANAEROBIC BACTERIA

Antimicrobial Therapy. Antimicrobial therapy is only one of a number of important considerations in the clinical management of diseases involving anaerobic bacteria. Surgical measures are often extremely important and include the debridement of necrotic tissue, drainage of abscesses, elimination of foreign bodies or malignant tumor, and removal of thrombi.[14,18,20] In most instances, it will be necessary to start antimicrobial therapy of anaerobe infections long before culture results are available. The clinical setting (e.g., necrotizing pneumonia, intra-abdominal abscess, etc.) and results of direct microscopic examination will often provide valuable information on the nature of the infectious process and the probable organisms involved (see pages 276–278). The first major goal of the laboratory should be to determine whether anaerobic bacteria are present. This involves observation of colonial morphology and Gram's-stain features or other available characteristics of isolates following incubation of primary isolation media. Prompt reporting of these preliminary results will allow the physician to adjust therapy within a relevant time frame. The second major goal should be to rapidly identify the most common and significant anaerobes, as discussed earlier in this chapter. In particular, the clinician needs to know if *Bacteroides fragilis* is present because of its predictable resistance to penicillin.

Tabulated results of susceptibility tests from the literature or from the clinician's own hospital must be available for reference. This assumes that the susceptibility patterns of most clinically encountered anaerobes are stable and predictable. Unfortunately, some anaerobes are becoming more resistant to certain antimicrobics, as recently reviewed by Finegold.[15] Although the "routine" susceptibility testing of anaerobe isolates is still probably not necessary for immediate patient care, we agree with Sutter and Washington,[41] who recommend that antimicrobial susceptibility of individual isolates be determined in serious infections such as endocarditis and brain abscess or in other anaerobic infections requiring prolonged antimicrobial therapy (e.g., chronic osteomyelitis). Susceptibility testing of anaerobes may also be indicated in instances when the patient has not responded to therapy.

As discussed in Chapter 11, other factors, besides the results of *in vitro* susceptibility tests, must be taken into account when antimicrobial agents are selected. These include pharmacological properties of the drug, penetration into and levels achievable in body fluids such as blood and CSF or tissue, possible alteration of the intestinal flora, toxicity, and the route and ease of administration.[15,18] Also, since most anaerobic infections are polymicrobial, the choice of antimicrobial agents to use will depend on the other bacteria involved.

Susceptibility of Anaerobes to Various Antimicrobial Agents. Table 10-19 summarizes current susceptibility data of anaerobes to various antimicrobial agents that are encountered in human infections. These data have been summarized from the reports of Martin, *et al.*[30] and Sutter and Finegold[39a] and tabulated to conform to Thornsberry's categories of susceptibility.[45] The Category Test of Thornsberry and interpretative criteria for determining susceptibility or resistance of anaerobe isolates are given on pages 314–315.

Penicillin G is highly active against most anaerobes with certain notable exceptions. These include most strains of the *B. fragilis* group, certain strains of *Clostridium ramosum, C. perfringens,* and most strains of *Fusobacterium mortiferum* and *F. varium.* Carbenicillin shows good activity against most anaerobes, except for about 10 to 20 per cent of *B. fragilis* strains, which are resistant.[19a]

Clindamycin shows excellent *in vitro* activity against most commonly encountered anaerobes, with the exception of occasional strains of *B. fragilis,*[36] many strains of *F. varium,* and certain species of Peptococcus and Clostridium which are resistant. Unfortunately, the use of clindamycin as a therapeutic agent has been deterred to some degree because of its association with pseudomembranous colitis.[43] Since the etiology of some cases of antibiotic-induced colitis has now been determined,[4] this complication may be prevented by also administering vancomycin or other antibiotics active against the toxigenic species of Clostridium that have been incriminated.

Chloramphenicol is another drug with good *in vitro* activity against anaerobes; however, its bone marrow toxicity is a well-documented complication. Many anaerobes are now resistant to tetracycline, particularly most *B. fragilis* isolates.

Metronidazole (Flagyl), long used in the treatment of *Trichomonas vaginalis* infections, has also been found to be active against most anaerobes. Exceptions include most species of Actinomyces, Arachnia, Propionibacterium, and about one third of Eubacterium species.[15] Flagyl has the advantage over clindamycin in the therapy of anaerobic infections of the central nervous system since it freely crosses the blood-brain barrier. In contrast to chloramphenicol, metronidazole shows bactericidal activity against *B. fragilis.* Metronidazole has not yet been approved by the Food and Drug Administration for parenteral use.

Antimicrobial agents that are inactive against most anaerobes include the aminoglycosides (gentamicin, amikacin, kanamycin, streptomycin) and polymyxin B and colistin (polymixin E). Vancomycin is not active against species of Bacteroides and Fusobacterium, but is quite active against most gram-positive anaerobes.[15] Most of the cephalosporins show poor activity against the *B. fragilis* group, but good activity against most other anaerobes. Cefoxitin, a cephalosporin that is not inactivated by β-lactamase, shows some promise in the treatment of anaerobic infections.[24]

Methods for Antimicrobial Susceptibility Testing of Anaerobes

Most of the methods for antimicrobial susceptibility testing discussed in Chapter 11 also apply to the anaerobes. However, the Bauer-Kirby technique should not be used despite its convenience.[44,45] Most anaerobes grow too slowly, the Bauer-Kirby interpretative charts were not designed for anaerobes, and there is poor correlation between zone size measurements and the results from MIC dilution tests. The modified disk agar diffusion technique developed by Sutter also cannot be recommended.

In 1972, a collaborative group formed as a subcommittee of the National Committee for Clinical Laboratory Standards (NCCLS) began work on developing a standardized method for the antimicrobial susceptibility testing of anaerobes. In 1976, their preliminary studies were presented. The agar dilution procedure which was developed through the studies of this committee, even if adopted by NCCLS as the reference method, would not necessarily have to be used in clinical laboratories; rather, it would serve as the standard for evaluating the

(Text continues on p. 316)

Table 10-19.   *Antimicrobial Susceptibility of Anaerobic Bacteria*
(Per Cent Susceptible Within Thornsberry Category I and II*)

Organism	Penicillin G			Carbenicillin			Clindamycin		
	I ≤0.25 μg./ml.	*II* 16 μg./ml.	*No.* *Strains* ( )	*I* ≤32 μg./ml.	*II* 64 μg./ml.	*No.* *Strains* ( )	*I* ≤2 μg./ml.	*II* 8 μg./ml.	*No.* *Strains* ( )
Bacteroides fragilis group†	<1	38	(271)	47	80	(76)	93	100	(271)
Bacteroides melaninogenicus	65	97	(95)	100		(66)	100		(95)
Other species of Bacteroides and Selenomonas	24	75	(72)	92	96	(72)	94	96	(72)
Fusobacterium nucleatum	81	100	(36)	100		(18)	100		(36)
Other species of Fusobacterium	50	89	(18)	94		(16)	94	94	(18)
Clostridium perfringens	84	100	(43)	100		(9)	93	100	(43)
Other species of Clostridium	31	100	(51)	94	100	(51)	71	90	(51)
Actinomyces species	69	100	(16)	100		(16)	100		(16)
Bifidobacterium species	20	100	(10)	100		(5)	100		(10)
Eubacterium species	18	100	(38)	100		(17)	93	100	(38)
Propionibacterium species	89	100	(28)	100		(12)	100		(28)
Peptococcus species (and Gaffkya)	78	100	(204)	100		(59)	82	98	(204)
Peptostreptococcus species	81	99	(101)	93	97	(29)	100		(101)
Gram-negative cocci	36	100	(39)	96		(26)	100		(39)

*These data as presented are meant to serve only as a guide to the susceptibility of the various anaerobes listed. The two concentrations of each antimicrobial given conform to Thornsberry's category I (very susceptible organisms—inhibited by the level of antimicrobial attained in the blood on usual dosage) and category II (moderately susceptible organisms—inhibited by the blood level of antimicrobial achieved on high dosage). The per cent of anaerobes given in each category for each antimicrobial is based on totals calculated from Martin, W. J., Gardner, M., and Washington, J. A., II: Antimicrob. Agents Chemother., 1:148–158, 1972, and Sutter, V. L., and

**Table 10-19. Antimicrobial Susceptibility of Anaerobic Bacteria
(Per Cent Susceptible Within Thornsberry Category I and II*) (Continued)**

Chloramphenicol			Tetracycline			Cephalothin			Erythromycin		
I ≤1 µg./ml.	II 8 µg./ml.	No. Strains ( )	I ≤2 µg./ml.	II 8 µg./ml.	No. Strains ( )	I ≤2 µg./ml.	II 16 µg./ml.	No. Strains ( )	I ≤2 µg./ml.	II 4 µg./ml.	No. Strains ( )
<1	94	(271)	35	41	(271)	<1	<4	(229)	60	85	(229)
48	100	(95)	78	92	(95)	66	89	(38)	95	97	(38)
38	96	(72)	43	60	(72)	41	51	(51)	86	96	(51)
69	100	(36)	92		(36)	82	86	(28)		68	(28)
83	100	(18)	94		(18)	83	100	(6)			
	100	(43)	72	81	(43)	100		(35)	94	100	(35)
	94	(51)	63	75	(51)	75	100	(24)	100		(24)
81	100	(16)	94		(16)	30	100	(10)	80	100	(10)
40	100	(10)	50	60	(10)		73	(30)		93	(30)
	100	(38)		76	(38)						
50	100	(28)	93		(28)	100		(24)	79		(24)
60	99	(204)	50	55	(204)	90	98	(187)	85	88	(187)
63	100	(101)	60	75	(101)	95	98	(86)	76	88	(86)
68	100	(39)	82		(39)	94	100	(32)	58	79	(32)

Finegold, S. M.: Antimicrob. Agents Chemother., 10:736–752, 1976, and does not account for possible differences in classification of isolates at the two laboratories or differences in the methods used for determining MIC's in the two reports.
†B. fragilis group includes a composite of data for B. fragilis, B. thetaiotaomicron, B. distasonis, B. vulgatus, and B. ovatus as reported by Sutter and Finegold, since these authors report that "little or no difference has been found in the susceptibility of these organisms to the antimicrobial agents in their study, but includes only Martin's strains (195) strains) designated Bacteroides fragilis.

Chart 10-1.   *Broth-Disk Procedure for Susceptibility Testing of Anaerobes*

Introduction	In this method antimicrobials are added to broth tubes using commercially available paper disks. The antimicrobials elute in the broth tubes to achieve the desired concentrations.
Principle	Anaerobic bacteria are either susceptible or resistant to the single selected concentration of each different antimicrobial. The antimicrobial concentration in the broth for each drug has been selected to approximate the concentration ordinarily achieved in a patient's blood.
Media and Reagents	1. *The medium* is PRAS brain-heart infusion broth supplemented with 0.0005% hemin, 0.0002% menadione, and 0.5% yeast extract. Add 5.0 ml. of this broth per tube under oxygen-free CO_2 and seal with butyl rubber stoppers.
	2. *Disks.* Use commercially available disks that are purchased for the Kirby-Bauer procedure (available from BBL, Difco, and Pfizer). Using a CO_2 gas cannula inserted in the neck of the tube, the disks are added to each tube with flamed forceps as indicated in the table below:

Antimicrobial	Disk Content in μg.	No. Disks per Tube	Final Concentration of Antimicrobial per ml. of Broth (μg.)
Penicillin G	10 U.	1	2 U.
Carbenicillin	100	5	100
Cephalothin	30	1	6
Tetracycline	30	1	6
Clindamycin	2	8	3.2
Chloramphenicol	30	2	12
Erythromycin	15	1	3
Control	0	0	0

Procedure	1. Inoculate each tube with 1 drop of an 18- to 24-hour PRAS chopped meat-glucose (CMG) broth culture using a Pasteur pipette. Prevent aera-

Chart 10-1. *Broth-Disk Procedure for Susceptibility Testing of Anaerobes* (Continued)

Procedure *(Continued)*	tion of each tube during inoculation again by the insertion of a cannula carrying oxygen-free CO_2. Reseal each tube with its rubber stopper. 2. Incubate for 18 to 24 hours at 37° C. in a conventional incubator.
Interpretation of Results	Compare the turbidity of the growth in each tube containing an antimicrobial with that of a growth control tube without antimicrobial. In tubes showing 50% or more of the turbidity of the growth control tube, the organism is considered resistant, while in antimicrobial tubes showing no turbidity or less than 50% of the control tube, the organism is considered susceptible. The susceptibility test is considered indeterminate if the turbidity is approximately 50% that of the control. In most instances there is simply turbidity equal to that of the control (resistant) or no turbidity (sensitive).
Comment	An acceptable alternative to the Wilkins and Thiel (1973) broth-disk method was reported by Kurzynski and coworkers (1976). In this modified broth-disk procedure thioglycollate broth (BBL 135C) is substituted for PRAS-BHI broth. The same number of disks (as for the Wilkins-Thiel procedure) are added to the thioglycollate broth (5 ml. of broth per 16 × 125-mm screwcap tube) which has been preboiled 5 mintes and cooled. The antimicrobials are allowed to elute into the broth by holding the tubes 2 hours at room temperature. Each tube is then inoculated with 2 drops of an overnight CMG culture. The caps are tightened and the tubes are incubated at 37°C. overnight in an ambient air incubator, or for 48 hours to permit adequate growth of slow-growing anaerobes. The tubes are read and interpreted similarly as for the Wilkins-Thiel procedure. This procedure is more practical and convenient for small clinical laboratories, since it does not require the use of anaerobe-grade CO_2 or a gassing apparatus and substitutes a less expensive, commercially available medium for PRAS-BHI broth.
Bibliography	1. Kurzynski, T. A., *et al.*: Aerobically incubated thioglycollate broth disk method for antibiotic susceptibility testing of anaerobes. Antimicrob. Agents Chemother., *10*:727–732, 1976. 2. Wilkins, T. D., and Thiel, T.: Modified broth-disk method for testing the antibiotic susceptibility of anaerobic bacteria. Antimicrob. Agents Chemother., *3*:350–356, 1973.

Chart 10-2. *The Category Susceptibility Test of Thornsberry*

Introduction and Principle	This test is an abbreviated version of the conventional broth dilution procedure (Stalons and Thornsberry, 1975) and tests anaerobes against only 2 or 3 concentrations of each antimicrobial. These concentrations generally conform to the clinical categories recommended by the International Collaborative Study (Ericsson and Sherris, 1971) and categorize the organisms as very susceptible, moderately susceptible, moderately resistant, or very resistant.
Medium	Prepare Schaedler broth (BBL or Difco) according to the manufacturer's directions and supplement with hemin (0.1 μg./ml.) and vitamin K_1 (5 μg/ml.). Dispense 2.7 ml. of the medium into each of several 13×100-mm. screwcap tubes. Prior to use, the medium must be boiled for at least 15 minutes, followed by cooling, or, alternatively, the tubes can be held in an anaerobic environment (e.g., a glove box or an evacuation/replacement jar containing 85% N_2, 5% CO_2, 10% H_2) for 3 hours or more to facilitate the development of reducing conditions within the medium.
Preparation of Antimicrobials	The concentrations of antimicrobial agents suggested for use in this test are given in the following table:

Antimicrobial	Concentrations to Test (μg./ml.)		
Penicillin G	0.25	16	128
Carbenicillin	32	64	256
Cephalothin	2	16	128
Tetracycline	2	8	32
Clindamycin	2	8	64
Chloramphenicol	1	8	
Erythromycin	2	4	64

Sources of the reference antimicrobial powders are given below:

Sources of Laboratory Reference Antimicrobial Powders

Penicillin G	Wyeth Laboratories P.O. Box 8299 Philadelphia, Pa. 19101
Carbenicillin	J. B. Roerig and Company Division of Pfizer Pharmaceuticals, Inc. 235 E. 42nd St. New York, N.Y. 10017
Clindamycin	The Upjohn Company 7171 Portage Road Kalamazoo, Mich. 49002
Chloramphenicol	Parke, Davis and Company P.O. Box 118 Detroit, Mich. 48232

Chart 10-2. *The Category Susceptibility Test of Thornsberry (Continued)*

Preparation of *Antimicrobials* *(Continued)*	Tetracycline	Lederle Laboratories Division of American Cyanamid Company P.O. Box 500 Pearl River, N.Y. 10965
	Cephalothin Erythromycin	Lilly Research Laboratories Division of Eli Lilly and Company P.O. Box 618 Indianapolis, Ind. 46206

Each drug is prepared in a stock solution at 1280 µg./ml. and stored at −70° C. Prior to performing the test, the stock solution is thawed and then diluted to 10 times the final desired concentrations shown in the above table, since there is a further 1:10 dilution in the test.

Disks containing the appropriate concentrations of each drug can be prepared by pipetting the proper quantity of drug onto sterile blank disks. The disks are then dried and stored with a desiccant at −70° C.

Procedure

1. The inoculum can be prepared by one of two ways. Grow the organism overnight in supplemented Schaedler broth (BBL or Difco) and adjust the turbidity to the turbidity of a 0.5 McFarland standard. Or, remove some growth from the surface of a Schaedler agar plate that has been incubated overnight and prepare a suspension in Schaedler broth. Adjust the turbidity to that of a 0.5 McFarland standard as described for the broth culture.
2. To each tube of 2.7 ml. of Schaedler broth add 0.3 ml. of the 10 × concentrated antimicrobial solution. Alternatively, if disks are used, add each disk to 3 ml. of the broth.
3. Inoculate each tube with 0.025 ml. of the adjusted inoculum using a sterile, disposable, calibrated dropper (e.g., Cooke Engineering Co., Alexandria, Va.). Replace the caps loosely.
4. Incubate the tubes in an anaerobe jar (GasPak or evacuation/replacement) or anaerobic glove box for 18 to 24 hours at 35° C.

Interpretation of
Results

Read the end point as the smallest concentration where no macroscopic growth (turbidity) occurred. There must be evidence of growth (good turbidity) in the control tube. The results can be reported as categories of susceptibility as shown in the following table.

	Presence or Absence of Growth in the Four Tubes			
Category of Susceptibility	Control Tube	Low Concentration	Medium Concentration	High Concentration
I	+	−	−	−
II	+	+	−	−
III	+	+	+	−
IV	+	+	+	+

(Continued)

Chart 10-2. *The Category Susceptibility Test of Thornsberry (Continued)*

Interpretation of *Results* *(Continued)*	*Category I.* Very susceptible organisms. Organisms would be inhibited by the level of antimicrobial attained in the blood on usual dosage. *Category II.* Moderately susceptible organisms. Organisms would be inhibited by the blood level of antimicrobial achieved on high dosage. *Category III.* Moderately resistant organisms. Organisms would be inhibited by levels achieved where the drug is concentrated (e.g., in urine). *Category IV.* Very resistant organism. The organisms would be resistant to usually achievable blood levels and would thus be considered resistant without qualification. An alternative way to report results by the category test is to report results only by MIC. For example, the results for penicillin G would be either susceptible to ≤ 0.25, 16, 128 μg./ml. or resistant to > 128 μg./ml. Use of a report form such as the one in Chapter 11, page 340, will assist the physician in interpreting MIC results and in determining the most appropriate dose and route of administration.

Bibliography	1. Ericsson, H. M., and Sherris, J. C.: Antibiotic sensitivity testing. Report of an international collaborative study. Acta. Pathol. Microbiol. Scand. Section B, Supplement 217, 1971. 2. Stalons, D. R., and Thornsberry, C.: A broth dilution method for determining the antibiotic susceptibility of anaerobic bacteria. Antimicrob. Agents Chemother., 7:15–21, 1975. 3. Thornsberry, C., Gavan, T. L., and Gerlach, E. H.: New developments in antimicrobial susceptibility testing. *In* Sherris, J. C. (ed.): Cumitech 6, Washington, D.C., American Society for Microbiology, 1977.

accuracy and precision of any other methods to be used. Details of the NCCLS reference method are to be published soon.

Conventional agar dilution and broth dilution procedures have been described by Sutter and Washington.[41] Some laboratories have employed a microdilution procedure that appears promising.[28] However, the details of these procedures are beyond the scope of this text.

Until the standard procedure for the antimicrobial susceptibility testing of anaerobes is established, the two procedures described in Charts 10-1 and 10-2 are relatively simple to perform and are now well established for use in the clinical laboratory.

REFERENCES

1. Allen, S. D.: Distribution of anaerobes isolated from human clinical specimens at Indiana University. *Unpublished data.*
2. Allen, S. D., *et al.*: Development and evaluation of an improved anaerobic holding jar procedure. Abstracts of the Annual Meeting of the American Society for Microbiology, Abstract C142, p. 59, 1977.
3. Aranki, A. S., *et al.*: Isolation of anaerobic bacteria from human gingiva and mouse cecum by means of a simplified glove box procedure. Appl. Microbiol., 17:568–576, 1969.
4. Barlett, J. G., *et al.*: Antibiotic associated pseudomembranous colitis due to toxin-producing clostridia. N. Engl. J. Med., 298:531–534, 1978.
5. Beerens, H., and Tahon-Castel, M.: Infections humaines a bacteries anaerobies non-toxigenes. Presse Acad. Europ., Bruxelles, 1965.
6. Breed, R. S., Murray, E. G. D., and Smith, N. R.: Bergey's Manual of Determinative Bacteriology. ed. 7. Baltimore, Williams & Wilkins, 1957.
7. Brock, T. D.: Biology of Microorganisms. Englewood Cliffs, N.J., Prentice Hall, 1974.

8. Buchanan, R. E., and Gibbons, N. E. (eds.): Bergey's Manual of Determinative Bacteriology. ed. 8. Baltimore, William & Wilkins, 1974.

9. Cato, E. P., and Johnson, J. L.: Reinstatement of species rank for *Bacteroides fragilis, B. ovatus, B. distasonis, B. thetaiotaomicron,* and *B. vulgatus:* designation of neotype strains for *Bacteroides fragilis* (Veillon and Zuber) Castellani and Chalmers and *Bacteroides thetaiotaomicron* (Distaso) Castellani and Chalmers. Int. J. System Bacteriol., *26:*230–237, 1976.

10. Dowell, V. R., Jr.: Anaerobic infections. *In* Bodily, H. L., Updyke, E. L., and Mason, J. O. (eds.): Diagnostic Procedures for Bacterial, Mycotic and Parasitic Infections. ed. 5, pp. 494–543. New York, American Public Health Assoc., 1970.

11. ———: Methods for isolation of anaerobes in the clinical laboratory. Am. J. Med. Technol., *41:*32–40, 1975.

12. Dowell, V. R., Jr., and Hawkins, T. M.: Laboratory Methods in Anaerobic Bacteriology. CDC Laboratory Manual. DHEW Publication (CDC) 74-8272. Atlanta, Ga., Center for Disease Control, 1974.

13. Dowell, V. R., Jr., and Lombard, G. L.: Presumptive Identification of Anaerobic Nonsporeforming Gram-Negative Bacilli. DHEW Publication. Atlanta, Ga., Center for Disease Control, 1977.

14. Finegold, S. M.: Anaerobic Bacteria in Human Disease. New York, Academic Press, 1977.

15. ———: Therapy for infections due to anaerobic bacteria: an overview. J. Infect. Dis. (March Suppl.), *135:*S25–S29, 1977.

16. Finegold, S. M., Attebery, H. R., and Sutter, V. L.: Effect of diet on human fecal flora: comparison of Japanese and American diets. Am. J. Clin. Nutr., *27:*1456–1459, 1974.

17. Finegold, S. M., Shepherd, W. E., and Spaulding, E. H.: Practical anaerobic bacteriology. *In* Shepherd, W. E. (ed.): Cumitech 5, Washington, D.C., American Society for Microbiology, 1977.

18. Finegold, S. M., *et al.:* Management of anaerobic infections. Ann. Intern. Med., *83:*375–389, 1975.

19. Georg, L.: The agents of human actinomycosis. *In* Balows, A., DeHaan, R. M., Guze, L. B., and Dowell, V. R., Jr. (eds.): Anaerobic Bacteria: Role in Disease. Springfield, Ill., Charles C Thomas, 1974.

19a. Gorbach, S. L.: Introduction. J. Infect. Dis. (March Suppl.) *135:*S52–S53, 1977.

20. Gorbach, S. L., and Bartlett, J. G.: Anaerobic infections. N. Engl. J. Med., *290:*1177–1184, 1237–1245, 1289–1294, 1974.

21. Hansen, S. L., and Stewart, J. B.: Comparison of API and Minitek to Center for Disease Control methods for the biochemical characterization of anaerobes. J. Clin. Microbiol., *4:*227–231, 1976.

22. Hanson, C. W., Welch, W. D., and Martin, W. J.: Comparison of Minitek and API methods on identification consistency of anaerobic bacteria. Abstracts of the Annual Meeting of the American Society for Microbiology, Abstract C171, p. 64, 1977.

23. Hauser, K. J., and Zabransky, R. J.: Evaluation of the API Anaerobe System (API 20A) for the identification of anaerobes in a clinical laboratory. Abstracts of the Annual Meeting of the American Society for Microbiology, Abstract C14, p. 28, 1976.

24. Hesseltine, P. N. R., *et al.:* Clinical experience with cefoxitin. Clin. Res., *24:*113A, 1976.

25. Holdeman, L. V., Cato, E. P., and Moore, W. E. C.: *Clostridium ramosum* (Vuillemin) comb. nov: amended description and proposed neotype strain. Int. J. System. Bacteriol., *21:*35–39, 1971.

26. ———: Current classification of clinically important anaerobes. *In* Balows, A., De Haan, R. M., Guze, L. B., and Dowell, V. R., Jr. (eds.): Anaerobic Bacteria: Role in Disease. Springfield, Ill., Charles C Thomas, 1974.

27. ———: Anaerobe Laboratory Manual. ed. 4. Blacksburg, Va., Virginia Polytechnic Institute and State University, 1977.

28. Jones, R. N., and Fuchs, P. C.: Identification and antimicrobial susceptibility of 250 *Bacteroides fragilis* subspecies tested by broth microdilution methods. Antimicrob. Agents Chemother., *9:*719–721, 1976.

28a. Killgore, G. E., Starr, S. E., Delbene, V. E., Whaley, D. N., and Dowell, V. R., Jr.: Comparison of three anaerobic systems for the isolation of anaerobic bacteria from clinical specimens. Amer. J. Clin. Pathol., *59:*552–559, 1973.

29. Lombard, G. L., *et al.:* The effect of storage of blood agar medium on the growth of certain obligate anaerobes. Abstracts of the Annual Meeting of the American Society for Microbiology, Abstract C95, p. 41, 1976.

30. Martin, W. H., Gardner, M., and Washington, J. A., II: *In vitro* antimicrobial susceptibility of anaerobic bacteria isolated from clinical specimens. Antimicrob. Agents Chemother. *1:*148–158, 1972.

31. Martin, W. J.: Practical method for isolation of anaerobic bacteria in the clinical laboratory. Appl. Microbiol., *22:*1168–1171, 1971.

32. Morello, J. A., and Graves, M. H.: Clinical anaerobic bacteriology, Lab management, *15:*20–25, 1977.

33. Prevot, A. R., and Fredette, V.: Manual for the Classification and Determination of the Anaerobic Bacteria. Philadelphia, Lea & Febiger, 1966.

34. Rosebury, T. S.: Microorganisms Indigenous to Man. New York, McGraw-Hill, 1962.

35. Rosenblatt, J. E., Fallon, A. M., and Finegold, S. M.: Comparison of methods for isolation of anaerobic bacteria from clinical specimens. Appl. Microbiol., *25:*77–85, 1973.

36. Salaki, J. S., *et al.: Bacteroides fragilis* resistant to the administration of clindamycin. Am. J. Med., *60:*426–428, 1976.

37. Smith, L.: The Pathogenic Bacteria. ed. 2. Springfield, Ill., Charles C Thomas, 1975.

38. Sommers, H. M., and Matsen, J. M.: Laboratory diagnosis of infectious disease. Workshop Manual for Course no. 704. Chicago, American Society of Clinical Pathologists, March, 1977.

39. Sutter, V. L., and Finegold, S. M.: Susceptibility of anaerobic bacteria to 23 antimicrobial agents. Antimicrob. Agents Chemother. *10:*736–752, 1976.

40. Sutter, V. L., Vargo, V. L., and Finegold, S. M.: Wadsworth Anaerobic Bacteriology Manual. Wadsworth Hospital Center, Veterans Administration, Los Angeles, 1975.

41. Sutter, V. L., and Washington, J. A., II: Susceptibility testing of Anaerobes. *In* Lennette, E. H., Spaulding, E. H., and Truant, J. P. (eds.): Manual of Clinical Microbiology. ed. 2. Washington, D.C., American Society for Microbiology, 1974.

42. Swenson, R. M., *et al.:* Incidence of anaerobic bacteria in clinical specimens from known or suspected human infections. Abstracts of the Annual Meeting of the American Society for Microbiology, Abstract M67, p. 84, 1973.

43. Tedesco, F. J.: Clindamycin and colitis: a review. J. Infect. Dis. (March Suppl.), *135:*S95–S98, 1977.

44. Thornsberry, C.: Factors affecting susceptibility tests and the need for standardized procedures. *In* Balows, A., DeHaan, R. M., Guze, L. B., and Dowell, V. R. (eds.): Anaerobic Bacteria: Role in Disease. Springfield, Ill., Charles C Thomas, 1974.

45. Thornsberry, C., Gavan, T. L., and Gerlach, E. H.: New developments in antimicrobial susceptibility testing. *In* Sherris, J. C. (ed.): Cumitech 6. Washington, D.C., American Society for Microbiology, 1977.

46. Whaley, D. N., and Gorman, G. W.: An inexpensive device for evacuating and gassing systems with in-house vacuum. J. Clin. Microbiol., *5:*668–669, 1977.

47. Wilson, G. S., and Miles, A. A.: Topley and Wilson's Principles of Bacteriology and Immunity. ed. 5. London, Arnold, 1964.

11 Antimicrobial Susceptibility Testing

Alexander Fleming[7] is credited with the discovery of penicillin. Fortuitously, one autumn day in 1928, he observed not only that a contaminant mold was growing in a culture dish that had been carelessly left open to the air, but that the staphylococcus colonies growing adjacent to the mold were undergoing lysis (Fig. 11-1). Fleming correctly concluded that the mold, later identified as a strain of *Penicillium notatum*, was producing a diffusible bacteriolytic substance capable of killing staphylococci. Fleming's unknown antibiotic was later called penicillin, heralding the advent of the modern antibiotic era.

In truth, the phenomenon of antibiosis had been observed on two other recorded occasions about 40 years prior to Fleming's discovery. In about 1880, Lord Lister, in search of new antiseptics, noted that bacterial growth was inhibited in some culture flasks that were contaminated with molds. Fleming later stated: "if fate had been so kind, medical history may have been changed and Lister might have lived to see what he was always looking for—a nonpoisonous antiseptic." In 1889, Doehle[6] published a paper together with a photograph illustrating the antibiotic action of an organism that he called *Micrococcus anthracotoxicus* because of its lytic action on colonies of anthrax growing in mixed culture on the same plate. Pasteur himself must also have recognized this phenomenon of bacterial antagonism since he coined the phrase "life hinders life."

Yet it may have been an unnamed ancient nomadic Egyptian physician, curing himself of some ailment by eating moldy bread, who was the first discoverer of antibiotic action. The moldy wheaten loaf was only one of a number of therapeutic substances known by ancient man to be effective against inflammatory diseases. Herbs and spices, such as hyssop and anointing oils, and plant resins such as the myrrh carried by the wise men were widely used for their medicinal effects. Soils of strong smell, as those collected from forest humus or from river banks and stagnant pools, were commonly ingested by people belonging to earth-eating cults. We now know that these soils are heavily contaminated with antibiotic-producing strains of streptomyces and other actinomycetes. To cure pus in the throat, the Talmud recommends: "Take some earth from the shaded side of the privy and mix with honey." The dosage is not stated!

ANTIBIOTIC SUSCEPTIBILITY TESTING—THE BEGINNINGS

More than a decade passed before Fleming's discovery had any practical application in the treatment of infectious disease, although the injection of antimicrobial chemicals into humans was not a new concept. Paul Ehrlich, after many years of study on the antibiotic effect of azo dyes, discovered his "magic bullet," salvarsan, in 1912. This was the first injectable substance effective *in vivo* against the spirochete of syphilis. The

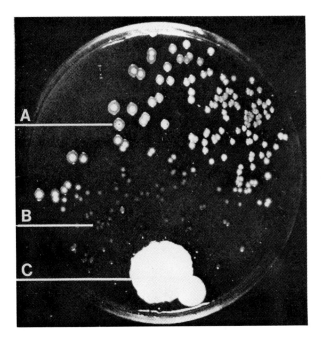

Fig. 11-1. Reproduced photograph of Fleming's discovery of the antibiotic action of Penicillium. Colonies of staphylococcus are seen growing at *A*; a contaminating colony of Penicillium is growing at *C*. The staphylococcus colonies around the fungus colony in area *B* are poorly developed and are undergoing lysis secondary to an antibiotic substance produced by the mold. This unknown substance was later called penicillin.

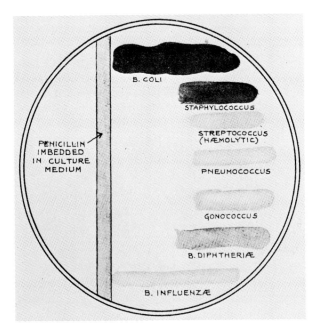

Fig. 11-2. Reproduction of Fleming's "ditch plate" antibiotic susceptibility test. A solution of penicillin was placed in the ditch. Several species of microorganisms were streaked perpendicular to the ditch. In this illustration, *B. coli* and *B. influenzae* are considered "resistant" to the action of the antibiotic since they grow up to the ditch; the other species are considered "sensitive" because their growth is inhibited in the zone adjacent to the ditch.

research on penicillin was spurred on by Domagk's 1932 discovery of Prontosil, a chemical analogue later found to be similar to sulfonamide. In 1939, Florey and Chain developed the practical technique by which the antimicrobial extract of penicillium molds could be obtained in sufficient purity and quantity for use in man.

The Fleming Ditch Plate Test

Fleming developed the first method for performing antibiotic susceptibility testing. His "ditch plate" technique is shown in Figure 11-2. A strip of agar in the form of a ditch (seen to the left in Fig. 11-2) is removed and replaced with medium containing the mold extract (penicillin). Multiple streak inocula of the organisms to be tested are made at right angles to the ditch. Note in Figure 11-2 that the strains labeled *Bacillus coli* and *Bacillus influenzae* (top and bottom streaks) appear "resistant" because they grow up to the ditch. In contrast, zones of growth inhibition are noted adjacent to the ditch for the strains labeled staphylococcus, streptococcus, pneumococcus, gonococcus, and *Bacillus diphtheriae.*

The need for antimicrobial susceptibility testing became evident soon after antibiotics became commercially available. Prior to World War II, penicillin production was limited and extremely expensive. Thus, it was necessary to develop some means for predicting when the use of penicillin would potentially cure an individual of an infectious disease.

During World War II, a number of different antibiotics were being discovered and patterns of susceptibility against various organisms were established. From his longtime interest in soil microbes, Waksman discovered streptomycin in 1943 and Dubos discovered gramicidin and tyrocidine soon thereafter. Duggar's research at Pearl River resulted in the manufacture of chlortetracycline (Aureomycin) by Lederele Laboratories in 1944. Although these new antibiotics were truly "wonder drugs" at the time of their introduction into medical practice, it was not long before resistant bacterial strains emerged and susceptibility testing became a practical necessity to guide physicians in the proper use of antibiotics.

The Broth Dilution Susceptibility Test

The broth dilution susceptibility test was among the first to be developed and still serves today as the reference method. Figure 11-3 shows 10 test tubes containing nutrient broth. To each have been added quantities of antibiotic, serially diluted from 100 μg./ml. to 0.4 μg./ml. Tube number 10 is free of antibiotic and serves as a growth control. Each of the ten tubes is inoculated with a calibrated suspension of the microorganism to be tested, followed by incubation at 35° C. for 18 hours. At the end of the incubation period, the tubes are visually examined for turbidity.

In Figure 11-3, note that the five tubes to the left are clear; the five to the right appear cloudy. Cloudiness indicates that bacterial growth has not been inhibited by the concentration of antibiotic contained in the turbid tubes. Figure 11-4 illustrates that the "break point" of growth inhibition is between tubes 5 and 6, or between 6.25 μg./ml. and 3.12 μg./ml. of antibiotic. This break point introduces the term, *minimal inhibitory concentration,* (MIC), defined as the lowest concentration of antibiotic in micrograms/ml. that prevents the *in vitro* growth of bacteria. Thus, in the example shown in Figures 11-3 and 11-4, the MIC lies somewhere between 6.25 μg./ml. and 3.12 μg./ml.; however, by convention, the MIC is interpreted as the concentration of the antibiotic contained in the first tube in the series that inhibits visible growth. Thus, in Figures 11-3 and 11-4, the MIC is 6.25 μg./ml.

What does the MIC mean to the physician? Simply that the MIC of the antibiotic being used must be achieved at the site of infection if bacterial growth is to be potentially inhibited. Generally, concentrations higher than the MIC are desirable because

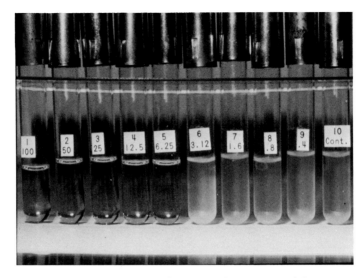

Fig. 11-3. Illustration of broth dilution antibiotic susceptibility test, in which the antibiotic to be tested is serially diluted in a range between 100 μg./ml. and 0.4 μg./ml. Tube number 10 serves as a positive growth control.

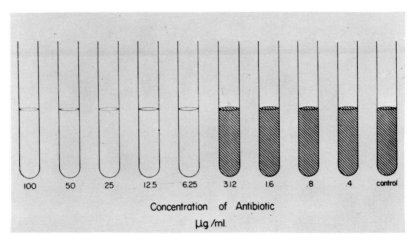

Concentration of Antibiotic
μg /ml

Fig. 11-4. Line drawing of the broth dilution susceptibility test shown in Figure 11-3. The minimal inhibitory concentration for the test illustrated here is 6.25 μg./ml.

other factors, such as the binding of the antibiotic to serum proteins or the presence of tissue inhibitors within the inflammatory exudate, may reduce the antibiotic action.

The *minimal bactericidal concentration* (MBC), in contrast to the MIC, is the least concentration in micrograms/ml. that will totally kill the bacterium being tested. Figure 11-5 illustrates an agar plate on which a series of subcultures were made from the visually clear tubes of broth shown in Figure 11-3. Note that the subcultures from tubes 3, 4, and 5 produced positive growth, indicating that although the antibiotic prevented growth in these tubes, it did not kill the bacteria present. Thus, the MBC is interpreted as tube number 3, or 50 μg./ml.

In practice, physicians use the MBC values in treating patients with infectious disease whose host defenses may be signifi-

Color Plate
11-1

Plate 11-1
Antimicrobial Susceptibility Testing

The agar diffusion disk technique is the method most commonly used in clinical laboratories for measuring the susceptibility of bacteria to various antibiotic agents. Filter paper disks impregnated with antibiotics are placed on the agar surface previously inoculated with the organism to be tested. The antibiotic from the disk diffuses into the agar, resulting in distinct zones of growth inhibition around those disks to which the bacterium being tested is susceptible.

A.
A 20 per cent agar hemoglobin reduction susceptibility test, illustrating distinct zones of growth inhibition around those disks to which the test bacteria are sensitive.

B.
A 20 per cent blood agar plate showing distinct zones of growth inhibition around the disks placed near the bottom of the photograph. The light zones around the susceptible disks represent the lack of hemoglobin reduction where the bacteria are not growing. This test, although rapid, is no longer used because it is difficult to reproduce the results.

Currently, the Bauer-Kirby disk diffusion technique is accepted as the standard antimicrobial susceptibility test. Frames C through L illustrate the sequence of steps necessary to perform the test.

C.
Frame C shows the selection of at least five well-isolated bacterial colonies from the surface of an agar plate. A Dacron tip swab, is used.

D.
Inoculation of trypticase soy broth with the swab containing the selected bacterial colonies.

E.
Broth tubes illustrating the procedure for the standardization of the bacterial inoculum. A MacFarland #1 turbidity standard is illustrated by the second tube from the left. The bacterial suspension in the left tube is too light in that the black line visualized through the suspension is more distinct than that of the turbidity standard. The broth suspension in the tube to the right is too dense in that the black line cannot be seen through the suspension at all. The third tube from the left in the series illustrates the adjusted bacterial suspension equal to the standard.

F.
Saturation of the Dacron tip of a swab with the standardized bacterial suspension in preparation of inoculation of the Mueller-Hinton agar plate included in the photograph.

G.
Inoculation of the Mueller-Hinton agar plate with the standard bacterial suspension. The swab is streaked back and forth across the agar, rotating the plate 60^0 and streaking again to give a uniform inoculum to the entire surface.

H.
Illustration of the modified Barry overlay technique for inoculating the susceptibility plate. The bacterial suspension is made in molten agar, which is then poured in a thin film to evenly cover the surface of the agar in the susceptibility plate.

I.
Positioning of the antibody-impregnated disks on the surface of a Mueller-Hinton agar plate that has been inoculated with the test organism. Forceps can be used as shown here; automatic dispensers are also available.

J.
Appearance of a Mueller-Hinton susceptibility test plate after proper performance of the procedure and incubation of the plate at 35^0 C. for exactly 18 hours. Note the distinct zones of growth inhibition around those disks containing antibiotics to which the test organism is susceptible.

K.
Zones of inhibition around the disks should be measured with a millimeter ruler or a caliper.

L.
The measurement of the zone of inhibition involves the taking of a reading across the center of the zone to include the diameter of the filter paper disk.

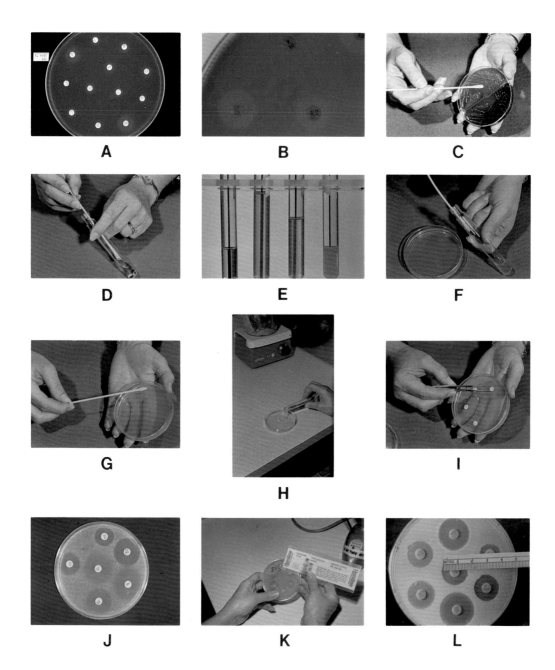

A

B

C

D

E

F

G

H

I

J

K

L

cantly lowered, particularly in patients receiving anticancer drugs or immunosuppressive agents. In patients with uncomplicated infections whose immune response is near normal, the MIC levels are generally sufficient to use as a guide for establishing antibiotic therapy.

Paper Disk Susceptibility Test: Early Developments

With the advent of several new antibiotics during the 1940's, tube dilution methods were no longer practical to meet the large volume of work required. For example, in a busy microbiology laboratory, 25 or more antibiotic susceptibility tests may be performed daily. Furthermore, each organism may be tested against 10 or more different antibiotics. To perform this volume of susceptibility tests, over 2500 tubes would be required, not to mention the drudgery and monotony of performing over 250 individual serial dilutions.

In 1943, Foster and Woodruff[8] first reported the use of antibiotic-impregnated filter paper strips in the performance of susceptibility tests. Their test was performed by placing a moistened strip to the surface of an agar plate that had been previously inoculated with the organism to be tested. Figure 11-6 simulates this test. Note the zone of growth inhibition adjacent to those strips containing the antibiotics to which the bacterium is sensitive. This procedure had the advantage over the tube dilution methods in that more than one antibiotic could be tested simply by placing multiple antibiotic-impregnated strips on the same agar plate.

Vincent and Vincent[16] introduced the use of paper disks in 1944, increasing even more the number of antibiotics that could be tested simultaneously. One year later, Morely[11] added another dimension by demonstrating that the paper disks could be dried after applying the antibiotic solution, thereby precluding the necessity of having fresh stock solutions available each time a test was to be performed.

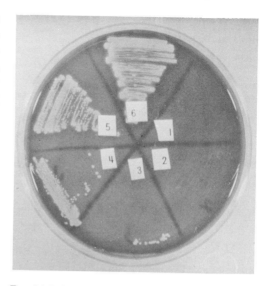

Fig. 11-5. Blood agar plate with subcultures from tubes 1 through 6 shown in Figure 11-3. Note that viable colonies were grown from tubes 3, 4, and 5. Since the microorganisms were killed only in tubes 1 and 2, the minimal bactericidal concentration is 50 μg./ml.

This discovery led shortly thereafter to the commercial manufacture of antibiotic susceptibility disks not unlike those used today. Bondi and associates[5] were the first to establish performance standards for the various concentrations of antibiotics to be used in different disks, from which were developed the first guidelines by which practical clinical applications could be made in the treatment of patients with infectious disease.

Figure 11-7 illustrates the basic principle of the disk diffusion method of antimicrobial susceptibility testing. As soon as the antibiotic-impregnated disk comes in contact with the moist agar surface, water is absorbed into the filter paper and the antibiotic diffuses into the surrounding medium. The rate of extraction of the antibiotic out of the disk is greater than its outward diffusion into the medium, so that the antibiotic concentration immediately adjacent to the disk may exceed that in the disk itself. However, as the distance from the disk

(Text continues on p. 326)

Fig. 11-6. Illustration of a simulation of the paper strip antibiotic susceptibility test of Foster and Woodruff.[8] Growth of the organism is inhibited around those strips containing antibiotics that are "sensitive."

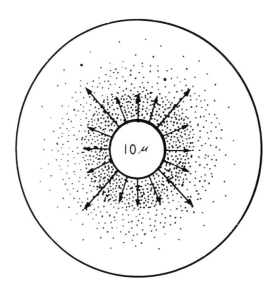

Fig. 11-7. Illustration of the principle of antibiotic diffusion in agar. The concentration of antibiotic decreases as the distance from the disk increases.

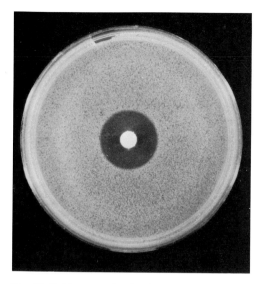

Fig. 11-8. Photograph of a disk antibiotic suscepti-bility plate showing the same principle as Figure 11-7. At the area where the concentration of anti-biotic is insufficient to prevent bacterial growth, a distinct margin can be seen.

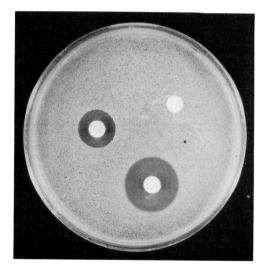

Fig. 11-9. Photograph of a disk antibiotic suscepti-bility plate on which have been placed three disks, each containing a different antibiotic. Note differ-ences in the sizes of the zones of growth inhibi-tion.

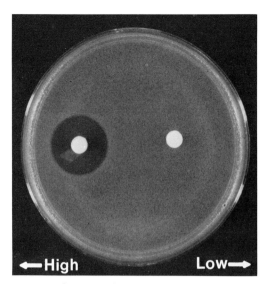

Fig. 11-10. Photograph of a disk antibiotic susceptibility plate on which have been placed a high- and a low-concentration disk. A "sensitive" organism will show a zone of growth inhibition around both disks; a "resistant" organism will be inhibited by neither disk; an organism of intermediate sensitivity is inhibited only by the high-concentration disk, as illustrated in this photograph.

increases, there is a logarithmic reduction in the antibiotic concentration until a point is reached at which the bacterial growth on the agar surface is no longer inhibited. The result is a sharply marginated zone of growth inhibition, as shown in Figure 11-8.

The so-called "zone vs. no zone" method was used for interpreting the results, meaning that the development of a zone of growth inhibition of any size around a disk indicated that the organism was "sensitive" to the antibiotic contained. "Resistant" bacteria grow right up to the margin of the disk.

Figure 11-9 illustrates an antibiotic susceptibility agar plate on which have been placed three disks. Note that there are differences in the sizes of the zones of inhibition around each of these three disks. Early in the interpretation of these tests, it was believed that zone sizes could be correlated with the relative sensitivity of the organism; that is, it was believed that the larger the zone, the more effective the anti-

biotic should be therapeutically. However, it was soon learned that zone sizes depend on certain physical-chemical properties of each antibiotic that influence *in vitro* diffusion rates in agar and do not necessarily correlate with *in vivo* antimicrobial activity.

The inability to produce quantitative results by the initial disk susceptibility test was considered a distinct disadvantage. In 1953, Schneierson[14] developed a semiquantitative two-tube broth test. One tube contained antibiotic in relatively high concentration, the other tube contained the antibiotic in low concentration. A highly "resistant" organism grew in both tubes; one of "intermediate" susceptibility grew only in the tube containing the broth with low antibiotic concentration; a highly "sensitive" organism grew in neither tube, being inhibited even by the low concentration of antibiotic.

This test was later modified into the high- and low-concentration two-disk test. Resistant organisms grew up to the margins of both disks, and sensitive organisms showed zones of growth inhibition around both disks. Organisms of intermediate sensitivity were inhibited only by the disk containing the higher concentration of antibiotic (Fig. 11-10).

The 18- to 24-hour delay required by the above methods before accurate results could be interpreted was considered to be too slow to be useful for application in some clinical situations. In 1956, Bass and coworkers[3] developed the hemoglobin-reduction disk susceptibility method by which results were available within 4 hours for the more rapidly growing organisms. A hemoglobin-reduction antibiotic susceptibility test plate is shown in Plate 11-1, *A*. Note that the color is a deeper red than most blood agar plates because the concentration of 20 per cent sheep blood was used instead of the usual 5 per cent concentration.

To perform this technique, a portion of a well-isolated colony of the organism to be tested is emulsified in approximately 3 ml. of molten brain-heart infusion agar that is allowed to cool to about 45° C., a temperature at which the agar is in a liquid state.

This "seed" culture is next evenly layered over the surface of the 20 per cent blood agar plate and the agar is allowed to harden. Several disks containing different antibiotics are then placed on the hardened seed layer. After 4 hours of incubation at 35° C., bright zones become visible around those disks that contain antibiotics to which the organism is "sensitive" (Plate 11-1, *B*). In the medium between the disks, or around those disks that contain antibiotics to which the organism is resistant, a dark color develops secondary to the accumulation of hemoglobin-reduction products from bacterial action. Although this method gave rapid results, quantitative interpretations were not possible and there was a relatively poor correlation of zone sizes for some antibiotics between the rapid test and those derived in the more conventional 18-hour test.

THE NEED FOR A STANDARDIZED ANTIMICROBIAL SUSCEPTIBILITY PROCEDURE

By the end of the 1950's, the status of antimicrobial susceptibility testing in microbiology laboratories throughout the world was in chaos, primarily because of the lack of an acceptable standard procedure. The antibiotic concentrations in disks varied considerably, a variety of media was being employed, methods of inoculation differed from laboratory to laboratory, the length of incubation time was not uniform, and results were being interpreted by several methods. A World Health Organization (WHO) Committee was formed[18] to investigate this problem, and the deliberations of this committee provided the groundwork leading to the development of first the Anderson and later the Bauer-Kirby standard techniques.

The Anderson Antibiotic Susceptibility Method

The several standardized steps that have been incorporated into the Bauer-Kirby antibiotic susceptibility technique were first developed by Anderson,[1] summarized as follows:

1. **Standardization of the antibiotic disks.** A single disk of known antibiotic concentration was used.
2. **Standardization of the medium.** Tryptic soy agar was the base medium.
3. **Standardization of the inoculum.** A concentration of approximately 10^8 organisms was required.
4. **Standardization of the incubation time.** The optimal time for the diffusion of antibiotic into the agar and reaction with the growing microorganisms was established to be 18 hours.
5. **Measurements of Zone Sizes.** The diameters of the zones of growth inhibition around each disk were carefully measured with a caliper or ruler and the results interpreted from a conversion table.

A number of modifications of the Anderson technique were later incorporated by Bauer, Kirby, and coworkers in the development of the currently accepted standard antimicrobial susceptibility test in the United States. Designating a national standard test not only permits more exacting quality control but also allows valid comparison of results between laboratories using the procedure.

The Bauer-Kirby Antibiotic Susceptibility Test[4]

The Bauer-Kirby susceptibility test differs from the Anderson procedure in only a few details, as described below. The sequence of the Bauer-Kirby method is shown in Plate 11-1, *C* through *L*.

Standardized Steps

The Medium. Mueller-Hinton agar has been selected as the standard culture medium instead of tryptic soy agar as used in the Anderson test. Mueller-Hinton agar supports the growth of most of the significant bacterial isolates encountered in clinical practice. Of more importance, the composition of the agar can be more uniformly controlled from batch to batch so that the microorganisms used in quality control testing give virtually identical reactivity from one lot of medium to the next.

It is also important that the medium be poured to a uniform depth of 4 mm. in the agar dish. If the medium is thinner than this, the antibiotics tend to diffuse from the disk to a greater extent in a lateral direction, increasing the zone sizes; agar deeper than 4 mm. results in more of the antibiotic diffusing downward, with a tendency to artificially narrow the zones of growth inhibition.

The Inoculum. Significant day-to-day variations in zone sizes of standard control organisms may occur if the concentration of bacteria in the inoculum is not controlled. The recommended procedure for inoculating susceptibility plates is as follows:

With a sterile polyester swab, touch the convex surfaces of 4 or 5 similar-appearing colonies that are well isolated on primary isolation medium, such as phenylethyl alcohol agar or MacConkey agar (Plate 11-1, *C*). Transfer the swab to 3 ml. of trypticase soy broth (Plate 11-1, *D*), rinse well in the fluid to remove the organisms from the inoculum, and then remove the swab. Place the culture tube into a 37° C. water bath for approximately 2 to 3 hours, or until the turbidity of the medium equals that of a MacFarland No. 1 standard. This is equivalent to a concentration of about 10^8 organisms. A MacFarland No. 1 standard is made by adding 0.5 ml. of barium sulfate to 99.5 ml. of 0.36 N sulfuric acid. The turbidity comparison between the standard and the test broth can best be made by viewing against a white card on which are scribed a series of horizontal black lines (see Plate 11-1, *E*).

If the suspension of organisms is lighter than the standard, the tube must be reincubated once again. If the turbidity of the organism suspension exceeds that of the standard, sterile saline can be added until the two match. When this has been achieved, a second, dry, sterile polyester swab is immersed into the bacterial suspension, expressing any excess fluid by rotating the swab against the inside wall of the tube immediately before it is removed (Plate 11-1, *F*).

The surface of a Mueller-Hinton agar plate that has been allowed to warm to room temperature is inoculated with this swab. While the plate is warming to room temperature, it is advisable to partially remove the lid to allow any excess moisture to evaporate from the agar surface. In order to evenly cover the entire surface of the plate, it is suggested that the swab be streaked in at least three directions, turning the plate at approximately 60-degree angles after each streak (Plate 11-1, *G*).

After allowing the inoculum to dry, the Mueller-Hinton plate is ready for the placing of the antibiotic disks on the agar surface, as described on page 329.

The Barry "agar overlay" method[2] provides an alternative procedure for inoculating the surface of Mueller-Hinton plates. Instead of 150-mm. plates, 100-mm. plates are used and the bacterial suspension is added to the agar medium while it is still molten, as described with the hemoglobin-reduction technique (page 326). The bacterial-agar suspension is next poured evenly over the agar surface to form a thin film (Plate 11-1, *H*). The advantages of this method are a more even distribution of organisms and the appearance of zones of growth inhibition that are more distinct and easy to interpret.

The Antibiotic Disks. The Bauer-Kirby method uses so-called "high-potency" disks. The concentrations of some of the more commonly used disks are tabulated below:

Ampicillin	10 μg.	Chloramphenical	30 μg.	Gentamicin	10 μg.
Carbenicillin	300 μg.	Polymyxin B	300 μg.	Kanamycin	30 μg.
Cephalothin	30 μg.	Tetracycline	30 μg.	Streptomycin	10 μg.
Methicillin	5 μg.			Tobramicin	10 μg.
Penicillin G	10 units	Clindamycin	2 μg.		
		Erythromycin	15 μg.	Naladixic Acid	30 μg.
		Lincomycin	2 μg.	Nitrofurantoin	300 μg.
				Sulfonamides	300 μg.

The term "high potency" refers to a relatively high concentration of antibiotic compared to the low concentration, often 1 to 2 μg. or less, in disks manufactured prior to the development of the standard Bauer-Kirby procedure. At low concentrations, antibiotics do not diffuse evenly into the agar, leading to spurious results. Therefore, the concentration of antibiotic in the disk must be sufficiently great to produce an even and easily reproducible diffusion in agar.

Manufacturers of antibiotic disks must carefully control the concentration of antibiotics within each disk, to within 60 to 120 per cent of the stated content, under guidelines established by the Food and Drug Administration. Disks may be purchased in individually packaged vials or in special cartridges designed to fit into automatic dispensers. All disks not in current use should be stored in a $-20°$ C. freezer; those currently in use should be kept in the refrigerator, preferably in a closed chamber with a desiccant. Disks should be allowed to warm to room temperature before placing on the agar surface.

In performing the Bauer-Kirby susceptibility test, antimicrobial-containing disks may be placed on the agar surface manually, using a pair of sterile forceps, as shown in Plate 11-1, *I*. Automatic dispensers that hold up to a dozen disk cartridges are also available. In either case, the disks must be placed at least 22 mm. from each other and 14 mm. from the rim of the Petri dish to avoid overlapping of the zones of growth inhibition or extension to the edge of the dish. Each disk should be gently pressed to the agar surface with the point of the forceps or an applicator stick to ensure that firm contact is made with the agar, taking care not to move the disk once it is in place.

Because the antibiotic concentration may be altered during storage, it is important that all disks be tested with a suitable quality control organisms of known reactivity each time the procedure is performed. The quality control procedure for susceptibility tests was discussed in Chapter 1.

Members of the penicillin group of drugs and cephalothin in particular are subject to deterioration if they are not stored properly. All vials and cartridges containing antibiotic disks must possess a clearly printed label indicating the exact concentration of antibiotic within the disks and the date of expiration. Laboratory personnel must take care not to use any disks that have exceeded their designated expiration dates.

Incubation. After the antibiotic susceptibility plates have been properly prepared, they should be placed in a 35° C. incubator, without increased atmospheric CO_2. Using incubators with CO_2 should be avoided because carbonic acid can form on the moisturized surface of the agar, resulting in a drop in *p*H. The growth of some organisms is inhibited in an acid *p*H, tending to falsely narrow the zone of growth inhibition. Also, the activity of different antibiotics may increase or decrease with a drop in *p*H, resulting in differences in diffusion rates and alterations of the zones of growth inhibition.

In laboratories with a small work load where only a CO_2 incubator is available, it is acceptable to place the susceptibility plates in a candle or anaerobe jar, sealing the lid to prevent access of the CO_2 within the incubator. The plates should be placed in the incubator upside down so that any moisture or condensation that collects under the lid does not fall onto the agar surface.

Although with some of the more rapidly growing organisms the zone of inhibition may be apparent within as early as 4 hours and reasonably accurate preliminary interpretations can be made, the recommended standard method requires that all final measurements should be made at exactly 18 hours. Eighteen hours has been established to be the time when the reactivity between the growing organisms and the inhibitor effects of the antibiotic are optimal and the zone margins of growth are most distinct. Even most of the slower-growing species have developed sufficiently by 18 hours that an accurate measurement can be made. If interpretation is delayed beyond 18 hours, alterations in the zone diameter may occur from drying of the agar, deterioration of the antibiotic, or overgrowth of the bacterial colonies.

Fig. 11-11. Photograph of an antibiotic susceptibility plate using a species of Proteus as the test organism. Note the swarming into the zone of inhibition at the peripheral margins. The second outer zone of growth inhibition should be used when measuring the width of the zone.

Measurement of Zone Diameters. After 18 hours of incubation, zones of growth inhibition will appear around those antibiotic disks to which the organism being tested is susceptible (Plate 11-1, *J*). These zone diameters must be carefully measured, viewing from the back of the Petri dish using a bright source of transmitted light. Sliding and graduated calipers, metric rulers that are scaled in millimeters, or specially prepared templates can be used. The reading of an inhibition zone with a metric ruler is illustrated in Plate 11-1, *L*. All measurements are made to the nearest millimeter, and the person performing the interpretation must take care to view the plate from a directly vertical line of sight to avoid any parallax that may result in misreading of the ruler markings.

Special templates with several circles with diameters indicating the resistant and susceptible zones of inhibition for many of the more commonly employed antibiotics are also available.

Atypical Results

Motile organisms, such as *Proteus mirabilis* or *Proteus vulgaris*, may swarm when growing on agar surfaces, resulting in a thin veil that may penetrate into the zones of inhibition around antibiotic susceptibility disks (Fig. 11-11). This zone of swarming should be ignored when making a reading, measuring the outer zone margin which is usually clearly outlined. Similarly, with sulfonamide disks, growth may not be completely inhibited at the outer margin of the zone, resulting in a faint veil where 80 per cent or more of the organisms are inhibited. Again, the outer margin of heavy growth inhibition should be used as the point of measurement.

The phenomenon shown in Figure 11-12 must be interpreted differently than that shown in Figure 11-11. Note that distinct colonies are present within the zone of inhibition. This does not represent swarming; rather, these colonies are either mutants that are more resistant than the major portion of the bacterial strain being tested, or the culture is not pure and the separate colonies are of a totally different species. A Gram's stain and/or subculture may be required to resolve this problem. If it is determined that the separate colonies represent a variant of a mutant strain, the bacterial species being tested must be considered "resistant," even though a wide zone of inhibition may be present for the remainder of the growth.

Figure 11-13 demonstrates the difficulty in measuring one zone diameter when there is overlapping with adjacent antibiotic zones, or when the zone extends beyond the margin of the Petri dish. Oval- or elliptical-shaped zones may occur, and it is difficult to determine whether to measure the short or long diameters. Unless the zones are very wide and the organism being tested is obviously sensitive, the test must be repeated with

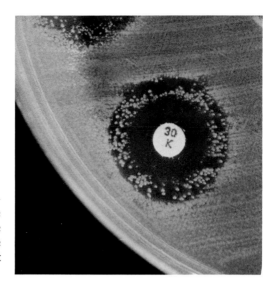

Fig. 11-12. Photograph of an antibiotic susceptibility plate in which two organisms in mixed culture are contained. The organisms growing within the zone of inhibition are "resistant," although the second species may be "sensitive." This test must be repeated.

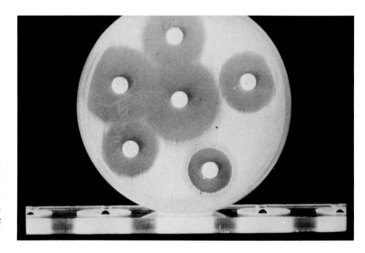

Fig. 11-13. Photograph of a poorly prepared antibiotic susceptibility plate showing objectionable overlapping of the zones of growth inhibition of adjacent disks.

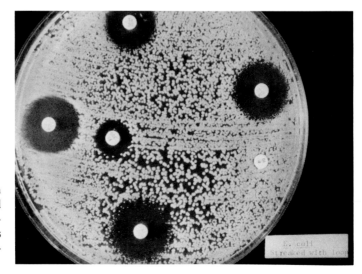

Fig. 11-14. Photograph of a poorly streaked antimicrobial susceptibility plate showing uneven growth. The zone margins are indistinct, compromising accurate measurements.

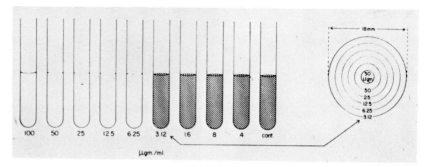

Fig. 11-15. Illustration showing the relationship between the minimal inhibitory concentration (MIC) of the tube dilution test and the diameter of growth inhibition by the disk diffusion method. The MIC "break point" in the tube dilution test of 6.25 μg./ml. corresponds with a zone diameter of 18 mm. in the disk diffusion test shown in this illustration.

more careful placement of the antibiotic disks so that overlapping will not occur.

Figure 11-14 illustrates a poorly prepared plate. The lines of streaking are irregular, leaving spaces between areas of colonial growth. The zone margins are not distinct, making it difficult to pick the exact points at which to make the measurements. Readings should not be attempted on a poorly inoculated plate such as this, and the test should be repeated for accurate results.

Interpretation of Results

It is important that physicians, microbiologists, and technologists have a clear understanding of the underlying meaning of the Bauer-Kirby susceptibility test and how the results should be interpreted. Figure 11-15 is an illustration that compares the results of a disk diffusion method with those of the broth dilution susceptibility test. As was discussed above, in the broth dilution test the concentration of antibiotic to be measured is serially decreased through a number of doubling dilutions. Note that the concentration declines by a logarithmic function. At some point, represented by the mean inhibitory concentration "break point," the antibiotic concentration is sufficient to inhibit the growth of the organism being tested, represented by the value of 6.25 micrograms/ml. in Figure 11-15.

The circular figure to the right in Figure 11-15 represents the diffusion of antibiotic into the agar away from the center antibiotic disk. This diffusion can be thought of as a series of concentric rings, with the antibiotic concentration decreasing as the distance from the disk increases. As a matter of fact, this decrease in antibiotic concentration also is a *logarithmic function*.

Thus, in a disk diffusion test, one can visualize a series of concentric rings extending out from the disk, each representing a declining concentration of antibiotic. Theoretically, there are concentric areas in the agar where the antibiotic concentration exactly matches those concentrations produced in the broth dilution test. It should now become clear that there is theoretically a point away from the disk in the diffusion test that exactly matches the "break point" in the broth dilution test. It is exactly this distance that is taken as the "sensitive" zone diameter in the interpretive charts against which Bauer-Kirby susceptibility test results are compared. In Figure 11-15, the distance of 18 mm. compares to 3.12 μg./ml., which is lower than the MIC. Therefore, any organism that gives a zone of 18 mm. or greater in the standard disk diffusion test can be considered "sensitive," or, extending the logic one step further, is within the MIC range of that antibiotic.

The fallacy of the "zone vs. no zone"

method of reading disk diffusion tests is that the presence of a zone of growth inhibition around a disk indicates that the organism is "sensitive"; however, if the zone is too small, the organism may be "sensitive" only at a very high concentration of antibiotic, beyond that which can be achieved in the blood or at the site of infection even using maximum dosages. In a practical sense, the organism is "resistant," even through a zone of growth inhibition is present.

The relationship between the size of zone diameters derived from the disk difussion method and the corresponding MIC break points for a number of commonly used antibiotics is shown in Table 11-1. Table 11-1 is a chart published in 1974 by the National Committee on Clinical Laboratory Standards (NCCLS), listing the commonly used antibiotics, the disk content, zone diameters for resistant, intermediate, and susceptible readings, and the approximate MIC correlations. The MIC conversion figures are meaningful in that the degree of reactivity is indicated, providing the physician with a better guideline for determining if a given antibiotic dosage will be sufficient. For example, a microorganism showing a 10-mm. zone of inhibition against ampicillin would be considered "resistant" at a level of 32 μg./ml. (See Table 11-1). However, if the infection is in the urinary tract, the concentration of ampicillin achievable in the urine with standard dosages is over 100 μg./ml., which exceeds the 32 μg./ml. MIC break point for ampicillin and represents a concentration sufficient to inhibit the growth of the microorganisms being tested. Thus, an antibiotic considered "resistant" by the disk diffusion method is in reality "sensitive" when one is dealing with an infection at a site where the concentration of antibiotic achievable exceeds the MIC break point.

Limitations of the Bauer-Kirby Test

Although the Bauer-Kirby test has been accepted as the standard technique for performing antibiotic susceptibility tests, giving useful information in most instances, there are a few distinct limitations:

1. Disk diffusion techniques are not applicable to slow-growing microorganisms. If prolonged incubation is required to achieve sufficient growth to produce a detectable zone of inhibition, there may be enough deterioration of the diffusing antibiotic to produce imprecise readings.

2. For antibiotics that diffuse slowly in agar, such as polymyxin B, rather large shifts in MIC values must occur before significant measurable changes are noted in the zone sizes. The high disk content of 300 μg/ml. counteracts the slow diffusability to some degree; however, results may be unreliable for slow-migrating antibiotics and controls must be compared.

3. Disk diffusion methods are not applicable to the determination of antibiotic susceptibility of anaerobes. The long period of incubation for the recovery of many anaerobes has made it difficult to establish reliable interpretive charts.

4. Interpretive charts, based on regression curves made from comparing disk diffusion zone diameter measurements with broth dilution MIC results, refer to antibiotic levels achievable in serum. As explained above, in the example of ampicillin and urinary tract infections, the antibiotic concentration at the site of infection may be totally dissimilar to that in the serum. In tissues or at an abscess site, the level of antibiotic may be so low as to be ineffectual, even though the disk diffusion test indicates a "sensitive" organism. On the other hand, a report of an organism "resistant" to an antibiotic by the disk diffusion method may not be valid since the concentration of the antibiotic in the disk (10 μg/ml., for example) itself may be far below the achievable levels in the urine, bile, or other secretions where the antibiotic in question may be concentrated.

5. Although certain disks, such as cephalothin, are considered "class disks" and theoretically produce similar results with any antibiotic in the generic group, the

(Text continues on p. 336)

Table 11-1. Zone Diameter Interpretive Standards and Approximate MIC Correlates*

Antimicrobial Agent	Disk Content	Zone Diameter, Nearest Whole mm.			Approximate MIC Correlates	
		Resistant	Intermediate	Susceptible	Resistant	Susceptible
Ampicillin[a] when testing gram-negative enteric organisms and enterococci	10 μg.	≤11	12–13	≥14	≥32 μg./ml.	≤8 μg./ml.
Ampicillin[a] when testing staphylococci and penicillin G-susceptible micro-organisms	10 μg.	≤20	21–28	≥29	≥32 μg./ml.; Penicillinase[b]	≤0.2 μg./ml.
Ampicillin[a] when testing Hemophilus species[†]	10 μg.	≤19	—	≥20	—	≤2.0 μg./ml.
Carbenicillin when testing Proteus species and Escherichia coli	100 μg.	≤17	18–22	≥23	≥32 μg./ml.	≤16 μg./ml.
Carbenicillin when testing Pseudomonas aeruginosa	100 μg.	≤13	14–16	≥17	≥250 μg./ml.	≤125 μg./ml.
Cephalothin[c]	30 μg.	≤14	15–17	≥18	≥32 μg./ml.	≤10 μg./ml.
Chloramphenicol	30 μg.	≤12	13–17	≥18	≥25 μg./ml.	≤12.5 μg./ml.
Clindamycin[d] (when reporting susceptibility to clindamycin only)	2 μg.	≤14	15–16	≥17	≥2 μg./ml.	≤1 μg./ml.
Colistin[e]	10 μg.	≤8	9–10	≥11	—[c]	—
Erythromycin	15 μg.	≤13	14–17	≥18	≥8 μg./ml.	≤2 μg./ml.
Gentamicin[†]	10 μg.	≤12	—	≥13	≥6 μg./ml.	≤6 μg./ml.
Kanamycin	30 μg.	≤13	14–17	≥18	≥25 μg./ml.	≤6 μg./ml.
Methicillin[f] when testing staphylococci	5 μg.	≤9	10–13	≥14	—	≤3 μg./ml.
Neomycin	30 μg.	≤12	13–16	≥17	—	≤10 μg./ml.
Penicillin G[g] when testing staphylococci	10 units	≤20	21–28	≥29	Penicillinase[b]	≤0.1 μg./ml.
Penicillin G[g] when testing other microorganisms	10 units	≤11	12–21[h]	≥22	≥32 μg./ml.	≤1.5 μg./ml.
Polymyxin B[e]	300 units	≤8	9–11	≥12	≥50 units/ml[e]	—

Antimicrobial agent	Disk content	≤	Intermediate	≥	≥	≤
Streptomycin	10 µg.	≤11	12–14	≥15	≥15 µg./ml.	≤6 µg./ml.
Tetracycline[i]	30 µg.	≤14	15–18	≥19	≥12 µg./ml.	≤4 µg./ml.
Tobramycin[†]	10 µg.	≤11	12–13	≥14	—	—
Vancomycin	30 µg.	≤9	10–11	≥12	—	≤5 µg./ml.
Sulfonamides[j]	250 or 300 µg.	≤12	13–16	≥17	≥350 µg./ml.	≤100 µg./ml.
Trimethoprim-sulfamethoxazole[j]	1.25 µg. 23.75 µg.	≤10	11–15	≥16	≥200 µg./ml.	≤35 µg./ml.
Nitrofurantoin[j]	300 µg.	≤14	15–16	≥17	≥100 µg./ml.	≤25 µg./ml.
Nalidixic acid[j]	30 µg.	≤13	14–18	≥19	≥32 µg./ml.	≤12 µg./ml.

*Table prepared by the Microbiology Resource Committee, National Committee for Clinical Laboratory Standards (NCLLS).

†To be considered tentative for 12 months from publication of this standard.

aClass disk for ampicillin, hetacillin, and amoxicillin. †When testing Hemophilus species an isolate equivocal results with the ampicillin disk and resistant results with the penicillin G disk (i.e., zones of ≤11 mm.) should be considered resistant to ampicillin class drugs, providing satisfactory controls are included.

bResistant strains of S. aureus produce penicillinase. There are significant reports of ampicillin-resistant strains which produce penicillinase.

cClass disk for cephalothin, cephaloridine, cephalexin, cephazolin, cephacetrile, cephradine, and cephaprin.

dThe clindamycin disk is used to test susceptibility to both clindamycin and lincomycin. Due to the greater activity of clindamycin and lincomycin, separate interpretive categories of zone diameters are recommended when reporting susceptibility to lincomycin as follows: 16 mm resistant: 17–20 mm-intermediate: ≥21 mm.- susceptible.

eColistin and polymyxin B diffuse poorly in agar, and thus the accuracy of diffusion tests is less than that found with other antimicrobics, and MIC correlates cannot be calculated reliably from regression analysis.

fClass disk for penicillinase-resistant penicillins (i.e., methicillin,).

gClass disk for penicillin G, phenoxymethyl penicillin, and phenethicillin.

hIntermediate category includes some microorganisms, such as enterococci, and certain gram-negative bacilli that may cause systemic infections treatable with high dosages of benzyl penicillin but not of phenoxymethyl penicillin or phenethicillin.

iClass disk for tetracyclines.

jUsed only for testing isolates from urinary tract infections.

[335]

Fig. 11-16. Photograph of the Steers replicator (see text).

results may not always be reliable for all antibiotics within the group. Antibiotic testing against specific drugs may be necessary in some instances.

Despite these limitations, the Bauer-Kirby technique represents forward progress in providing standardized results that can be compared between laboratories. Two stable organisms, *Staphylococcus aureus* (ATCC 25923) and *Escherichia coli* (ATCC 25922), allow each laboratory using the test to achieve a high degree of day-to-day precision. However, disk diffusion methods are starting to give way to broth dilution susceptibility tests now that automatic dispensers are available by which MIC results can be conveniently derived against multiple antibiotics.

METHODS FOR PERFORMING MINIMUM INHIBITORY CONCENTRATION (MIC) ANTIBIOTIC SUSCEPTIBILITY TESTS

Three methods are in use for determining the MIC antibiotic susceptibility of bacteria: (1) the broth dilution technique; (2) the agar dilution technique; and (3) the microtube broth dilution technique.

Broth Dilution Technique[9]

As discussed on page 321, the broth dilution technique has been used since the early 1940's to derive MIC susceptibility results and has been the standard reference method for comparing disk diffusion tests. However, the procedure is cumbersome and not applicable to the volume necessary for routine use in clinical laboratories.

Agar Dilution Technique[17]

The agar dilution technique is used with success in large-volume laboratories where the number of antibiotic susceptibility tests exceeds 20 per day. This procedure involves the use of a series of agar plates, each containing a different concentration of antibiotic, encompassing the different levels within the achievable therapeutic range. For example, if the therapeutic range for a given antibiotic is between 2 and 12 μg./ml., a series of agar plates containing 1, 4, 8, 12, and 16 μg/ml. of antibiotic might be used to determine the susceptibility of the organism being tested. If the organism grew in the first three plates but not in the plates containing 12 and 16 μg/ml. of antibiotic, an MIC value

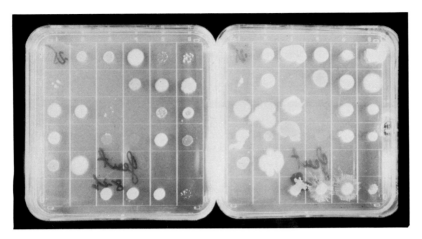

Fig. 11-17. Photograph of agar dilution antimicrobial susceptibility plates that had previously been inoculated with several species of bacteria from a Steers replicator. "Sensitive" organisms are inhibited by the concentration of antibiotic contained in the plate: "resistant" organisms appear as distinct colonies of bacterial growth.

of somewhere between 8 and 12 μg/ml. can be established, similar to the interpretation of the "break point" in the broth dilution technique.

In order to facilitate the testing of a large volume of cultures, an instrument known as the Steers replicator is used (Fig. 11-16 and Plate 6-3, *A* and *B*). The main feature of the instrument is a spring-loaded head fitted with 32 to 36 flat-surfaced inoculating pins, each about 3 mm. in diameter. The head is attached to a piston and cylinder mechanism by which it can be moved up and down in a vertical plane. The counterpart is an aluminum seed plate containing 32 to 36 wells. These wells are tooled in such a manner that when the seed plate is properly aligned within the guide at the base of the replicator, each of the inoculating pins on the movable head fits exactly into the wells. Each well in the seed plate provides a receptacle into which different bacterial suspensions can be placed.

The agar dilution susceptibility test is performed by placing the seed tray with its 32 to 36 bacterial suspensions directly under the inoculating head. The head is lowered so

that the pins extend fully into each of the wells, thereby sampling approximately 0.001 ml. of each bacterial suspension on the surface of each inoculating pin. The head is raised and the seed plate removed. Next, one of the antibiotic-containing agar plates is placed beneath the inoculating head, which in turn is lowered so that the flat surface of each inoculating pin just touches the agar surface. The head is again raised, and the inoculated agar plate is removed and replaced with the aluminum seed tray. The procedure is repeated for all of the antibiotic plates to be tested. After all the plates have been inoculated, they are placed in a 35° C. incubator for 18 hours.

An agar dilution plate ready for interpretation is shown in Figure 11-17. Note that those microorganisms that are "sensitive" to the concentration of antibiotic contained in any given agar plate do not produce a button of growth at the inoculum site, whereas those that are "resistant" appear as circular colonies. The agar plates are marked with a grid so that each microorganism can be identified by a number and the results entered on a worksheet.

The advantages of this method are:

1. A total of 32 to 36 organisms (depending on the size of the inoculating head), including three or more control strains, can be tested in each inoculation cycle.

2. The cost is relatively low.

3. Each plate can be subjected to strict quality control, using three or more test organisms with known reactivity.

4. Results can be interpreted as MIC values in μg./ml.

5. The method is well adapted to automation, data processing and statistical analysis of results. Electronic reading devices are currently on the market by which results can be automatically read and interpreted by computer.

The Microtube Broth Dilution Technique

The microtube dilution procedure is similar in principle to the broth dilution method above, except that the susceptibility of microorganisms to antimicrobial agents is determined in a series of "microtubes" molded into a plastic plate. Each plate may contain 80, 96, or more wells, depending upon the number of vertical and horizontal rows. A microtube plate containing 80 wells is shown in Figure 11-18, arranged in 10 vertical and 8 horizontal rows. This permits the testing of 10 different antibiotics, each of which can be diluted through 8 doubling dilutions. The number at the top of each vertical row represents each antibiotic being tested; the letter to the left of each horizontal row indicates the antibiotic concentration contained within each well.

The microtube plate is prepared by adding 50 microliters of the different concentrations of antibiotic solutions to the appropriate wells. Automatic dispensers and dilutors are available, or, in some locales, preprepared frozen plates ready for immediate use can be purchased. The user can select the number and types of antibiotics to be tested and the exact dilution concentrations desired.

Most commonly one plate is used to test the susceptibility pattern of one organism against several antibiotics, although more than one organism could be tested against fewer antibiotics at the discretion of the user. If a frozen plate is used, it is taken from the freezer just prior to inoculation and allowed to thaw and warm to room temperature. A bacterial suspension is prepared by inoculating a loopful of the colony to be tested in ½ to 1 ml. of nutrient broth and incubating at 35° C. for 4 to 6 hours (producing a bacterial concentration of about 10^9 organisms). Fifty microliters of the suspension are added to 25 ml. of water, resulting in a final concentration of about 10^5 organisms/ml. Fifty microliters of this diluted suspension are added to each of the 80 wells.

After the bacterial suspension has been added to each well, a plastic sheet or other suitable material is placed over the surface of the plate to prevent drying during incubation. In laboratories where multiple plates are inoculated, it is possible to stack one on top of the other, covering only the top plate with the plastic sheet. All plates are placed in a 35° C. incubator for 15 hours.

The interpretation of the microtube plate shown in Figure 11-18 is as follows: assume that row 1 contains ampicillin, serially diluted from 64 μg/ml. in row A to 0.5 μg/ml. in row H. Note that a button of bacterial growth is present within the centers of all wells in row number 1. This indicates that the organism being tested is resistant to ampicillin at all concentrations tested; or, the MIC is greater than 64 μg./ml. (the highest concentration tested).

Assume that row number 2 contains carbenicillin, serially diluted from concentrations of 128 μg./ml. in row A to 1 μg./ml. in row H. Note in row number 2 that bacterial growth is absent from the wells in rows A, B, and C. Well D in row number 2 is the last one in which bacterial growth is observed; or, the "break point" is between wells C and D. The concentration of antibiotic in well C is 32 μg./ml. (well A = 128 μg/ml, well B = 64 μg/ml., etc) and in well D is 16 μg./ml. The MIC is interpreted as the first well showing no growth; or, as illustrated in Figure 11-18, the MIC of the organism being

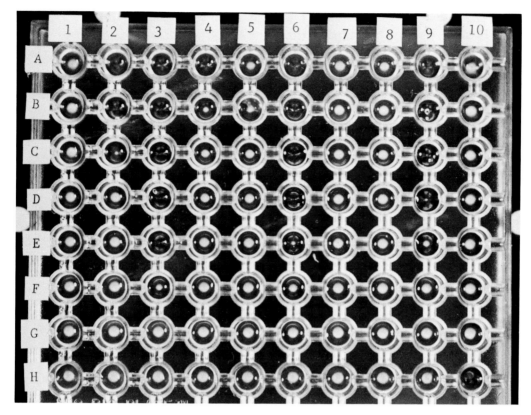

Fig. 11-18. Photograph of a microtube broth dilution antibiotic susceptibility plate. The numbers across the top row of microtubes indicate the different antibiotics being tested; the letters along the left vertical row of microtubes reflect the concentration of antibiotics contained within each well. The appearance of a button of bacterial growth in any well indicates "resistance" to that concentration of antibiotic.

tested against carbenicillin (row number 2) is 32 µg./ml. (well C).

If row number 3 contains cephalothin, serially diluted from 64 µg./ml. (well A) to 0.5 µg./ml. (well H), the MIC of the organism is 4µg./ml., the concentration of antibiotic in well E, the first one in row number 3 showing no bacterial growth. The remaining antibiotics can be interpreted in the same manner. One must be careful to note the exact dilution within each of the wells because the concentration may vary with each antibiotic used.

In order to assist the physician in the clinical applications of the MIC susceptibility test, a report form similar to that shown in Table 11-2 should be used. Note that the theoretical blood and urine levels in µg./ml.

attainable for several antibiotics at different dosage schedules and routes of administration are clearly listed. By knowing the blood and urine levels theoretically achievable for each antibiotic and the MIC results, the physician not only can select the antibiotic most suitable, but can select the dosage schedule and most appropriate route of administration as well.

Microtube dilution MIC tests may be completed in just over 4 hours for many of the more rapidly growing microorganisms by using a small amount of tetrazolium dye as an indicator of bacterial growth. This modification of the technique involves incubating the inoculated microtube plate for about 3½ hours, after which 50 microliters of a solution of oxidized tetrazolium salt is added to

Table 11-2. *Sample Form for Reporting of MIC Susceptibility Results*

ANTIBIOTIC SUSCEPTIBILITY REPORT

The microdilution susceptibility test gives the minimum inhibitory concentration (MIC) of antibiotics against the organism being tested. The MIC can be thought of as the *least amount of antibiotic required to inhibit bacterial growth*. The amount of antibiotic in serum (blood) should be equal to or greater than the MIC at all times. The peak serum level should be *at least 2 to 8 times the MIC* because of "peak and valley" fluctuations following administration of the medication.

Below is a list of antibiotics and attainable MIC blood and urine levels that can be expected for different dosages via several routes of administration.*

Antibiotic	Route of Administration	Therapeutic Dose (Adults)	Dose Schedule	Attainable Blood Levels μg./ml.	Attainable Urine Levels μg./ml.
Ampicillin	Oral	250 mg.	q6h	1.5–2.5	50–100
	I.M.	500 mg.	q6h	2.5–4.0	200–400
	I.V.	1.0–1.5 g.	q4h	15–20	200–400
Carbenicillin	Oral	1.0 g.	q6h	2–10	250–1400
	I.V.	4.0–5.0 g.	q4h	100–150	1000–7000
Cephalexin	Oral	0.25 g.	q6h	2–6	300–400
	Oral	1.0 g.	q6h	20–30	1000–2000
Cephalothin	I.M.	0.5–1.0 g.	q6h	10–20	1000–2000
	I.V.	1.0–2.0 g.	q6h	30–50	1000–2000
Chloramphenicol	Oral	250–500 mg.	q6h	2–5	200–800
	I.V.	500 mg.–1.0 g.	q6h	10–20	500–1400
Clindamycin	Oral	150–300 mg.	q6h	2–3.5	30–90
	I.V.	300–600 mg.	q6h	4–8	45–240
Erythromycin	Oral	500 mg.	q6h	3–5	Very low
Gentamicin	I.M. or I.V.	1.0–2.0 mg./kg.	q6–8h	5–10	Up to 250
Kanamycin	I.M. or I.V.	500 mg.	q8–12h	15–20	100–600
Methicillin	I.M.	1.0 g.	q4h	10–15	700–1000
	I.V.	2.0 g.	q4h	20–25	1000–2000
Nitrofurantoin	Oral	50–100 mg.	q6h	Insignificant	200+
Oxacillin	Oral	0.25–1.0 g.	q6h	1–8	100–700
	I.M.	1.0 g.	q4h	10–15	700–1000
	I.V.	2.0 g.	q4h	20–25	1000–2000
Penicillin	Oral	250 mg.	q6h	1.0–1.5	
	I.V.	1.0 Mil. Units	q4h	15–20	1000–3000
Tetracycline	Oral	250–500 mg.	q6h	2–5	600–1200
T/S (trimethoprim [TMP], sulfamethoxazole) [SMX]	Oral	160 mg. TMP 800 mg. SMX	1–2 tab. q12h	1–3 TMP 40–60 SMX	100–200 TMP 200–300 SMX
Tobramycin	I.M. or I.V.	3.5 mg./kg.	q8h	4–8	50–150

Interpretation:

Date Reported _____

*If Colistin or Poly B susceptibility tests are desired, please notify laboratory.

each well. The plate is returned to the incubator for an additional 30 minutes. In those wells in which the organism is growing, the dye will be reduced to a visible red formazan, producing a red button at the bottom of the well. Where no growth has occurred the solution remains clear.

Tests of Therapeutic Effect

In patients with severe bacterial infections, or those in renal failure receiving a nephrotoxic drug, it may be important to know the level of antibiotic that is present in the serum or body fluid. A sufficient concentration of antibiotic to inhibit the growth of or even kill the causative microorganism must be achieved at the site of infection, yet toxic levels must be prevented. Serum antibiotic levels can be determined by either assessing the antimicrobial activity of the serum itself or by measuring the concentration of antibiotic by drug assay methods.

Serum Antimicrobial Activity

The antimicrobial activity of serum, also known as the Schlicter serum killing power test,[13] measures the ability of serial dilutions of serum to inhibit the growth of or kill the microorganism causing the infection. It is important to understand that the microorganism causing the infection in the patient must be isolated and subjected to this test. Usually paired serum samples are obtained from the patient, about ½ hour before and ½ hour after the dose of antibiotic has been administered. This allows an assessment of both the "valley" and the "peak" serum antibacterial activity. A full description of the Schlicter test is given in Chart 11-1.

The serum antimicrobial activity test is most useful in guiding antibiotic therapy in patients with subacute bacterial endocarditis. Assessing the overall antimicrobial activity of serum has advantages over measuring the levels of individual antibiotics in that the innate antibacterial properties of the serum itself (from opsonins, antibodies, and bacte-

riolysins) are taken into account, the effects of protein drug binding are discounted, and the effects of multiple antibiotics are included in the end point measurement.

Serum Drug Assays

Serum drug assays have primary value in monitoring serum levels of potential hepatotoxic and nephrotoxic antibiotics that are being administered to patients with liver or kidney failure. The aminoglycosides, in particular gentamicin, are the drugs most commonly assayed because of their toxic effects.[10] Since the aminoglycosides are excreted almost exclusively by the kidney, patients in renal failure may develop high serum levels of gentamicin at standard or less than standard dosages, increasing the potential for ototoxicity or other toxic effects. Therefore, monitoring of serum levels of gentamicin and other aminoglycosides has become a practical necessity in many hospitals, necessitating the implementation of drug assay procedures in clinical microbiology laboratories. These procedures have been extended to include assays of antibiotics in fluids other than serum and even tissue levels as well.

Several methods are currently being used for antibiotic assays, some of which utilize readily available materials and are relatively easy to perform. The method most commonly used for the assay of gentamicin is an agar diffusion procedure[12] in which the inhibition of *Bacillus subtilis, Staphylococcus epidermidis*, or other bacterial strains selectively susceptible to gentamicin is evaluated. An example of the procedure is as follows:

1. A control organism, such as *Bacillus subtilis* (ATCC 6633), *Staphylococcus epidermidis* (ATCC 27626), or *Staphylococcus aureus* (ATCC 25923), is inoculated into approximately 3 ml. of brain-heart infusion broth and incubated at 35° C. long enough to produce a faintly visible turbidity.

2. An amount of 10 microliters of this bacterial suspension is mixed with 10 ml. of

(Text continues on p. 344)

Chart 11-1.   *Schlichter's Serum Antibacterial Potency Test (Modified)*

Introduction	In the treatment of patients with septicemia or bacterial endocarditis, it is often important to know whether the prescribed dosages of antibiotics are achieving blood levels sufficiently high to kill the causative organism. In 1947, Schlichter and MacLean described a direct technique for assessing the antibacterial potency of serum obtained from patients with endocarditis who were undergoing penicillin therapy.
Principle	The Schlichter test involves the determination of the exact serum dilution capable of killing the causative organism previously recovered in culture from a patient with septicemia. The bacteriostatic level is the dilution that prevents visible bacterial growth in an organism-serum suspension; the bactericidal level is that serum dilution that effects complete bacterial killing. Schlichter indicated that bacterial killing at a dilution of 1:2 signifies adequate antibiotic dosage; currently, bactericidal levels of 1:8 are considered adequate in most cases. The test results are used by physicians to adjust the antibiotic dosage appropriately.

Media and Reagents

1. Patient serum. Blood may be drawn from the patient at any time interval; generally, samples are obtained immediately before administering the next dose of antibiotic in order to evaluate the lowest level of serum potency.
2. Bacterial suspension. It is essential that a pure culture of the organism previously recovered from the patient be used. The final bacterial concentration in the incubation suspension should equal between 10^5 and 10^6 colony forming units (c.f.u.).
3. Suspension diluent
   A. Nutrient broth. Trypticase soy, Mueller-Hinton, Columbia, brain-heart infusion, and glucose phosphate broths may be used.
   B. Pien and associates recommend the use of heat inactivated human serum as the diluent in order to take into account antibiotic-protein-binding properties. One- to threefold higher bactericidal dilution levels have been demonstrated in broth than in serum, particularly with semisynthetic penicillins.
   C. Stratton and Reller advocate the use of Mueller-Hinton broth (MHB) supplemented with 50 mg./L. of Ca^{++} and 20 mg./L. of Mg^{++} (MHBS). Optionally, small aliquots of MHB-S broth can be mixed in a 1:1 ratio to pooled human serum (Flow Laboratories, Inglewood, California) for use as the diluent (MHB-S/HS). MHB-S broth can be prepared as follows:
      Add 21 g. of Mueller-Hinton powder to 997.6 ml. of distilled water. Boil to dissolve. Add 2 ml. of 10 % $CaCL_2 \bullet 2 H_2O$ (Calcium Chloride Injection, 10%, Invenex Pharmaceuticals, Chagrin Falls, Ohio) and 0.4 ml. of 50 % $MgSO_4 \bullet 7 H_2O$ (Magnesium Sulfate Injection, U.S.P., the Vitarine Co., New York, N.Y.) Dispense in 10 ml. aliquots in screwcap tubes and autoclave for 15 minutes at 250° C. and 21 ppi.

Procedure

1. Place twelve 75 × 10 mm. sterile test tubes (or Kahn tubes) in a rack. Label the tubes 1 through 12.
2. Place 1 ml. of the patient's serum to be tested in Tube 1. Tube 1 will also serve as a serum sterility control.

Chart 11-1.   *Schlichter's Serum Antibacterial Potency Test (Modified) (Continued)*

Procedure *(Continued)*	3. To Tubes 2 through 12, add 0.5 ml. of nutrient broth. Alternately, inactivated serum, MHB-S, or MHB-S/HS may be selected, as discussed above.
	4. With a sterile pipette, transfer 0.5 ml. of serum from Tube 1 to Tube 2. Mix well.
	5. Perform serial twofold dilutions of the serum by sequentially transferring 0.5 ml. of the mixture from Tube 2 to Tube 3, from Tube 3 to Tube 4, and so on, through Tube 10. Discard from Tube 10 0.5 ml. of the mixture. The final dilution of serum in Tube 10 is 1:512. Further dilutions can be carried out if the organism being tested is particularly sensitive.
	6. Prepare a suspension of the patient's organism in nutrient broth (or inactivated serum, MHB-S, or MHB-S/HS). Incubate the suspension for 4–6 hours to equal a MacFarland 0.5 standard (approximately 10^8 to 10^9 c.f.u./ml.).
	7. Make a 1:100 dilution of the suspension by adding 1.0 ml. of the suspension to 9.9 ml. of the same diluent used in Step 6. Mix well. This diluted suspension should contain between 10^5 and 10^6 c.f.u./ml. This may be checked by performing a standard colony count using a 10^{-3} ml. standard inoculating loop.
	8. Add 0.5 ml. of this diluted suspension to Tubes 2 through 11. Tube 11, which does not contain serum, serves as a positive growth control. Do not add the bacterial suspension to Tube 12. Tube 12 contains only the broth or serum diluent and serves as the negative growth control.
	9. Incubate all tubes for 18 hours at 35° C. in an incubator with no added CO_2.
	10. Tubes 1 and 12 should be sterile after incubation, while Tube 11 should be turbid. The test must be repeated if the controls were not correct.
	11. Observe all tubes after incubation. Using a 10^{-2} ml. calibrated inoculating loop, subculture the suspensions that appear visually clear from all tubes to sheep blood agar plates. Incubate these plates for an additional 18 to 24 hours.
Interpretation	The minimal inhibitory or bacteriostatic dilution is the first dilution of patient's serum that appears visually clear. The minimal bactericidal dilution is the first dilution that shows a 99.9% kill of the organism. This is determined by observing the sheep blood agar subcultures. The first dilution showing growth of fewer than 10 colonies is taken as the end point. Usually a dosage of antibiotic that gives a 1:8 or 1:16 bactericidal dilution is considered adequate.

Bibliography	Pien, F. D., Williams, R. D. and Vosti, K. L. Comparison of broth and human serum as the diluent in the serum bactericidal test. Antimicrob. agents chemoth., 7:113–114, 1975.
	Stratton, C. W. and Reller, L. B. Serum dilution test for bactericidal activity. I. Selection of a physiologic diluent. J. Infect. Dis., 136:187–195, 1977.
	Schlichter, J. G. and MacLean, H. A method of determining the effective therapeutic level in the treatment of subacute bacterial endocarditis with penicillin. Am. Heart J., 34:209–211, 1947.
	Schlichter, J. G., MacLean, J. and Milzer, A. Effective penicillin therapy in subacute bacterial endocarditis and other chronic infections. Am. J. Med. Sci., 217:600–608, 1949.

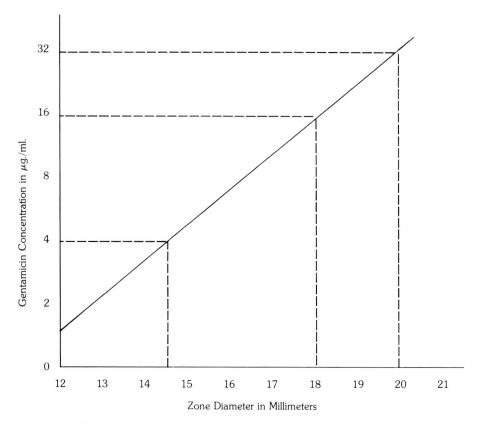

Fig. 11-19. Illustration of a regression curve by which the concentration of drug (gentamicin) is related to the zone diameter in millimeters in the agar diffusion drug assay test. Antibiotic concentrations in µg./ml. are read along the ordinate corresponding to various zone diameters along the abscissa by projecting vertical and horizontal lines to the standard curve as illustrated.

molten Mueller-Hinton agar cooled to 50° C. The bacterial broth suspension is thoroughly mixed with the Mueller-Hinton agar and quickly poured into a 150-mm. Petri dish. This produces a relatively thin agar depth and is used if impregnated filter paper disks are to be placed on the surface.

Alternatively, 40 ml. of the bacterial suspension-agar mixture can be poured into the plate if the agar well diffusion method is used.

3. After the bacterial agar suspension has been allowed to harden and cool, 3 filter paper disks are placed on the surface to serve as controls, and additional disks equal in number for the unknown assays to be performed. With the agar well method, 3-mm. holes are punched into the agar, gently aspirating the agar plugs.

4. An amount of 20 microliters of antibiotic standards diluted in antibiotic-free human serum, usually in concentrations of 2, 8, and 16 µg./ml. is added to the three control disks. To the additional paper disks are added 20 microliters of serum or body fluid to be tested. In the agar well method, only 10 microliters of control and test sera are added to each of the wells.

5. The plates are placed into a 35° incubator for 15 hours. At the end of the incubation period the zones of inhibition are carefully measured with a millimeter caliper or ruler for the standards and unknown tests.

6. A standard curve is prepared as shown

in Figure 11-19. The antibiotic concentrations in μg./ml. are listed along the ordinate as a logarithmic function, using semilog paper, while the zone diameters in millimeters are plotted along the abscissa as a geometric function.

For example, assume on the day that the curve shown in Figure 11-19 was constructed the standard disks produced the following zones of bacterial growth inhibition:

2 μg./ml. = 13.0 mm.
4 μg./ml. = 14.7 mm.
8 μg./ml. = 16.0 mm.
16 μg./ml. = 18.0 mm.

These points were plotted at the appropriate intersects and the best straight line drawn as shown in Figure 11-19. A new standard curve must be constructed each time the test is run to take into account variations in medium, reagents, and technique.

7. The zone diameter of each unknown can be plotted on the standard curve and the antibiotic concentration can be determined by finding the appropriate intersect with the ordinate.

For example, Figure 11-20 is a photograph of an agar diffusion gentamicin assay plate on which are placed three disks inoculated with serum samples containing unknown concentrations of gentamicin. Assume that the zone diameters are 14.5, 18, and 20 mm., respectively. Using the regression curve shown in Figure 11-19, the gentamicin concentrations for each of these unknowns can be found by projecting a vertical line from the appropriate point on the mm. scale on the abscissa until it intersects the regression curve. The gentamicin concentration can then be read directly from the scale on the ordinate by drawing a horizontal line from the points of intersect on the regression curve as illustrated in Figure 11-19. Thus, the gentamicin concentrations for the serum unknowns shown in Figure 11-20 are 4, 16, and 32 μg./ml., respectively.

By knowing the exact serum antibiotic levels as determined by the drug assay procedure described above, the physician can adjust the antibiotic dosage up or down, depending upon the level desired. If the

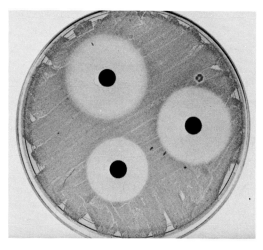

Fig. 11-20. Photograph of an agar diffusion drug assay plate illustrating serum samples applied to disks with varying concentrations of antibiotic. The higher the antibiotic concentration, the larger the zone of growth inhibition, as shown. These zone diameters are then interpreted from a regression curve similar to that shown in Figure 11-19.

example cited above were true, the two samples of 16 μg./ml., and 32 μg./ml. are considered to be within the toxic range (concentrations over 8 μg./ml. are considered toxic) and the dosage schedule would have to be reduced downward.

The disk diffusion procedure should always be determined in triplicate on three separate seed plates, and the results arithmetically averaged before the standard curve is constructed and the measurements of unknown samples interpreted.

If the patient is receiving penicillin or cephalosporin in addition to gentamicin at the time the test is to be performed, penicillinase in a concentration of 20 Levy units/ml. or cephalosporinase in the same relative concentration should be added to the Mueller-Hinton bacterial suspension. Alternatively, the serum to be tested can be inactivated by adding beta-lactamase.

Two other procedures, radioimmunoassay and an enzymatic method[15] in which a specific radioactive product is measured after treatment of the test sample with a mixture of ^{14}C-labeled adenosine 5'-triphosphate, buffer, and a gentamicin adenylating enzyme, are also in current use. Both of these methods have the advantage of being rapid, with test results available within one or two hours, and gentamicin is specifically assayed, an important feature when other antibiotics are also present in the serum being tested.[15] However, preparation of the radioimmunoassay antisera or enzyme solutions, the need for expensive radioactive counting equipment, and excessive expenditure of technical time are drawbacks for the general implementation of either of these methods in clinical laboratories. In addition, assays applying high-pressure liquid chromatography appear quite promising, particularly for use in larger laboratories.

REFERENCES

1. Anderson, T. G.: An evaluation of antimicrobial susceptibility testing. In Antimicrobial Agents Annual. New York, Plenum Press, 1961.
2. Barry, A. L.: The agar overlay technique for disc susceptibility testing. *In* Balows, A. (ed.): Current Techniques for Antibiotic Susceptibility Testing. Pp. 17–25. Springfield, Ill. Charles C Thomas, 1974.
3. Bass, J. A., *et al.*: Evaluation of a rapid (hemoglobin reduction) method for determining antibiotic susceptibility of microorganisms. Randolph Air Force Base, Texas, Air University School of Aviation Medicine, United States Air Force, 1956.
4. Bauer, A. W. *et al.*: Antibiotic susceptibility testing by a standardized single disc method. Am. J. Clin. Pathol. *45:*493, 1966.
5. Bondi, A. *et al.*: A routine method for rapid determination of susceptibility to penicillin and other antibiotics. Am. J. Med. Sci., *213:*221–225, 1947.
6. Doehle, : Beobachtungen über einen Anta-

gonisten des Milzbrandes. Kiel, Schmidt and Klaunig, 1889.
7. Fleming, A.: On the antibacterial action of cultures of a penicillium with special reference to their use in the isolation of *B. influenzae*. Br. J. Exp. Pathol., *10:*226–236, 1929.
8. Foster, J. W., and Woodruff, H. B.: Microbiological aspects of penicillin. J. Bacteriol., *46:*187–202, 1943.
9. Gavan, T. L., and Town, M. A.: Microdilution method for antibiotic susceptibility testing. Bact. Proc., p. 73, 1969.
10. Giamarellou, H., et al.: Assay of aminoglycoside antibiotics in clinical specimens. J. Infect. Dis. *132:*399–406, 1975.
11. Morley, D. C.: A simple method of testing the sensitivity of wound bacteria to penicillin and sulfathiazole by use of impregnated blotted paper discs. J. Pathol. Bacteriol., *57:*379–382, 1945.
12. Sabath, L. D., *et al.*: Rapid microassay of gentamicin, kanamycin, neomycin, streptomycin and vanomycin in serum or plasma. J. Lab. Clin. Med., *78:*457–463, 1971.
13. Schlichter, J. G., and MacLean, H.: A method of determining the effective therapeutic level in treatment of subacute bacterial endocarditis with penicillin. Am. Heart J., *34:*209–211, 1947.
14. Schneierson, S. S.: A simple rapid disk-tube method for determination of bacterial sensitivity to antibiotics. Antibiot. Chemother., *4:*125–132, 1954.
15. Smith, A., et al.: Comparison of enzymatic and microbiological gentamicin assays. Antimicrob. Agents Chemother., *6:*316–319, 1974.
16. Vincent, J. G., and Vincent, H. W.: Filter paper modification of the Oxford cup penicillin determination. Proc. Soc. Exp. Biol. Med., *55:*162–164, 1944.
17. Washington, J. A., II.: The agar dilution technique. *In* Balows, A. (ed.): Current Techniques for Antibiotic Susceptibility Testing. Pp. 127–141. Springfield, Ill. Charles C Thomas, 1974.
18. W.H.O. Technical Report Series 210. Standardization of Methods for Conducting Microbic Sensitivity Tests. Geneva, 1961.

12 Mycobacteria

The diagnosis of tuberculosis depends upon the recovery and identification of the organism from the patient. For many years both the diagnosis and treatment of tuberculosis depended on clinical symptoms, x-ray evidence of disease, and the presence of acid-fast bacilli in the sputum. Today the causative organism should be isolated, identified, and in many instances tested for susceptibility to antimycobacterial drugs.

The isolation of mycobacteria from sputum and other clinical specimens poses a special problem to the laboratory. Mycobacteria require a prolonged time for replication, in the order of 20 to 22 hours for *M. tuberculosis,* while the generation time of other bacteria that may be present in the specimen may be as short as 20 to 30 minutes. This disproportionate rate of growth between mycobacteria and other bacteria may result in rapid accumulation of metabolic waste products which make the culture medium unsatisfactory for the growth of mycobacteria. For this reason, the successful isolation of mycobacteria is in large part dependent upon the selective suppression of contaminating bacteria.

SPECIMEN COLLECTION FOR THE ISOLATION OF MYCOBACTERIA

Mycobacteria can be isolated from a variety of clinical specimens, including sputum, urine, and cerebrospinal fluid, and from tissue biopsies, including liver, bone marrow, and lymph nodes, or from any source suspected of being the site of tuberculous infection.[8] Specimens which may contain a mixed bacterial flora should be processed promptly to reduce the problem of culture contamination. Expectorated sputum or sputum collected by ultrasonic nebulization should be obtained shortly after the patient awakens in the morning. Although some studies suggest that a 24-hour sputum sample may result in an increased recovery of mycobacteria, the incidence of contamination and the necessity to discard cultures is also greater.[7,9] Nebulized early-morning sputum specimens not only result in more rapid growth of mycobacteria in cultures, but less contamination when compared to 24-hour pooled sputum specimens.

The irregular ulceration and release of acid-fast bacilli from subepithelial bronchial foci of tuberculous infection can result in a variable pattern of recovery. Cultures may be positive on one day but negative the next. For this reason, a minimum of 3 to 5 early-morning specimens should be collected from patients suspected of having either pulmonary or renal tuberculosis. All specimens should be transported promptly to the laboratory and refrigerated if processing of them is delayed.

SPECIMEN PROCESSING

The high cell wall lipid content of most mycobacteria make them more resistant to killing by strong acidic and alkaline solutions than other types of bacteria that may be present in sputum, urine, and other types of clinical specimens. For this reason, specimens likely to contain a mixed bacterial flora are treated with a decontaminating agent to reduce undesirable bacterial overgrowth and to liquify mucous. After treatment with the decontamination agent for a carefully controlled period of time, the acid or alkali used is neutralized and the mixture is then centrifuged at high speed to concentrate the mycobacteria.

Some decontaminating solutions, such as 6 per cent sodium hydroxide, are so strong that they may kill or seriously injure mycobacteria in the specimen so that they grow very slowly if at all. Decreasing the strength of the acidic or alkaline decontamination solution has resulted in an improved recovery of mycobacteria by culture but frequently at the price of a higher incidence of contamination. Exposure of specimens to strong decontaminating agents such as 4 per cent sodium hydroxide, 5 per cent oxalic acid, or 3 per cent sodium hydroxide must be carefully timed to prevent excessive chemical injury.

The use of mild decontaminating agents, such as trisodium phosphate or trisodium phosphate combined with benzalkonium chloride (Zephiran), is popular in some laboratories. Specimens containing large numbers of *M. tuberculosis* can withstand the action of these agents as long as an overnight exposure, and careful timing is not required.[10] Specimens treated with TSP-Zephiran should be inoculated to egg base culture media to neutralize the growth inhibition of the Zephiran. If agar base media are to be used, neutralization of the Zephiran can be accomplished by adding lecithin.

Following the introduction of successful chemotherapy and the subsequent need to recover mycobacteria for susceptibility testing, a concentrating solution containing 2 per cent NaOH and n-acetyl-l-cysteine (NALC) was described. NALC is a mucolytic agent without antibacterial activity which liquifies mucus by splitting disulfide bonds. The mycobacteria are released when

the mucus is liquified and can be more readily sedimented by high-speed centrifugation (3,000 relevant centrifuge force [RCF] or higher if an angle-head centrifuge is used).

The 2 per cent NaOH in the solution serves as a decontaminating agent. Occasionally the concentration of NaOH must be increased to 3 per cent during warm weather or in treating specimens from patients with large pulmonary cavities associated with persistent bacterial contamination.

Usually, concentrated HCl or concentrated NaOH is employed to neutralize the decontaminating agents. Because of the strength of these solutions, a neutral end point is sometimes difficult to achieve. One advantage of the NALC procedure is that the addition of a large volume of phosphate buffer makes shifts in *p*H less likely. The buffer serves to "wash" the specimen, and dilutes toxic substances and decreases the specific gravity of the specimen so that centrifugation is more effective.

Table 12-1 lists commonly used agents for decontaminating and concentrating specimens, along with comments concerning their use. Each laboratory should select the agents to employ on the basis of the number and types of specimens received and the time and technical staff available to process specimens. The needs of a laboratory receiving specimens from hospitalized patients will differ from those of one serving outpatient clinics.

Mycobacterium tuberculosis is a facultative intracellular parasite that may be present in macrophages of the bone marrow, liver, and lymph nodes of patients with disseminated infections. Since tissue biopsies are usually not contaminated with other microorganisms, they can be homogenized and inoculated directly to culture media without the use of a decontaminating solution. Draining sinuses or other cutaneous lesions suspected of being tuberculous are best cultured by obtaining a small portion of infected tissue or drainage.

Culture swabs ordinarily are not recommended for recovery of mycobacteria because the hydrophobic nature of the lipid-

Table 12-1. *Commonly Used Agents for Decontamination and Concentration of Specimens*

Agent	Comments
n-Acetyl-l-cysteine plus 2% NaOH	Mild decontamination solution with mucolytic agent NALC to free my-cobacteria entrapped in mucus
Dithiothreitol plus 2% NaOH*	Very effective mucolytic agent used with 2% NaOH. Trade name of dithio-threitol is Sputolysin. Reagent is more expensive than NALC.
Trisodium phosphate, 13%, plus benzalkonium chloride (Zephi-ran)	Preferred by laboratories that cannot carefully control time of exposure to decontamination solution. Zephiran should be neutralized with lecithin if not inoculated to egg base culture medium.
NaOH, 4%	Traditional decontamination and concentration solution. Time of exposure must be carefully controlled. NaOH, 4%, will effect mucolytic action to promote concentration by centrifugation.
Trisodium phosphate, 13%	Can be used for decontamination of specimens when exposure time cannot be completely controlled. It is not as effective as TSP-Zephiran mixture.
Oxalic acid, 5%	Most useful in the processing of specimens that contain *Pseudomonas aeru-ginosa* as a contaminant
Cetylpyridium chloride, 1%, plus 2% NaCL	Effective as a decontamination solution for sputum specimens mailed from outpatient clinics. Tubercle bacilli have survived 8-day transit without sig-nificant loss.

*See reference 23.
†See reference 24.

containing cell wall of the bacteria inhibits transfer of organisms from the swab to the aqueous culture medium. If a swab has been used and submitted for culture, the tip should be placed directly on the surface of the culture medium or into a tube of liquid medium and incubated for 4 to 8 weeks. Mycobacteria, if present, may be found forming colonies in the fibers of the swab at the junction with the culture medium.

Safety Precautions

An approved microbiological safety hood should be used to transfer sputum and other clinical specimens from collection containers to centrifuge tubes, prepare smears for acid-fast staining, inoculate cultures, and transfer isolated mycobacteria colonies for further studies. Personnel should wear either disposable or sterilizable caps, gowns, masks, and gloves when working with cultures or potentially infectious specimens.

Perhaps the greatest potential for infectious hazard to laboratory personnel exists from the aerosolization of specimens that can occur from broken tubes in a rapidly spinning centrifuge. It is strongly recom-mended that laboratories obtain either 50 ml. or 250 ml. centrifuged cups with screw-cap tops that can be adapted to hold 50 ml. centrifuge tubes. If these are not available, the centrifuge should be exhausted to the outside although the variability that occurs with storms and wind currents makes this alternative less desirable.

CULTURE MEDIA USED IN THE RECOVERY OF MYCOBACTERIA

Recovery of mycobacteria from agar culture media was poor when first tried late in the 19th century. However, it was found that a culture medium of whole eggs, potato flour, glycerol, and salts, solidified by heating to 85 to 90° C. for 30 to 45 minutes (inspissation), was more effective for the isolation of *M. tuberculosis*. Later it was found that the use of aniline dyes such as malachite green or crystal violet in the inspissated medium was helpful in the control of contaminating bacteria. It should be noted that although increasing the aniline dye content of the culture medium reduces the amount of contamination, it may also inhibit the growth of mycobacteria.

Table 12-2. *Nonselective Mycobacterial Isolation Media*

Medium	Components	Inhibitory Agent
Löwenstein-Jensen	Coagulated whole eggs, defined salts, glycerol, potato flour	Malachite green, 0.025 g./100 ml.
Petragnani	Coagulated whole eggs, egg yolks, whole milk, potato, potato flour, glycerol	Malachite green, 0.052 g./100 ml.
American Thoracic Society medium	Coagulated fresh egg yolks, potato flour, glycerol	Malachite green, 0.02 g./100 ml.
Middlebrook 7H10	Defined salts, vitamins, cofactors, oleic acid, albumin, catalase, glycerol, dextrose	Malachite green, 0.0025 g./100 ml.
Middlebrook 7H11	Defined salts, vitamins, cofactors, oleic acid, albumin, catalase, glycerol, 0.1% casein hydrolysate	Malachite green, 0.0025 g./100 ml.

Nonselective Culture Media

Numerous egg base culture media for isolation of mycobacteria are currently in use. Löwenstein-Jensen medium is the most commonly used of these (Table 12-2). Petragnani medium is more inhibitory and should be used only with specimens that contain large numbers of contaminants. The American Thoracic Society (ATS) medium is a less inhibitory egg base medium. This medium is helpful in the primary isolation of mycobacteria from specimens such as cerebrospinal fluid, pleural fluid, and tissue biopsies where contamination with other bacteria is less likely. Plate 12-1, *A* illustrates examples of Löwenstein-Jensen and ATS media.

Media of Cohen and Middlebrook

During the 1950's, Cohen and Middlebrook developed a series of defined culture media for use both in the research and clinical laboratories. These media were prepared from defined salts and organic chemicals, with some containing agar, but all were found to require the addition of albumin for optimal growth of mycobacteria. The Middlebrook media which contain agar are transparent and allow the detection of growth after 10 to 12 days instead of the 18 to 24 days of incubation required with other media. Experienced mycobacteriologists can often make a preliminary identification of *M. tuberculosis* and other groups of mycobacteria within 10 days by examining early microcolonies on Middlebrook agar and observing certain well-defined morphologic features.[19]

Not many of the earlier Cohen and Middlebrook culture media are used today. However, 7H9 is a popular liquid medium, and both 7H10 and 7H11 agar media are widely used for isolation and susceptibility testing. The medium 7H11 differs from 7H10 only in containing 0.1 per cent casein hydrolysate, an additive found to improve the rate and amount of growth of mycobacteria resistant to isoniazid (INH).[2] Both 7H10 and 7H11 contain malachite green but in much smaller quantities than usually used in egg base media, in part explaining the higher incidence of contamination than on egg base media.

Although essentially all culture media yield more growth and larger colonies of mycobacteria when incubated in 5 to 10 per cent CO_2, the Middlebrook media absolutely require capneic incubation for proper performance. Exposure of 7H10 or 7H11 to strong light or storage of the media at 4° C. for more than 4 weeks may result in deterioration and release of formaldehyde,[15] which is very inhibitory to the growth of mycobacteria.

Both 7H10 and 7H11 are used for mycobacterial drug susceptibility testing. The antimycobacterial agents should be incorporated into the medium just before it solidifies to reduce the loss of activity that is known to occur with some drugs during the

Table 12-3. *Selective Mycobacterial Isolation Media*

Medium	Components	Inhibitory Agents
Gruft modification of Löwenstein-Jensen	Coagulated whole eggs, defined salts, glycerol, potato flour, RNA—5 mg./100 ml.	Malachite green, 0.025 g./100 ml. Penicillin, 50 units per ml. Nalidixic acid, 35 μg. per ml.
Mycobactosel* Löwenstein-Jensen	Coagulated whole eggs, defined salts, glycerol, potato flour	Malachite green, 0.025 g./ml. Cycloheximide, 400 μg./ml. Lincomycin, 2 μg./ml. Nalidixic acid, 35 μg./ml.
Middlebrook 7H10	Defined salts, vitamins, cofactors, oleic acid, albumin, catalase, glycerol, dextrose	Malachite green, 0.0025 g./100 ml. Cycloheximide, 360 μg./ml. Lincomycin, 2 μg./ml. Nalidixic acid, 20 μg./ml.
Selective 7H11 (Mitchison's medium)	Define salts, vitamins, cofactors, oleic acid, albumin, catalase, glycerol, dextrose, casein hydrolysate	Carbenicillin, 50 μg./ml. Amphotericin B, 10 μg./ml. Polymyxin B, 200 units/ml. Timethoprim lactate, 20 μg./ml.

*Bioquest, Division of BBL Laboratories, Baltimore, Maryland

long heating period used in preparing inspissated egg base media. The names and components of a number of nonselective culture media for recovery of mycobacteria are listed in Table 12-2.

Selective Culture Media

Culture media containing antimicrobial agents to suppress bacterial and fungal contamination have been used for many years. Although certain antimicrobial agents are known to reduce contamination, they may also inhibit the growth of mycobacteria. Despite inhibition of some mycobacterial species, the use of selective media can result in greatly improved recovery of mycobacteria. Table 12-3 lists the names and components of several selective media.

Currently the medium described by Gruft[5] which consists of Löwenstein-Jensen medium with penicillin, naladixic acid, and RNA is one of the more commonly used of the selective media. Petran[18] subsequently described a selective medium containing cyclohexamide, lincomycin, and naladixic acid to control fungal and bacterial contaminants. By varying the concentrations of these agents, the medium can be prepared with either Löwenstein-Jensen or 7H10 base (see Table 12-3).

Selective 7H11 is a modification of an oleic acid agar medium first described by Mitchison.[16] The medium was originally designed for use with sputum specimens without the use of a decontaminating agent. Mitchison's medium contains carbenicillin, polymyxin, trimethoprim lactate, and amphotericin B. McClatchy[14] suggested reducing the concentration of carbenicillin from 100 to 50 μg./ml. and the use of 7H11 medium instead of oleic acid agar. He called this modification Selective 7H11.[14] Reports comparing the use of Selective 7H11 medium with Löwenstein-Jensen and 7H11 have shown recovery of mycobacteria is definitely improved, particularly when the Selective 7H11 medium is used with the NALC-2 per cent NaOH decontamination and selective procedure.[14] Plate 12-1, *G*, illustrates the effect of Selective 7H11 medium on the growth of mycobacteria colonies.

STAINING OF ACID-FAST BACILLI

The cell walls of mycobacteria, because of their high lipid content have the unique capability of binding the dye carbolfuchsin so that it resists destaining with acid alcohol. This "acid-fast" staining reaction of mycobacteria along with their characteristic size and shape is a valuable aid in the early detection of infection and in the monitoring of therapy for mycobacterial disease. The presence of acid-fast bacilli in the sputum, combined with a history of cough, weight loss, and chest x-ray showing a pulmonary infiltrate is still presumptive evidence of active tuberculosis.

It has been estimated that when using standard concentrating techniques, approximately 10,000 acid-fast bacilli per milliliter of sputum are required to be detected microscopically. Patients with extensive disease shed large numbers of mycobacteria, and there is a good correlation between a positive smear and a positive culture. Many patients have minimal or less advanced disease, and the correlation of positive smears to positive cultures in this group may be only 60 to 70 per cent.

Acid-fast stained smears are also useful in following a patient's response to treatment. After antimycobacterial drugs are started, cultures become negative before the smears do, suggesting that the organisms are injured sufficiently to prevent replication but not the binding of the stain. With continued treatment, more organisms are killed and fewer are shed, so that assessing the number of organisms in the sputum during treatment can provide an early objective measure of response. Should the number of organisms fail to decrease after therapy is started, the possibility of drug resistance must be considered. Additional cultures and susceptibility studies should be obtained.

Two types of acid-fast stains are commonly used:

---

1. Carbolfuchsin stains: a mixture of fuschin with phenol (carbolic acid)
   A. Ziehl-Neelsen ("hot stain")
   B. Kinyoun ("cold stain")
2. Fluorochrome stain: auramine 0, with or without a second fluorochrome, rhodamine

---

Both staining techniques are said to bind to mycolic acid in the mycobacterial cell wall. Smears stained with the carbolfuchsin technique must be scanned with an oil immersion objective. This limits the total area of a slide that can be viewed in a given unit of time. In contrast to the carbolfuchsin stains, smears stained by the auramine procedure can be scanned with a 25 × objective, thereby increasing the field of view and reducing the time needed to scan a given area of the slide. Fluorochrome-stained smears require a strong light source, either a 200-watt mercury vapor burner or a strong blue light with a fluoricein isothiocyanate (FITC) filter.

Fluorochrome-stained bacteria are bright yellow against a dark background, allowing the slide to be scanned under lower magnification without losing sensitivity. The sharp contrast between the brightly colored mycobacteria and the dark background offers a distinct advantage in scanning the slide, because a smaller number of cells per slide is required for detection than with the carbolfuchsin stain (Plate 12-3, *I*). Modifications of the auramine fluorochrome stain include the addition of rhodamine, giving a golden appearance to the cells, or the use of acridine orange as a counterstain, resulting in a red to orange background.

Although workers in various laboratories may be partial to one staining method over another, the specificity for mycobacteria of the two seems about the same, with the possible exception of *Mycobacterium fortuitum.* In a recent study of 15 strains of *M. fortuitum,* Joseph[6] reported that five of the strains did not stain with auramine but all 15 stained with carbolfuchsin.

Since a significantly larger area of the smear can be scanned per unit of time with the auramine fluorochrome stain than with the carbolfuchsin method, the fluorochrome stain offers the advantage for greater sensitivity. Some workers use the fluorochrome method for scanning purposes and then confirm their findings by reexamining the preparation after destaining and restaining

Table 12-4. *Acid-Fast Staining Procedures*

Ziehl-Neelsen Procedure	Kinyoun's "Cold" Procedure	Auramine Fluorochrome Procedure
Carbolfuchsin: dissolve 3.0 g. of basic fuchsin in 10.0 ml. of 90–95% ethanol. Add 90 ml. of 5% aqueous solution of phenol.	*Carbolfuchsin:* dissolve 4.0 g. of basic fushsin in 20 ml. of 90–95% ethanol and then add 100 ml. of a 9% aqueous solution of phenol (9.0 g. of phenol dissolved in 100 ml. of distilled water).	*Phenolic auramine:* dissolve 0.1 g. of auramine 0 in 10 ml. of 90—95% ethanol and then add to a solution of 3 g. of phenol in 87.0 ml. of distilled water. Store the stain in a brown bottle.
Acid-alcohol: add 3.0 ml. of concentrated HCL *slowly* to 97.0 ml. of 90–95% ethanol, in this order. Solution may get hot!	*Acid-Alcohol:* add 3 ml. of concentrated HCL *slowly* to 97.0 ml. of 90–95% ethanol, in this order. Solution may get hot!	*Acid-alcohol:* add 0.5 ml. of concentrated HCL to 100 ml. of 70% alcohol.
Methylene blue counterstain: dissolve 0.3 g. of methylene blue chloride in 100 ml. of distilled water.	*Methylene blue counterstain:* dissolve 0.3 g. of methylene blue chloride in 100 ml. of distilled water.	*Potassium permanganate:* dissolve 0.5 g. potassium permanganate in 100 ml. of distilled water.
Procedure: 1. Cover a heat-fixed, dried smear with a small rectangle (2 × 3 cm.) of filter paper. 2. Apply 5–7 drops of carbolfuchsin stain to thoroughly moistened filter paper. 3. Heat the stain-covered slide to steaming but do not allow to dry. Heating may be done by gas burner or over an electric staining rack. 4. Remove paper with forceps, rinse with water, and allow to drain. 5. Decolorize with acid-alcohol until no more stain appears in the washing (2 minutes). 6. Counterstain with methylene blue (1–2 minutes). 7. Rinse, drain, and air dry (1 to 2 minutes). 8. Examine with 100X oil immersion objective. Bacilli are stained red and the background light blue.	*Procedure:* 1. Cover a heat-fixed, dried smear with a small rectangle (2 × 3 cm.) of filter paper. 2. Apply 5–7 drops of carbolfuchsin to thoroughly moisten filter paper. Allow to stand for 5 minutes. Add more stain if paper dries. Do not steam! 3. Remove paper with forceps, rinse with water and allow to drain. 4. Decolorize with acid-alcohol until no more stain appears in the washing (2 minutes). 5. Counterstain with methylene blue (1–2 minutes). 6. Rinse, drain, and air dry (1 to 2 minutes). 7. Examine with 100X oil immersion objective. Bacilli are stained red and the background light blue.	*Procedure:* 1. Cover a heat-fixed, dried smear with carbol auramine and allow to stain for 15 minutes. Do not heat or cover with filter paper. 2. Rinse with water and drain. 3. Decolorize with acid-alcohol (2 minutes). 4. Rinse with water and drain. 5. Flood smear with potassium permanganate for 2 and not more than 4 minutes. 6. Rinse with tap water. Drain. 7. Examine with 25X objective using a mercury vapor burner and BG-12 filter or a strong blue light. Mycobacteria are stained yellowish-orange against a dark background.

with a carbolfuchsin method. After laboratory workers have become familiar with the auramine fluorochrome method, they prefer it over the carbolfuchsin procedure. Table 12-4 lists the procedures for preparing carbolfuchsin and fluorochrome stains.

It should be emphasized that neither the auramine nor the auramine-rhodamine fluorchrome stain is a fluorescent antigen-antibody technique. Fluorescent-tagged antibodies to aid in the identification of various species of mycobacteria have been described but are not in widespread use, nor are they commercially available.

INCUBATION OF CULTURES

Different species of mycobacteria show a striking dependence on the temperature of incubation for optimal growth. Species having a predilection for causing infection of the skin, such as *Mycobacterium marinum* or *M. ulcerans,* grow best at skin temperatures (30 to 32° C.) and very poorly or not at all at 37° C. *Mycobacterium tuberculosis* grows best at 37° C. and poorly or not at all at 30° C. or 42 to 45° C. *Mycobacterium avium,* the major cause of tuberculosis in birds, grows best at 42 to 45° C., which is the body temperature of birds. *Mycobacterium xenopi,* a species not commonly found as the cause of infection in humans, grows best at 42° C. and has been implicated as an environmental contaminant in the hot-water system of a large hospital.[4]

Optimal recovery of different mycobacterial species will depend on the incubation of at least part of the concentrated specimen at a temperature most likely to promote growth of that species. An incubator set at 30° C. should be used for all specimens from suspected skin or subcutaneous infections. If a 30° C., incubator in a temperature-monitored closed box placed in a sheltered area away from warm or cool drafts. The temperature in the box should be monitored and recorded daily. An incubator maintained at 42° C. can be helpful in the recovery of *Mycobacterium xenopi,* but human infections with this organism are uncommon.

Mycobacteria grow best in an atmosphere of 3 to 11 per cent CO_2. The use of CO_2 is mandatory if Middlebrook 7H10 or 7H11 medium is used. For reasons that are not well understood, mycobacteria do not grow well in candle extinction jars. The CO_2 concentration in incubators should be monitored daily and a record kept of both the incubator temperature and the CO_2 content.

IDENTIFICATION OF MYCOBACTERIA

The identification of mycobacteria is not difficult, but requires patience, familiarity with the end points of different identification characteristics, and a collection of control organisms.[25] Difficulties in identification have arisen because in many parts of the country acute-care general hospitals are having to assume the responsibility for provid-

ing laboratory services formerly carried out in tuberculosis sanitariums, many of these having been closed due to the lack of patients. Thus, the experience and competence of special mycobacterial laboratories have been either lost or assigned to other tasks.

Not every laboratory needs the ability to differentiate into species all mycobacterial isolates. The number of patients with tuberculosis in any given hospital is usually not large, and certain tests and procedures can be best performed by reference laboratories equipped to perform the tests required for definitive identification, as well as the susceptibility testing of isolates. Suggestions for the *extent* or *level* of services offered by individual laboratories have been published by both the College of American Pathologists and the American Thoracic Society (see the list on the facing page). Most clinical laboratories fall into extent levels 2 or 3 as outlined in the following list, and the functions described for level 4 are carried out by specialized reference laboratories or laboratories interested in, or assigned to, the care of patients with mycobacterial disease.

Inasmuch as *Mycobacterium tuberculosis* is the most common cause of mycobacterial disease in man, most laboratories strive to maintain proficiency in the identification of this organism. *Mycobacterium tuberculosis* can be identified using a few simple tests that can be performed by any interested laboratory. The recommended procedures are:

1. Determining the optimal temperature for isolation and rate of growth
2. Pigmentation of colonies
3. Niacin accumulation
4. Reduction of nitrates to nitrites
5. Catalase production
6. Growth inhibition by thiophene-2-carboxylic acid hydrazide

A brief review of each of these procedures is presented in the following paragraphs, with details given in Charts 12-1 through 12-5. These procedures are necessary to derive a definitive species identification, though an assessment of colonial morphology is helpful in making a presumptive identification of

Color Plates
12-1 *to* 12-3

Plate 12-1
Identification of Mycobacteria
Colonial Characteristics

A.

American Thoracic Society (ATS) medium (top) and Lowenstein-Jensen medium (bottom) with colonies of *Mycobacterium tuberculosis*. Note the larger size of the colonies on the ATS medium, which is less inhibitory than Lowenstein-Jensen medium.

B.

Rough, dysgonic colonies of *M. tuberculosis* viewed through a dissecting microscope. Colonies of *M. tuberculosis* have been described as resembling "bread crumbs."

C.

7 H1O agar plate with hypermature colonies of *M. tuberculosis* strain H37Rv. Note the thin, spreading colonies with aerial extensions where the colonies meet. The colony in the lower left portion of the plate shows fragmentation of bacterial growth resulting from an inoculating spade moved through several colonies.

D.

Buff-colored, rough, dysgonic, mature colony of *Mycobacterium bovis* as visualized through a dissecting microscope.

E.

Lowenstein-Jensen medium (top) and ATS medium (bottom) with colonies of *Mycobacterium kansasii*. Both cultures were inoculated from the same broth. The larger size of the colonies on ATS medium reflects less inhibitory properties of this medium than Lowenstein-Jensen. Although supporting growth better than Lowenstein-Jensen, ATS medium is more easily contaminated when used for the inoculation of clinical specimens containing mixtures of bacteria. ATS medium has its best application for the culture of specimens from body sites that are normally sterile but may contain small numbers of mycobacteria.

F.

Dissecting microscopic view of nonpigmented colonies of *M. kansasii*. Growth in the center of the colonies may appear rough, but with progression to the periphery, the borders become smooth. These colony types have been termed "semirough."

G.

7 H11 medium (right) and Selective 7 H11 medium (left) with colonies of *Mycobacterium scrofulaceum*. Note both the marked inhibition of growth on the selective medium as well as the lack of pigment production. Pigment formation may take place on the selective medium as the colonies mature but will be less intense than on the non-antibiotic-containing medium.

H.

Lowenstein-Jensen medium (top) and ATS medium (bottom) with colonies of *Mycobacterium gordonae*. Different species of mycobacteria may show significant stimulation or inhibition of growth with slight variations in culture media components, illustrating the need to use at least two or preferably three types of culture media for the inoculation of all specimens suspected of containing mycobacteria.

I.

7 H10 agar plate with colonies of *Mycobacterium avium*. Note the light yellow pigmented colonies, which are large, smooth, and butyrous. Many of the *M. avium-intracellulare* strains produce colonies that are lightly pigmented, but the intensity of the pigment does not increase with exposure to light; hence, the term "non-photochromogen".

J.

Dissecting microscopic view of *M. avium-intracellulare* illustrating two colony types. The small,

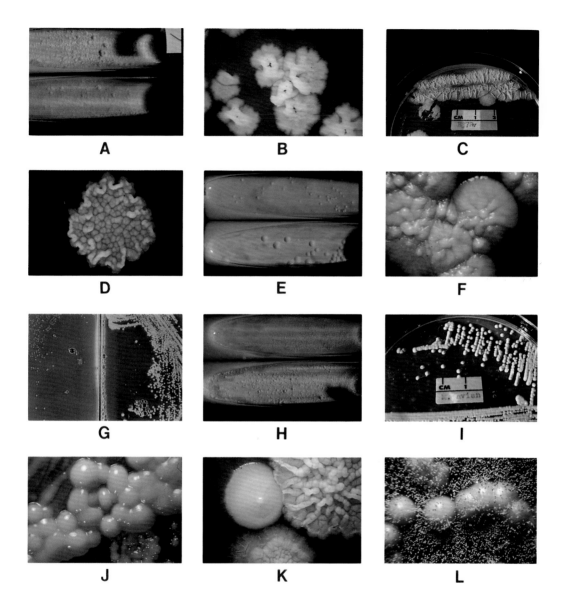

A.   B.   C.

D.   E.   F.

G.   H.   I.

J.   K.   L.

flat, spreading colonies are typical of those isolated from patients with infection with this organism. The large, round, smooth, butyrous colonies are considered to be mutants that can develop on subculture. The large size of the mutant colonies and their relatively rapid growth can mislead laboratory workers into classifying the organism as one of the Group IV, rapidly growing mycobacteria.

K.

Mycobacterium fortuitum illustrating three colony types with smooth and rough variants. The colony at the bottom of the picture demonstrates an irregular border with peripheral extensions. These can also be seen in 1 to 3-day-old cultures of *M. fortuitum* and are helpful in distinguishing *M. fortuitum* from *M. chelonei*. The extensions seen here are suggestive of microrhizoids.

L.

Colonies of *M. avium-intracellulare* badly overgrown with Aspergillus species. Fungal contamination of mycobacterial cultures can be a serious problem when a nonselective medium is used. Contamination can be controlled by sealing plates in polyethylene bags before placing them in the incubator.

A.

Lowenstein-Jensen medium with colonies of *M. tuberculosis* (left), *M. kansasii* grown in the light (center), and *M. kansasii* grown in the dark and not exposed to light (right). Only young, actively metabolizing mycobacteria can make pigment after exposure to light; old cultures are, therefore, not photochromogenic.

B.

Lowenstein-Jensen medium with colonies of *Mycobacterium szulgai*. One culture tube was wrapped in aluminum foil, while the other was exposed to ambient light during incubation at 37° C. Both cultures produced the same amount of pigment and therefore are "scotochromogenic" after incubation for 2 weeks.

C.

Lowenstein-Jensen medium with colonies of *M. szulgai* incubated at 24° C. The top tube was wrapped in foil while the lower tube was exposed to ambient light during a 3-week incubation. Note that the photochromogenic nature of this organism was not apparent when the cultures were incubated at 37° C. (Frame *B*).

D.

Proskauer-Beck broth containing cultures of *M. tuberculosis*. Aniline and cyanogen bromide were added to the tube on the right. The appearance of a yellow color indicates the accumulation of niacin, a helpful characteristic to use to identify *M. tuberculosis*.

E.

Niacin accumulation test. Niacin test using a reagent-impregnated paper strip, illustrating the yellow color in a positive test as seen in the tube on the left. The test is performed by placing an aqueous or saline extract from the surface of a mature culture slant of *M. tuberculosis* into the bottom of a small, screw-capped culture tube. The reagent-impregnated strip is then placed in the tube and the top carefully tightened. Yellow color in the fluid within 12 to 15 minutes indicates the accumulation of niacin.

F.

Nitrate reduction test. The appearance of a red color after addition of reagents to the medium signifies that nitrates have been reduced to nitrites. The two tubes on the left have been inoculated with *M. tuberculosis*, strains H37Rv and H37Ra, both capable of reducing nitrates, while the tubes on the right were inoculated with *M. avium-intracellulare* and *M. bovis*, respectively, strains devoid of nitroreductase activity.

G.

Semiquantitative catalase test. The tube on the left shows no catalase activity while the tube on the right shows a column of gas bubbles greater than 50 mm. in height. The gas bubbles are liberated from hydrogen peroxide by the colonies growing on the surface of the medium. Intermediate and weak catalase activity with bubble columns less than 50 mm. are illustrated in the middle two tubes.

H.

Tubes illustrating the hydrolysis of Tween 80® when mycobacteria are added to the left tube (pink color). The red color is that of the neutral red indicator at the *p*H of the buffer in the substrate while the straw color in the tube on the right represents a change in optical rotatory characteristics upon addition of Tween 80. Hydrolysis of the Tween 80 results in a return of the color from straw-yellow to pink. The color change does not indicate a change in *p*H, but the hydrolysis (destruction) of Tween 80.

I.

Substrate tubes illustrating a positive arylsulfatase reaction in the left tube (red color), compared to a negative reaction in the right tube. Arylsulfatase is an enzyme produced by some species of mycobacteria, notably *Mycobacterium fortuitum* and *M. chelonei*, which may give a positive reaction within 3 days. The action of the enzyme converts phenolphthalein disulphate (colorless) to free phenolphthalein, which is red after the addition of alkaline sodium carbonate to the substrate.

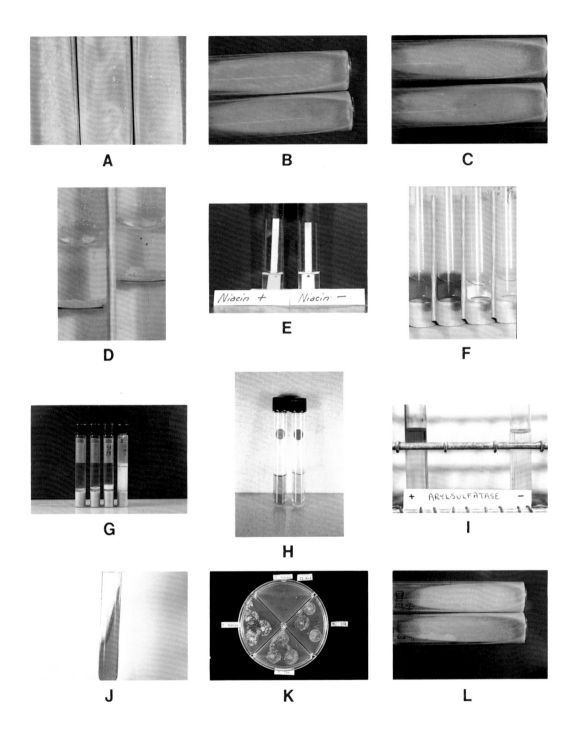

A

B

C

D

E

F

G

H

I

J

K

L

Plate 12-2 (Continued)

J.

Some species of mycobacteria produce urease, which can convert urea medium to a red color. The test is best performed in a urea broth substrate.

K.

7 H11, four-quadrant agar plate. The upper and lower quadrants contain thiophene-2-carboxylic acid hydrazide (T_2H). The top and left quadrants have been inoculated with *M. bovis;* the lower and right quadrants, with *M. tuberculosis.* Note the inhibition of *M. bovis* in the top quadrant in the presence of T_2H, while *M. bovis* grows well in the quadrant without the inhibitor. *M. tuberculosis*

grows in the quadrants both with and without the inhibitor. The inability to grow in the presence of T_2H is a valuable characteristic for the identification of *M. bovis.*

L.

Illustration of the iron uptake test. Lowenstein-Jensen medium with colonies of *Mycobacterium fortuitum* incubated at 37^0 C. for two weeks. A solution of ferric chloride was added to the bottom culture tube after 7 days of incubation and this culture was reincubated for an additional 7 days. The incorporation of iron salts in the mycobacterial cells, as illustrated by the rusty color in the bottom tube, is a helpful characteristic in the identification of *M. fortuitum.*

Plate 12-3
Antimicrobial Susceptibilty Testing
of Mycobacteria

The antimicrobial susceptibility testing of mycobacteria is performed in quadrant petri dishes containing 7 H11 agar. One quadrant is used as a control and each of the other three contain a different antimicrobial drug. In performing the test, all quadrants are inoculated with a standardized concentration of the organism to be tested. The susceptibility of the organism to each drug is determined by comparing the number of colonies growing on the drug-containing quadrants with the control.

A.

Petri plate with control quadrant in the upper right. The quadrant in the upper left contains isonicotinic acid hydrazide (INH), 1.0 ug./ml., the lower-left quadrant, streptomycin, 10 ug./ml., and the lower-right quadrant, paraminosalicylic acid (PAS), 10 ug./ml. The inoculum to each quadrant was the same as that in the control quadrant. The media containing streptomycin and PAS have completely inhibited the bacterial growth, while approximately 50 colonies are present in the INH quadrant. The number of colonies growing in the control quadrant are too numerous to count; therefore, the percentage of colonies resistant to INH relative to the number inoculated cannot be determined.

B.

Quadrant Petri plate inoculated with a 100-fold dilution of the inoculum used in Frame A. Between 50 and 100 colonies are growing in the control quadrant, but the INH quadrant is free of growth, indicating that any spontaneous resistance in the parent strain is less than 1 per cent. The organism being tested in Frames A and B is *M. kansasii.* The red color in the streptomycin quadrant is Congo red dye, previously, but no longer, used as a marker for the drug contained in that quadrant.

C and D.

Quadrant Petri plates inoculated with 10^{-2} (Frame C) and 10^{-4} (Frame D) dilutions of a 10-day broth culture of *M. tuberculosis.* In Frame C, confluent growth is seen in both the control and INH quadrants, with smaller numbers of inhibited colonies in the PAS-containing quadrant. In Frame D, equal numbers of colonies are seen in the control and INH quadrants (about 50-100) while in the PAS quadrant, 14 colonies are present. This number indicates the resistance is greater than 1 per cent, contraindicating clinical use.

E and F.

Quadrant Petri plates inoculated with *M. tuberculosis.* In Frame E a 10^{-2} dilution of a 7-day broth culture was used as the inoculum,

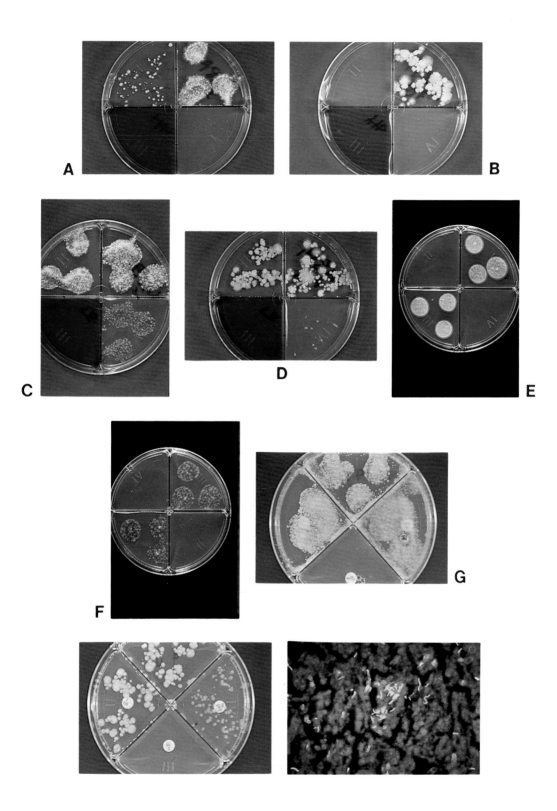

Plate 12-3 (Continued)

and in Frame *F*, a 10^{-4} dilution of the same culture was used. Growth is confluent in both the control and streptomycin quadrants in Frame *E*, and is equivalent at about 100-200 colonies in Frame *F*. These plates indicate susceptibility to INH and PAS but complete resistance to streptomycin.

G and H.

Quadrant Petri plates using drug-containing filter paper disks. The plates were inoculated from a broth culture of *M. kansasii* diulted 10^{-2} (Frame *G*) and 10^{-4} (Frame *H*) and incubated for 21 days.

In Frame *G* confluent growth is seen on the INH, PAS, and control quadrants and no growth is seen in the streptomycin quadrant. In the 10^{-4} dilution, individual colonies can be seen in Frame *H*.

I.

Photomicrograph of an auramine-stained smear of a lymph node. The yellow color of the mycobacterial organisms stands in bright contrast to the green-staining lymphocytes against the dark, unstained background. Different counterstains and filter systems may result in various color combinations.

Extent of Services (College of American Pathologists)	Areas of Proficiency (American Thoracic Society)
Extent Level: 1. No mycobacteriologic procedures performed 2. Acid-fast stain of exudates, effusions, and body fluids, etc., with inoculation and referring of cultures to reference laboratories for further identification 3. Isolation of mycobacteria; identification of *Mycobacterium tuberculosis* and preliminary identification of the atypical forms as photochromogens, scotochromogens, nonphotochromogens, and rapid growers 4. Definitive identification of mycobacteria isolated to the extent required to establish a correct clinical diagnosis and to aid in the selection of safe and effective therapy. Drug susceptibility testing may or may not be performed.	1. Collection and transportation of specimens; preparation and examination of smears for acid-fast bacilli 2. Detection, isolation and identification of *M. tuberculosis* 3. Determination of drug susceptibility of mycobacteria 4. Identification of mycobacteria other than *M. tuberculosis*

some Mycobacteria species. The characteristic morphology of several mycobacteria species are shown in Plate 12-1.

Optimal Temperature for Isolation and Rate of Growth of Mycobacteria

The general principle of incubation temperature and growth rates were discussed above (see page 354). *Mycobacterium tuberculosis* grows maximally at 37° C. with very poor or minimal growth at 30° C. or 42° C. Although colonies may be detected on egg base medium as early as 12 days after inoculation, the average recovery time is 21 days. Six weeks or more may be required for detectable colonies of occasional strains to appear.

For assessing growth rate, any standard, nonselective culture medium can be used, either in tubed slants or in Petri plates. Petri plates are preferable because developing colonies can be studied with a dissecting microscope or low-power microscopy.

A well-isolated colony of the test organism is subcultured to a 7H9 broth containing Tween 80 and incubated several days or until the medium is faintly turbid. The broth is then diluted 1:100 and isolation streaks are made to the test medium to obtain

isolated colonies. To accurately determine growth rate it is necessary to use an inoculum sufficiently dilute to produce individual colonies. An inoculum of large numbers of slowly growing mycobacteria may form a visible colony within a few days and give an erroneous impression of the growth rate.

Pigment Production by Mycobacteria

Pigment production of mycobacteria is an important identification characteristic depending on whether color develops upon incubation in the dark (scotochromogenic) or is stimulated only after exposure to light (photochromogenic). *M. tuberculosis* fails to produce any pigment, except a light buff color, even after exposure to bright light. Details of the procedures for determination of pigment production are given in Chart 12-1. Examples of photoreactivity studies are illustrated in Plate 12-2, *A, B,* and *C.*

Niacin Accumulation

M. Tuberculosis, in addition to a few other mycobacteria species listed in the introduction to Chart 12-2, is unable to convert free niacin to niacin ribonucleotide. Thus, the accumulation of water-soluble

Chart 12-1.   *Determination of Photoreactivity of Mycobacteria*

Introduction	The pigmentation of young colonies of mycobacteria, either after having been grown in the dark or following induction by exposure to light, can be an important characteristic in the identification of certain species of mycobacteria.
Principle	Yellow pigment production in photochromogenic mycobacteria is the result of yellowish orange carotene crystals produced by actively metabolizing mycobacteria following exposure to a bright light. The type of pigment produced by mycobacteria when grown in the dark is not known.
Media and Reagents	Any noninhibitory medium known to support the growth of the test organism is acceptable. Selective or antimicrobial containing media should not be used since these may interfere with pigment formation. Löwenstein-Jensen and ATS media are both good for determining photochromogenicity.
Procedure	A broth culture of test organisms, diluted sufficiently so that isolated colonies can be obtained, is inoculated to three Löwenstein-Jensen culture tubes. Two of the tubes are wrapped with aluminum foil and the third is left exposed to the ambient light in the incubator. Cultures thought to be photochromogenic should be incubated at 30° and 37° C. while those considered to be scotochromogenic should be inoculated in duplicate and incubated at both 24 to 26° C. and 37° C.
	Several days after growth is first noted on the light-exposed control tube, the foil-wrapped tubes are examined for growth. If there is evidence of colony formation or early growth, one of the two foil-wrapped tubes is exposed to strong light. A 100-watt tungsten bulb or fluorescent equivalent is adequate. The cap of the culture tube should be loosened during the exposure time of 3 to 5 hours.

niacin in an egg base culture medium is a valuable differential characteristic in the identification of *M. tuberculosis*. As discussed in Chart 12-2, reagent-impregnated filter paper strips have been developed which eliminates the necessity for using cyanogen bromide, a highly toxic substance required for performance of the test as it was originally described. The development of a yellow color in the medium indicates a positive test, as illustrated in Plate 12-2, *D* and *E*.

Reduction of Nitrates to Nitrites

M. tuberculosis, among a few other species of mycobacteria listed in the introduc-

tion to Chart 12-3, produces the enzyme nitroreductase which catalyzes the reduction of nitrates to nitrites. The development of a red color upon addition of the reagents indicates a positive test (Plate 12-2, *F*).

For reasons that are not well understood, the test for nitroreductase is not highly reproducible between laboratories. This lack of reproducibility is disappointing because the nitrate test is a key characteristic in the identification of *M. szulgai*. In one comparative study, almost half of over two hundred laboratories examining the organism in a proficiency test survey reported the organism to be negative.[26] The International Working Group of Mycobacterial Taxonomy also found by interlaboratory compari-

Chart 12-1. *Determination of Photoreactivity of Mycobacteria (Continued)*

Procedure *(Continued)*	Following exposure of the culture to the light, it is returned to the incubator and inspected after 24 and 48 hours for the development of a yellow or orange pigment.
Interpretation	Color changes, especially subtle changes, are compared in the culture tube exposed to ambient light and the foil-wrapped culture tube that had not been exposed to light. Mycobacteria that are scotochromogenic, that is, that produce pigment when incubated in the dark, include *M. scrofulaceum, M. gordonae, M. flavescens, M. xenopi,* and *M. szulgai,* the latter only when incubated at 37° C. (Plate 12-2, *B* and *C*) Species that are photochromogenic include *M. kansasii, M. marinum, M. simiae,* and *M. szulgai,* the latter only when incubated at 22 to 24° C. Many species are normally lightly pigmented, such as *M. intracellulare-avium.* Runyon has termed these organisms "nonphotochromogenic," meaning that exposure to light does not make the pigment more intense. Nonpigmented or only lightly pigmented mycobacterial species include *M. tuberculosis, M. bovis, M. ulcerans, M. fortuitum, M. chelonei,* and certain of the species of the group III nonphotochromogenic mycobacteria.
Controls:	Photochromogenic control: *M. kansasii* Scotochromogenic control: *M. scrofulaceum* Photochromogenic and scotochromogenic control: *M. szulgai* at 24° and 37° C. Negative control: *M. tuberculosis*
Bibliography	Wayne, L. G., and Doubek, S. R.: The role of air in the photochromogenic behavior of *M. kansasii.* Am. J. Clin. Pathol., *42*:431–435, 1964.

son that the nitroreductase test was unreliable.[33] Until the problems associated with the test are better understood, three control cultures should be used with the test, one giving a strong positive reaction, one giving a weak reaction, and one being a negative control.

Catalase Activity

Most of the mycobacteria produce catalase; however, there are different forms of catalase, some being inactivated if the test medium is heated to 68° C. for 20 minutes. Thus, measuring the quantity of catalase produced by a given strain of mycobacteria before and after heating is helpful differen-

tial test. This characteristic is particularly useful for the identification of isoniazid (INH) resistant strains of *M. tuberculosis* which do not produce catalase. The details of the semiquantitative catalase test are presented in Chart 12-4 (Plate 12-2, *G*).

Growth Inhibition of Mycobacteria by Thiophene-2-Carboxylic Hydrazide (T₂H)

Thiophene-2-carboxylic acid hydrazide selectively inhibits the growth of *Mycobacterium bovis,* while most other mycobacteria can grow on medium containing this compound. This characteristic can be particularly useful in differentiating *M. tuberculosis*
(Text continues on p. 364)

Chart 12-2. *Niacin Accumulation*

Introduction

Niacin is formed as a metabolic by-product by all mycobacteria, but most species possess an enzyme that converts free niacin to niacin ribonucleotide. *M. tuberculosis*, *M. simiae*, occasional strains of *M. marinum* and *M. chelonei*, as well as a number of strains of M. bovis lack this enzyme and will accumulate niacin as a water-soluble by-product in the culture medium. The amount of niacin present in a culture slant is in part a reflection of the number of colonies on the slant and the age of the culture.

Principle

In the chemical test for niacin described by Runyon, nicotinic acid reacts with cyanogen bromide in the presence of a primary amine (aniline) to form a yellow compound. Other variations of the test have been described but are generally considered to be less sensitive than the Runyon test.

Media and Reagents

1. Löwenstein-Jensen culture medium or 7H10 or 7H11 media supplemented with 0.1% potassium aspartate
2. Sterile water or 0.85% saline
3. Aniline, 4%
   A. Colorless aniline, 4 ml.
   B. Ethyl alcohol, 95% 96 ml.
   Store in a brown bottle in refrigerator. If solution turns yellow discard and prepare a fresh solution.
4. Cyanogen bromide, 10%
   Cyanogen bromide, 5 g.
   Dissolve in 50 ml. of distilled or deionized water. Store in refrigerator in brown bottle with tightly fitting cap. If precipitate forms, warm to room temperature and redissolve before use. Solution is volatile and may loose strength on storage. (Note: cyanogen bromide is a tear gas and should be used in a safety cabinet vented to the outside. In acid solutions, cyanogen bromide may hydrolyze to hydrogen cyanide, a very toxic gas. Always discard solution containing cyanogen bromide into germicides mixed with 2 to 4% NaOH.)

Procedure

The test is performed by adding 1.0 ml. of either sterile water or sterile saline to the slant of a mature Löwenstein-Jensen culture or to the surface of 7H10 or 7H11 culture medium with a number of well-formed colonies. Since niacin must be extracted from the culture medium, the culture should not cover the entire surface of the medium. If this has happened, scrape part of the bacterial growth to one side, allowing the extracting fluid to be in direct contact with the culture medium. Allow the extracting fluid to cover the surface of the culture medium for 15 to 30 minutes and then remove 0.5 ml. of the fluid and place it in a clean, screwcap test tube. Add 0.5 ml. of aniline and then 0.5 ml. of cyanogen bromide. If niacin has been extracted from the culture, a yellow color will appear in the extract within a few minutes.

Chart 12-2. *Niacin Accumulation (Continued)*

Procedure *(Continued)*	Reagent-impregnated filter paper strips have been developed that simplify the test after the extraction has been made. These strips are sensitive and work well when instructions are followed carefully. When using the paper strips, be sure to place both the extract from the culture medium and the reagent-containing paper strip into a screwcapped tube and tighten the top firmly. The reaction consists of the evoluation of cyanogen chloride gas given off from the top of the paper strip, which reacts with the niacin in the extracting fluid at the bottom of the tube. The tube containing the extracting fluid and test strip must be tightly closed to prevent any loss of the cyanogen chloride gas. Do not use strips that have become discolored because the reagents may have undergone some degree of deterioration and will not be reliable.
Interpretation	The appearance of a yellow color in the extracting fluid indicates the presence of niacin (Plate 12-2, *D* and *E*). Negative tests on niacin-positive organisms can occur if there is insufficient or early growth prior to the accumulation of sufficient niacin in the medium to be detected. In cultures where there seems to be an insufficient amount of growth, reincubate the culture for an additional 2 to 4 weeks or make a subculture that will have a minimum of 50 colonies or more. A negative test can also occur if there is bacterial growth over the entire surface of the culture medium, preventing extraction of niacin from the culture medium. This can be controlled by scraping part of the confluent bacterial growth to one side with an inoculating needle or spatula.
Controls	Reagent controls should be added to a culture tube of uninoculated medium. Positive control: *Mycobacterium tuberculosis.* Negative control: Mycobacterium intracellulare.
Bibliography	Gangadharam, P. R., and Droubi, D. S.: A comparison of four different methods for testing the production of niacin by mycobacteria. Am. Rev. Respir. Dis., *104*:434–437, 1971. Kilburn, J. O., and Kubica, G. P.: Reagent impregnated paper for detection of niacin. Am. J. Clin. Pathol., *50*(4):530–531, 1968. Konno, K.: New chemical method to differentiate human type tubercle bacilli from other mycobacteria. Science, *124*:985, 1956. Konno, K., *et al.:* Niacin metabolism in mycobacteria. Am. Rev. Respir. Dis., *93*:41–46, 1966. Runyon, E. H., Selin, M. J., and Hawes, H. W.: Distinguishing mycobacteria by the niacin test. Am. Rev. Tuberc., *79*:663–665, 1959. Young, W. D., Jr., *et al.:* Development of paper strip test for the detection of niacin produced by mycobacteria. Appl. Microbiol., *20*:939–945, 1970

Chart 12-3. *Reduction of Nitrates to Nitrites*

Introduction

The presence of the enzyme nitroreductase is an important identification characteristic. *M. tuberculosis, M. kansasii, M. szulgai,* and *M. fortuitum* all reduce nitrates to nitrites. Other species that are positive for nitroreductase include *M. flavescens, M. terrae, M. triviali,* and *M. chelonei.*

Principle

Mycobacteria containing nitroreductase can derive oxygen from nitrates and other reduction products. The chemical reaction is

$$NO_3 + 2\bar{e} + H_2 \longrightarrow NO_2 + H_2O$$

Nitrate   Nitrite

The presence of nitrite in the test medium is detected by the addition of sulfanilamide and n-naphthylethylenediamine in an acid pH. If nitrites are present, a red diazonium dye is formed.

Reagents

1. M/100 $NaNO_3$ in pH 7.0, M/45 phosphate buffer

$NaNO_3$	0.085 g.
KH_2PO_4	0.117 g.
Na_2HPO_4 $12H_2O$	0.485 g.
Distilled water	100.0 ml.

2. 1:1 dilution of concentrated HCl (Add 10 ml. of HCl to 10 ml. of H_2O; always add concentrated acid to water.)
3. Dissolve 0.2 g. of sulfanilamide in 100 ml. of water.
4. Dissolve 0.1 g. of n-naphthylethylenediamine dihydrochloride in 100 ml. of distilled water.

Store reagents at 4° C. and discard if any change in color occurs. A reagent-containing filter paper strip has been developed for determining the presence of nitroreductase (Quigley). The test is not very sensitive, and, although valid when positive, should be repeated by the standard Virtanen procedure (see below) when negative.

Procedure

1. Place several drops of sterile distilled water in a sterile, screwcapped test tube.
2. Emulsify a loopful of actively growing mycobacteria from the test culture in the water.
3. Add 2 ml. of the buffered sodium nitrate solution to the emulsified organisms, mix by shaking, and incubate at 37° C. for two hours.

Chart 12-3. *Reduction of Nitrates to Nitrites (Continued)*

Procedure *(Continued)*	4. Acidify the test culture by adding 1 drop of the 1:1 dilution of HC1. 5. Add 2 drops of the sulfanilamide solution. 6. Add 2 drops of the n-naphthylethlenediamine solution.

Interpretation

A positive test is indicated by the development of a red color within 30 to 60 seconds (Plate 12-2, *F*). The color may vary from pink to deep red. Quantitation can be made on comparison with color standards. If no color develops, confirm as a negative test by adding a small amount of powdered zinc from the tip of an applicator stick. If a red color develops after adding the zinc, the test was a true negative. If no color develops after adding the zinc, the test was positive, with the nitrates in the solution reduced beyond nitrites into colorless compounds. Since reduction to gaseous products such as N_2O is not common by mycobacteria, all false negative tests should be repeated.

Controls

Each new group of reagents should be compared with established and validated reagents before using with clinical isolates. The following three organisms should be used each time the test is determined on recent isolates.
M. tuberculosis H37R: strongly positive
M. kansasii: selected to be weakly positive
M. intracellulare: negative
 Occasionally colors may be pale and hard to interpret. When this happens, color standards can be prepared to help with the interpretation of the test.

Bibliography

Quigley, H. J., and Elston, H. R.: Nitrite test strips for detection of nitrate reduction by mycobacteria. Am. J. Clin. Pathol., 53:663–665, 1970.
Vestal, A. L.: Procedures for the Isolation and Identification of Mycobacteria. DHEW Publication No. (CDC) 75-8230, 1975.
Virtanen, S.: A study of nitrate reduction of mycobacteria. Acta Tuberc. Scand. (Suppl.), 48:119, 1960.
Wayne, L. C., and Doubek, S. R.: Classification and identification of mycobacteria. II. Tests employing nitrate and nitrite as substrates. Am. Rev. Respir. Dis., 91:738–745, 1965.

Chart 12-4. *Catalase Activity*

Introduction
Most mycobacteria produce the enzyme catalase, but they vary in the quantity produced. Also some forms of catalase are inactivated by heating at 68° C. for 20 minutes and others are stable. The semiquantitation of catalase and susceptibility to heating at 68° C. at pH 7.0 are both useful characteristics in the identification of mycobacteria. A third procedure, the qualitative "spot test," is sometimes used to screen for catalase activity.

Principle
Organisms producing the enzyme catalase have the ability to decompose hydrogen peroxide into water and free oxygen.
$$2H_2O_2 \longrightarrow 2H_2O + O_2$$
Catalase

The test for mycobacterial catalase differs from the test used for the detection of catalase in other types of bacteria by using 30% hydrogen peroxide (Superoxal) in a strong detergent solution (10% Tween 80) instead of the usual 3% hydrogen peroxide solution. The detergent helps to disperse the hydrophobic, tightly clumped mycobacteria from large clumps to individual bacilli, maximizing the detection of catalase.

Media and Reagents
Media:
1. For spot test, growth on egg or agar base media is recommended.
2. For semiquantitative and heat-stable catalase tests, growth on egg base media only is recommended.
Reagents:
1. Hydrogen peroxide, 30% (Superoxol)
2. Tween 80, 10%, sterilized at 121° C. for 10 minutes and stored at 4° C. Swirl before using if settling has occurred.
3. Just before use, mix equal amounts of 30% hydrogen peroxide and Tween 80 in the amounts needed. Discard any Tween-peroxide mixture left, because it is unstable and should not be reused.
4. M/15 phosphate buffer
   A. Stock solutions:

(1) Anhydrous Na_2HPO_4	9.47 g.
Distilled water	1000.00 ml.
(2) Potassium phosphate KH_2PO_4	9.07 g.
Distilled water	1000.00 ml.

   B. Phosphate buffer, pH 7.0:
   Mix 61.1 ml. of solution A (1) with 38.9 ml. of solution B (2). Confirm pH with meter.

Procedure
Spot Test:
Add 1 to 2 drops of a freshly mixed Tween-peroxide solution to a colony of mycobacterial growth on a plate or tube of culture medium. Observe for 4 to 5 minutes for evolution of bubbles. Appearance of bubbles may be rapid (strongly positive) or slow (weakly positive). The absence of any bubbles is a negative test for catalase.

Chart 12-4. *Catalase Activity (Continued)*

Procedure *(Continued)*	*Semiquantitative Test:* 1. Inoculate the surface of a tube of Löwenstein-Jensen medium prepared as a "deep" with 0.1 ml. of a 7-day liquid culture to the test organism. 2. Incubate at 37° C. for two weeks. Caps on the culture tube must be loose to permit adequate exchange of air. 3. Add 1.0 ml. of freshly prepared Tween-peroxide solution and leave upright for 5 minutes. 4. Measure the height of the column of bubbles above the surface of the culture medium and record on work sheet (Plate 12-2, G). *Test for Heat-Stable Catalase—pH 7/68° C. for 20 Minutes* 1. Using a small test tube emulsify several colonies of the test organism in 0.5 ml. of M/15 phosphate (pH 7.0) buffer. 2. Place the tube in a water bath or a constant-temperature block at 68° C. for 20 minutes. 3. Remove the tube and allow to cool to room temperature. 4. Add 0.5 ml. of freshly prepared Tween-peroxide mixture. 5. Watch for bubbles on the surface of the fluid. Do not discard as negative until after 20 minutes. 6. Development of bubbles is a positive reaction and a negative test will be devoid of any bubbles. Do not shake the tube because a false impression of bubbles can develop from the presence of the detergent in the mixture.
Interpretation	In each of the tests, the presence of catalase is indicated by bubbles. The spot test is a quick and easy method for detecting the presence or absence of catalase, but only gives a broad guide as to the amount present. The determination of heat-stable catalase ia a very helpful characteristic in the identification of the nonpigmented mycobacteria. Heat-labile catalase is a characteristic of *M. tuberculosis, M. bovis, M. gastri,* and occasional strains of the *M. intracellulare-avium* complex. 　The semiquantitative test for catalase is useful in distinguishing strains that are strongly positive for catalase from those with small amounts of catalase and is particularly helpful for the identification of INH resistant *M. tuberculosis* (negative). 　A column of bubbles 5 to 45 mm. in height is considered weakly positive. If the column of bubbles is greater than 45 mm. the test is strongly positive. Absence of bubbles indicates a negative catalase test.
Controls	*Mycobacterium kansasii* for strongly positive semiquantitative and heat-stable catalase *Mycobacterium tuberculosis* H37Rv for weakly positive semiquantitative and heat-labile catalase *Mycobacterium tuberculosis* INH-resistant for negative spot test and semiquantitative catalase
Bibliography	Kubica, C. P., et al.: Differential identification of mycobacteria. I. Tests on catalase activity. Am. Rev. Respir. Dis., *94*:400–405. 1966.

Chart 12-5.   *Growth Inhibition by Thiophene-2-Carboxylic Hydrazide (T₂H)*

Introduction	The distinction between *Mycobacterium tuberculosis* and *M. bovis* can sometimes be difficult since up to 30% of *M. bovis* BCG strains may accumulate small amounts of niacin while other strains may exhibit weakly positive tests for nitrate reduction. A characteristic that has been found to be helpful in differentiating between these two species is the ability of thiophene-2-carboxylic acid hydrazide to inhibit the growth of *M. bovis* but not the other species of mycobacteria. This characteristic is especially helpful in differentiating *M. bovis* from *M. tuberculosis*.
Media and Reagents	Thiophene-2-carboxylic acid hydrazide (T_2H) is incorporated into 7H10 or 7H11 agar in concentration of 1 and 5 µg./ml. (available from Aldrich Chemical Co., Milwaukee, Wis.). The medium can be dispensed into plastic biplates or quadrant plates. The use of 10 µg./ml. as previously recommended was found to inhibit some strains of *M. tuberculosis*.
Procedure	Inoculate media containing 0, 1, and 5 µg. of T_2H per ml. with a loopful of a barely turbid broth culture of the organism to be tested and streak to obtain isolated colonies. Incubate in 5 to 8% CO_2 for 14 to 21 days and examine for growth.
Interpretation	A positive test will show good growth of the organism on the medium without the drug and lack of growth on the medium containing T_2H (Plate 12-2, *J*). Although the results of an international collaborative study have suggested that the test is highly reproducible with the concentration of 1.0 µg./ml., many laboratories find it advisable to use a second concentration of 5 µg./ml. to confirm the results obtained with the lower concentration.
Controls	Positive control: *Mycobacterium bovis* Negative control: *Mycobacterium tuberculosis*, H37Rv
Bibliography	Harrington, R., and Karlson, A. G.: Differentiation between *M. tuberculosis* and *M. bovis* by in vitro procedures. Am. J. Vet. Res., *27*:1193–1196, 1967. Wayne, L. G., *et al.*: Highly reproductible techniques for use in systematic bacteriology in the Genus *Mycobacterium*. Tests for niacin and catalase and for resistance to isoniazid, thiophene 2-carboxylic acid hydrazide, hydroxylamine and p-nitrobenzoate. Int. J. System. Bacteriol., *26*:311–318, 1976.

from those strains of *M. bovis* that may have other characteristics that are similar (Plate 12-2, *J*). For example, 30 per cent of *M. bovis* BCG strains may be weakly niacin-positive, while others may be weakly nitrate-positive.[33] The details of this characteristic are presented in Chart 12-5.

Identification of Mycobacteria Other than *M. tuberculosis*

As presented in the list on page 355, extent level-4 reference laboratories (number 4 in the list under Extent of Services) are available for definitive identification and suscep-

Table 12-5. *Runyon Classification of Atypical Mycobacteria*

Runyon Group	Characteristics	Organisms
Group I: Photochromogens	Production of a yellow carotene pigment when viable colonies are exposed to a strong light (photochromogenic)	*M. kansasii* *M. simiae* *M. marinum*
Group II: Scotochromogens	Production of bright yellow pigmented colonies when grown either in the light or dark. In some species the pigment may be intensified on exposure to light.	*M. scrofulaceum* *M. gordonae* *M. flavescens*
Group III: Nonphotochromogens	Although some of these species may produce small amounts of pale yellow pigment, exposure of colonies to light does not intensify the color; hence they are designated "nonphotochromogens."	*M. intracellulare-avium* complex *M. terrae* *M. gastri* *M. nonchromogenicum-triviale* complex
Group IV: Rapid growers	Ability to grow much more rapidly than the other three groups, often showing mature colonies in 3 to 5 days. Some rapid growers produce an intense yellow pigment; however, the two species listed here, known to cause infection in man, are nonpigmented. These species may be associated with disease of the skin or eye, or multiple organ dissemination in immune supressed patients	*M. fortuitum* *M. chelonei*

tibility testing of mycobacteria. However, as general knowledge of mycobacteria has increased and better standardized identification procedures have become available, more and more laboratories at extent level 3 have required the ability to accurately identify most of the species of Mycobacteria of medical importance.

Classification of Mycobacteria

With the introduction of streptomycin in 1945 and subsequently other chemotherapeutic agents, it became highly desirable to isolate the microorganism from each patient in case susceptibility testing was needed. As a greater variety of organisms were recovered and identified, "atypical" strains were noted which differed in colonial appearance and biochemical reactions from *M. tuberculosis*. In 1954, after studying a series of these atypical strains, Timpe and Runyon[28] proposed classifying them into four groups on the basis of their rate of growth at 37° C. and the presence or absence of pigmented colo-

nies when grown in the dark and then exposed to light. This classification was subsequently refined and is presented in Table 12-5.[20]

Although Runyon's classification was helpful in the subgrouping of atypical mycobacteria isolates, it is now apparent from a clinical standpoint that it is necessary to identify all isolates in order to properly manage the treatment of patients. The confusion and misunderstanding which can result from the use of terms such as photochromogen and scotochromogen should be avoided. For example, *M. szulgai*, a mycobacterial species recently recognized in association with human disease, does not fit the Runyon classification.[22] *M. szulgai* is pigmented (scotochromogenic) when incubated at 37° C, but photochromogenic when grown at 22 to 24° C.

The species of mycobacteria currently considered to be of clinical and laboratory significance and their differential characteristics are shown in Table 12-6. Descriptions of the characteristics other than those pre-

(Text continues on p. 368)

Table 12-6. *Identification Characteristics of Mycobacteria**

| Organism | Optimum Isolation— Temperature and Rate of Growth | Pigmentation Growth in | | Niacin Test | Nitrate Reduction |
		Light	Dark		
M. tuberculosis	37° C. 12–25 days	Buff	Buff	+	3–5 +
M. africanum	37° C. 31–42 days	Buff	Buff	V	V
M. bovis	37° C. 24–40 days	Buff	Buff	V	−
M. ulcerans	32° C. 28–60 days	Buff	Buff	−	−
M. kansasii	37° C. 10–20 days	Yellow	Buff	−	1–5 +
M. marinum	31–32° C. 5–14 days	Yellow	Buff	V	−
M. simiae	37° C. 7–14 days	Yellow***	Buff	+	∓
M. szulgai	37° C. 12–25 days	Yellow to orange	Yellow—37° C. Buff—25° C.	−	±
M. scrofulaceum	37° C. 10 + days	Yellow	Yellow	−	−
M. gordonae	37° C. 10 + days	Yellow to orange	Yellow	−	−
M. flavescens	37° C. 7–10 days	Yellow	Yellow	−	+
M. xenopi	42° C. 14–28 days	Yellow	Yellow	−	−
M. intracellulare-avium complex	37° C. 10–21 days	Buff to pale yellow	Buff to pale yellow	−	−
M. gastri	37° C. 10–21 days	Buff	Buff	−	−
M. terrae complex	37° C. 10–21 days	Buff	Buff	−	1–5 +
M. triviale	37° C. 10–21 days	Buff	Buff	−	1–5 +
M. fortuitum	37° C. 3–5 days	Buff	Buff	−	2–5 +
M. chelonei					
ss chelonei	37° C. 3–5 days	Buff	Buff	V	−
ss abscessus	37° C. 3–5 days	Buff	Buff	V	−
M. smegmatis	37° C. 3–5 days	Buff to yellow	Buff to yellow	−	1–5 +

*Reproduced with permission, American Society of Clinical Pathologists. From Sommers, H. M.: The identification of mycobacteria. Technical Improvement Service, vol. 28, 1977.
Key: + = 84% of strains +; ± = 50-84%; ∓ = 16–49%; − = 16% of strains +; V = variable; blank spaces = little or no data.

Identification Characteristics of Mycobacteria* (Continued)

Catalase Semi quantitative†	pH 7.0 68° C.	Tween 80 Hydrolysis— 10 days	Arylsulfatase— 3 days	Urease	Resistance to T₂H— 1 µg./ml.	Growth on 5% NaCL	Iron Uptake
<40‡	−	∓	−	+	+	−	−
<20	−	−	−		+	−	−
<20	−	−	−	+	−	−	−
>50	+	−	−		+		
>50	+	+**	−	+	+	−	−
<40	∓	+	∓		+	−	−
>50	+	−	−		+		
>50	+	∓	∓	+	+	−	
>50	+	−	−	+	+	−	−
>50	+	+	−	−	+	−	−
>50	+	+	−	+	+	+	−
<40	+	−	±	−	+	−	−
<40	+	−	−	−	+	−	−
<40	−	+	−	+	+	−	−
>50	+	+	−	−	+	−	−
>50	+	+	∓		+	+	+
>50	+	±	+	+	+	+	+
>50	+	−	+	+	+	−	−
>50	+	−	+	+	+	+	−
>50	±	+	−		+	+	+

† Numbers indicate millimeters of bubbles.
‡ NH = resistant strains may be negative.
** Positive (most) in 24–48 hours.
*** Photochromogenicity unstable with repeated subcultures.

Chart 12-6.   *Hydrolysis of Tween 80*

Introduction

Ability to hydrolyze Tween 80 is an important characteristic for differentiation of mycobacteria. With rare exceptions, strains that hydrolyze Tween 80 are clinically insignificant (for example, the "tap water" bacilli, *M. gastri*, *M. terrae* complex, and *M. triviale*), while the clinically important species (*M. scrofulaceum* and members of the *M. intracellulare-avium* complex) are Tween 80-negative.

Principle

Tween 80 is the trade name for the detergent polyoxyethylene sorbitan monooleate. Certain mycobacteria possess a lipase that splits Tween 80 into oleic acid and polyoxyethylated sorbitol, which modifies the optical characteristics of the test solution from a straw yellow (produced by light passing through the intact Tween 80 solution) to pink. Although pink is the color of the neutral red indicator, the color change is not the result of a pH shift since the oleic acid formed is neutralized by the buffer solution. The color change directly indicates the hydrolysis or destruction of the Tween 80 molecule.

Media and Reagents

1. Phosphate buffer, 0.067 M, pH 7.0   100 ml.
   61.1 ml. $^m/_{15}$ Na_2HPO_4 (9.47 g./l.)
   38.9 ml. $^m/_{15}$ KH_2PO_4 (9.09 g./l.)
2. Tween 80   0.5 ml.
3. Neutral red, 0.1% aqueous   2.0 ml.
   Note: it is important to prepare the neutral red solution on the basis of dye activity. Commercial products are often less than 100% active. For example, if the actual dye content is 85%, dissolve 0.1 g. in 85 ml. of water rather than in 100 ml. of water in order to achieve a 0.1% solution.

---

viously reviewed in Charts 12-1 though 12-5 are discussed in the paragraphs that follow and in Charts 12-6 and 12-7.

Mycobacteria species are listed in Table 12-6 and a more detailed discussion of this test is presented in Chart 12-6 (Plate 12-2, *H*).

Tween 80 Hydrolysis

Tween 80 is the trade name of a detergent than can be used in identifying those mycobacteria that possess a lipase that splits the compound into oleic acid and polyoxyethylated sorbitol. This test is helpful in the identification of *M. kansasii,* which can produce a positive result in as short a time as 3 to 6 hours, and in differentiating between *M. gordonae* (positive) and *M. scrofulaceum* (negative) in the standard 10-day test. The Tween 80 hydrolysis reaction of the other

Determination of Arylsulfatase Activity

Determination of the activity of the enzyme arylsulfatase in mycobacteria is helpful in the identification of certain species, notably the rapidly growing members of the *M. fortuitum-chelonei* complex (Plate 12-2, *I*). Small quantities of this enzyme can also be produced by *M. marinum, M. kansasii, M. szulgai, M. xenopi,* and other less frequently encountered species as shown in Table 12-6. Many of these species grow so slowly that insufficient enzyme is produced to give a

Chart 12-6. *Hydrolysis of Tween 80 (Continued)*

Media and Reagents (Continued)	Mix the three reagents and dispense 3 to 5 ml. amounts into screwcap test tube. Autoclave and store in the refrigerator in a light-proof container to protect from spontaneous hydrolysis. This substrate is stable for only 2 to 4 weeks. A Tween 80 hydrolysis test concentrate is commercially available and is reported to be equivalent to the standard substrate (Kilburn).
Procedure	1. Place a 3 mm. loopful of actively growing mycobacterial bacilli in the Tween 80 substrate. (Inasmuch as there is no nitrogen source in the substrate, the organism used for testing must be mature and actively metabolizing.) 2. Incubate at 35 to 37° C. for 10 days. 3. Observe for color change initially in 3 days and daily thereafter. (The exception is when *M. kansasii* is suspected, which may produce a positive reaction within 3 to 6 hours.)
Interpretation	A positive test is shown by a change in the color of the substrate from straw yellow to pink (Plate 12-2, *H*).
Controls	Rapid positive: *M. kansasii* Delayed positive: *M. gastri* Negative: *M. scrofulaceum*
Bibliography	Kilburn, J. O., et al.: Preparation of a stable mycobacterial Tween hydrolysis test substrate. Appl. Microbiol., *26*:826, 1973. Wayne, L. G., Doubek, J. R., and Russell, R. L.: Classification and identification of mycobacteria. I. Tests employing Tween 80 as substrate. Am. Rev. Respir. Dis., *90*:588–597, 1964.

consistently positive reaction. Therefore, the most useful laboratory application of the standard 3-day test is for the differentiation of the members of the *M. fortuitum-chelonei* complex from the group III nonphotochromogenic mycobacteria. The arylsulfatase test is discussed in detail in Chart 12-7.

The Urease Test

The test for urease activity was described in Chart 2-11 in Chapter 2. As applied to mycobacteria, this test provides a simple means to differentiate *M. scrofulaceum* (positive) from *M. gordonae* (negative). It is also helpful in separating *M. gastri* (positive) from other members of the group III non-

photochromogenic mycobacteria, as shown in Table 12-6. The test for mycobacteria can be performed either by inoculation of the organism into distilled water containing urea base concentrate, or by use of filter paper disks containing urea which are added to distilled water.[17,30] The test is interpreted as positive if a pink to red color develops after three days of incubation at 37° C.

Growth in 5 Per Cent NaCl

The ability to grow on an egg base culture medium containing 5 per cent NaCL is shared by *M. flavescens, M. triviale,* and the rapidly growing mycobacteria with the exception of *M. chelonei* ss *chelonei* (Table

Chart 12-7. *Arylsulfatase Test*

Introduction	Arylsulfatase is an enzyme secreted by certain mycobacteria which can aid in their identification. Members of the *M. fortuitum-chelonei* complex are unique among the mycobacteria in that they produce sufficient arylsulfatase to produce a positive test within 3 days. Small quantities may also be produced by *M. xenopi, M. szulgai, M. marinum, M. kansasii, M. gordonae,* and members of the *M. avium-intracellulare* complex, among others (see Table 12-6); however, the test is less helpful in identifying these species because they commonly grow too slowly to produce consistently reliable results. Thus, the test is best used primarily to identify the *M. fortuitum-chelonei* complex of organisms.
Principle	Arylsulfatase is an enzyme that splits free phenolphthalein from the tripotassium salt of phenolphthalein disulfate. In the test described by Wayne, the phenolphthalein salt is incorporated into oleic acid agar and the test is performed by adding a small amount of an alkaline solution of sodium carbonate to the surface of a 3-day-old culture. A positive test is indicated by the development of a purple color. The test can also be performed by using 0.001 M solution of the phenolphthalein salt in Dubos or Middlebrook 7H9 broth. Use of 0.003 M tripotassium phenolphthalein disulfate solution has been recommended to detect smaller quantities of the enzyme, with incubation periods increased to as much as 14 days (Kubica and Vestal).

Media and Reagents	1. Substrate:
	Tripotassium phenolphthalein disulfate 65 g.
	Glycerol 1 ml.
	Dubos oleic agar base 100 ml.
	   Add the phenolphthalein and glycerol to 100 ml. of melted Dubos agar and dispense in 2-ml. amounts into 16×125 mm. screwcap culture tubes. Let harden in upright position and store in the refrigerator. This medium is commercially available as Wayne Arylsulfatase Agar.*
	2. Sodium carbonate, 1 M.:
	   Add 5.3 g. to 100 ml. of water

12-6). Other mycobacteria do not tolerate this increased salt concentration. Egg base slants of Löwenstein-Jensen medium containing 5 per cent NaCL are commercially available. The test can not be performed using agar base medium.[11]

Iron Uptake by Mycobacteria

Members of the *M. fortuitum-chelonei* complex have many similarities and it is necessary only occasionally to distinguish between them. *M. fortuitum* has the ability to take up soluble iron salts from the culture medium, producing a rusty-brown appearance upon addition of an aqueous solution of 20 per cent ferric ammonium citrate. *M. chelonei* lacks this property[31] (Plate 12-2, *K*).

CLINICAL SIGNIFICANCE OF THE MYCOBACTERIA

The clinical significance of the different species of mycobacteria depends in large part on the state of the host's natural resistance. *M. tuberculosis* is almost always asso-

Chart 12-7.   *Arylsulfatase Test* (Continued)

Procedure	1. Prepare a suspension of the organism to be tested and inoculate 1 or 2 drops into a vial of substrate. 2. Incubate at 35 to 37° C. for 3 days. 3. Add 1 ml. of sodium carbonate solution at the end of the incubation period and observe for a color change.
Interpretation	The development of a pink color in the substrate near the bacillary growth after adding the sodium carbonate indicates the release of free phenolphthalein which is a positive test for arylsulfatase (Plate 12-2, *I*).
Controls	Positive control: *M. fortuitum* Negative control: *M. tuberculosis*
Bibliography	Kubica, G. P., and Ridgon, A. L.: The arylsulfatase activity of mycobacteria: III. Preliminary investigation of rapidly growing mycobacteria. Am. Rev. Respir. Dis., *83*:737–740, 1961. Kubica, G. P., and Vestal, A. L.: The arylsulfatase activity of acid-fast bacilli: I. Investigation of stock cultures of mycobacteria. Am. Rev. Respir. Dis., *83*:728–732, 1961. Wayne, L. G.: Recognition of *Mycobacterium fortuitum* by means of a three-day phenophthalein sulfatase test. Am. J. Clin. Pathol., *36*:185–187, 1961.

*Bioquest, Division of BBL Laboratories, Baltimore, Maryland

ciated with infection and tuberculosis is known to be a highly communicable disease. Outbreaks in closed populations, such as schools, ships, or crowded family groups are all too common. Greater than 95 per cent of *M. tuberculosis* strains isolated from untreated patients are susceptible to antituberculosis drugs and will respond promptly to treatment with two or preferably three drugs.

Certain other species of mycobacteria can also cause human disease. It is important to be able to differentiate these in the laboratory from those which are virtually never associated with disease. Table 12-7 lists various species of Mycobacteria, their relative pathogenicity for man, their legitimate as well as their commonly accepted names and names given them without legitimate standing, and their equivalent Runyon group.

Infections with *M. kansasii* are not as

communicable as those with *M. tuberculosis* and, although the organism is moderately resistant to many of the first line antituberculosis drugs *in vitro,* the infections usually respond to agressive therapy with three or more agents. In contrast, many *M. intracellulare* infections respond poorly to drug therapy, frequently requiring four or more drugs to achieve a positive response. A disappointing number of patients with *M. intracellulare* infections do not show a satisfactory response to drug therapy at all.

The apparent lack of communicability of mycobacterial infections involving species other than *M. tuberculosis* has led to the impression that these species are "opportunistic" pathogens. For example, there is a greater incidence of chronic lung disease preceding infection with *M. kansasii* and *M. intracellulare* than with *M. tuberculosis.*[1] For

(Text continues on p. 374)

Table 12-7. Nomenclature of Mycobacteria*

Legitimate Name	Relative Pathogenicity for Man	Equivalent Runyon Group	Acceptable Common Name	Names Without Legitimate Standing and Comments
M. africanum	+ + +			Intermediate form between M. bovis and M. tuberculosis. It is found in North and Central Africa.
M. asiaticum	+ +	Group I photochromogen		Similar to M. simiae but differs antigenically
M. bovis	+ + +		Bovine tubercle bacillus	Causes bovine and human tuberculosis; avirulent strains are used for BCG vaccines.
M. chelonei	+	Group IV rapid grower		May cause occasional skin disease. Includes two subspecies, ss. chelonei and ss. abscessus.
M. flavescens	0	Group II scotochromogen, sometimes placed with group IV rapid growers		Grows rapidly. It should be differentiated from M. scrofulaceum.
M. fortuitum	+	Group IV rapid grower		M. ranae; M. minetti—skin and lung infections. It may cause disease in immunosuppressed host.
M. gastri	0	Group III nonphotochromogen		Not known to be pathogenic for man. It may be found in gastric aspirates.
M. gordonae	0	Group II scotochromogen	Tap water scotochromogens	M. aquae—rarely, if ever, pathogenic for man.
M. intracellulare-avium complex	+ + +	Group III nonphotochromogen	Battey bacillus	M. batteyi, M. battey—frequently drug-resistant

*Sommers, H. M.: The Clinically Significant Mycobacteria. Chicago, American Society of Clinical Pathologists, 1974.

Table 12-7. *Nomenclature of Mycobacteria* (Continued)*

Legitimate Name	Relative Pathogenicity for Man	Equivalent Runyon Group	Acceptable Common Name	Names Without Legitimate Standing and Comments
M. kansasii	+ + +	Group I photochromogen		Rare, nonpigmented, scotochromogenic, and niacin-positive strains
M. marinum	+ + +	Group I photochromogen		*M. balnei, M. platypeocilus* associated with skin infections
M. scrofulaceum	+ +	Group II scotochromogens		*M. marinum*—may cause cervical lymphadenitis
M. simiae	+ +	Group I photochromogen		*M. habana* facultatively pathogenic; photoreactivity may be unstable; it is niacin-positive.
M. szulgai	+ + +	Group I photochromogen at 25° C. Group II scotochromogen at 37° C.		Associated with chronic pulmonary and extrapulmonary disease; distinctive lipid composition of cell walls
M. terrae	Rare	Group III nonphotochromogen	"Radish bacillus"	May be closely related to *M. triviale*
M. triviale	0	Group III nonphotochromogen	"V" bacillus	Has been called "atypical-atypical" mycobacterium
M. tuberculosis	+ + +	Human tuberculosis	Human tubercle bacillus	Causes human tuberculosis—highly contagious
M. ulcerans	+ + +			Associated with skin infections in tropics—*M. buruli*
M. xenopei	+ +	Group III nonphotochromogens (Scotochromogen)		*M. littorale, M. xenopei* grows slowly; best at 42° C.; may contaminate hot-water system.

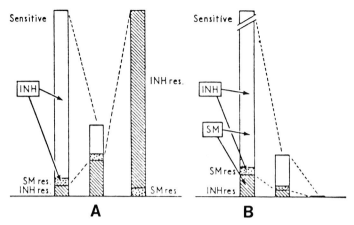

Fig. 12-1. Emergence of mycobacterial antimicrobial resistance with one and two drug therapy. In Fig. A, the patient represented by the diagram is treated with only INH. Although the small number of streptomycin resistant mutants are inhibited by the INH, the INH resistant mutants are refractory and in time make up the majority of the population. This represents drug failure.

The patient represented by Fig. B is treated with both streptomycin and INH. The streptomycin resistant mutants are inhibited by the INH and the isoniazid resistant mutants are inhibited by streptomycin. Thus, neither of these mutant strains can overgrow and drug therapy is successful. (From Crofton, J.: Some principles in the chemotherapy of bacterial infections. Br. Med. J., *2:*209–212, 1969.)

this reason, patients with mycobacterial infections other than *M. tuberculosis* infections should be investigated for possible defects in host resistance.

SUSCEPTIBILITY TESTING OF MYCOBACTERIA

In determining the pattern of drug susceptibility of mycobacteria, several principles must be clearly understood.[13] First, random drug resistance of mycobacteria is independent of exposure to the agents. The frequency of drug-resistant mutants in a culture of tubercle bacilli has been estimated to be about one in 10^5 bacteria for isoniazid and one in 10^6 for streptomycin. If two drugs, isoniazid and streptomycin, are both given, the incidence of resistance will be the product of the two separately, one in every 10^{11} organisms. Knowledge of the incidence of mutants becomes important because it has

been determined that patients with an open pulmonary cavity may have a total bacillary population of 10^7 to 10^9 bacteria. Therefore, if these patients are treated with a single antituberculous agent, their cultures may soon show only resistant organisms to that agent, and treatment failure. For this reason, patients with tuberculosis must always be treated with two and preferably three drugs. From the above it can be seen that failure of patients to take only one of the drugs may lead to the rapid emergence of drug-resistant tubercle bacilli (Fig. 12-1).

A second principle of mycobacterial drug susceptibility testing is based in the *in vivo* correlation between the clinical response to an antimycobacterial agent and the result of *in vitro* susceptibility testing. It has been found that if more than 1 per cent of a patient's tubercle bacilli are resistant to a drug *in vitro,* therapy with that drug will not be clinically useful. Therefore, the method

Table 12-8. *In Vitro Concentrations of Antimycobacterial Agents in 7H11 Agar Used for Susceptibility Testing**

Agent	Concentration in (μg./ml.)	
	Low	High
Isoniazid	0.2	1.0
Streptomycin	2.0	10.0
Para-aminosalicylic acid	8.0	
Ethambutol	7.5	15.0
Riframpin	1.0	5.0
Ethionamide	10.0	15.0
Cycloserine	30.0	60.0
Capreomycin	10.0	20.0
Kanamycin	6.0	12.0
Viomycin	10.0	15.0
Pyrazinamide†	50.0	

*Concentrations used at the National Jewish Hospital, Denver, Colorado, from J. K. McClatchy, Ph.D. *Personal communication.*
†Tested in 7H10 medium at pH 5.5. See reference 27.

for drug susceptibility testing of mycobacteria must enable the determination of the proportion of bacilli susceptible and resistant to a given drug. To achieve this, the inoculum should be adjusted so that the number of spontaneously resistant mutants will not mislead the laboratory worker to interpret the culture as resistant. By the same token, there must be a sufficient number of colonies on the plate so that the incidence of drug resistance in the range of 1 per cent can be determined. This is best accomplished when 100 to 300 colony-forming units are present on each quadrant of a 4-quadrant Petri plate. To determine the incidence of resistance, it is usually necessary to inoculate two sets of susceptibility test plates, the second set of plates with a 100-fold dilution of the inoculum used for the first set. This procedure is known as the proportional susceptibility testing method.

At the present time, there are eleven drugs used in the treatment of tuberculosis. Five are considered "primary" drugs and include streptomycin, isoniazid, para-aminosalicylic acid, rifampin, and ethambutol, while the remaining six, ethionamide, capreomycin, kanamycin, cycloserine, viomycin, and pyrazinamide, are considered secondary drugs and are used only when resistance develops to the primary drugs. The suggested concentrations of the drugs used

for mycobacterial susceptibility testing are listed in Table 12-8.

The test is performed using plastic Petri plates divided into 4 quadrants (Plate 12-3). An amount of 5 ml. of agar is placed in each quadrant, the first without any antimycobacterial agent to act as a growth control, the other three containing dilutions of the drugs to be tested. Although drugs have been incorporated in inspissated egg base media in the past, most laboratories now prefer using either 7H11 or 7H10 as a base medium, adding the drugs after cooling the agar to 45° C. and thereby decreasing the loss of activity that can occur during inspissation. An additional loss of drug activity may occur in egg base media as the result of some agents binding to albumin and other proteins.

A simplified method for preparing drug susceptibility plates has been described which utilizes filter paper disks containing the primary antituberculous drugs.[32] Appropriate disks are available from standard reagent supply sources. Medium containing the desired concentration of drug is prepared by placing a disk in one quadrant of the plate and adding 5 ml. of 7H10 or 7H11 agar. Since each disk is labeled with the name and concentration of the drug it contains, labeling errors are eliminated as well as errors that can occur in weighing, dilut-

Table 12-9. *Distribution of Drug-Containing Disks for Susceptibility Tests**

Plate Number	Quadrant Number		Amount of Drug per Disk (μg.)	Final Drug Concentration (μg./ml.)
1	I	Control No. 1	—	0
	II	Isoniazid	1	0.2
	III	Isoniazid	5	1.0
	IV	Rifampin	5	1.0
2	I	Streptomycin	10	2.0
	II	Streptomycin	50	10.0
	III	Ethambutol	25	5.0†
	IV	Ethambutol	50	10.0†
3	I	Para-aminosalicylic acid	10	2.0
	II	Para-aminosalicylic acid	50	10.0
	III	Control No. 2	—	0
	IV	—	—	—

*From Kubica, G. P., *et al.*: Laboratory services for mycobacterial diseases. Am. Rev. Respir. Dis., *112*:773–787, 1975.
†Improved correlation with clinical response has been noted with ethambutol concentrations of 7.5 and 15 μg./ml. These can be achieved by using 1½ 25-μg. disks in each quadrant for the 7.5 μg./ml. concentration and a 25- and 50-μg. disk for the 15 μg./ml. concentrations.

ing, and measuring of drug solutions. Table 12-9 gives a suggested schedule for the use of drug-containing paper disks.

A set of susceptibility testing media is inoculated with 3 drops per quadrant of a 10^{-2} dilution and a second set of media is inoculated with 3 drops per quadrant of a 10^{-4} dilution of a barely turbid broth culture. Plates are incubated in CO_2 at 37° C. and read after 2 to 3 weeks of incubation. Incubation of ethambutol-containing media for more than three weeks can result in the appearance of microcolonies as a result of inactivation of the drug. Interpretation of the results should include either an estimate or direct count of the number of colonies on the control and on the drug-containing media. All colonies, even those showing inhibition of growth on drug-containing media, should be counted and compared to the number of colonies on the control quadrant. Since the control quadrant should reflect the number of organisms added to the test media, the percentage of resistant colonies can be readily calculated from the two sets of plates inoculated with the two dilutions of the culture 100-fold apart (10^{-2} to 10^{-4}). A number of drug susceptibility plates are

illustrated in Plate 12-3. They show how the results can be interpreted.

Mycobacterial susceptibility plates can also be inoculated directly with digested and concentrated sputum found to be positive for acid-fast bacilli ("direct test"). The method described above using a broth culture derived from colonies of a primary culture is referred to as an "indirect test." The direct test gives good results if there are large numbers of mycobacteria in the specimen. The advantage of the direct susceptibility test is an earlier report of drug resistance or susceptibility (3 to 4 weeks) in contrast to the indirect test which may take up to 6 to 8 weeks. The disadvantage of the direct susceptibility test is that it usually requires a fairly large number of mycobacteria for successful growth and may often be overgrown with contaminating bacteria.

Drug susceptibility studies are not indicated for all mycobacterial isolates since the incidence of primary drug resistance (defined as a drug-resistant organism isolated from a previously untreated patient) is less than 3 to 5 per cent. Even if there is primary resistance to one drug, such as isoniazid, treatment with the recommended triple-drug therapy will provide adequate coverage. Susceptibility tests are necessary on bacterial isolates from patients who have shown relapse while on drug therapy. The probability of induced resistance in this group of patients is high. Other indications for performing antimycobacterial drug susceptibility tests are given in the following list. If susceptibility tests are not performed on isolates obtained from patients not previously seen by a laboratory, at least one culture should be saved for six months or a year should the patient not respond to therapy. Control strains for susceptibility studies should be run with each set of isolates tested. The controls should include a susceptible, an intermediate susceptible (e.g., *M. kansasii* resistant to 0.2 µg./ml. of isoniazid but susceptible to 1.0 µg./ml. of isoniazid), and a resistant strain.

Indications for Testing the Susceptibility of Mycobacterial Isolates*

---

1. May not be indicated for previously untreated patients
2. Should be obtained under the following conditions:
   A. Sputum-positive patients who have had previous chemotherapy (relapsed or retreatment cases)
   B. Patients on therapy whose sputum reverts to positive
   C. Patients whose sputum smears do not convert to negative within two to three months of treatment
   D. Patients whose cultures do not convert to negative within four to six months treatment
   E. Patients with increasing numbers of tubercle bacilli seen in smears after an initial decrease
   F. Patients with mycobacterial infections other than *Mycobacterium tuberculosis*
   G. Patients suspected of having primary resistance:
      (1) Tuberculosis patients who have lived abroad
      (2) Persons possibly infected by patients with drug-resistant tuberculosis

---

*Adapted from Pennsylvania State Public Health Laboratories and J. K. McClatchy. *Personal Communication.*

REFERENCES

1. Ahn, C. H., *et al.:* Ventilatory defects in atypical mycobacteriosis. Am. Rev. Respir. Dis., *113:*273–279, 1976.
2. Cohn, M. L., *et al.:* The 7H11 medium for the culture of mycobacteria. Am. Rev. Respir. Dis., *98:*295–296, 1968.
3. David, H. C.: Bacteriology of the Mycobacterioses. DHEW, Public Health Service Publication No. (CDC) 76-8316, 1976.
4. Gross, W., Hawkins, J., and Murphy, B.: *Mycobacterium xenopi* in clinical specimens. I. Water as a source of contamination (abstract). Am. Rev. Respir. Dis., *113* (Part 2):78, 1976.
5. Gruft, H.: Isolation of acid-fast bacilli from contaminated specimens. Health Lab. Sci., *8:*79–82, 1971.
6. Joseph, S. W.: Lack of auramine-rhodamine fluorescence of Runyon group IV mycobacteria. Am. Rev. Respir. Dis., *95:*114–115, 1967.
7. Kestle, D. G., and Kubica, G. P.: Sputum collection for cultivation of mycobacteria. An early morning specimen or the 24 to 72-hour pool? Am. J. Clin. Pathol., *48:*347–351, 1967.
8. Krasnow, I.: Primary isolation of mycobacteria. *In* Baer, D. M. (ed.): Technical Improvement Service P. 68. Chicago, Commission on Continuing Education, American Society of Clinical Pathologists, No. 25, 1976.
9. Krasnow, I., and Wayne, L. G.: Comparison of methods for tuberculosis bacteriology. Appl. Microbiol., *18:*915–917, 1969.
10. ———:Sputum digestion. I. The mortality rate of tubercle bacilli in various digestion systems. Am. J. Clin. Pathol., *45:*352–355, 1966.
11. Kubica, G. P.: Differential identification of mycobacteria. VII. Key features for identification of clinically significant mycobacteria. Am. Rev. Respir. Dis., *107:*9–21, 1973.
12. Kubica, G. P., *et al.:* Laboratory services for mycobacterial diseases. Am. Rev. Respir. Dis., *112:*783–787, 1975.
13. McClatchy, J. K.: Susceptibility testing of mycobacteria. *In* Baer, D. M., (ed.); Technical Improvement Service. Chicago, Commission on Continuing Education, American Society of Clinical Pathologists, No. 29, 1977.
14. McClatchy, J. K., *et. al.:* Isolation of mycobacteria from clinical specimens by use of selective 7H11 medium. Am. J. Clin. Pathol., *65:*412–415, 1976.
15. Miliner, R., Stottmeir, K. D., and Kubica, G. P.: Formaldehyde: a photothermal acti-

vated toxic substance produced in Middle-brook 7H10 medium. Am. Rev. Respir. Dis., *99:*603–607, 1969.

16. Mitchison, D. A., *et al.: A selective oleic acid albumin agar medium for tubercle bacilli. J. Med. Microbiol., *5:*165–175, 1972.

17. Murphy, D. B., and Hawkins, J. E.: Use of urease test discs in the identification of mycobacteria. J. Clin. Microbiol., *1:*465–468, 1975.

18. Petran, E. I., and Vera, H. D.: Media for selective isolation of mycobacteria. Health Lab. Sci., *8:*225–230, 1971.

19. Runyon, E. H.: Identification of mycobacterial pathogens utilizing colony characteristics. Am. J. Clin. Pathol., *54:*578–586, 1970.

20. ———:Anonymous mycobacteria in pulmonary disease. Med. Clin. North Am., *43:*273–290, 1959.

21. Runyon, E. H., *et al.: Mycobacterium. In* Lennette, E. H., Spaulding, E. H., and Truant, J. P. (eds.): Manual of Clinical Microbiology. P. 148–174. Washington, D. C., American Society for Microbiology, 1974.

22. Schaefer, W. B., *et al.: Mycobacterium szulgai*—a new pathogen. Am. Rev. Respir. Dis., *108:*1320–1326, 1973.

23. Shah, R. R., and Dye, W. E.: The use of dithiothreitol to replace N-acetyl-L-cysteine for routine sputum digestion-decontamination for the culture of mycobacteria. Am. Rev. Respir. Dis., *94:*454, 1966.

24. Smithwick, R. W., *et al.: Use of cetylpyridium chloride and sodium chloride for the decontamination of sputum specimens that are transported to the laboratory for the isolation of *Mycobacterium tuberculosis.* J. Clin. Microbiol., *1:*411–413, 1975.

25. Sommers, H. M.: The identification of myco-

bacteria. *In* Baer, D. M. (ed.): Technical Improvement Service. P. 1, Chicago, Commission on Continuing Education, American Society of Clinical Pathologists, No. 28, 1977.

26. ———: Special Mycobacterial Survey, Skokie, Ill. College of American Pathologists: Critique for Specimen E-4 *(M. szulgai),* 1976.

27. Stottmeier, K. D., Beam, E., and Kubica, G. P.: Determination of drug susceptibility of mycobacteria to pyrazinamide in 7H10 agar. Am. Rev. Respir. Dis., *96:*1072–1075, 1967.

28. Timpe, A., and Runyon, E. H.: The relationship of "atypical" acid-fast bacteria to human disease. J. Lab. Clin. Med., *44:*202–209 1954.

29. Vestal, A. L.: Procedures for the Isolation and Identification of Mycobacteria. DHEW Public Health Service Publication No. (CDC) 75-8230, 1975.

30. Wayne, L. G.: Simple pyrazinamidase and urease tests for routine identification of mycobacteria. Am. Rev. Respir. Dis., *109:*147–151, 1974.

31. Wayne, L. G., and Doubek, J. R.: Diagnostic key to mycobacteria encountered in clinical laboratories. Appl. Microbiol., *16:* 925–931, 1968.

32. Wayne, L. G., and Krasnow, I.: Preparation of tuberculosis testing mediums by means of impregnated discs. Amer. J. Clin. Pathol., *45:*769–771, 1966.

33. Wayne, L. G., *et al.:* Highly reproducible technique for use in systematic bacteriology in the genus *Mycobacterium:* tests for niacin and catalase and for resistance to isoniazid, thiophene 2-carboxylic acid hydrazide, hydroxylamine, and p-nitrobenzoate. Int. J. System. Bacteriol., *16:*311–318, 1976.

13 Mycology

This chapter will focus on the role of the laboratory microbiologist or mycologist in the diagnosis of fungal disease in man. It should be stressed that the laboratory should not become an island and microbiologists must not isolate themselves from the arena of clinical practice if the interests of the patient with mycotic disease are to be best served.

Many primary-care physicians are still lacking in their understanding of the basic principles of clinical mycology and how to properly approach the diagnosis of a patient with potential fungal disease. Physicians must be reminded of the clinical signs and symptoms of mycotic disease and how to properly collect specimens and have them transported to the laboratory in an optimal condition for the recovery of fungi. Clinical pathologists, microbiologists, and medical technologists should participate frequently in infectious disease conferences, teaching rounds, or other activities where the clinical and laboratory aspects of mycotic disease can be discussed.

CLINICAL AND LABORATORY APPROACHES TO THE DIAGNOSIS OF MYCOTIC DISEASES

There are three major approaches to the diagnosis of fungal disease.[18] First there is the clinical setting, where the physician encounters a patient with certain symptoms, takes a history, performs a physical examination, requests x-rays, orders laboratory tests, and obtains a specimen for culture. If the physician is not alert to the signs and symptoms of fungal infection, fails to obtain a proper specimen, or sends material to the laboratory in improper containers or in a formalin fixative, the diagnosis may be missed.

In another setting, the diagnosis of fungal disease may first come from the surgical pathologist who recognizes in a stained smear or in histologic tissue sections an inflammatory reaction or the hyphae, yeast forms, or spore structures suspicious for fungi. It is essential that surgical pathology laboratories be organized in such a way that selected portions of tissues are either preserved in a frozen state or appropriate fragments are submitted to the microbiology laboratory for culture. Too often attempts to make a definitive diagnosis of fungal infection on the basis of tissue sections alone are unsuccessful, whereas the diagnosis is easily confirmed if provisions are made to culture the infected material. The observation of a granulomatous inflammatory reaction in a frozen section preparation should always prompt a request for culture.

The microbiology laboratory is the third area in which the diagnosis of fungal disease can be made. A presumptive diagnosis can often be made immediately by the microscopic examination of infected specimens. In other instances, the definitive diagnosis will only be made after isolation and identification of a fungus isolated in culture. It is the responsibility of the microbiologist to select

Table 13-1. *Classification of the More Commonly Encountered Human Mycoses*

Deep-Seated Mycoses	Opportunistic Mycoses	Subcutaneous Mycoses	Superficial Mycoses
Blastomycosis	Actinomycosis*	Actinomycosis*	Dermatomycosis
Coccidioidomycosis	Aspergillosis	Chromomycosis	Tinea capitis
Cryptococcosis	Candidiasis	Maduromycosis	Tinea corporis
Histoplasmosis	Geotrichosis	Nocardiosis*	Tinea cruris
Paracoccidioidomycosis		Sporotrichosis	Tinea pedis
(South American blasto-			Tinea versicolor
mycosis)			

*The Actinomycetaceae, including the genera Actinomyces and Nocardia are currently considered to be bacteria and are classified with Schizomycetes.

the appropriate culture media to bring about optimal recovery and perform those laboratory procedures which will allow exact identification of the species isolated.

The effectiveness with which a patient with potential mycotic disease is evaluated and properly treated is dependent upon how well communications are maintained between these three areas of activity. The physician should inform the laboratory if he suspects a certain mycotic disease based on the history and physical examination of the patient so that specific procedures can be carried out; the surgical pathologist must learn to recognize the tissue alterations associated with mycotic infection and select appropriate material for culture before the specimen is placed in formalin; and the microbiologist must contact physicians immediately at the earliest appearance of growth in cultures that suggest a pathogenic fungus to resolve quickly whether or not the isolate may or may not be clinically significant and if further studies are required.

CLINICAL PRESENTATION OF MYCOTIC DISEASES

In the past fungal infections were classified as superficial, subcutaneous, and deep-seated mycoses. With the advent of broad-spectrum antibiotic therapy and the treatment of patients with chronic metabolic and neoplastic diseases with immunosuppressive and cytotoxic agents, the distinction between a pathogenic and "contaminant" fungus is much less clear, and this former classification of mycotic disease must be viewed with a different perspective. Table 13-1 presents a revision of this older classification.

Fungi causing deep-seated mycoses are virtually always pathogenic and potentially can result in serious or life-threatening disease. Fungal agents such as *Aspergillus fumigatus,* members of the genera Zygomycetes (Phycomycetes), Nocardia, and Candida, formerly considered as laboratory contaminants of little clinical significance, are now known to cause disseminated and even fatal disease in the immunosuppressed host. Other environmental fungi, such as Scopulariopsis, Fusarium, and Cladosporium, previously not recognized as causing disease, are now considered the etiologic agents of occasional cases of endocarditis, mycotic keratitis, and other localized infections, or are incriminated in allergic bronchopulmonary disease, including the farmer's lung syndrome. Certain of these species also produce aflatoxins which can cause gastrointestinal upset or neurologic manifestations when ingested by humans and animals.[22]

It is beyond the scope of this text to discuss in detail the various mycotic diseases; however, the reader is referred to texts by Emmons and coworkers,[9] Conant and associates,[5] and Rippon[22] for more in-depth coverage. Only a brief overview of the clinical manifestations of the different types of mycotic disease will be presented here so that laboratory microbiologists can gain some facility in the identification of several species of fungi that may be recovered in the laboratory.

General Symptoms of Fungal Infection

The patient's complaints may be vague and nonspecific. Low-grade fever, night sweats, weight loss, lassitude, easy fatigability, cough, and chest pain are often the presenting symptoms. In this regard, deep-seated or disseminated fungal diseases may mimic other infections such as tuberculosis, brucellosis, syphilis, and sarcoidosis, or disseminated carcinomatosis.

Laboratory tests may reveal nonspecific inflammatory responses. The erythrocyte sedimentation rate may be increased, and elevated levels of serum enzymes or gamma globulin and low-grade neutrophilia or monocytosis may be present. Even positive x-ray findings, such as pulmonary infiltrates or inflammatory processes in other organs, are often not specific. In some patients with severely compromised host resistance, general symptoms or laboratory and x-ray findings may be absent altogether.

The initial clues to fungal disease in a patient may be dependent upon the recognition of more specific signs and/or symptoms or from information derived from the clinical history. It is important that the physician elicit any relevant historical information from the patient, such as past or recent travel into geographic areas known to be endemic areas of fungal infection, or recent exposure to soil, dust, birds, or other sources having a high probability of fungal contamination.

Specific Signs and Symptoms

Pulmonary Signs and Symptoms

The respiratory tract is considered the primary route of infection for most fungal infections contracted from exogenous sources, such as by inhalation of dust-laden spores. In the acute phase of infection, a transient influenza-like syndrome may develop. Cough that may or may not be productive of sputum, chest pain that is frequently pleuritic in nature, dyspnea, tachypnea, and, less commonly, hemoptysis, are common respiratory symptoms in pulmonary infections. Cavity formation in the lung is only infrequently encountered. Peripherally located, small calcified nodules or "coin" lesions are usually manifestations of chronic healed forms of the disease. Allergic bronchopulmonary disease is a manifestation of hypersensitivity to fungal spores or products commonly of Aspergillus species. The formation of fungus balls in old tuberculous cavities is a special type of pulmonary lesion in which congenital bronchial cysts or cavitary lesions caused by tuberculosis may become colonized with a fungal species, usually Aspergillus or one of the Zygomycetes.

Cutaneous Signs and Symptoms

Primary cutaneous fungal infection caused by one of the dimorphic pathogenic species is rare, but may occur secondary to skin inoculation with contaminated soil or vegetative matter that may gain entrance to the site of a traumatic injury. Nonhealing pustules, ulcers, or draining sinuses may be the initial presenting sign of a disseminated mycotic infection. Nonhealing ulcers with or without regional lymph node involvement are the common presenting signs of sporotrichosis or agents causing mycetomas such as *Petriellidium (Allescheria) boydii* or Nocardia species. Chromomycosis is a rarely encountered cutaneous fungal disease in the United States and is caused by a variety of slow-growing, dematiacious molds. The scaling, itching lesions of tinea infections ("athletes foot," tinea capitis, tinea barbae, *etc.*) and typical ringworm infections of tinea corporis are common manifestations of infections with the superficial dermatophytic fungi. Hair and nail infections may also be caused by dermatophytic fungi, although *Candida albicans* must always be considered in patients with nail infections.

Central Nervous System Manifestations

Presenting symptoms of meningitis usually point to infection with *Cryptococcus neoformans,* while brain abscess may be caused by one of the members of the Zygomycetes group. Symptoms may be insidious or abrupt in onset, including headaches which increase in frequency and severity, ataxia, vertigo, vomiting, memory lapses, and in some cases seizures of the Jacksonian type. Vomiting, hallucinations, and drowsiness or coma are manifestations of advanced central nervous system disease.

Manifestations of Disseminated Disease

Fever, anemia, leukopenia, weight loss, and lassitude often indicate the dissemination of a mycotic disease beyond the primary organ of infection. Virtually any of the body organs may become involved, resulting in specific symptoms related to that organ. For example, involvement of the adrenal glands may lead to Addison's disease and potentially a fatal outcome due to adrenal gland insufficiency.

Signs and Symptoms of Miscellaneous Fungal Infections

Conjunctivitis, corneal infections, and keratoconjunctivitis may be caused by a variety of environmental fungi, including species of Aspergillus, Cladosporium, Cephalosporium, and Fusarium. Intraocular infections are most commonly caused by *Candida albicans,* species of Aspergillus, or members of the Zygomycetes (Phycomycetes), commonly following eye surgery or trauma.

Aspergillus niger is a common fungal agent that causes otomycosis, or "swimmers ear."

Endocarditis may be caused by a variety of fungi. Kaufman[15] has recently reviewed 46 previously reported cases. Of these, thirty-eight were caused by species of Candida, six by species of Aspergillus, and one each by a species of Hormodendrum and Paecilomyces.

Orbitocerebral phycomycosis is a potentially serious disease, particularly in diabetic patients whose diabetes is out of control. The disease most commonly begins as an infection in one of the nasal sinuses, progressing with invasion of the ocular orbit, optic nerve, or meninges. The disease may be rapidly fatal through invasion of cerebral blood vessels resulting in cerebral infarction or hemorrhage.

Rhinosporidium seeberi causes rhinosporidiosis, a chronic granulomatous infection of the nasal passages with the production of hyperplastic nasal polyps.

Lumpy jaw and deeply penetrating sinus tracts of the thorax or abdomen are caused by the Actinomycetes, notably *A. israelii.*

A good review of the opportunistic mycotic diseases and miscellaneous conditions less commonly caused by some of the environmental saprobic fungi can be found in the text by Rippon.[22]

HISTOPATHOLOGY OF FUNGAL INFECTIONS

Although microbiologists are infrequently called upon to interpret tissue sections that reveal structures suspicious for fungi, a few basic principles will be presented here.

The invasion of organs or tissues by fungi produces a variety of inflammatory reactions. *Blastomyces dermatitidis* and *Nocardia asteroides* commonly produce an acute suppurative inflammation, that is, the production of numerous polymorphonuclear leukocytes that form abscesses. *Histoplasma capsulatum* and *Coccidioides immitis* usually produce a "round cell" or monocytic cellular infiltrate that often takes on the appearance of granulomatous inflammation similar to that produced by *Mycobacterium tuberculosis.* Tuberclelike lesions, including granulomas with caseous necrosis and the accumulation of Langhans' giant cells, may be observed.

Aspergillus fumigatus and Zygomycetes (Phycomycetes) species often cause necrotizing inflammation with infarcts of organs and tissues due to their propensity to invade and thrombose blood vessels. *Cryptococcus neoformans* may not produce any observable inflammatory reaction, and the production of abundant capsular polysaccharide in the tissues at times produces a picture difficult to distinguish from mucin-secreting carcinoma. When inflammation does occur

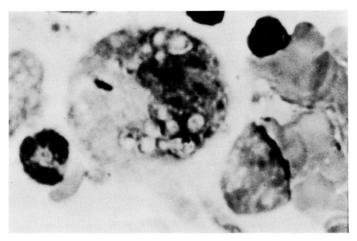

Fig. 13-1. Wright-Giemsa-strained photomicrograph of bone marrow histiocyte containing multiple pseudoencapsulated yeast forms of *Histoplasma capsulatum* (oil immersion).

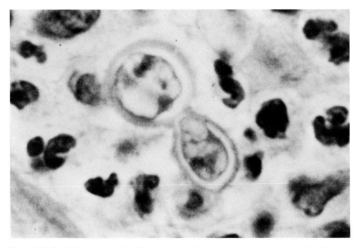

Fig. 13-2. Hematoxylin-Eosin-stained photomicrograph of *Blastomyces dermatitidis* yeast form within a purulent exudate. Note the thick wall and the single broad-based bud (oil immersion).

with *C. neoformans,* it can be acute and suppurative in some cases or chronic and granulomatous in others, indistinguishable from the reactions caused by other fungi.

Although fungal hyphae, spores, and other fungal forms are often distorted in tissues by the inflammatory response and may be difficult to identify, two basic structures can usually be distinguished: (1) a mycelial form, characterized by the presence of filamentous structures called hyphae or pseudohyphae; and (2) a yeast or "tissue" form in which only yeast cells can be distinguished.

Some fungi produce only a yeast form in tissues, others only a mycelial form. Rarely do the two forms occur together within the same organ. The dimorphous fungi, so called because they are in the mycelial form

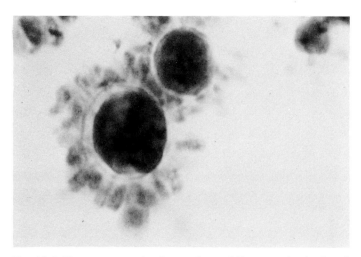

Fig. 13-3. Photomicrograph of yeast form of *Paracoccidioides brasiliensis,* showing a large central yeast cell and multiple peripheral buds, simulating a "mariner's wheel" (oil immersion).

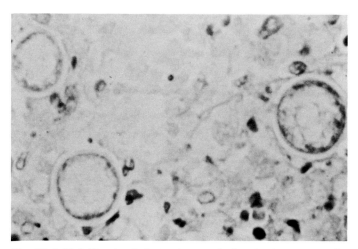

Fig. 13-4. H & E-stained photomicrograph showing immature spherules of *Coccidioides immitis,* simulating the yeast forms of *Blastomyces dermatitidis* (oil immersion).

when incubated at room temperature and in the yeast form at 37° C., are virtually always in the yeast form when seen in tissues.

Yeast forms in tissue sections can be presumptively identified by observing: (1) the size of the individual cells; (2) their arrangement and location; and (3) the number and modes of attachment of buds (blastospores).

Yeast cells that are tiny (1 to 4 μ.) suggest *Histoplasma capsulatum,* particularly if they are located intracellularly (Fig. 13-1), *Torulopsis glabrata,* the nonencapsulated yeast cells of *Cryptococcus neoformans,* or the endospores of *Coccidioides immitis.* Larger yeast cells (8 to 20 μ. or larger) suggest *Blastomyces dermatitidis* (Fig. 13-2) or *Paracoccidioides brasiliensis* (Fig. 13-3). *B. dermatitidis,* which produces only single buds with a broad base can be distinguished from *P.*

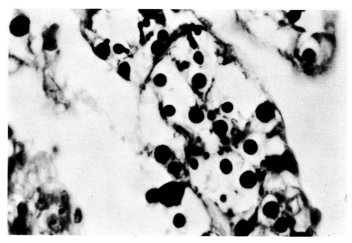

Fig. 13-5. Silver-strained photomicrograph of *Cryptococcus neoformans* showing irregular size of yeast cells and abundant capsular material (oil immersion).

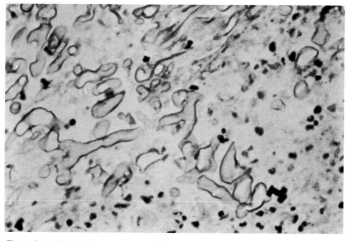

Fig. 13-6. H & E tissue section showing irregular-sized, broad, aseptate hyphal forms of one of the species of Zygomyces[Phycomyces] (oil immersion).

brasiliensis which produces multiple buds forming a structure simulating a "mariner's wheel." The spherules of *Coccidioides immitis* (Fig. 13-4) can be confused with the yeast cells of *B. dermatitidis*, particularly when they are immature and devoid of endospores. Encapsulated forms of *Cryptococcus neoformans* (Fig. 13-5), although similar in size to *B. dermatitidis*, can usually be distinguished because of the thick polysaccharide capsule and the irregular size of the yeast cells, ranging from 3 to 10 μ. in diameter.

Fungi that produce hyphae in tissues can be presumptively identified observing (1) the breadth of the hyphal strands; (2) the presence or absence of septa; and (3) the presence or absence of a brown pigmentation which suggests a member of the dark or dematiacious group of fungi.

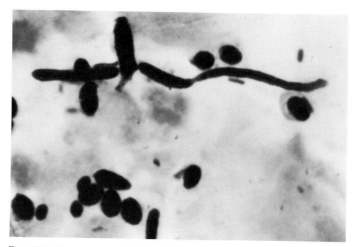

Fig. 13-7. Gram-stained photomicrograph of sputum showing pseudo-hyphae and budding blastospores characteristic of *Candida albicans* (oil immersion).

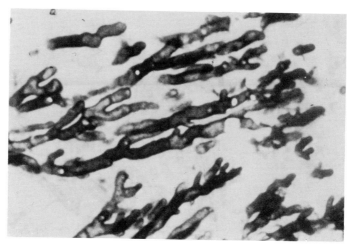

Fig. 13-8. Silver-stained photomicrograph of lung tissue showing infiltration with regular, dichotomous branching septate hyphae of Aspergillus species (oil immersion).

If the hyphae appear irregular in diameter, ranging between 10 and 60 μ., are ribbonlike, branch irregularly, and are devoid of septa (Fig. 13-6), one of the fungal species belonging to the Zygomycetes (Phycomycetes) class should be suspected. The use of silver stains, helpful in the demonstration of most fungi, may be of little value with this group of fungi because the stain does not readily penetrate the hyphae.

All other filamentous fungi produce septate hyphae in tissue sections. The observation of regular points of constriction along the hyphal strands, simulating link sausages, suggests the pseudohyphae of Candida species (Fig. 13-7). Pseudohyphae are thought to be elongated blastospores. This presumption can be further substantiated if typical budding yeasts (blastospores) are also observed.

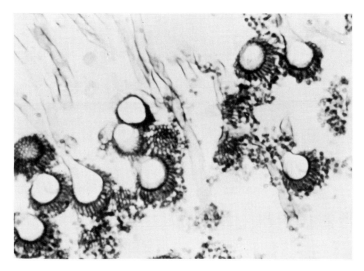

Fig. 13-9. H & E-stained photomicrograph of material from an aspergillus "fungus-ball" infection, showing the club-shaped vesicles with sporulation only from the top half of the vesicles, characteristic of *A. fumigatus* (high power).

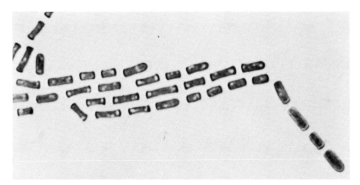

Fig. 13-10. Lactophenol (aniline) blue stain of regularly staining, rectangular arthrospores, characteristic of species of Geotrichum (oil immersion).

Species of Aspergillus, the most common fungal agents that are usually observed in tissue sections, produce septate hyphae that are uniform in diameter, ranging between 5 and 10 μ. (Fig. 13-8). These hyphae characteristically branch at 45 degree angles (dichotomous). The diagnosis of aspergillus infection can occasionally be confirmed in tissue sections if the characteristic spore-bearing vesicles are identified (Fig. 13-9). Vesicles are most commonly seen in "fungus-ball" infections, presumably because the fungal growth within the cavity is exposed to the air through connection with an open bronchus. If the vesicle is club-shaped and sporulation is seen only over the top half, a presumptive identification of *A. fumigatus* can be made.

The observation in tissue sections of septate hyphae that break into distinct thickened arthrospores (Fig. 13-10) suggests possible infection with *Geotrichum candidum.* *Coccidioides immitis* is the most commonly encountered pathogenic fungus that pro-

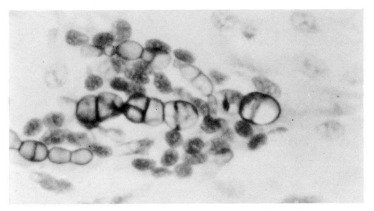

Fig. 13-11. H & E-stained photomicrograph of a subcutaneous exudate showing tiny multicelled granules characteristic of chromomycosis (oil immersion).

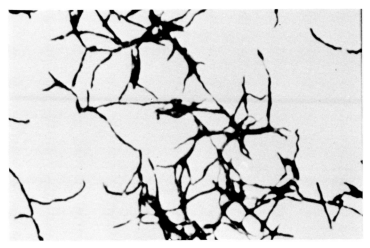

Fig. 13-12. Gram-stained photomicrograph of delicate branching filaments, characteristic of species of Actinomyces (oil immersion).

duces arthrospores; however, this form is not found in the tissues except under unusual circumstances (cavitary lesion). Species of dermatophytes may also produce arthrospores; however, these infections are limited to the superficial corneum of the skin, and tissue biopsies are rarely obtained for the diagnosis of these infections.

Septate hyphae with a distinctly dark yellow or brown appearance in tissue sections suggest infection with one of the dematiacious fungi, specifically those causing chromomycosis. The hyphae are generally short, fragmented, and distorted by the inflammatory tissue reactions. The presumptive diagnosis of chromomycosis in tissue sections can be strengthened by detecting the characteristic spherical, multicelled brown-staining sclerotic bodies (Fig. 13-11).

The presence of thin, delicate, freely branching filaments that are no more than 1 μ. in diameter is highly suspicious of infection with a species of Actinomyces, Nocardia or Streptomyces (rare occurrences; see Fig. 13-12). Nocardia species can be presumptively identified in tissue sections if the filaments are acid-fast, using 3 per cent hydrochloric acid or 1 per cent sulfuric acid as the decolorizing agent in place of acid alcohol.

It is strongly recommended that tissue sections possessing fungal structures be made available in mycology laboratories to serve as a reference with which to compare suspicious forms that may be seen in new cases and as a valuable teaching aid for self study or for instructing new students.

THE LABORATORY DIAGNOSIS OF FUNGAL INFECTIONS

Specimen Collection and Transport

Physicians, nurses, and laboratory personnel must work together to ensure that specimens optimal for the recovery of fungi are properly collected and promptly submitted to the laboratory for culture. Specimens should be transported promptly to the laboratory in sealed, sterile containers. If transport is delayed, the specimen should be refrigerated at 4° C. to prevent overgrowth by contaminating bacteria or yeasts that may also be present. It is recommended that unprocessed specimens not be shipped in the mail; however, if this cannot be prevented, 50,000 units of penicillin, 100,000 µg. of streptomycin, or 0.2 mg. of chloramphenicol can be added per ml. of specimen to inhibit overgrowth with contaminating microorganisms. Directions for proper packaging and labeling of specimens for shipping and mailing were discussed in Chapter 1.

The general principles of specimen collection discussed in Chapter 1 also apply to specimens for the culture of fungi. When obtaining specimens from cutaneous sources, it is recommended that the area to be sampled is first swabbed with 70 per cent alcohol to remove bacterial contaminants. Typical "ringworm" lesions should be sampled from the erythematous, peripheral growing margin. Infected nails should be sampled from beneath the nail plate to obtain softened material from the nail bed. If this is not possible, superficial portions of the nail should be scraped away, using a scal-pel blade to obtain subsurface material where the infective organisms are more likely to be present.

Specimen Processing

Once a specimen is received in the laboratory, it should be promptly examined and immediately inoculated to appropriate culture media.

Direct Examination

It is recommended that a direct microscopic examination be made of most specimens submitted for fungal culture. This can aid in the selection of appropriate culture media and can also provide the physician with a rapid presumptive identification.

Specimens that are fluid in consistency and relatively clear can be examined directly under the microscope after a small portion is mixed with a drop of water or saline on a glass slide. Skin scales, nail scrapings, hairs, or other materials that are thick in consistency or opaque should first be emulsified in a drop of 10% potassium hydroxide on a glass slide. The mixture can be gently warmed over the flame of a Bunsen burner, a coverslip applied, and the mixture examined after about 10 to 15 minutes, the time necessary to allow the background material to clear. Hyphae and yeast cells, which resist digestion by the potassium hydroxide, can then be seen clearly against a homogeneous background. Potassium hydroxide acts by dissolving keratin and intensifying the contrast of fungal structures with other material in this microscopic mount.

The use of India ink for the identification of the irregularly sized, encapsulated yeast cells of *C. neoformans* in cerebrospinal fluid, is one of the more commonly used procedures for direct examinations. To prepare an India ink mount, the spinal fluid is centrifuged for 10 minutes at about 3000 r.p.m. and a drop of sediment is mixed with a small drop of India ink on a microscope slide. India ink preparations may also be used to detect other fungal forms in other fluid specimens, such as pleural, synovial, peritoneal and pericardial fluids.

Table 13-2. *Presumptive Identification of Fungi Based on Direct Microscopic Examination of Material from Clinical Specimens*

Direct Microscopic Observations	Presumptive Identification
Hyphae relatively small (6–10 μ.) and regular in size, dichotomously branching at 45 degree angles with distinct cross septa	Aspergillus species
Hyphae irregular in size, ranging from 6 to 50 μ. ribbonlike, and devoid of septa	Zygomycetes (Phycomycetes) species Rhizopus-*Mucor-Absidia*
Hyphae small (2–3 μ.) and regular, some branching, with rectangular arthrospores sometimes seen; found only in skin, nail scrapings, and hair	Dermatophyte group: Microsporum species Trichophyton species Epidermophyton species
Delicate branching filaments (1 μ. or less in diameter), often contained within "sulfur granules"; gram-positive in Gram's stain. Species of Nocardia are partially acid-fast.	Actinomycetes group: Actinomyces species Nocardia species Streptomyces species
Hyphae, distinct points of constriction simulating link sausages (pseudohyphae), with budding yeast forms (blastospores) often seen	Candida species
Yeast forms, cells spherical and irregular in size (6–15 μ.), classically with a thick polysaccharide capsule (not all cells are encapsulated), with one or more buds attached by a narrow constriction	*Cryptococcus neoformans* Cryptococcus species nonencapsulated
Yeast forms, large (8–15 μ.), with cells appearing to have a thick, double-contoured wall, with a single bud attached by a broad base	*Blastomyces dermatitidis*
Large, irregularly sized (10–50 μ.), thick-walled spherules, many of which contain small (2–4 μ.), round endospores	*Coccidioides immitis*

A phase contrast microscope is a valuable adjunct in the direct examination of specimens.[26] The advantages include: (1) mounts can be made and examined quickly; (2) there is no need for direct staining; (3) the objects can be visualized with clarity. Presumptive identification of fungi on the basis of direct microscopic examination is shown in Table 13-2.

Selection of Culture Media and Inoculation of Specimens

The culture media used for the recovery of fungi from clinical specimens need not be elaborate.[24] Although the recovery rate may be somewhat enhanced by using a variety of isolation media, considerations of cost, lack of adequate storage and incubator space, and insufficient technologist time generally dictate a more conservative approach in most laboratories.

Two general types of culture media are essential for the primary recovery of fungi

from clinical specimens. One medium should be nonselective, that is, one which will permit the growth of virtually all fungal species. Sabouraud's dextrose agar is the nonselective medium most commonly used. The low *p*H serves to inhibit the growth of many contaminating bacteria that may be present in the specimen. Some mycologists prefer to use the Emmons modification of Sabouraud's agar, in which the dextrose content is decreased from 4 g. to 2 g. and the *p*H is adjusted to 6.9. This modification enhances sporulation and is particularly useful for the subculture of fungi where identification may be difficult due to poor development of the characteristic fruiting structures.

Sabouraud's Dextrose Agar	
Dextrose	40 g.
Peptone	10 g.
Agar	15 g.
Distilled water	1 liter
Final *p*H = 5.6	

The second type of medium useful for primary fungal isolation is one that is selective for fungi. This is commonly accomplished by adding antibiotics to a nonselective basal medium to inhibit bacterial contamination. The combinations of penicillin (20 units/ml.) and streptomycin (40 units/ml.) or gentamicin (5μg/ml.) and chloramphenicol (16 μg./ml.) can be used to inhibit the growth of bacteria. Cycloheximide (Actidione), in a concentration of 0.5 mg./ml. may be added to prevent the overgrowth of some of the more rapidly growing molds that may contaminate culture plates.[24] It should be noted that some of the pathogenic fungi, including *C. neoformans* and *A, fumigatus,* may also be particularly or totally inhibited by cycloheximide as well.

The antibiotic combinations may be added to a variety of culture bases, including Sabouraud's dextrose agar. The commercial products Mycosel (BBL) and Mycobiotic (Difco) agars are prepared with Sabouraud's dextrose agar with chloramphenicol and cycloheximide ("C & C"). These media are particularly useful for the recovery of dermatophytes from cutaneous specimens.

For the recovery of more fastidious dimorphous fungi such as *B. dermatitidis* or *H. capsulatum,* an enriched agar base such as brain-heart infusion must be used. Antibiotic combinations may be added to prevent overgrowth with bacteria or contaminating molds since the incubation of the plates or tubes may require one month or more. For optimal recovery of these organisms, the addition of 5 to 10 per cent sheep blood is also recommended; however, if blood is used, isolates often must be subcultured to a less enriched medium such as Sabouraud's dextrose agar or plain brain-heart infusion agar, on which characteristic sporulation is more likely to take place.

It is currently recommended that all fungal cultures be incubated at 25 to 30° C.[18,24] At one time it was also recommended that a second set of media be incubated at 35 to 37° C. in order to recover the yeast forms of the dimorphous fungi. However, it has been recognized that the recovery of these fungi is not enhanced by the higher incubation temperature, and the relatively low frequency of recovery in most laboratories of these species does not warrant the added cost and time required to process the extra media. Any mold recovered in primary culture that is suspicious of one of the dimorphous fungi can be subcultured to another plate or tube and then secondarily incubated at 37° C. to convert it to the yeast phase.

All fungal cultures should be incubated for a minimum of 30 days before discarding as negative.[24] It is important that media be held for this length of time even though some of the more rapidly growing molds may appear earlier. The recovery of one mold does not preclude the recovery of one of the slower-growing pathogenic species in the same culture dish or tube.

The decision should be made whether to use Petri dishes or culture tubes for the primary recovery of fungi. Both are satisfactory, but certain advantages and disadvantages of each should be considered before making a final decision. For low-volume laboratories, staffed with personnel who may not be familiar with the handling of fungi or where equipment may be less than optimal, the use of culture tubes is recommended. These require less space in storage, the volume of media required is less, the media are stable for a longer period of time, and the cultures can be handled more safely. Large culture tubes (150 × 25 mm., for example) with tightly fitting screwcaps are recommended. The media should be poured in thick slants to prevent dehydration during the prolonged incubation period. After inoculation, the caps should be screwed on but left lightly loosened so that the culture can "breathe."

In large-volume laboratories doing a substantial number of fungal cultures, Petri dishes are generally used. Petri dish cultures provide better aeration and examination of the culture surface, and it is considerably easier to subculture colonies in them than in culture tubes. One major advantage is that fungal colonies in mixed culture are easier to recover, identify, and subculture. A major

Table 13-3. *Processing and Inoculation of Fungal Specimens from Various Clinical Materials*

Clinical Material	Processing and Inoculation Techniques	Recommended Media
Cerebrospinal fluid	Filter 1 to 3 ml. of freshly collected cerebrospinal fluid through a 0.45 μ. Swinnex filter (Millipore Corporation) attached to a sterile syringe. Remove the filter and place it on the agar surface so that the side containing the concentrate touches the agar surface. Examine daily and move the filter pad to another location. If less than 2 ml. of sample is received, centrifuge for 10 min. and apply 1-drop aliquots of sediment to several areas on the agar surface.	Brain-heart infusion agar Chocolate agar Sabouraud's dextrose agar *Note:* media containing cycloheximide should not be used since some important fungi such as *C. neoformans* may be inhibited.
Blood[27]	Using aseptic technique, draw 10 ml. of blood from the patient and add to the blood culture bottle. The bottle should be vented throughout the duration of incubation using a sterile cotton-plugged needle. Examine daily for growth. In small laboratories, it may be preferable to inoculate 5 to 10 ml. of blood directly to the surface of appropriate agar.	Biphasic blood culture bottle containing a brain-heart infusion agar slant bathed in brain-heart infusion broth. Flood the agar surface daily with the broth by tipping the bottle gently. For plate techniques, Sabouraud's dextrose agar or brain-heart infusion agar are satisfactory.
Urine	All urine samples should be centrifuged and the sediment inoculated onto an appropriate medium. Streak the specimen over the agar surface with a loop to ensure adequate isolation of colonies.	Sabouraud's dextrose agar Brain-heart infusion agar *Note:* the addition of antibiotics (see text) is recommended because specimens are often contaminated with gram-negative bacteria.
Respiratory secretions: Sputum Bronchial washing Transtracheal aspirations	Respiratory samples that are thick, purulent, or flecked with blood are most likely to produce positive fungal cultures. The sputum grading procedure described in Chapter 1 is not applicable to the processing of specimens for fungal culture. As much of the specimen as possible should be inoculated onto the surface of an appropriate medium. Cultures should be incubated at 30° C. and examined every other day for the visual presence of growth.	Since respiratory secretions are commonly contaminated with bacteria and rapidly growing molds which may suppress the slower-growing pathogenic fungi, media containing antibiotics should be used: 1. Sabouraud's dextrose agar with chloramphenicol and cycloheximide 2. Brain-heart infusion agar with chloramphenicol and cycloheximide or gentamicin (Cycloheximide is inhibitory to some pathogenic fungi.)
Tissue, bone marrow, and body fluids	All biopsy tissue should be minced with a sharp scalpel blade before being cultured. Grinding is discouraged since some of the hyphal forms (particularly those of the Phycomyetes) may be damaged Five to 10 ml. of tissue homogenate, bone marrow sample, or fluid specimen sediment should be placed onto the surface of appropriate media. Examine cultures daily for the presence of growth.	Sabouraud's dextrose agar Brain-heart infusion agar with antibiotics

Table 13-3. *Processing and Inoculation of Fungal Specimens from Various Clinical Materials*
(Continued)

Clinical Material	Processing and Inoculation Techniques	Recommended Media
Corneal scrapings and ear cultures	As much of the specimen as possible should be inoculated onto the surface of appropriate medium. Examine cultures daily for visual evidence of growth	Mycotic keratitis and external otomycosis are most often caused by the rapidly growing saprobic molds; therefore, media used should not contain antifungal antibiotics (such as cycloheximide.)
Oral mucosa	As much of the specimen as possible should be inoculated onto the surface of an appropriate medium. Cultures should be incubated for a minimum of 30 days because *H. capsulatum* is commonly recovered from lesions of the oral mucosa.	Sabouraud's dextrose agar Brain-heart infusion agar with chloramphenicol and cycloheximide
Skin scrapings, nails, and hair	Place skin scales, nail scrapings, or hairs directly on the surface of the medium. A few fragments should be submerged beneath the surface with a straight inoculating wire to produce maximal contact with the medium. Examine periodically for visual evidence of growth and hold all cultures for a minimum of 30 days.	Sabouraud's dextrose agar with chloramphenicol and cycloheximide (Mycosel or Mycobiotic agars)

disadvantage of plates is their tendency to become dehydrated during prolonged incubation. This can be minimized by pouring sufficient agar to a depth of 6 to 8 mm., taping down the lids in at least two places to prevent inadvertent opening, placing the plates into plastic wrappers that are lightly sealed, and using an incubator with a humidity of at least 30 to 40 per cent. The placement of a flat pan of water on the bottom shelf of the incubator can serve to raise the ambient humidity.

Recommended procedures for the inoculation of material from specimens obtained from a variety of clinical sources and the media to select are listed in Table 13-3.

The Identification of Fungal Cultures

The limited space in this text allows only a brief overview of the laboratory approach to the identification of fungal species that commonly cause disease in man or are recovered with some frequency in clinical laboratories. A more complete description of the approach outlined in this text can be found in texts by Koneman and Roberts[18] and by Koneman, Roberts, and Wright.[19]

After proper specimen collection, processing, and inoculation techniques have been carried out as discussed above, the challenge remains with the microbiologist to make a prompt and accurate identification of any fungi recovered in culture. The immediate task is to determine as quickly as possible the potential pathogenicity of the fungus recovered, in particular, whether it belongs to the dimorphic group. The following are visual criteria helpful in excluding a fungal colony recovered in culture from the dimorphic group.[16]

1. *Rapid growth*, that is, development of a mature colony within 5 days. With the exception of some strains of *Coccidioides immitis* which may grow in 5 days, the dimorphic molds are slower growing, often requiring two weeks or more to develop mature colonies.

2. *The appearance of brightly colored surface pigmentation.* The rapidly growing molds commonly produce brightly colored

spores; the dimorphic fungi are white, gray, or brown, but never assume pastel hues.

3. *Production of water-soluble pigments* that diffuse into the agar, producing a brightly colored reverse surface of the colony. The dimorphic fungi may produce brownish to black discoloration of the reverse side of the colony. Other colors are not produced.

4. *Growth inhibition of agar containing antifungal agents* such as cycloheximide. The dimorphic molds are not inhibited.

The presumptive identification of a dimorphic mold must be confirmed by demonstrating conversion to the yeast form when a subculture of the colony is incubated at 37° C. on an appropriate medium. More than one transfer may be required for some strains of *B. dermatitidis* or *H. capsulatum* before complete transformation into the yeast form can be accomplished. Cottonseed medium, containing dextrose, 20 g., agar, 10 g., Pharmamedia (obtained from Traders Protein Division, Fort Worth, Texas), and distilled water, 1000 ml., has been recommended[24] to aid in the more rapid conversion to the yeast form of some of the dimorphous molds, particularly *B. dermatitidis*.

In most instances, it is not difficult to determine by visual examination of a fungal colony whether it is a mold or a yeast. This is an important initial assessment in that one of two well-defined pathways must be followed if the identification is to be rapidly made. Each of these two approaches are described below.

Identification of the Filamentous Molds

Although colonial morphology in some instances may be helpful in the identification of a filamentous mold recovered in culture, particularly of certain frequently recovered strains known to be endemic in a given area, there is enough variation depending upon the conditions of culture and type of media used that colony characteristics are unreliable. For example, Plate 13-2, *A* illustrates strikingly different colonial manifestations of the same species of Aspergillus that grow on three different types of culture media. In like manner, *Histoplasma capsulatum* often appears yeastlike when grown on a blood-enriched medium. On Sabouraud's dextrose agar or brain-heart infusion agar without blood, *H. capsulatum* presents a fluffy white or tan colony. Therefore, microscopic examination is generally required before a definitive identification can be made.

The tease mount, Scotch tape preparation, and the microslide technique are three commonly used methods for the microscopic examination of filamentous molds.[16,19] In each instance a portion of the colony is mounted in a drop of lactophenol aniline blue stain on a microscope slide. A coverslip is positioned over the drop and gently pressed to disperse the sample more evenly throughout the mounting fluid to facilitate microscopic examination.

1. *The Tease Mount.* With a pair of dissecting needles or pointed applicator sticks, dig out a small portion of the colony to be examined, including some of the subsurface agar. Place on a microscope slide in a drop of lactophenol aniline blue, tease the colony apart with the dissecting needles, and overlay with a coverslip. Examine microscopically first under the low-power (10×) objective, using the high-power objective (40×) if suspicious fungal structures are seen. Unfortunately, the delicate nature of many of the filamentous molds does not allow one to observe the characteristic spore arrangements, and often an identification cannot be made using this technique.

2. *Scotch Tape Preparation.* The Scotch tape method of preparing cultures for microscopic examination is often helpful in that the spore arrangements of the more delicate filamentous molds are better preserved. Using a nonfrosted, clear brand of Scotch tape, press the sticky side gently but firmly to the surface of the colony, picking up a portion of the aerial mycelium. Immediately place the sticky side down in a small drop of lactophenol cotton blue on a microscope slide and examine microscopically in the same manner as described above for the tease mount preparation. This method is inexpensive,

rapid, simple to perform, and with few exceptions allows one to make an accurate identification.

3. *The Microslide Culture Technique.* In those instances where neither the tease mount nor the Scotch tape preparations are able to establish an accurate identification or if permanent slide mounts are desired for future study, the microslide culture technique is recommended. Although somewhat tedious to perform, high-quality preparations in which the spore structures and arrangements are beautifully preserved can be made. The procedure is as follows:

A. Place a 1-cm. square block of Sabouraud's dextrose agar or potato dextrose agar on the surface of a sterile 3 × 1-inch microscope slide.

B. Inoculate each corner of the block with a small portion of the colony to be studied, using a straight inoculating wire or the tip of a dissecting needle.

C. Gently heat a coverslip passing it quickly through the flame of a Bunsen burner and immediately place it directly on the surface of the inoculated agar block. The gentle heating produces a firm seal with the agar, the surface of which is briefly melted by the heated glass.

D. Place the slide mount into a Petri dish on top of glass rods or applicator sticks to elevate it above the surface of a moistened filter paper disk placed in the bottom of the dish.

E. Place the lid on the dish and incubate at room temperature or 30° C. for 3 to 5 days.

F. The mount can be periodically examined under the scanning lens of a microscope to determine when sporulation is optimal to harvest the culture. When the culture is mature, the coverslip is gently lifted from the surface of the agar block. Portions of the mycelium adhere to the under surface of the coverslip.

G. Place the coverslip on a drop of lactophenol aniline blue mounting fluid. The agar block can be removed from the original microscope slide and the mycelium adhering to its surface also stained with the lactophenol aniline blue fluid. A coverslip is placed over the mount.

H. Each mount can be preserved by rimming the margins of the coverslip with mounting fluid or clear fingernail polish.

These techniques allow one to observe the microscopic morphologic features of the filamentous fungi. This need not be difficult if a logical approach is followed, based on a few preliminary observations of certain characteristics of gross colony morphology and microscopic features.

Three broad groups can be immediately separated, based on the following criteria:

1. *The Zygomycetes (Phycomycetes).* Microscopically the hyphae are broad, ribbon-like, twisted, and devoid of cross walls called septa. The colonial morphology is characteristic since growth is very rapid, completely filling the Petri dish or culture tube with a gray-white, fluffy mycelium within two or three days (Plate 13-1, *A*).

2. *The Dark or Dematiacious Saprobes.* This group can be immediately suspected because of the dark brown or black pigmentation of the colony which can be observed both on the surface and the reverse side of the colony (Plate 13-1, *B* through *F*). Microscopically the hyphae show distinct septa and have a yellow to brown intrinsic pigmentation that can be observed without staining.

3. *The Hyaline Saprobes.* The hyphae are distinctly septate but have a transparent or hyaline appearance. A variety of colonial types may be observed, depending upon the species, environmental conditions, and media used. Frequently a variety of bright colors are observed due to the production of pigmented spores (Plates 13-2 and 13-3). These features will be described below.

General Microscopic Features of Filamentous Fungi

Before discussing the specific characteristics of various species of filamentous molds, some of the general microscopic features of fungi will be reviewed.

The fundamental microscopic units of

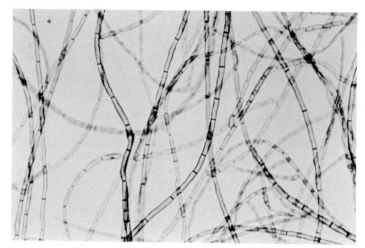

Fig. 13-13. Photomicrograph of septate hyphae (high power).

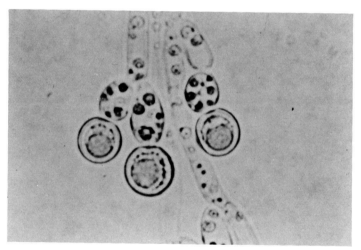

Fig. 13-14. Photomicrograph of chlamydospores (oil immersion).

fungi are the threadlike structures called "hyphae" a number of which combine to form a "mycelium." Hyphae that are subdivided into individual cells by transverse walls or septa are septate (Fig. 13-13); those missing walls are aseptate. The portion of the mycelium that extends into the substrate or culture medium is the vegetative mycelium; the portion that projects above the substrate is the aerial or the reproductive mycelium, in which spores are often produced.

The identification and classification of fungi is based primarily on the morphologic differences in their reproductive structures. Fungi reproduce by the production of spores which are borne within a variety of specialized structures called "fruiting bodies." Three general types of reproduction or spore formation may be observed: (1) vegetative; (2) aerial (conidial); and (3) sexual.

Vegetative Sporulation. Three types of spores may form directly from the vegetative

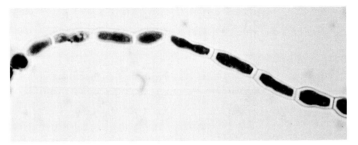

Fig. 13-15. Photomicrograph of arthrospores (high power).

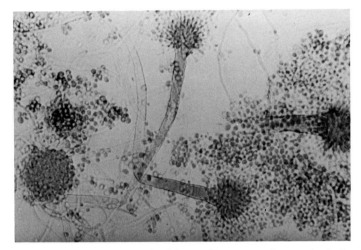

Fig. 13-16. Photomicrograph of Aspergillus species showing both the asexual conidial-bearing fruiting heads and a sexually derived cleisto-thecium (left margin of photograph, high power).

mycelium, blastospores; chlamydospores; and arthrospores. Blastospores are the familiar budding forms characteristically produced by yeasts (Fig. 13-7). Chlamydospores are spherical swellings within the hyphae that become surrounded by a thick wall (Fig. 13-14). This type of sporulation is characteristically exhibited by *Candida albicans*. Arthrospores are thickened segments of septate hyphae that become detached at their points of septation (Fig. 13-15). This type of sporulation is characteristic of *Coccidioides immitis* and *Geotrichum* species.

Aerial Sporulation. Aerial sporulation is generally more elaborate and spores are borne within a variety of fruiting bodies. The formation of enclosed saclike structures called sporangia within which are produced sporangiospores is characteristic of the Zygomycetes. Spores produced directly from the surface of a variety of fruiting structures are called "conidia" (Latin for "dust"). Conidia may be borne singly, in chains, or in clusters from special hyphal segments called condidiophores. Tiny conidia (1 to 2 units), usually single-celled, are called "microconidia" (or "microaleuriospores"). Larger, multicelled spores are called "macroconidia" (or "macroaleuriospores").

Sexual reproduction is not a common feature of those fungi of medical importance that are most commonly encountered in clinical laboratories. Fungi that reproduce sexually are called "perfect fungi," while those

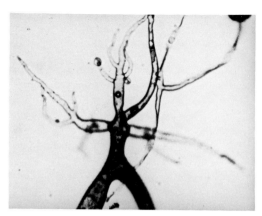

Fig. 13-17. Photomicrograph of Rhizopus sp. illustrating a rhizoid (oil immersion).

species in which a sexual stage of reproduction has not been demonstrated are classified with the *Fungi imperfecti.*

The species of the genus Aspergillus, one of the more commonly encountered molds in the laboratory, are capable of sexual reproduction, particularly members of the *A. glaucus* and the *A. nidulans* groups (Fig. 13-16). The sexually derived reproductive structure most commonly observed is the large baglike cleistothecium (closed) or perithecium (small opening) within which are produced ascospores. It is believed by most mycologists that virtually all fungi have a sexual stage, even though this has not been demonstrated in culture for most fungi of medical importance. One example is the fungus *Allescheria (Petriellidium) boydii,* one of the causes of subcutaneous mycetoma in man, which is now known to be the perfect stage of *Monosporium apiospermum,* formerly considered to be a distinct species within the *Fungi imperfecti.*

This brief orientation will serve as a point of reference for discussing individual groups of fungi in the following paragraphs, the classification of which is made on the basis of specific differences in the structure of fruiting bodies and modes of sporulation.

Features of the Zygomycetes

The general microscopic features of the Zygomycetes include aseptate, broad, ribbonlike hyphae, the formation of special saclike fruiting structures called sporangia within which are borne small (2 to 3 μ.) spherical yellow-brown sporangiospores, and the formation of rudimentary rootlike structures called rhizoids (Fig. 13-17). Three genera, Rhizopus, Mucor, and Absidia, cause most human infections and should be distinguished in the laboratory. Three other genera, Syncephalastrum, Circinella, and Cunninghamella, are rarely encountered as laboratory "contaminants" and are not of clinical interest.

Both Rhizopus and Absidia produce rhizoids; however, these two genera can be distinguished by observing the relationship between the rhizoids and the derivation of the sporangiophores, the specialized hyphal extensions that support the sporangia. With Rhizopus, the sporangiophores are derived immediately adjacent to the rhizoids (Fig. 13-18), a so-called "nodal" derivation, while with Absidia the sporangiophores are derived intranodally from hyphal segments between the rhizoids (Fig. 13-19). Mucor species do not form rhizoids (Fig. 13-20).

Any one of these three genera may cause zygomycosis (phycomycosis, mucormycosis) in humans with compromised host resistance, particularly in patients with diabetes whose disease is out of control. The disease commonly begins as an infection of the nasal sinus, with a propensity to invade the ocular orbit or extend deeply into the meninges or brain. Pulmonary infections have been reported in immunosuppressed hosts, and the disseminated form is often fatal.

Features of the Dematiacious Saprobes

The dematiacious saprobes can be suspected in culture by observing the dark gray, brown, or black hairy or velvety colonies which also pigment the reverse side. Young colonies are often yeastlike in consistency before the low aerial mycelium begins to develop. Presumptive identification on the basis of colony morphology can be confirmed by microscopically demonstrating the dark yellow or brown septate hyphae. The following is a list of the dematiacious saprobes more commonly encountered in clinical laboratories:

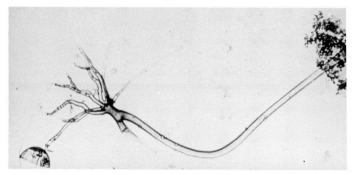

Fig. 13-18. Photomicrograph of Rhizopus species illustrating the characteristic nodal derivation of the conidiophore (high power).

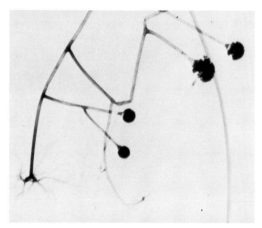

Fig. 13-19. Photomicrograph of Absidia species illustrating internodal derivation of the conidiophores (high power).

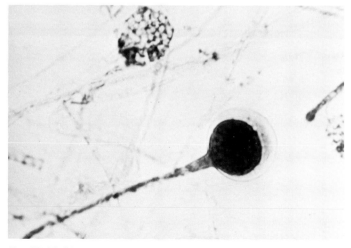

Fig. 13-20. Photomicrograph of Mucor species illustrating only sporangia and absence of rhizoids (oil immersion).

Table 13-4. *Cultural Features of the Dematiacious Molds*

Genus	Colonial Morphology	Microscopic Features	Illustration
Curvularia	Dense, cottony, well-developed aerial mycelium. Initially gray-white, soon turning dark brown to red purple. Margins entire and sharply demarcated. Reverse is red-purple to black.	Hyphae distinctly septate and yellow-brown. Conidiophores twisted and roughened at points of conidial attachments. Dark brown macroconidia are divided into 4 to 6 cells by transverse septa having a curved or boomerang appearance.	
Helminthosporium	Colony is similar in appearance to Curvularia.	Hyphae distinctly septate and yellow-brown. Conidiophores twisted and roughened at points of conidial attachments. Elongated, cylindrical, smooth-walled, dark brown macroconidia divided into many cells by thick transverse septa. In direct mounts, macroconidia often appear vacuolated.	
Heterosporium	There are two colonial types: 1. Colony similar in appearance to Curvularia 2. Low velvety mycelium with a light gray to gray-brown coloration	Conidiophores similar to those of Helminthosporium, with roughening at points of conidial attachments. Conidia are oval to elliptical, divided into 3 to 5 cells by transverse septa, and when mature are covered by fine hairlike echinulations simulating cocoons.	
Alternaria	Colony is similar in appearance to Curvularia.	Hyphae distinctly septate and yellow-brown. Macroconidia are dark brown, multicelled, with septa both transverse and longitudinal, drumstick or beak-shaped, arranged in tandem in long chains.	
Stemphylium	Colonies spreading and covered with a low, well-developed aerial mycelium. Gray-white at onset with development of irregular, varigated dark brown to black pigmentation. Reverse of colony is dark brown to black.	Hyphae distinctly septate and yellow-brown. Conidiophores are often very short, bearing single, large, multicellular macrocondia, oval or round, divided by transverse and longitudinal septa.	

Table 13-4. *Cultural Features of the Dematiacious Molds (Continued)*

Genus	Colonial Morphology	Microscopic Features	Illustration
Epicoccum	Colonies spreading but retain a distinct, serpiginous border. The aerial mycelium is well developed, presenting a cottony surface which develops a play of colors with maturity, including black, yellow, orange, red, and brown.	Hyphae distinctly septate and yellow-brown. Irregularly sized, spherical to club-shaped macrocondia are borne in clusters directly from the hyphae and are divided into multiple cells by both transverse and longitudinal septa.	
Nigrospora	Colonies spreading, gray-white, and covered by a well-developed fluffy mycelium. Darkening occurs only with maturity.	Hyphae initially hyaline and septate. Yellow-brown pigmentation occurs only with age. Conidiophores are short, somewhat helical, with a swollen urnlike tip within which are borne large, subspherical jet-black conidia, appearing as miniature cockhats.	
Cladosporium	Colonial types varying from deep brown to black, smooth, leathery, and rugose, to velvety, deep green variant covered by a low, hairlike mycelium. Early colonies may be smooth and yeastlike in nature.	Hyphae distinctly septate, yellow-brown. Conidiophores are freely branching, having the appearance of a brush from the tips of which are borne long chains of small, dark, yellow-brown oval or elliptical conidia.	
Aureobasidium (Pullularia)	Colonies grow slowly and are initially white to gray, yeastlike and glabrous, turning dark brown to jet black with age. Aerial mycelium never develops unless the colony becomes sterile.	Hyphae are broad, separated into distinct segments by thick-walled septa simulating arthrospores, giving rise to myriads of tiny elliptical non-pigmented microconidia.	

Curvularia	Stemphylium
Helminthosporium	Nigrospora
Heterosporium	Cladosporium
Alternaria	Aureobasidium

Selected colonies are illustrated in Plate 13-1, *C* through *F*. The colonial and microscopic features for identifying these dematiacious molds are listed in Table 13-4. These fungi rarely cause invasive disease in humans, although they are thought to be associated with allergic upper and lower respiratory diseases. Slower-growing counterparts, belonging to the Phialophora, Fonsecaea, and Cladosporium genera, cause chromomycosis and some forms of mycetoma in humans, diseases rarely encountered in the United States. A discussion of these diseases is beyond the scope of this text; however, a recent review has been reported by Roberts.[23]

Table 13-5. *Characteristics of Three Species of Aspergillus*

Species	Colonial Morphology	Microscopic Features	Illustration
Aspergillus fumigatus	Mature colonies have a distinct margin and are some shade of green, blue-green, or green-brown. Surface has a powdery or granular appearance from profuse production of pigmented spores. A white apron usually is seen at the edge in the zone of active growth.	Hyphae are hyaline and distinctly septate. Conidiophores are long, terminating in a large club-shaped vesicle. Chains of 2- to 3-μ. spherical conidia are borne from a single row of sterigmata that are produced only from the top half of the vesicle surface.	
Aspergillus niger	Colonies are initially covered with a white, fluffy, aerial mycelium. As colony matures, a salt-and-pepper effect is noted, with the surface ultimately covered with black spores. The reverse of the colony remains a light tan or buff color, which separates A. niger from the dematiacious molds.	Hyphae are hyaline and distinctly septate. Conidiophores are long and vesicle is usually not seen because it is covered with a thick ball of spores that are derived from the entire surface. Where vesicles can be seen, they have a concave undersurface simulating a mushroom. Spores are 2 to 3μ. spherical, and black.	
Aspergillus flavus	Colonies have a distinct margin, are covered by a fluffy, well-developed aerial mycelium, and when mature have a yellow or yellow-brown color.	Spherical 2- to 3-μ. spores are borne in short chains from the entire circumference of the vesicle. Vesicles are spherical and give rise to a double row of sterigmata from which the spores are borne. Hyphae are hyaline and distinctly septate.	

Features of Hyaline Saprobes

Included in the hyaline saprobes are the following genera:

Aspergillus	Gliocladium
Penicillium	Trichoderma
Paecilomyces	Cephalosporium
Scopulariopsis	Fusarium

These eight fungi account for more than 95 per cent of the hyaline saprobes that will be encountered in clinical laboratories. The colonial morphology may vary considerably, depending upon environmental conditions and the culture media used. The colonies are generally rapidly growing and mature within 5 days, generally forming a distinct border and often a brightly pigmented surface, ranging in pastel shades of lavender, blue, green, yellow, and brown (Plates 13-2 and 13-3).

Of these hyaline saprobes, species of Aspergillus are not only the most frequently encountered fungi in the laboratory but also are the most significant clinically. Although over 700 species have been recognized by Raper and Fennell,[20] only three are of diagnostic medical significance and need to be specifically identified in the laboratory: *Aspergillus fumigatus, Aspergillus flavus*, and *Aspergillus niger*.

Aspergillus fumigatus is the species most

Color Plates
13-1 *to* 13-6

Plate 13-1
The Dematiaceous Molds

The dematiaceous (dark) molds are characterized by the development of deep green, brown, or black colonies, with black pigmentation of the reverse surface. Most species are rapidly growing (mature colonies within 5 days), although some of the pathogenic species causing mycetomas and chromomycosis may require two weeks or more for growth. Representative colonies of the dematiaceous molds growing on Sabouraud's dextrose agar are illustrated in this plate.

A and B.
Diffuse growth over the agar surface with a dark, fluffy mold representative of one of the Zygomyces (Phycomyces) species (Frame A). Although the surface of the colony appears dark, simulating one of the dematiaceous molds, it cannot be one of these molds because of the lack of black pigmentation on the reverse side of the colony (Frame B).

C and D.
Colony of Helminthosporium species illustrating the surface-black pigmentation of the mycelium (Frame C). Note the deep black pigmentation of the reverse side of the colony (Frame D).

E.
Olive green, granular, rugose colony of Cladosporium species, one of the saprobic, dematiaceous molds more commonly encountered in the clinical laboratory.

F.
Black colony of *Cladosporium carrionii*, one of the slower growing, pathogenic molds causing chromomycosis.

G.
Gray colony of Heterosporium species, a dematiaceious mold closely related to Cladosporium species.

H.
Epicoccum species, illustrating the characteristic variegated play of yellow, orange, and black colors within different portions of the mycelium.

I.
Aureobasidium pullulans, illustrating the flat, yeast-like colony with a late growth of a low white mycelium located centrally. *Aureobasidium pullulans* should be initially considered when a black, yeast-like colony is recovered in cultures. It should be remembered that some strains of the Phialophora group may also produce colonies that initially are yeast-like in appearance.

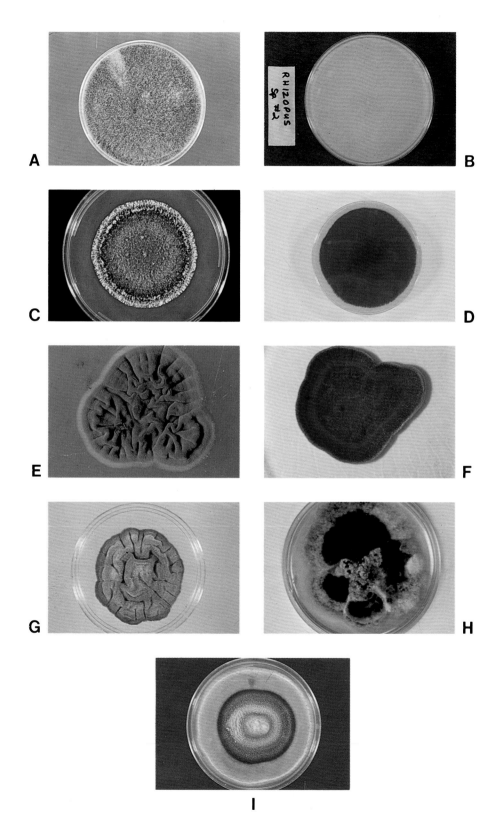

Plate 13-2
Identification of *Asperigillus fumigatus*,
Aspergillus flavus, and *Aspergillus niger*

The Aspergilli, of which there are over 700 identified species, are among the most commonly encountered molds in the clinical laboratory. In most instances they grow in culture media as contaminants; however, three species, *A. fumigatus*, *A. flavus*, and *A. niger* may be the agents of primary mycotic diseases in man, particularly in immunosuppressed or compromised hosts. It is desirable that at least these three species of Aspergillus be recognized in the clinical microbiology laboratory.

A.

Photograph of Sabouraud's dextrose agar (upper left), 20 per cent maltose agar (upper right) and Czapek's agar (lower) inoculated simultaneously with the same strain of Aspergillus species, illustrating the differences in growth rate and colonial appearance that occurs when different media are used. When describing a "classic" mold colony, the type of media employed and conditions of incubation must always be cited.

B,* C and D.

Cultural variants of *Aspergillus fumigatus*. Most strains produce green, green-brown, or green-blue colonies, as illustrated in these frames. Rugal folds often develop. The white apron commonly seen at the periphery of the colony (Frames *B* and *C*) is the zone of active growth, with the pigmented spores developing behind the sterile hyphae at the colony margin. Note differences in colonial morphology on Czapek's (left) and maltose (right) agars.

E, F* and G.

Colonial variants of *Aspergillus flavus*. As the species name implies, *A. flavus* forms colonies that are some shade of yellow; however, as illustrated in Frame *G* (photograph of Czapek's and 20 per cent maltose agar plates), *A. flavus* can appear green and granular, closely simulating *A. fumigatus*. Microscopic study must always be carried out to confirm presumptive identification of colonies.

H.*

Appearance of *Aspergillus niger*, illustrating the dense salt and pepper effect from profuse proliferation of black spores. This surface appearance may suggest one of the dematiaceous molds; however, the reverse of *A. niger* colonies is light buff and never black.

*Reproduced from Dolan, C. T., *et. al.* Atlas of Clinical Mycology. Chicago, American Society of Clinical Pathologists, 1976.

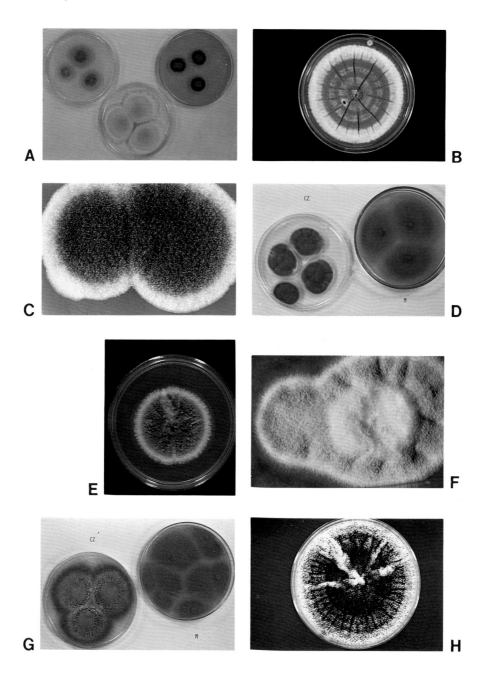

Plate 13-3
Identification of the Hyaline Molds

The hyaline molds, so called because they produce a mycelium composed of transparent hyphae without dark pigmentation, include a number of species that are commonly encountered as contaminants in the clinical laboratory. They are further characterized by relatively rapid growth of colonies in culture (mature colonies in 3 to 7 days) that develop a variety of colors due to the production of pigmented spores. In rare instances one of these species may cause mycotic disease in compromised humans.

A, B and C.
Colonial variants of Penicillium species. Most strains are some shade of green and simulate the colonies of *Aspergillus fumigatus.* Yellow or brown varients may be encountered, as illustrated in Frame C. The surface of the colony is quite granular due to the dense proliferation of spores, and radial rugal folds are generally present.

D.
Green, granular colony of Paecilomyces species, a mold closely resembling Penicillium species, both in its colonial and microscopic morphology.

E* and F.
Colonial variants of Scopulariopsis species. This hyaline mold produces colonies that are always some shade of buff or brown, green or blue pastels do not develop. The colony surface is extremely granular due to dense spore production, and irregular rugal folds are usually produced.

G.*
Green, granular colony without a margin covering the surface of the agar as a "lawn." This colonial appearance is suggestive of either Gliocladium species or Trichoderma species and microscopic study of the fruiting structures is required to make a definitive identification.

H.
Cephalosporium species. The colony is not distinctive. Many strains of Cephalosporium species produce light green, blue, and yellow pastel colors, although off-white variants, as seen here, are also common. Due to the delicate nature of the aerial mycelium, Cephalosporium species usually develop colonies with a low, flat mycelium, appearing almost yeast-like in nature, as illustrated in this photograph.

I.
Fusarium species, illustrating the classic fluffy mycelium with a deep pigmentation ranging from rose red, to lavender, to deep purple. Fusarium species are one of the more common causes of mycotic keratitis in humans. This species has also been incriminated in aflotoxin disease.

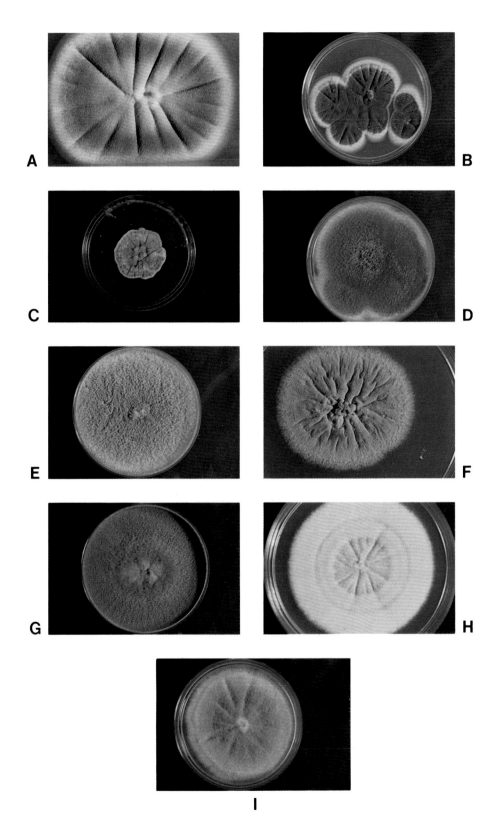

Plate 13-4
Identification of the Dermatophytes

The colonies representative of the dermatophytic molds on artificial culture media are not distinctive. There is sufficient strain variation and differences in appearance of the same strain, depending on the type of culture medium used and environmental conditions during incubation, that colonial morphology is not a reliable criterion for identification. Microscopic studies are almost always necessary before a definitive classification can be made. The photographs on this plate are representative of several of the more commonly encountered dermatophytes; however, the strain selected in any given photograph does not necessarily reflect the marked variation that may occur under different cultural conditions.

A.
Colony of *Microsporum audouinii*, illustrating the delicate velvety aerial mycelium and gray-to-buff pigmentation of the colony. *M. audouinii* grows more slowly than the other species of dermatophytes and does not grow on rice medium at all, two helpful characteristics in the identification of colonies of this species.

B and C.
Surface (Frame *B*) and reverse (Frame *C*) views of *Microsporum canis* colonies, illustrating the yellow-orange pigmentation commonly produced by this species. a lemon-yellow apron at the growing peripheral margin of the colony is a helpful identification feature.

D.
Colony of *Microsporum gypseum*, illustrating the granular surface due to the dense production of macroaleuriospores and the characteristic cinnamon-brown pigmentation. The irregular, fluffy margin, although not distinctive, is frequently produced by *M. gypseum*, as illustrated in this photograph.

E and F.
Granular (Frame *E*) and fluffy (Frame *F*) colonies of *Trichophyton mentagrophytes*. Although there are many strain variants of *T. mentagrophytes*, they are broadly grouped into granular and fluffy types, as illustrated here. The colonies are not distinctive and further microscopic and biochemical studies are required before a species identification can be made.

G and H.
Surface (Frame *G*) and reverse (Frame *H*) views of colonies of *Trichophyton rubrum*. As with *T. mentagrophytes*, both granular and fluffy colonial variants occur. The deep red pigmentation (Frame *H*) produced in the reverse side of the colony, particularly when seen with colonies growing on cornmeal agar, is a helpful characteristic in the identification of *T. rubrum*, although some strains of *T. mentagrophytes* can produce red pigment as well. Further microscopic and biochemical studies are required to make a definitive species identification.

I.
Young colony of *Epidermophyton floccosum*, illustrating the yellow-khaki color of the surface mat. As the colony matures it becomes covered with a floccose aerial mycelium. Microscopic studies are required to make a definitive identification.

J.
Colony of *Trichophyton tonsurans*, illustrating the somewhat creamy-tan mat with the characteristic radial rugal folds. The colony grows somewhat more slowly than either *T. mentagrophytes* or *T. rubrum*. Nutritional studies utilizing selective trichophyton agars supplemented with thiamine, nicotinic acid, inositol, and other enrichments are helpful in the identification of *T. tonsurans*. This species grows poorly on agar containing only casein (Trichophyton #1 agar) but grows well in

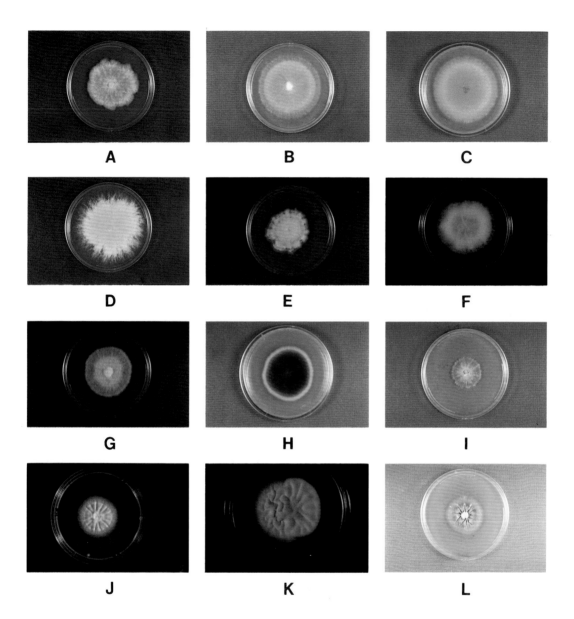

A B C

D E F

G H I

J K L

agar containing casein and thiamine (Trichophyton #4 agar).

K.
Trichophyton verrucosum, illustrating a colony not unlike that produced by *T. tonsurans.* The colonies of *T. verrucosum* are also slow growing. Most strains grow on casein agar (Trichophyton #1 agar), although some strains require both thiamine and inositol for growth.

L.
Trichophyton violaceum, illustrating the small, waxy colony with a violet-orange pigmentation. The colony is extremely slow growing and the waxy appearance is due to the inability of this species to produce an aerial mycelium or to sporulate in culture medium. This species is rarely encountered in the United States and is the cause of *tinea capitis* (favus) infections in Northern Europe and Scandanavia.

Plate 13-5
Identification of the Dimorphic Fungi

The dimorphic fungi, so called because they grow in a mold form at 25⁰ C. (room temperature) and in a yeast form at 37⁰ C., are pathogenic in man, and are the cause of the deep-seated mycoses. Although the definitive species identification depends upon microscopic study, confirmation that a given species belongs to the dimorphic group is dependent upon demonstrating both the mold and yeast forms in the laboratory.

A.*
Two tubes of brain heart infusion agar; the one on the left contains cycloheximide and chloramphenicol, and the one on the right is free of antibiotics. Both tubes of media support the growth of a fluffy, gray-white mold. The ability of a fungus to grow in the presence of cycloheximide is presumptive evidence that it belongs to the dimorphic group. Rapidly growing saprobes are inhibited in this antibiotic medium. The cultures shown here are those of *Coccidioides immitis.*

B.*
Sabouraud's dextrose agar plate with the cottony, white mold form of *C. immitis.* Laboratory workers should be wary of cottony molds such as illustrated here, particularly if their growth is delayed (5 to 10 days). The mold form of *C. immitis* is highly infectious and handling of these cultures must be carried out under a properly functioning bacteriological safety hood.

C.*
Brain heart infusion agar plate with a colony of *Histoplasma capsulatum,* illustrating the delicate,

silky nature of the mycelium. Note the tendency for the colony to turn a gray or tan color upon maturity, as seen in the central portion.

D and E.*
Colonies of *Blastomyces dermatitidis,* illustrating dimorphism with both the mold and yeast forms visible. Frame *E* in particular shows the incomplete conversion to the yeast form as a residual fluffy white mycelium is still observed on top of the smooth, yellow yeast colonies, even after incubation at 37⁰ C. for several days.

F.*
Colonies of a dimorphic mold illustrating the prickly stage of yeast conversion. Both *H. capsulatum* and *B. dermatitidis* colonies may show this prickly stage in the process of yeast conversion.

G and H.*
Colonial variants of yeast forms of the dimorphic fungi. Frame *G* is the yeast form of *B. dermatitidis* after complete conversion following incubation at 37⁰ C. for several days. Frame *H* is the yeast colony of *Sporothrix schenckii.*

I.
Tubes of brain heart infusion agar, one containing cycloheximide with the yeast (top) and mold (bottom) forms of *S. schenckii.* Because of the delicate nature of the hyphae and fruiting structures of this organism, the mold colony may also appear yeast-like. The mold colony tends to darken with maturity, as illustrated in the bottom tube of this photograph.

*Reproduced from Dolan, C. T., *et. al.:* Atlas of Clinical Mycology. Chicago, American Society of Clinical Pathologists, 1976.

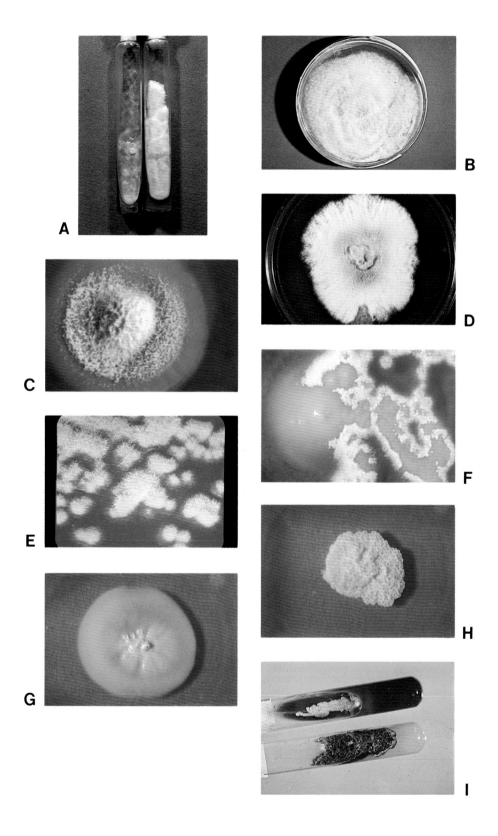

Plate 13-6 (Part I)
Identification of Yeasts

The laboratory identification of medically important yeasts involves the visual assessment of colonial characteristics, the microscopic study of the type and arrangement of blastospores of colonies growing on cornmeal agar, as outlined in Table 13-10, and the interpretation of the results of carbohydrate fermentation and assimilation tests and other biochemical reactions. Select identification characteristics outlined in Figure 13-22 are illustrated in the frames in this plate.

A.*
Fermentation studies illustrating four carbohydrate broths that have been inoculated with the yeast species to be identified. The yellow color in the dextrose and maltose broth media indicates acid production from these carbohydrates as compared to the red color in the tubes showing no acid formation. The accumulation of gas within the Durham vials in the dextrose and maltose tubes is the endpoint of yeast fermentation.

B.
Carbohydrate assimilation studies illustrating a yeast nitrogen base agar plate on which have been placed multiple filter paper disks impregnated with various 1 per cent carbohydrate solutions. The endpoint of the assimilation of a given carbohydrate by the yeast is the visual presence of colonial growth around the disk, as illustrated in this photograph.

C.
Sabouraud's dextrose agar plate on which are growing golden yellow, somewhat mucoid colonies of Rhodotorula species. Some strains of Cryptococcus species may also produce a yellow pigmentation, and further fermentation or assimilation studies are required before a definitive identification can be made.

D.
Surface of a niger seed (birdseed) agar plate with colonies of *Cryptococcus neoformans*. The maroon-red pigmentation of colonies on this medium is characteristic of *C. neoformans*, while other species of Cryptococcus are not capable of producing the pigmentation.

E.*
Photomicrograph of an acid fast stain performed on a smear preparation made from a colony of Saccharomyces species. Note the deep red color of the ascospores, diagnostic of this genus of yeasts, and the blue-staining blastospores in the background.

*Reproduced from Dolan, C. T., et., al.: Atlas of Clinical Mycology. Chicago, American Society of Clinical Pathologists, 1976.

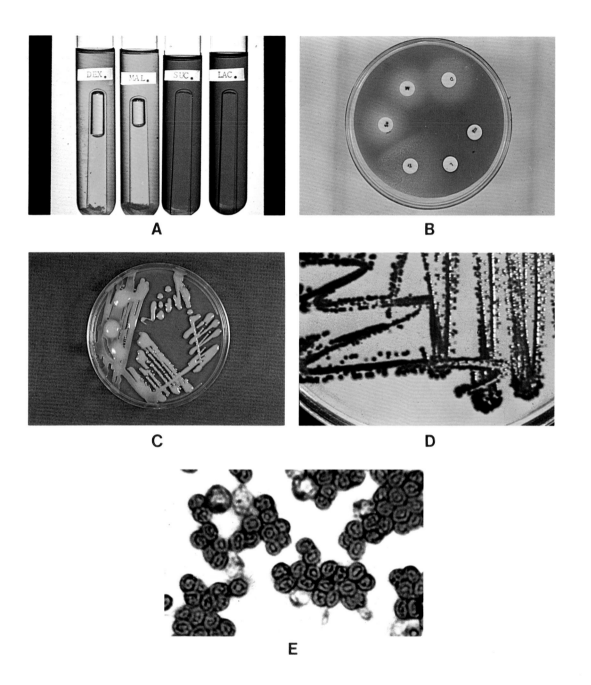

A

B

C

D

E

Plate 13-6 (Part II)
Identification of Yeasts

A, B and C.
Photographs of the API 20C Yeast System.* The test strips include a series of plastic, reagent, or media-containing cupulae in which can be performed a variety of yeast carbohydrate assimilation and fermentation studies. The portion of the strip illustrated in Frame *B* is designed for the determination of yeast fermentations (positive reactions are indicated by the production of acid *and* gas within the cupula) and the portion illustrated in Frame *C* is utilized for the determination of yeast assimilations. In the assimilation studies, a positive reaction is determined by visually observing the turbidity of the yeast suspension against the background horizontal brown lines within each of the cupulae.

D.
Photograph of a Corning Uni-Yeast-Tek "wheel" composed of a central chamber containing cornmeal agar on which a microscopic examination of the yeast colony can be made surrounded by 11 triangular chambers. Ten biochemical characteristics can be determined, based on the development of diagnostic color reactions. One of the reaction chambers serves as a growth control.

Plate 13-6 (Part II)
Identification of the Aerobic Actinomycetes

E.
Photomicrograph of a smear preparation stained with the modified acid-fast technique illustrating delicate, branching, acid-fast filaments characteristic of Nocardia species.

F.
Sabouraud's dextrose agar plate with a wartlike, brittle, yellow colony of *Nocardia asteroides*. Most strains of this species produce colonies with yellow or organge pigmentation, although chalky-white variants may also be encountered. The detection of a pungent earthy odor is an additional characteristic helpful in the presumptive identification of the aerobic actinomycetes.

G.
Chalky-white, folded, brittle-appearing colony of Streptomyces species. Most strains of Streptomyces produce gray or white colonies; however, some variants are yellow and further studies may be required to differentiate species of Streptomyces from Nocardia.

*This API 20C Yeast System has recently been revised to include only carbohydrate assimilation tests.

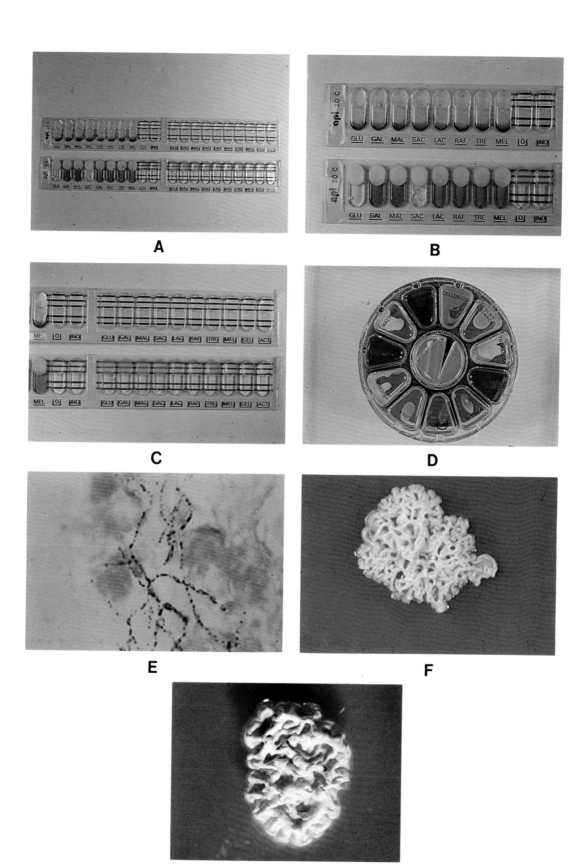

A

B

C

D

E

F

G

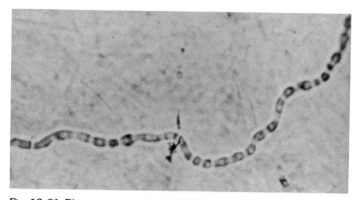

Fig. 13-21. Photomicrograph of a KOH preparation of skin scales illustrating a hyphal segment of one of the dermatophytic fungi. Note that the hyphal fragment is breaking up into tiny arthrospores (oil immersion).

likely to cause invasive pulmonary or disseminated aspergillosis in man, particularly in compromised hosts. *A. fumigatus* and *A. flavus* are most commonly associated with allergic bronchopulmonary disease while *A. niger* may be associated with fungus-ball infections of the nasal sinuses or lung, and is the agent most commonly recovered from cases of otitis externa ("swimmers ear").[8] Typical colonies of these species of Aspergillus are illustrated in Plate 13-2. The colonial and microscopic features by which these species can be distinguished in laboratory cultures are listed in Table 13-5.

The distinguishing laboratory characteristics of the other hyaline saprobes listed above are reviewed in Table 13-6. The colonies are usually pigmented with a variety of pastel hues as the growth matures and spores are produced. Representative colonies are illustrated in Plate 13-3. Microscopically, Penicillium, Paecilomyces, and Scopulariopsis can be distinguished because they form brushlike phialides with conidia borne in long chains, in contrast to Gliocladium, Trichoderma, and Cephalosporium which bear conidia in compact clusters. Fusarium is distinctive by the production of large, sickle-shaped, multicelled macroconidia. These colonial and microscopic features are outlined in Table 13-6.

The hyaline saprobes are rarely incriminated as the primary agents of human disease. These infrequent mycoses have been reviewed by Rippon[22]. Mycotic keratitis is one of the more common infections caused by this group of fungi, and members of the genus Fusarium are most frequently involved.

Laboratory Identification of the Dermatophytic Molds

The dermatophytes are a distinct group of fungi that infect the skin, hair, and nails of man and animals, producing a variety of cutaneous diseases colloquially known as "ringworm." The term "tinea" (Latin, "grub" or "worm") also refers to these diseases. Tinea barbae, tinea corporis, tinea cruris, and tinea pedis (athlete's foot) designate specific skin sites that are involved.

With the advent of Griseofulvin and topical antifungal compounds, the laboratory identification of the dermatophytes is less frequently required than previously. The clinical suspicion of one of the tinea infections can be confirmed by observing the delicate hyphae, often forming arthrospores, in a KOH mount of skin scales or nail scrapings (Fig 13-21). Therapy can then be instituted without obtaining a culture.

Table 13-6.   *Characteristics of the Hyaline Saprobes*

Genus	Colonial Morphology	Microscopic Features	Illustration
Penicillium	Colony is initially white and fluffy, soon turning shades of green or green-blue as pigmented spores are produced. Yellow or tan variants are occasionally seen. Radial rugae are often formed.	Hyphae are hyaline and septate. Conidiophores give rise to branching phialides forming a brush or "penicillus." Spherical or Oval 1- to 2-μ. conidia are borne in long chains from sterigmata, the tips of which are blunt and appear cut off at right angles.	
Paecilomyces	Colonies are usually powdery or granular and develop light pastel, yellow-green, green-blue, or buff as spores are produced. Margins are often not distinct.	Hyphae are hyaline and septate. Conidiophores branch freely into a brush-like structure. Oval 1- to 2-μ. conidia are borne in chains from the tips of sterigmata that are long and tapering.	
Scopulariopsis	Colonies are characteristically powdery, buff to brown in color, and develop shallow radial grooves.	Hyphae are hyaline and septate. Conidiophores branch to form penicillus; 3- to 4-μ. conidia are borne in chains. Conidia are lemon-shaped and with age develop surface echinulations.	
Gliocladium	Colonies develop diffusely over the surface as a green granular lawn. A distinct margin does not form.	Hyphae are hyaline and septate. Conidiophores branch into a brushlike structure; 2- to 3-μ. conidia are borne in clusters which obscure the tips of the sterigmata.	

Any mold recovered in culture from skin, nail, or hair samples should be suspected of being one of the dermatophyte species. With few exceptions, to be described below, the appearance of the colonies is not a reliable characteristic for making a species identification. Representative colonies are illustrated in Plate 13-4, *A* through *L*. A microscopic evaluation of the types and arrangements of macro- and microaleuriospores is required.

Microscopically, one of the dermatophyte species can be suspected when delicate, hyaline, septate hyphae are observed, accompanied by a variety of vegetative structures such as chlamydospores, favic chandeliers, pectinate bodies, and nodular organs.

These vegetative structures are nonspecific, and identification must be based on conidial morphology.

There are three genera of dermatophytes that require recognition: Microsporum, Trichophyton, and Epidermophyton. An unknown dermatophyte can be assigned to one of these genera by looking at the following characteristics.[3,17,19]:

1. *Genus Microsporum.* Production of large, multiseptate, rough-walled macroaleuriospores (Table 13-7). They are usually cylindrical or spindle-shaped and are borne singly on short conidiophores directly from the hyphae. Microaleuriospores are generally few in number and have no specific morphologic features.

Table 13-6.   *Characteristics of the Hyaline Saprobes* (Continued)

Genus	Colonial Morphology	Microscopic Features	Illustration
Trichoderma	Colony is similar to that of Gliocladium forming a diffuse yellow or yellow-green lawn covering the entire surface of the agar. Colony surface is granular to fluffy.	Hyphae are hyaline and septate. Conidiophores generally are short and give rise to blunt sterigmata with tapered points. Clusters of 1 to 2 μ. in diameter, spherical to elliptical conidia form in compact clusters, held together by a thin mucinous secretion.	
Cephalosporium	Colonies are often white and covered with a fluffy, well-developed aerial mycelium. Light pastel yellow or orange colors develop with some strains.	Hyphae are quite delicate, hyaline, and septate. Conidiophores are long and slender, giving rise to elongated, elliptiform conidia clustered in a mosaic pattern simulating the cortical surface of a brain.	
Fusarium	Colonies are initially white and covered by a well-developed fluffy aerial mycelium. With maturity delicate lavender to purple-red pigment develops both over the surface and on reverse side.	Hyphae are hyaline and septate. Microconidia are 2 to 3 μ. in diameter and elliptical, form clusters simulating those of Cephalosporium. Identification is made by demonstrating pointed, banana-shaped or sickliform multicelled macroconidia.	

2. *Genus Trichophyton.* Macroaleuriospores are sparse or absent. When present, they are long and pencil-shaped and have a smooth, thin wall (Table 13-8). Microaleuriospores are generally numerous and are borne either laterally along the hyphae (en thyrses) or in clusters (en grappe). (see Table 13-7).

3. *Genus Epidermophyton.* Microaleuriospores are never produced. Genus differentiation is made by observing the large, clavate, smooth-walled macroaleuriospores divided into 2 to 5 cells by transverse septa (Table 13-8). These macroaleuriospores are either borne singly from short conidiophores or in clusters of two or three.

A practical laboratory approach to the differentiation of the dermatophytes has been published by Koneman and Roberts.[17] Although well over 30 different species of dermatophytes are discussed in detail by Beneke,[3] it is estimated that over 95 per cent of dermatophyte infections in the United States are caused by only seven of these species.[17] Laboratory personnel should learn how to identify these accurately. On those occasions when one of the less common species is encountered, reference textbooks[3,5,9] may be consulted in making a final species identification.

The three commonly encountered species of Microsporum are *M. audouinii*, *M. canis*, and *M. gypseum*. It is important epidermiologically to separate these three species. *M. audouinii* causes contagious inflammatory ringworm infection of the scalp in children, that can be directly transmitted from one child to the next. *M. canis* infection is zoophilic, that is, it can be derived from animals, particularly from cats and dogs. The family pet must also be treated if recurrence of the infection is to be prevented. *M.*

Table 13-7. *Characteristics of Three Microsporum Species*

Species	Colony Morphology	Microscopic Features	Other Features	Illustration
M. audouinii	Colonies are moderately slow growing (7 to 14 days), producing a velvety aerial mycelium that is light tan or buff in color. The reverse appears salmon pink.	Macroaleuriospores are rarely produced; if present, they are bizarre-shaped. Microaleuriospores are usually rare. Terminal chlamydospores, favic chandeliers, and pectinate bodies usually abound.	No growth on rice grain medium	
M. canis	Colonies produce a granular to fluffy white to buff surface. A bright, lemon-yellow apron at the peripheral growing margin is typical. Colony reverse is usually yellow-orange.	Macroaleuriospores are thick-walled, spindle-shaped, multiseptate, and echinulate. Many have a characteristic curved tip. Microaleuriospores are generally sparse and laterally attached to the hyphae.	Grow well on rice grain medium. There are no other specific features.	
M. gypseum	Colonies are generally granular due to production of numerous aleuriospores. Surface is often cinnamon colored and the reverse is light tan.	Macroaleuriospores are thick-walled, multiseptate, and echinulate. They generally are longer and less spindle-shaped than *M. canis*, with rounded rather than pointed tips which do not tend to curve.	Grow well on rice grain medium. There are no other specific features.	

gypseum, on the other hand, is geophilic, derived from soil, and animal quarantine will be of no aid in preventing spread of dermatophyte infections with this species. The features by which these three species can be recognized in the laboratory are listed in Table 13-7. *M. distortum, M. nanum,* and *M. vanbreuseghemii* are other species of Microsporum that uncommonly cause infections in man.

The three species of Trichophyton most commonly isolated from human infections are *T. mentagrophytes, T. rubrum,* and *T. tonsurans. T. mentagrophytes* and *T. rubrum* are the two most frequently recovered spe-

cies from athlete's foot infections. *T. tonsurans* causes a particular type of tinea capitis infection known as "black dot ringworm," so called because the hairs of the scalp are infected in such a way that they break off near the scalp surface and exhibit a black dot appearance. The differential features by which these three species can be separated in the laboratory are listed in Table 13-8.

T. verrucosum, T. schoenleinii, and *T. violaceum* are three other species of Trichophyton uncommonly encountered in the United States. *T. schoenleinii* and *T. violaceum* are common causes of favus infections, severe ringworm infections of the scalp that

Table 13-8. *Characteristics of Three Trichophyton Species and Epidermophyton floccosum*

Species	Colony Morphology	Microscopic Features	Other Features	Illustration
T. mentagrophytes	There are two distinct colony types, fluffy, and granular. Color is usually white to pinkish. Reverse is buff to reddish brown. Red-brown pigment is produced by some strains, usually never as intense as with T. rubrum.	Microaleuriospores are usually produced in abundance, and are globose and arranged in pine-tree or grapelike clusters. Spiral hyphae are seen in 30% of isolates. Macroaleuriospores are rarely seen, are thin-walled, smooth, and pencil-shaped.	Positive urease test within two days[13] Produce conical-shaped areas of invasion of hair shafts in hair-baiting test (positive test)[2]	
T. rubrum	Colonies are generally white and downy in consistency. May be pinkish or reddish. Granular colony variants are found with strains that sporulate heavily. Reverse is often wine-red to red-yellow, particularly on corn meal agar.	Microaleuriospores are usually produced in profusion and are tear-shaped and borne laterally and singly from the hyphae. Macroaleuriospores are usually absent or are thin-walled, smooth, and pencil-shaped.	Urease not rapidly produced (Faint positive test may be seen in 7 days.) Hair baiting test negative	
T. tonsurans	Colonies are generally tan, brown, or creamy red in color. Mycelium is usually low, giving a velvety to powdery surface. Rugal folds are common, with heaped sunken center. Reverse is yellow to tan.	Macroaleuriospores are rarely produced and are bizarre-shaped when present. Microaleuriospores are characteristically tear-shaped or club-shaped with flat bottoms and larger than other dermatophytes. Occasionally there are balloon forms.	Cannot grow on trichophyton No. 1 agar which contains only casein; good growth on trichophyton No. 4 agar which contains casein plus thiamine	
E. floccosum	Colonies are generally white and floccose; they tend to turn khaki green-brown with age. Center of colony is often folded. Reverse is yellow brown with observable folds.	Microaleuriospores are not produced. Macroaleuriospores are large, smooth-walled, clavate, and divided into 2 to 5 cells. They are borne singly or in clusters of two or three.	No special features; may be confused with M. nanum; however, macroaleuriospores of this species are thick-walled and echinulate.	

are difficult to treat and appear to have a familial predisposition, in northern and southern Europe, respectively.

Epidermophyton floccosum is the other commonly encountered dermatophyte in the United States. This organism is frequently recovered from cases of groin (tinea cruris) infection or athlete's foot. *E. floccosum* never infects the hair. The identification features for *E. floccosum* are also listed in Table 13-8. Representative colonial types of dermatophytes are illustrated in Plate 13-4.

The Dimorphic Molds

Dimorphism is the ability of some species of fungi to grow in two forms, depending upon environmental conditions: (1) as a mold when incubated at 25° to 30° C. and (2) in a yeast form when incubated at 35 to 37° C. It is not clearly established whether or not there is a causal relationship between dimorphism and pathogenicity; however, the following dimorphic molds are all overt pathogens for man:

Blastomyces dermatitidis
Paracoccidioides brasiliensis
Histoplasma capsulatum
Coccidioides immitis
Sporothrix schenckii

Since it is currently recommended that primary fungal cultures be incubated only at 25 to 30° C., the mold form is initially isolated in the laboratory. It is also the mold form that is infective for man. In order to confirm that an unknown mold is one of the dimorphic pathogens, it must be converted into its yeast form. There are a number of species of environmental saprobic molds that may be confused with the dimorphic pathogens. Conversion to the yeast form is accomplished by transferring an inoculum of the unknown mold to a moist slant of brain-heart infusion agar containing 10 per cent sheep blood. A few drops of brain-heart infusion broth are also added to provide moisture during incubation. The screwcap of the tube is first tightened, then slightly

loosened to allow the culture to "breathe." The tube is incubated at 35 to 37° C. and inspected daily until yeast conversion has been accomplished. More than one transfer may be necessary before total yeast conversion can be completed.

With the dimorphic fungi, it is the yeast form that causes disease in man. Only in rare instances will the mold form be found in human tissues. An immediate presumptive diagnosis of disease caused by one of the dimorphic pathogens can be made by observing yeast forms in direct mounts of infected material. The mold form, on the other hand, is the infective form for man and the disease is usually contracted through the inhalation of airborne spores.

The cultural features of the dimorphic molds that cause disease in man are listed in Table 13-9. Included in this table are illustrations of both mold and the yeast forms of these fungi. Color illustrations of the various colonies are included in Plate 13-5.

It is beyond the scope of this text to discuss in any detail the diseases caused by these fungi. The reader is referred to texts by Emmons *et al.*,[9] Conant and associates,[5] and by Wilson and Plunkett,[30] or to the overview by Koneman and Roberts.[17] The following section gives a brief summary of the clinical diseases caused by the dimorphic fungi.

Blastomycosis

Blastomycosis is a fungal disease that may occur in two forms. (1) Pulmonary blastomycosis is most commonly a self-limited disease, manifesting as a viruslike respiratory disease in the acute form. Pulmonary spread or dissemination to other organs occurs only rarely. (2) Primary cutaneous blastomycosis (Gilchrist's disease) is rare, manifesting as nonhealing, suppurative ulcers of unclothed skin surfaces of the hands, face, or forearms. In most instances the disease is self-limited; however, rarely, the lesion may spread, forming an unsightly wartlike growth with a raised, verrucous advancing border. The disease is endemic in the Mississippi and Ohio River valleys and is found particularly in individuals having close contact with the soil.

Paracoccidiodomycosis (South American Blastomycosis)

This disease is not found in the United States; rather, it is limited to South America, with the highest prevalence rates in Brazil, Venezuela, and Columbia. The disease is primarily limited to the nasal or oral mucosa, with ulcers that spread over the mucous membranes and form raised, mulberrylike, erythematous lesions. Pulmonary disease occurs in a high percentage of patients but disseminated disease is rare, with secondary involvement of the lymphatic system, spleen, intestines, and liver.

Histoplasmosis

Histoplasmosis most commonly involves the lungs, usually as a self-limited influenza-like syndrome that lasts two or three weeks. In the chronic form, cavitary lesions of the lungs may form, or localized granulomas may progress to thick, laminated, calcified "histoplasmomas" that may be surgically removed. The organisms primarily reside within reticuloendothelial cells, and in the chronic disseminated form hepatosplenomegaly and diffuse lymphadenopathy are usually present. Any of the organs may be involved, and a fatal outcome may ensue particularly if the adrenal glands are involved, leading to Addison's disease. The disease is endemic in the Mississippi River valley and along the river's tributaries, and infections in man commonly occur after inhaling dust laden with excreta of chickens, turkeys, pigeons, or bats ("cave fever").

Coccidioidomycosis

Coccidioidomycosis is endemic in the dry desert regions of the southwestern United States and in Mexico. Man acquires the infection from inhalation of arthrospores from infected top soil. Over 90 per cent of individuals living in endemic areas have positive skin tests, indicating that most infections are self-limited. In the acute phase, an influenza like respiratory infection is experienced. Only 2 per cent of infected individuals develop complications, most of whom present with "coin lesions" of the lung that radiologically simulate metastatic tumors. Most coin lesions are removed surgically, and in most instances the causative organisms can be cultured from the tissue specimen. Disseminated disease rarely occurs, but when it does it manifests by destructive lymphadenopathy, osteomyelitis, invasion of virtually any organ, and, in rare cases, central nervous system disease with meningitis. Cutaneous coccidioidomycosis is most commonly caused by secondary spread of the organism and generally indicates that disseminated disease is already present.

Sporotrichosis

Sporotrichosis is the so-called "rose gardener's" disease because the primary lesion often presents as a nonhealing ulcer of the skin of the fingers, hand, forearm, usually developing within one or two weeks of skin puncture by a rose thorn or other sharp piece of vegetative matter that is infected with spores. Masonry workers who handle old bricks may also become infected through cracks or fissures in the skin. The lesions typically spread through the regional lymphatics, producing a series of subcutaneous nodules that eventually suppurate and ulcerate. Cases of pulmonary sporotrichosis have been reported, although disseminated disease is quite rare. The disease is most common in the midwestern United States, particularly in the states bordering the Missouri and Mississippi River valleys where the organism is commonly found in the soil and on vegetative matter.

The *Atlas of Clinical Mycology,* recently *(Text continues on p. 413)*

Table 13-9. Characteristics of the Dimorphic Molds

Species	Mold Form				Yeast Form	
	Colonial Morphology	Microscopic Features	Illustration	Colonial Morphology	Microscopic Features	Illustration
Blastomyces dermatitidis	Growth in 7 days to 4 weeks. On blood agar, colonies are cream to tan, soft, wrinkled, and appear waxy. On BHI or SAB agar, colonies appear fluffy and white to tan.	Hyphae delicate, hyaline and septate. Round to oval conidia are borne singly from the tips of conidiophores of irregular length that are borne laterally from the hyphae. They have the appearance of "lollipops."		Colonies are tan or cream in color, and very wrinkled and waxy in appearance when grown at 37° C.	Large thick-walled yeast cells having a single bud attached to the parent cell by a thick "collar" or wall.	
Paracoccidioides brasiliensis	Growth in 21 or more days. On BHI or SAB agar the aerial mycelium is white to tan-brown. Center of colony may become heaped with a crater cut into the agar surface.	Mycelium tends to be sterile and many chlamydospores may be seen. Occasional round or oval conidia similar to those of *B. dermatitidis* may be seen.		Colonies are tan to cream in color, and may become wrinkled and pasty in appearance when grown at 37° C.	Large, thick-walled yeast cells similar to those of *B. dermatitidis* except there are multiple daughter buds, forming structures simulating a mariner's wheel	
Histoplasma capsulatum	Growth in 7 to 45 days. Growth on blood agar appears moist, waxy, and cerebriform, and ranges from pink to tan in color. On BHI or SAB agar, colonies are cottony to silky and are white or turning brown with age.	Hyphae are small, hyaline, and septate. Round to tear-drop microaleuriospores are borne on short lateral branches. Macroaleuriospores spherical to pyriform and tuberculated, are the diagnostic forms.		Initial growth appears as a rough, mucoid, cream-colored colony. It turns smooth and brown with age.	Small, oval, budding cells are seen. If observed during yeast conversion phase, cells are larger and some resemble arthrospores.	

	Colony Morphology	Microscopic Morphology (Mold)	Yeast/Conversion	Tissue Form
Coccidioides immitis	Growth in 5 to 21 days. Young colonies are moist and adhere to blood or SAB agar. Older colonies develop cottony aerial mycelium which becomes unevenly distributed over the agar surface in a "cobweb" appearance. It is white at first, becoming brown with age.	Early cultures have septate hyphae, and many raquet hyphae; as the culture ages, hyphae become enlarged and dissociate through points of septation into barrel-shaped arthrospores that stain alternating dark and clear, with dead cells inbetween. 	No yeast form in routine culture; remains in mold form even at 37° C. incubation	10-60 μ. in diameter spherules containing 2–4 μ. in diameter endospores seen only in tissues. 
Sporothrix schenckii	Growth in 3 to 5 days. Early colony is smooth and white to cream colored. With age, surface becomes wrinkled, turning brown to black. Surface remains smooth and devoid of an aerial mycelium.	Hyphae are hyaline septate, and small in diameter. Branched slender conidiophores arise at right angles from hyphae. Small pyriform conidia arranged in "flowerettes" at tips of conidiophores are diagnostic. Conidia are attached by delicate thread.	Colonies are cream to white in color and soft and creamy in consistency, resembling the typical yeast colony of many other species.	Elongated yeast cells resembling cigars with delicate buds are typically seen. Occasional yeast cells may appear more oval and bear multiple delicate buds. 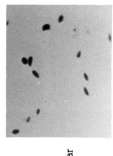

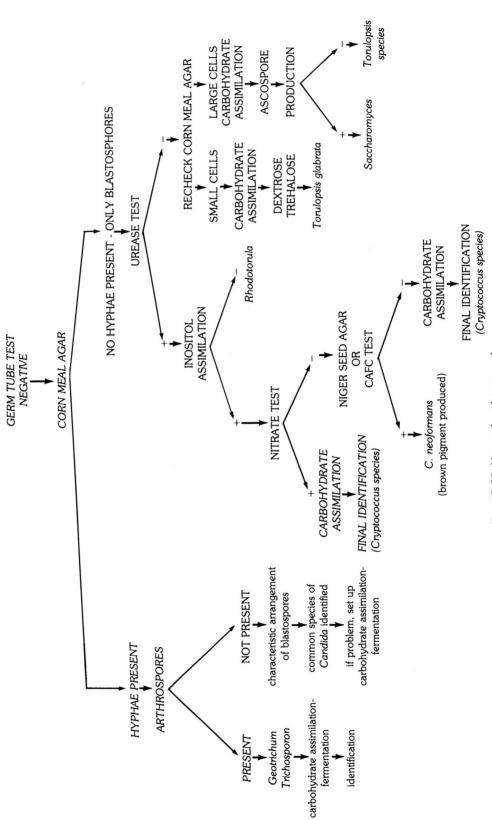

Fig. 13-22. Yeast identification schema.

published by the American Society of Clinical Pathologists[8] includes a number of color photogrpahs depicting the various clinical manifestations of the five dimorphic mycoses described above.

Laboratory Identification of the Yeasts

Each laboratory must decide to what extent they need to differentiate yeasts. Since yeasts are considered to be normal flora in the oropharynx and gastrointestinal tract, their recovery from sputum, throat swabs, bronchial washings, gastric washings, and stool specimens is of questionable clinical significance. Yeasts may also be recovered from urine cultures, nail scrapings, and vaginal specimens. Often the significance of the recovery of yeasts in these circumstances must be determined by the clinical history. The repeated isolation of yeasts from a series of clinical specimens from the same patient usually indicates infection with the organism recovered and identification of the isolates is necessary. Other clinical situations in which species identification is justified is the recovery of yeasts from normally sterile body fluids such as blood, cerebrospinal fluid, or fluids aspirated from joints, the pleural cavity, or pericardial sac.

A number of approaches may be taken in the laboratory for the identification of yeasts; however, the algorithm originally developed by Dolan[6] and revised by Roberts[7] as shown in Figure 13-22 is well within the capability of most clinical laboratories and sufficiently accurate for most clinical applications.

Since *Candida albicans* is the species of yeast most frequently cultured from clinical specimens, initial laboratory studies should be directed to its identification before additional costly tests are performed. Demonstrating the spiderlike colonies on eosin-methylene blue agar or observing the production of chlamydospores on corn meal agar are acceptable methods for identifying *C. albicans;* however, the rapid germ tube test is preferred by most laboratories. Ex-

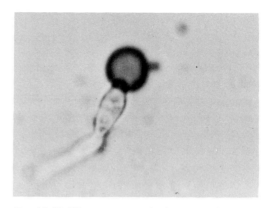

Fig. 13-23. Photomicrograph of a germ tube, characteristic of *Candida albicans* (oil immersion).

cept for strains of *Candida stellatoidea* (which is though to be a variant of *C. albicans*) and rare strains of *C. tropicalis,* only *C. albicans* will form germ tubes under the conditions of the test.

> **Germ Tube Test**
> 1. A very small portion of an isolated colony of the yeast to be tested is suspended in a test tube containing 0.5 ml. of rabbit or human serum.
> 2. The test tube is incubated at 37° C. for no longer than 3 hours.
> 3. A drop of the yeast-serum suspension is placed on a microscope slide, overlaid with a coverslip, and examined microscopically for the presence of germ tubes (see Fig. 13-23).

If germ tubes are observed, a presumptive identification of *C. albicans* can be reported and further testing is not necessary. If germ tubes are not observed, a portion of the unknown colony should be inoculated to a corn meal agar plate and to the surface of a Christensen's urea agar slant. Tween 80 (polysorbate) in a final concentration of 0.02 per cent should be added to the corn meal agar to reduce surface tension and enhance the optimal formation of hyphae and blastospores.

The corn meal Tween 80 agar is inoculated by making three parallel cuts one-half inch apart into the agar, holding the inocu-

Table 13-10. *Morphology of Clinically Important Yeasts on Corn Meal Tween 80 Agar*

Yeast Species	Microscopic Morphology	Illustration
Candida albicans	Chlamydospores may be numerous, borne singly or in clusters. Chlamydospores may not develop at 35° C. incubation.	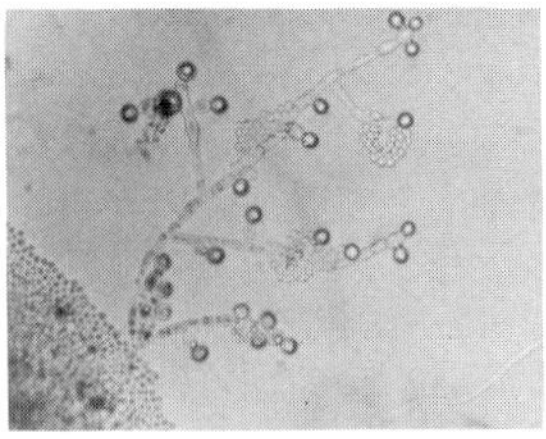
	Blastospores are usually numerous. They often are aggregated in dense clusters at regular intervals along the pseudohyphae.	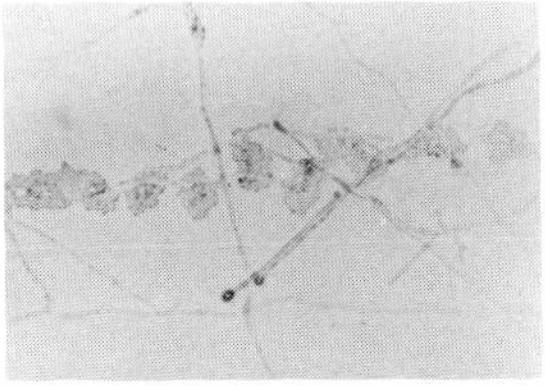
Candida tropicalis	Blastospores are fewer in number than with *C. albicans,* usually borne singly or in small clusters, and more sparsely spaced than with *C. albicans.* This feature is illustrated in the low-power and high-power photo micrographs to the right.	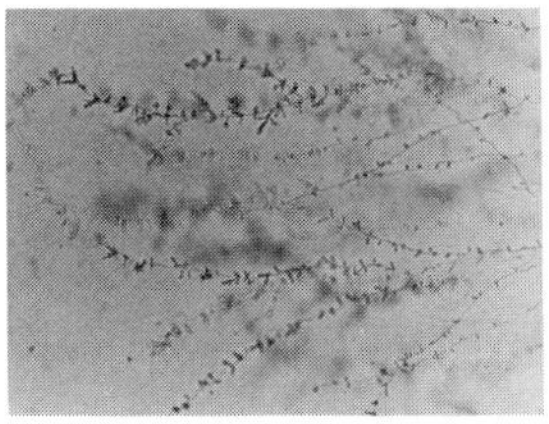
	Mycelium is generally delicate, with the formation of satellite colonies with a sage brush or cross-matchstick appearance. Blastospores are sparse and borne singly or in short chains.	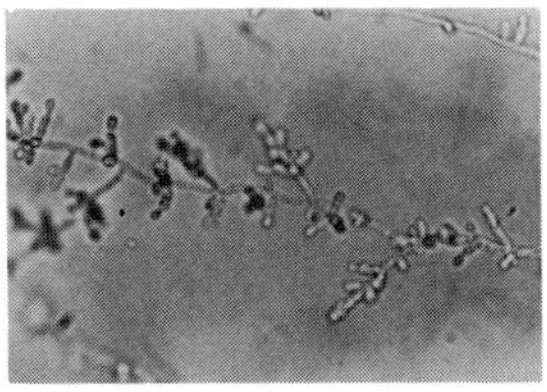
Candida parapsilosis	Mycelium is generally delicate, with the formation of satellite colonies with a sage brush or cross-matchstick appearance. Blastophores are sparse and borne singly or in short chains.	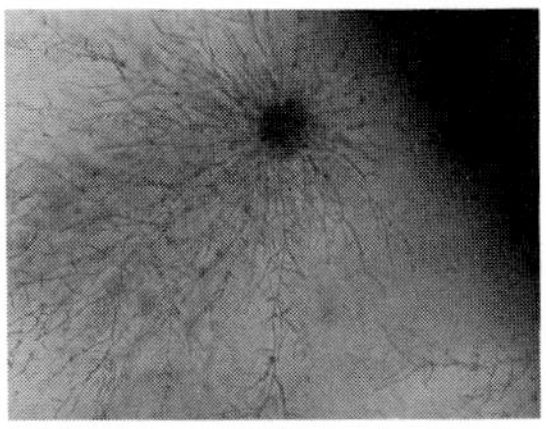
	Another characteristic of *C. parapsilosis* is the formation of giant hyphae, as shown to the right.	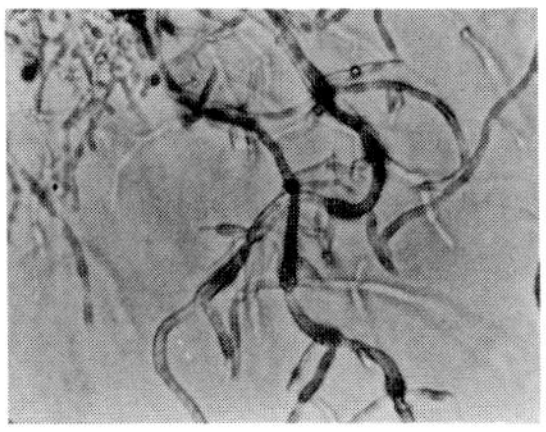

Table 13-10. *Morphology of Clinically Important Yeasts on Corn Meal Tween 80 Agar*
(Continued)

Yeast Species	Microscopic Morphology	Illustration
Candida pseudotropicalis	There are elongated blastospores that readily dissociate from the pseudohyphae, tending to lie in parallel fashion, in a "logs-in-stream" arrangement.	
Cryptococcus species	Hyphae do not form on the corn meal agar. The irregularly sized yeast cells form tall mosaic clusters adjacent to the streak line. If capsules are formed, the yeast cells may appear widely separated from one another.	
Torulopsis glabrata	Hyphae do not form on corn meal agar. The yeast cells form compact masses adjacent to the streak line on corn meal agar, appearing similar to those of species of Cryptococcus, except they are smaller and more tightly arranged.	
Geotrichum species	They form arthrospores on corn meal agar. The microscopic morphology may be difficult to differentiate from Trichosporon, unless the formation of blastospores from the corner of the arthrospores, a characteristic feature of Geotrichum is seen.	

lating wire at about a 45-degree angle. A coverslip is laid on the surface of the agar, covering a portion of the inoculation streaks. The inoculated plates are incubated at 30° C. for 24 to 48 hours and then examined microscopically, preferably through the coverslip to prevent inadvertent contamination of the tip of the microscope objective into the agar.

Interpretation of Corn Meal Agar Plates

The corn meal agar preparations should be examined for the presence of hyphae, blastospores, chlamydospores, or arthrospores.

If hyphae are present, observe for the presence of arthrospores. Trichosporon and Geotrichum both produce arthrospores on corn meal agar. Many species of Trichosporon produce urease; species of Geotrichum, however, are negative for this characteristic. Carbohydrate fermentation and assimilation studies (see below) may be required to differentiate species within these two genera if the results of the urease test are equivocal.

If arthrospores are not present, but pseudohyphae and blastospores are observed, the unknown yeast belongs to the genus Candida. It may be possible in many instances to make a species identification

Chart 13-1. *Carbohydrate Fermentation Test for Yeast Identification*

Introduction	Carbohydrate fermentation tests are useful for supplementing carbohydrate assimilation test results in the definitive identification of species of yeasts recovered from clinical specimens.
Principle	Yeast fermentation in appropriate culture medium containing a single carbohydrate source is detected by the production of gas, and even though present, acid production (carbohydrate utilization) is not an indication of fermentation. For accurate results, it may be necessary to "starve" the yeast by passing it one or two times through carbohydrate-free broth to remove any carbohydrate assimilated during growth on primary recovery medium.

Media and Reagents

1. Yeast Fermentation Broth:

Distilled water	1000 ml.
Peptone	10 g.
NaCL	5 g.
Beef extract	3 g.
1N NaOH	1 ml.

2. Indicator (Bromcresol Purple):

   Dissolve 0.04 g. of bromcresol purple in 100 ml. of distilled water. Add a small amount of 1N NaOH to make an alkaline solution and let stand overnight. After the dye is in solution, add 1N HCL until the neutral pH is reached. Add 100 ml. of indicator to 1 liter of fermentation broth.

3. Carbohydrate Solutions:

   Stock solutions are prepared by adding 20 g. of carbohydrate to 100 ml. of distilled water. Dissolve by placing in a 56° C. water bath for a few minutes. Sterilize by filtration through 0.45 Millipore or Nucleopore filters.

---

based on the morphology and specific arrangement of the blastospores, as reviewed in Table 13-10. The production of chlamydospores is diagnostic of *C. albicans. C. albicans* may also be tentatively identified if compact clusters of blastospores are formed at regular intervals along the hyphae. Smaller numbers of blastospores, widely spaced singly or in small clusters along the hyphae, are more consistent with *C. tropicalis;* satelliting of "spider colonies" and formation of giant hyphae are suggestive of *C. parapsilosis,* and a "log-in-stream" arrangement of the blastospores is the characteristic leading to a presumptive identification of *C. pseudotropicalis.*[6,7]

If there is difficulty in deriving a presumptive identification based on these growth patterns on corn meal agar, carbohydrate fermentation or assimilation tests may be required for final species identification. The carbohydrate fermentation test which assesses the ability of various species of yeast to form gas from different carbohydrates is discussed in detail in Chart 13-1. (see Plate 13-6, *A*). The principle of the carbohydrate assimilation test is to determine the ability of different species of yeasts to grow in various single carbohydrate substrates, and is reviewed in Chart 13-2.[1] The specific fermentation and assimilation reactions for several species of Candida when tested with a number of carbohydrates are listed in Table 13-11 (see Plate 13-6, *B*).

Chart 13-1.   *Carbohydrate Fermentation Test for Yeast Identification (Continued)*

Media and Reagents
(Continued)
	4. Preparation of Wickerham Tubes:
	    A. Use 18 × 150- mm. tubes containing 9 ml. of broth indicator and an inverted Durham tube. The broth indicator medium is prepared by adding one part of the indicator solution to 10 parts of broth.
	    B. Sterilize the broth indicator medium by autoclaving at 121° C. for 15 minutes.
	    C. Add 0.5 ml. of stock carbohydrate solution to the broth indicator media tubes just before use.

Procedure	1. To each of the yeast fermentation tubes (dextrose, maltose, sucrose, and lactose are most frequently used), add 0.2 ml. (5 drops) of a yeast suspension in saline equivalent to a McFarland No. 4 standard. 2. Do not screw caps on tubes tightly. Incubate at 37° C. for 48 hours. Leave all negative tests in incubator for 6 to 10 days before discarding.
Interpretation	The presence of bubbles in the inverted Durham tube or a drop in the liquid level of the Durham tube is indicative of fermentation (see Plate 13-6, *A*). Development of a yellow color indicates carbohydrate utilization, not fermentation.
Controls	A positive- and negative-reacting yeast species must be included for each of the carbohydrates tested.
Bibliography	Adams, E. D., Jr., and Cooper, B. H.: Evaluation of a modified Wickerham medium for identifying medically important yeasts. Am. J. Med. Technol., *40:*377–388, 1974. Wickerham, L. J., and Burton, K. A.: Carbon assimilation tests for the classification of yeasts. J. Bacteriol., *56:*363–371, 1948.

Yeasts That Do Not Form Hyphae in Corn Meal Agar

If hyphae are not observed in a corn meal preparation inoculated with an unknown yeast, the appropriate portion of the scheme outlined in Figure 13-22 must be followed. The yeast cells that are present (blastospores) should be carefully examined for the presence of a capsule. If present, one must be highly suspicious for the presence of *Cryptococcus neoformans*. A positive urease test also suggests one of the species of Cryptococcus although species of Rhodotorula also produce urease. However, species of Rhodotorula, as the name would indicate (*rhodo,* red), produce orange or orange-red colonies (Plate 13-6, *C*), and inositol is not assimilated, two characteristics that are usually sufficient to separate them from Cryptococcus.

If a species of Cryptococcus is suspected in the identification of an unknown yeast, *C. neoformans* must be specifically tested for because of its potential pathogenicity for man. Cryptococcal meningitis is the most common manifestation, although virtually every organ can be involved in disseminated cases, often with a fatal outcome.

Two identification characteristics are helpful in making a presumptive identification of *C. neoformans:* (1) the inability of this species to reduce nitrates to nitrites (see Chart 2-2 in Chap. 2 for a review of the nitrate reduction test); and (2) the production of a reddish color on birdseed agar (see

Chart 13-2. *Carbohydrate Assimilation Test for Yeast Identification*

Introduction	Carbohydrate utilization tests are widely used methods for the definitive identification of clinically important yeasts and yeastlike organisms. Several methods have been described (see Bibliography). The method of utilizing carbohydrate-impregnated disks will be described here.
Principle	The disk plate method for assessing the ability of yeasts to utilize carbohydrates is based on the use of carbohydrate-free yeast nitrogen base agar and observing for the presence of growth around carbohydrate-impregnated filter paper disks after an appropriate period of incubation.
Media and Reagents	Yeast Nitrogen Base Agar: 1. Prepare a 2% agar solution (20 g. of agar per liter of distilled water). Autoclave for 15 minutes at 121° C./15 psi. 2. Dissolve 6.7 g. of yeast nitrogen base (BBL) in 100 ml. of distilled water. Adjust *pH* to 6.2 to 6.4 by adding 1N NaOH. Discard 12 ml. of solution. 3. Sterilize by filtration through an appropriate Millipore or Nucleopore filter. 4. Add 88 ml. of yeast nitrogen base and 100 ml. of the filter-sterilized bromcresol purple indicator as prepared for the carbohydrate fermentation test to 1 liter of the 2% agar solution. Pour into sterile, plastic Petri dishes, approximately 20 ml./plate. Carbohydrate-Impregnated Filter Paper Disks: These may be purchased commercially (BBL, Difco), or can be prepared by soaking filter paper disks 10 mm. in diameter in 1% carbohydrate solutions and drying before use.

Plate 13-6, *D*). Rhodes and Roberts[21] have recently described a modification of the nitrate reduction test, utilizing semisolid indole nitrate medium and a heavy inoculum of the organism. The production of a red color after adding sulfanilic acid and alpha-naphthylamine to a 48- to 72-hour culture is indicative of a positive test.

A rapid nitrate reduction test has also been described by Hopkins and Land,[11] utilizing a cotton swab which has been impregnated with an inorganic nitrate substrate. The tip of the swab is swept across two or three colonies of the yeast to be tested. The inoculated swab is then swirled against the bottom of an empty test tube to embed the yeast cells within the fibers of the swab. The tube and swab are incubated for 10 minutes at 45° C. The swab is removed and two drops each of alpha-naphthylamine and sulfanilic acid reagents are added to the tube. The swab is replaced in the tube to allow it to absorb the reagents and the immediate development of a red color is indicative of a positive reaction.

The niger seed (bird seed) agar test of Staib and Senska[28] is quite useful for detecting and identifying *C. neoformans.*

The plate is heavily inoculated with the test organism and incubated at 37° C. for at least one week. The production of a red-brown or maroon pigment is characteristic of *C. neoformans* (see Plate 13-6, *D*).

Hopfer and Gröschel[10] recently described a more rapid test utilizing the active ingredient of niger seed (caffeic acid) as a substrate. Yeasts, notably *C. neoformans,* capable of producing phenoloxidase enzymes, also produce a brown pigment in media containing the caffeic acid substrate.[29]

Chart 13-2. *Carbohydrate Assimilation Test for Yeast Identification (Continued)*

Procedure
1. With a sterile transfer pipette, flood the surface of a yeast nitrogen base plate with a suspension of yeast in saline equivalent to a McFarland standard No. 4.
2. Aspirate the excess suspension using the same pipette. Let surface of agar dry for about 5 minutes.
3. Place carbohydrate disks on the agar and press down firmly with flamed forceps. The disks should be placed on four quadrants and in the center to form a cross configuration.
4. Incubate plate at 30° C. for 24 hours and record results.

Interpretation
Growth around a carbohydrate disk indicates that the sugar contained in that disk has been assimilated by the yeast species under study, and is a positive test (Plate 13-6, *B*). If growth on the plate is not sufficiently concentrated to enable a reading, reincubate for 24 hours and read again.

Controls
Each newly prepared batch of yeast nitrogen base agar and new carbohydrate disks must be tested with positive and negative-reacting yeasts for each carbohydrate to be used.

Bibliography
Bowman, P. I., and Ahern, D. G.: Evaluation of commercial systems for the identification of clinical yeast isolates. J. Clin. Microbiol., 4:49–53, 1976.
Huppert, M. G., *et al.*: Rapid methods for identification of yeasts. J. Clin, Microbiol., 2:21–34, 1975.
Roberts, G. D.: Laboratory diagnosis of fungal infections. Hum. Pathol., 7:161–168, 1976.

Staib's Medium, Used for the Detection and Identification of *C. neoformans*.

Pulverized *Guizotia abyssinicia* seed*	50 g.
Dextrose	1 g.
KH$_2$PO$_4$	1 g.
Creatinine	1 g.
Agar	15 g.

The seed is added to 100 ml. of distilled water and ground in a Waring blender. Boil mixture for ½ hour in a liter of water and strain through a cloth to remove the water extract from the seed. Adjust volume of the seed extract to 1000 ml. with distilled water.

Add remaining ingredients and boil until dissolved. Place in a flask and autoclave at 121° C for 15 minutes. Final pH = 5.5. Dispense in sterile, 100-mm Petri dishes (35 ml./plate).

*Available from Philadelphia Seed Co., Philadelphia, Pa.

If the results of these two tests are equivocal, carbohydrate fermentation and assimilation tests may be necessary to identify members of the Cryptococcus genus. The reactions for several species of Cryptococcus and the more commonly used carbohydrates are listed in Table 13-12.

Yeasts that do not produce hyphae in corn meal-Tween 80 agar and are urease-negative belong either to the Torulopsis or Saccharomyces genera. If the yeast cells appear quite small in the corn meal preparation, Torulopsis species can be suspected. *Torulopsis glabrata* is the only species currently known to be pathogenic for man, most commonly associated with urinary tract infections. *T. glabrata* assimilates both dextrose and trehalose, characteristics not shared by other members of the genus.

Saccharomyces species are not considered

(Text continues on p. 422)

Table 13-11. Characteristics of Those Candida Species Most Often Isolated from Clinical Specimens

Organism	Assimilations												Fermentations						Other Reactions					
	Dextrose	Maltose	Sucrose	Lactose	Galactose	Melibiose	Cellobiose	Inositol	Xylose	Raffinose	Trehalose	Dulcitol	Dextrose	Maltose	Sucrose	Lactose	Galactose	Trehalose	Urease	KNO₃	Pseudohyphae	Growth at 37° C.	Germ Tubes	India Ink Capsule
Candida albicans	+	+	+	−	+	−	−	−	+	−	+	−	AG	AG	A*	−	AG	AG	−	−	+	+	+	−
C. stellatoidea†	+	+	−	−	+	−	−	−	+	−	+*	−	AG	AG	−	−	− or A*	AG or A*	−	−	+	+	+	−
C. parapsilosis	+	+	+	−	+	−	−	−	+	−	+	−	AG or A	−	−	−	AG or A*	AG or A*	−	−	+	+	−	−
C. tropicalis	+	+	+	−	+	−	+*	−	+	−	+	−	AG	AG	AG	−	AG	AG	−	−	+	+	−	−
C. pseudotropicalis	+	−	+	+	+	−	+	−	+*	+	−	−	AG	−	AG	AG	AG	− or A*	−	−	+	+	−	−
C. krusei	+	−	−	−	−	−	−	−	−	−	−	−	AG	−	−	−	−	−	+*	−	+	+	−	−
C. guillermondii	+	+	+	−	+	+	+	−	+	+	+	+	AG	−	AG	−	AG or A*	AG or A*	−	−	+	+	−	−
C. rugosa	+	−	−	−	−	+	−	−	+	−	−	−	−	−	−	−	−	−	−	−	+	−	−	−

Key: + = assimilation, growth density greater than 1 + turbidity by the Wickerham card; AG = acid and gas; A = acid only produced in fermentation broth; − = negative.
* Strain variation.
† Report as C. albicans when C. stellatoidea is identified.

Table 13-12. Characteristics of Those Cryptococcus Species Most Commonly Encountered in Clinical Specimens

Organism	Assimilations												Fermentations						Other Reactions					
	Dextrose	Maltose	Sucrose	Lactose	Galactose	Melibiose	Cellobiose	Inositol	Xylose	Raffinose	Trehalose	Dulcitol	Dextrose	Maltose	Sucrose	Lactose	Galactose	Trehalose	Urease	KNO₃	Pseudohyphae	Growth at 37° C.	Germ Tubes	India Ink Capsule
Cryptococcus neoformans	+	+	+	−	+*	−	+	+	+	+*	+	+	−	−	−	−	−	−	+	−	R	+*	−	+
C. uniguttulatus	+	+	+	−	−*	−	−*	+	+	+	+*	−	−	−	−	−	−	−	+	−	−	+	−	+
C. albidus var. albidus	+	+	+	+*	−*	−*	+	+	+	+*	+	+*	−	−	−	−	−	−	+	+	+*	+	−	+
C. laurentii	+	+	+	+	+	+*	+	+	+	+*	+	+*	−	−	−	−	−	−	+	−	−	+	−	+
C. luteolus	+	+	+	−	+	+	+	+	+	−*	+	+*	−	−	−	−	−	−	+	−	−	+	−	−
C. albidus var. diffluens	+	+	+	−	−*	+*	+	+	+	−	+	+*	−	−	−	−	−	−	+	+	−*	+	−	+
C. terreus	+	+*	−*	−*	+*	−	+	+	+	−	+	+*	−	−	−	−	−	−	+	+	−*	+	−	+
C. gastricus	+	+	−*	−	+	−	+	+	+	−	+	−	−	−	−	−	−	−	+	−	−*	−	−	+

*Strain variation in assimilation.

Key: + = positive; R = occasional to rare hyphae; − = negative.

pathogenic for man. They produce large yeast cells in corn meal preparations and also produce ascospores when grown on ascospore agar medium. This medium contains potassium acetate, 10.0 g., yeast extract, 2.5 g., dextrose, 1.0 g., agar 30.0 g., and distilled water, 1000 ml. The unknown colony is inoculated to the surface of an ascospore agar plate and incubated at 30° C. for up to 10 days. Ascospores are acid-fast and appear as large, thick-walled structures when stained with the Kinyoun stain (see Plate 13-6, *E*).

Packaged Yeast Identification Kits

The scheme (shown in Fig. 13-22) described here is only one of many that may be useful for the identification of yeasts, and each laboratory supervisor must determine the approach best suited for the needs of the laboratory involved. Packaged kits are currently available for the identification of yeasts. An evaluation of three of these kits has recently been reported by Bowman and Ahearn.[4] A brief review of these three systems is as follows:

*The API 20C Yeast System** consists of a strip of 20 small plastic tubules containing dehydrated carbohydrate assimilation and fermentation media (see Plate 13-6, *F*, *G*, and *H*). The fermentation tubules contain glucose, galactose, maltose, lactose, raffinose, trehalose, and melibiose, with bromcresol purple as the *p*H indicator. The assimilation tubules include inositol, glucose, galactose, maltose, sucrose, lactose, raffinose, trehalose, melibiose, cellobiose, and a growth control containing only basal medium. To use, each tubule is filled with a suspension of the unknown yeast to be tested, prepared in a molten agar medium supplied for use with the kit. After incubation at 30 or 35° C. for 48 to 72 hours in a moisturized plastic chamber supplied with the kit, the fermentation tubules are examined for the presence of bubbles and the assimilation tubules for turbidity indicating growth (Plate 13-6, *F* and *G*).

The Uni-Yeast-Tek system† consists of a

multicompartmented plastic dish, each compartment containing a different carbon assimilation agar medium, a central well containing corn meal-Tween 80 agar for determining mycelial and chlamydospore production, urea agar, nitrate assimilation agar, and additionally a 0.05 per cent glucose-2.6 per cent beef extract broth for performing the germ tube test (see Plate 13-6, *I*). The carbohydrate concentrations vary from 1 to 4 per cent and include sucrose, lactose, maltose, cellobiose, soluble starch, trehalose, and raffinose. Each of the compartments is inoculated through a small portal on the side of each well with one drop of a distilled water suspension of the yeast to be tested, equal in turbidity to a McFarland no. 4 standard. The corn meal agar is inoculated by making two or three parallel slashes into the agar with an inoculating needle containing a small portion of the yeast colony to be tested. The plastic dish is incubated for 2 to 7 days at 30 or 35°C. The carbohydrate assimilation compartments are observed for a blue to yellow (positive) color change, the nitrate medium for a color change from blue to blue-green (nitrate reduction), and the corn meal agar for the presence of hyphae and blastospores.

The Micro-Drop test system‡ includes plastic Petri dishes 150 mm. in diameter, bottles containing 70 ml. of sterile basal agar with bromcresol purple, and 11 sealed dispensing cartridges that hold 2 ml. of sterile solutions of basal substrates (glucose, galactose, lactose, maltose, inositol, and KNO_3). The assimilation medium is first melted and cooled to about 50° C. and inoculated with 5 ml. of a saline suspension of the yeast to be tested (equal to a McFarland no. 5 standard). After the agar solidifies, one drop of each carbohydrate solution is dispensed onto the plate, duplicating a template provided with the kit. The plates are incubated at room temperature for 18 to 24 hours and observed for a yellow color change (posi-

---

*Analytab Products Inc., Plainview, N.Y., 11803
†Corning Medical Diagnostics, Roslyn, N.Y., 11576
‡Clinical Sciences, Inc., Whippany, New Jersey, 07981

tive) indicating utilization of the various carbon sources.

Bowman and Ahearn[4] conclude in their review that reasonably accurate results can be obtained with each of these systems, ranging from 83 per cent for the Micro-Drop system to 99 per cent for the Uni-Yeast-Tek system. All systems, however, require technical skills and sufficient familiarity to enable accurate interpretation of the various reactions. Before deciding to implement the use of one of these systems in a laboratory, such factors as cost, stability, and adaptability to individual needs must be taken into consideration.

SEROLOGIC DETECTION OF FUNGAL DISEASES

It is not possible here to discuss in detail the serologic diagnosis of fungal infections. This subject has recently been reviewed by Roberts.[25] Although viewed with some skepticism in the past, the value of serologic tests has been clearly demonstrated in the diagnosis of certain mycotic infections.[14]

These tests are not without pitfalls, primarily because most antigens are crude extracts of fungal organisms and cross-reactivity between genera is common. Usually the antibody titer is greater with the antigen from the causative organism and positive results can be best interpreted when correlated with the clinical signs and symptoms of the patient.

A brief overview of the current status of fungal serologic tests for those mycotic diseases for which they have been found useful in making a diagnosis is presented in Table 13-13. The specific techniques most commonly employed include immunodiffusion using Ouchterlony agar dishes, complement fixation, and latex agglutination tests.

Most of these techniques require careful technical performance and should not be attempted in all laboratories. Many reagents are not commercially available and considerable skill is required to produce antigens and antibodies of sufficient quality to give reliable results.

Commercial test kits are now available, making it possible for virtually any laboratory to perform serologic tests for some of the fungal diseases. The kit for the detection of cryptococcal antigen (International Biological Laboratories, Rockville, Md. 20850) in particular can be recommended, and this procedure can be quite helpful in the testing of the cerebrospinal fluid of patients with cryptococcal meningitis from whom cultures have been negative.

LABORATORY IDENTIFICATION OF THE AEROBIC ACTINOMYCETES

The family Actinomycetales includes several genera, including the anaerobic Actinomyces, Arachnia, and Bifidobacteria, which are discussed in Chapter 10, and the aerobic Nocardia and Streptomyces, which will be briefly discussed here.

Nocardia species and Streptomyces species are bacteria not fungi, although tests for their identification are performed by mycology laboratories in most hospital and reference centers. They are taxonomically more closely related to the Mycobacteria, particularly the Nocardia species which are partially acid-fast.

Human nocardiosis generally begins as a chronic pulmonary disease manifested by a nonresolving pulmonary infiltrate on x-ray, a cough, and a productive sputum. In rare instances, the disease may become disseminated to involve the kidneys, spleen, liver, and adrenal glands. Nearly one third of patients with progressive pulmonary nocardiosis develop metastatic brain abscesses which cause headache and local sensory or motor disturbances.

Several species of Nocardia and, less commonly, species of Streptomyces, may cause actinomycotic mycetomas of the skin, mucous membranes, and subcutaneous tissue. These lesions are characterized by slowly healing, deeply penetrating sinus tracts that discharge purulent material within which may be demonstrated yellow, gray, or white grains or granules.

The laboratory identification of Nocardia

Table 13-13. *Serologic Tests Useful in the Diagnosis of Mycotic Disease*

Disease	Antigens	Interpretation
Aspergillosis	Aspergillus fumigatus Aspergillus flavus Aspergillus niger	*Immunodiffusion:* one or more precipitin bands is suggestive of active disease. Precipitin bands can be found in 95% of patients with fungus-ball infections and in 50% of patients with bronchopulmonary disease. Positive cultures are required before the presence of precipitin bands can be considered clinically diagnostic.
Blastomycosis	Blastomyces dermatitidis Yeast form	*Complement fixation:* titers of 1:8 to 1:16 are highly suggestive of active infection; titers of 1:32 or greater are considered diagnostic. Cross-reactions in patients with histoplasmosis or coccidioidomycosis; 75% of patients with blastomycosis have negative titers. *Immunodiffusion:* preliminary results indicate an 80% positive detection rate, considerably more sensitive than the complement fixation test.
Candidiasis	Candida albicans	*Immunodiffusion:* precipitins occur in 20 to 30% of normal individuals, making interpretation difficult. Clinical correlation is necessary to interpret positive serologic test results.
Coccidioidomycosis	Coccidioidin	*Complement fixation:* titers of 1:2 to 1:4 have been seen in active infection; repeat test should be done at 2- to 3-week intervals. Rising titer or titer greater than 1:16 is usually indicative of active infection. Cross-reactions are seen in patients with histoplasmosis, and false negative results in patients with solitary pulmonary "coin lesion." *Latex agglutination:* occurrence of precipitins during first three weeks of infection is diagnostic. This is useful as a screening test in early infections.
Cryptococcosis	No antigen: latex particles are coated with hyperimmune anticryptococcal globulin.	*Latex agglutination:* presence of cryptococcal polysaccharides in spinal fluid or other body fluids is indicative of cryptococcosis. Positive CSF tests seen in 95% of patients with cryptococcal meningitis, often when cultures fail to recover the organism. Serum antigen is less frequently detected; however, positive serum test often indicates disseminated disease.

Histoplasmosis	Histoplasmin
	Yeast form of *Histoplasma capsulatum*
		Complement fixation: titers of 1:8 to 1:16 are highly suspicious of infection; titers of 1:32 are usually diagnositic. Cross-reactions occur with patients with aspergillosis, blastomycosis, and coccidioidomycosis, although titers are usually low. Rising titers indicate progressive infection; decreasing titers indicate regression.
		   Recent skin tests in individuals previously exposed to *H. capsulatum* will often result in increased titers.
		   The yeast antigen is positive in 75 to 80% of cases; histoplasmin antigen detects only 10 to 15% of cases.
		Immunodiffusion: appearance of H and M bands together is indicative of active infection. M band alone may appear in early infection or may appear following skin testing. The H band appears later than the M band, but disappears earlier, often indicating regression of disease.
		Latex agglutination: Test is unreliable because false positive and false negative test results frequently occur. Screening test only and positive tests should be confirmed by complement fixation.

Sporotrichosis	Yeast form of *Sporothrix schenckii*	*Agglutination:* titers of 1:80 or more usually indicate active infection. Most positive tests are seen in extracutaneous infections; titers are negative in many cases of primary skin infections.

and Streptomyces can be made using the following approach. A Gram's stain of a portion of a suspected colony will show delicate, branching filaments no more than 1 micron in diameter (Plate 13-6, *J*). Nocardia species are partially acid-fast (that is, they do not decolorize when treated with 1% H_2SO_4 or 3% HCl instead of the more active acid-alcohol decolorizer used in the Ziehl-Nielsen or Kinyoun stains), while Streptomyces species are not. If the unknown organism is not acid-fast, a report of "non-acid-fast aerobic actinomycete" can be made.

The typical colonies of species of Nocardia and Streptomyces are dry to chalky in consistency, usually heaped or folded, and range in color from yellow to gray white (Plate 13-6, *K* and *L*). Nocardia species are more commonly some shades of yellow whereas Streptomyces species are gray-white. Both groups produce colonies with a pungent "musty basement" odor.

Nocardia and *Streptomyces* species can be differentiated by inoculating plates of casein, xanthine, and tyrosine agars, incubating them for up to two weeks at 30° C., and observing for hydrolysis (clearing of the medium around the colonies). The reactions of several species of Nocardia and Streptomyces on these media are listed in Table 13-14. An example of casein hydrolysis by a colony of Streptomyces is shown (Fig. 13-24).

If the results of these are equivocal, thin-layer chromatography may be helpful. Dif-

Fig. 13-24. Photograph of a casein agar plate illustrating the hydrolytic action of two species of Streptomyces.

ferences in the cell wall composition of the carbohydrates arabinose, galactose, xylose, and madurose and the type of diamino-pimelic acid (DAP) are used in making a species identification. Streptomyces species have levo-DAP and no diagnostic carbohydrates in their cell wall whereas Nocardia species possess meso-DAP and various combinations of the carbohydrates listed above. Nocardia asteroides, the species that causes most human infections, can be identified by its inability to hydrolyze casein, xanthine, or tyrosine, and by demonstrating meso-DAP and arabinose and galactose in chromatographic analysis of the cell wall.[24]

Table 13-14. **Morphologic and Physiologic Characteristics of Common Aerobic Actinomycetes and *Mycobacterium fortuitum***

Organism	Casein Hydrolysis	Xanthine Hydrolysis	Tyrosine Hydrolysis	Acid Fastness	Branching Filaments
Nocardia asteroides	−	−	−	+	+
Nocardia brasiliensis	+	−	+	+	+
Nocardia caviae	−	+	−	+	+
Streptomyces species	±	+	+	−	+
Actinomadura madurae	+	−	±	−	+
Actinomadura dassonvillei	+	+	+	−	+
Actinomadura pelletierii	+	−	+	−	+
Mycobacterium fortuitum	−	−	−	+	−

REFERENCES

1. Adams, E. D., Jr., and Cooper, B. H.: Evaluation of a modified Wickerham medium for identifying medically important yeasts. Am. J. Med. Technol., *40:*377–388, 1974.
2. Ajello, L., and Georg, I. K.: *In vitro* hair cultures for differentiation between atypical isolates of *Trichophyton mentagrophytes* and *Trichophyton rubrum*. Mycopathol. Mycol. Appl., *8:*3–11, 1957.
3. Beneke, E. S., and Rogers, A. L.: Medical Mycology Manual. ed. 3. Minneapolis, Burgess Publishing, 1970.
4. Bowman, P. I., and Ahearn, D. G.: Evaluation of commercial systems for the identification of clinical yeast isolates. J. Clin. Microbiol., *4:*49–53, 1976.
5. Conant, N. F., *et al.:* Manual of Clinical Mycology. ed. 4. Philadelphia, W. B. Saunders, 1971.
6. Dolan, C. T.: A practical approach to identification of yeast-like organisms. Am. J. Clin. Pathol., *55:*580–590, 1971.
7. Dolan, C. T., and Roberts, G. D.: Identification procedures. *In* Washington, J. A., II (ed.): Laboratory Procedures in Clinical Microbiology. Pp. 153–164. Boston, Little, Brown, 1974.
8. Dolan, C. T., *et al.:* Atlas of Clinical Mycology. Chicago, American Society of Clinical Pathologists, 1976.
9. Emmons, C. W., *et al.:* Medical Mycology. ed. 3. Philadelphia, Lea & Febiger, 1977.
10. Hopfer, R. L., and Gröschel, D.: Six hour pigmentation test for the identification of *Cryptococcus neoformans*. J. Clin. Microbiol., *2:*96–98, 1975.
11. Hopkins, J. M., and Land, G. A.: Rapid method for determining nitrate utilization by yeasts. J. Clin. Microbiol., *5:*497–500, 1977.
12. Huppert, M., *et al.:* Rapid methods for identification of yeasts. J. Clin. Microbiol., *2:*21–34, 1975.
13. Kane, J., and Fisher, J. B.: The differentiation of *Trichophyton rubrum* and *T. mentagrophytes* by use of Christensen's urea broth. Can. J. Microbiol., *17:*911–913, 1971.
14. Kaufman, L.: Serodiagnosis of fungal diseases. *In* Rose, N. R., and Friedman, H. (eds.): Manual of Clinical Immunology. Pp. 363–381. Washington, D.C., American Society for Microbiology, 1976.
15. Kaufmann, S. M.: Curvularia endocarditis following cardiac surgery. Am. J. Clin. Pathol., *56:*466–470, 1971.
16. Koneman, E. W.: Laboratory Identification of Saprobic Fungi. Check Sample MB-82. Chicago, American Society of Clinical Pathologists, 1976.
17. Koneman, E. W., and Roberts, G. D.: The Dermatomycoses. Check Sample MB-66. Chicago, American Society of Clinical Pathologists, 1973.
18. ——: Clinical and Laboratory diagnosis of mycotic disease. *In* Davidsohn, I., and Henry, J. B. (eds.): Clinical Diagnosis by Laboratory Methods. ed. 16, Chap. 50. Philadelphia, W. B. Saunders, *in press.*
19. Koneman, E. W., Roberts, G. D., and Wright, S. F.: Practical Laboratory Mycology. ed. 2. Baltimore, Williams & Wilkins, 1978.
20. Raper, K. B. and Fennell, D. L.: The Genus *Aspergillus*. Baltimore, Williams & Wilkins, 1965.
21. Rhodes, J. C., and Roberts, G. D.: Comparison of four methods for determining nitrate utilization by cryptococci. J. Clin. Microbiol., *1:*9–10, 1975.
22. Rippon, J. W.: Medical Mycology: The Pathogenic Fungi and the Pathogenic Actinomycetes. Philadelphia, W. B. Saunders, 1974.
23. Roberts, G. D.: Laboratory Identification of the Dematiacious Fungal Etiologic Agents of Chromoblastomycosis and Eumycotic Mycetoma. Check Sample AMB-88. Chicago, American Society of Clinical Pathologists, 1977.
24. ——: Laboratory diagnosis of fungal infections. Hum. Pathol., *7:*161–168, 1976.
25. ——: Fungal Serologic Tests. Check Sample AMB-16. Chicago, American Society of Clinical Pathologists, 1976.
26. ——: Detection of fungi in clinical specimens by phase-contrast microscopy. J. Clin. Microbiol., *2:*261–265, 1975.
27. Roberts, G. D., and Washington, J. A., II: Detection of fungi in blood cultures. J. Clin. Microbiol., *1:*309–310, 1975.
28. Staib, F., and Senska, M.: Der Braunfarbeffkt (BFE) bei *Cryptococcus neoformans* auf Guizzotia abyssinica-kreatinin-agar in abhangigkeit vom ausgangs pH-wert. Zentralbl. Bakteriol. (Orig. A), *225:*113–124, 1973.
29. Wang, H. S., Zeimis, R. T., and Roberts, G. D.: Evaluation of a caffeic acid-ferric citrate test for rapid identification of *Cryptococcus neoformans*. J. Clin. Microbiol., *6:*445–449, 1977.
30. Wilson, J. W., and Plunkett, O. A.: The Fungous Diseases of Man. Berkeley, University of California Press, 1965.

14 Parasitology

INTRODUCTION: HISTORICAL PERSPECTIVES

Because the relatively large size of many of the more common animal parasites makes them visible to the eye, ancient man undoubtedly knew something of their structures and the diseases which they cause. Parasites such as ascaris (roundworms), enterobius (pinworms), and tapeworms were known in ancient Egypt, Greece, and Rome. Mosaic Law prohibited the ingestion of certain animal foods (pork for example), perhaps not so much because of religious taboos but because a high percentage of people who transgressed this law developed parasitic infection. On the other hand, the ancient Chinese are purported to have believed that a man should harbor at least three worms to remain healthy; and in Europe as late as the 18th century, the presence of worms in children was considered beneficial.[7]

Parasitology as a science was quite well developed by the middle of the 19th century. By the mid 17th century, Redi, who has been called the father of parasitology, wrote the first illustrated text on the subject. Andry published his first text in 1699, and by the 1781 edition he had added a number of detailed illustrations. During the late 18th century, Gotz wrote the most comprehensive text on the parasites of man that had been published up to that time.

The name of Gotz, however, brings up the controversy of spontaneous generation that at one time was as hotly debated in reference to the parasites as it was with the origin of bacteria. Andry, for one, did not believe in spontaneous generation and in 1699 stated that the presence of intestinal worms was due to the ingestion or the inhalation of their own seeds. Gotz, however, in company with Bloch, strongly argued for the case of spontaneous generation of parasites, and he and Bloch were awarded gold and silver medals by the Royal Society of Science of Copenhagen for their essays on the subject. Brera, on the contrary, stated in 1798 that worms developed from eggs taken in the food. He stated: "Perhaps in time more happy observers will discover the eggs of the principal human worms in the bowels of the animals from which we take our food. . . ."

In the 1860's the life cycles of many of the parasites were clearly established, and parasitologists of the day had traced various parasites through their adult, egg, host, and back to adult stages, essentially putting the question of spontaneous generation to rest. Many excellent texts were written by the mid 19th century, and the nature of many human parasites was well established.

There were a number of major advances during the late 19th century, but only a few can be mentioned here. Much of this history has been reviewed in an interesting text by Foster.[4] The names of Bilharz, Cobbold, Looss, and Manson stand out for their work on schistosomiasis. Leuckart and Looss developed the basic knowledge required to understand hookworm infestation. Wucherer and Bancroft have been immortalized in the name of the filarial worm causing elephantiasis. Gruby, Lewis, Evans, and notably Major David Bruce were all instrumental in defining the disease and the life cycle of the protozoa causing trypanosomiasis. Theobald Smith is famous for his investigation of Babesia as the cause of Texas fever in cattle. Laberan, Ronald Ross, Manson, and Grassi made major contributions in piecing together the puzzle of the epidemiology and transmission of the plasmodia protozoa responsible for malaria.[4]

The basic understanding of parasitology has not significantly advanced in the 20th century. The life cycles of virtually all parasitic diseases of man have been well established and preventive measures are in effect in most of the developed countries in the world. This does not negate the fact that parasitic diseases still account for inestimable loss of life, widespread morbidity, and in many countries the retardation of economic development. With the increasing travel and mobility of individuals into all parts of the world, it is necessary that personnel working in clinical microbiology laboratories remain adept in the recognition of the various parasites that may be encountered in clinical materials submitted to virtually any diagnostic laboratory.

THE RISK OF PARASITIC INFECTION AND ITS PREVENTION

The factors to consider in assessing the risk of acquiring parasitic infections during travel to infested areas of the world and the prophylactic measures necessary to prevent the acquisition of these diseases have been reviewed by Warren and Mahmoud.[17] At lowest risk is the businessman who stays in first-class hotels in large cities of developed countries for short periods of time. At the opposite end of the spectrum are volunteers or missionaries who live in tents or native dwellings in rural settings or less-developed countries for long periods of time.

Most parasitic diseases are contracted either through the ingestion of contaminated food or water or through the sting or bite of an arthropod vector. Drinking of untreated water is particularly hazardous. Also, since most intestinal parasites withstand freezing, contaminated ice water is equally unsafe. Hot tap water is relatively safe in that the infective forms of most intestinal parasites are heat-sensitive. Ingesting fresh milk should be avoided in endemic areas. Carbonated bottled beverages are usually safe.

Undercooked meats or raw fresh-water fish can transmit liver flukes and tapeworms. Raw vegetables are relatively safe if peeled before eating; however, lettuce is particularly difficult to rid of infectious eggs, cysts, or infectious bacteria.

Precautions should be taken to reduce insect bites in tropical areas. The use of screens, bug bombs, insect repellents, and long-sleeved protective clothing is highly recommended. Travelers to foreign countries, particularly to underdeveloped regions in the tropical or subtropical climates, should consult local health authorities about an appropriate immunization program. Travelers to areas where malaria is endemic should receive chloroquine prophylaxis.

Travelers to tropical regions should be warned against swimming in fresh water. The infective larvae of Schistosoma abound in many fresh-water rivers, lakes, and canals and can easily penetrate unbroken skin. Chlorinated water used in swimming pools is safe and larvae are not found in sea water.

The examining physician should make an effort to obtain any history of recent travel into regions where parasitic diseases are endemic and should question the patient carefully about the conditions under which he lived. The laboratory should be informed of any suspected parasitic disease so that relevant specimens can be collected and the proper procedures carried out for optimal recovery of the diagnostic forms.

CLINICAL MANIFESTATIONS OF PARASITIC DISEASE

Specific signs and symptoms of parasitic diseases will be briefly reviewed in the discussions of the various groups of parasites that follow below. The most common symptom of intestinal parasitic infestation is diarrhea, which may be bloody and/or purulent. Cramping abdominal pain may be a prominent feature of parasitic diseases in which the bowel mucosa or wall is invaded, such as infestations with hookworms, mansonian or oriental schistosomes, or intestinal flukes. Heavy infection with *Ascaris lumbricoides* can result in small bowel obstruction. Patients with tapeworms may be asymptomatic except for weight loss despite increased appetite and food intake.

Hepatosplenomegaly is a common manifestation of liver fluke infestation. Portal hypertension in particular can be caused by *Schistosoma mansoni,* and jaundice is a common manifestation. Space-occupying cystic lesions of the liver and other organs can be found in amebiasis, echinococcosis, and *Taenia solium* cysticercus infections.

Suprapubic pain, frequency of urination, and hematuria are highly suggestive of *Schistosoma haematobium* infection. Transient pneumonitis may be experienced during the larval migratory phases of ascaris or hookworm infections. Cough, chest pain, and hemoptysis, together with the formation of parabronchial cysts, are common manifestations of *Paragonimus westermani* lung fluke infestation. Low-grade fever, weight loss, and skeletal muscle pain point to possible infection with *Trichinella spiralis.* Focal itching of the skin may occur at the sites of penetration of hookworm or schistosome larvae.

Peripheral blood eosinophilia (15 to 50%) is one of the more important signs that a parasitic infestation may be present. An increased concentration of eosinophiles may also be observed in various body secretions such as sputum, diarrheal stools, suppurative exudates, or fluids from pseudocysts or various body cavities. A generalized urticarial skin rash may also point to parasitic infection, thought to be a hypersensitivity reaction secondary to the metabolic products or lytic products of dead organisms that are absorbed into the circulation.

Generalized constitutional symptoms are more commonly experienced secondary to

infections with the blood parasites. Fever, chills, night sweats, lassitude, and weight loss are common manifestations of malaria, leishmaniasis, and trypanosomiasis. Varying degrees of hepatosplenomegaly and lymphadenopathy are also seen with these diseases. Neurologic signs and symptoms ("sleeping sickness") are commonly seen with African trypanosomiasis; cardiac myopathy is one of the more serious complications of South American trypanosomiasis *(Trypanosoma cruzi).* Huge swellings of the legs, arms, and scrotum (elephantiasis) are seen in filariasis due to chronic scarring and destruction of lymphatics. Localized subcutaneous nodules or serpiginous inflammatory areas in the skin may be seen in diseases such as onchocerciasis, dracunculiasis, or in the cutaneous manifestations of visceral larva migrans.

These are only a few of the general and specific clinical manifestations of some of the classic parasitic diseases. One of the most interesting aspects of parasitic infections is the propensity of certain parasites, even to the point of being species-specific in some instances, to invade and infest a specific organ or tissue. This so-called organotropism, in which there are complex larval migrations from the site of primary inoculation to a specific distant end organ where the adult forms develop and mature, is poorly understood. Many of the parasites can complete their life cycles only in specific animal or arthropod hosts, a better understanding of which has been important in the epidemiology of parasitic diseases and in developing a prophylactic approach to the management of them.

LIFE CYCLES OF PARASITES OF IMPORTANCE TO MAN

The question, "Do we have to learn all these complicated life cycles?" is often asked by new students learning the rudiments of parasitology. Except under conditions of extremely conservative tutelage, which fortunately is rapidly disappearing, the answer is a qualified "no." However,

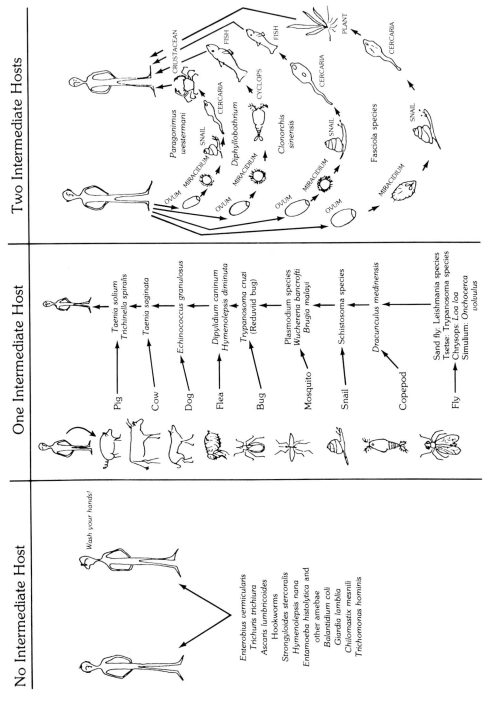

No Intermediate Host

Wash your hands!

Enterobius vermicularis
Trichuris trichiura
Ascaris lumbricoides
Hookworms
Strongyloides stercoralis
Hymenolepsis nana
Entamoeba histolytica and
    other amebae
Balantidium coli
Giardia lambia
Chilomastix mesnili
Trichomonas hominis

One Intermediate Host

Pig → *Taenia solium*
      Trichinella spiralis
Cow → *Taenia saginata*
Dog → *Echinococcus granulosus*
Flea → *Dipylidium caninum*
       Hymenolepsis diminuta
Bug → *Trypanosoma cruzi*
      (Reduvid bug)
Mosquito → *Plasmodium* species
           Wuchereria bancrofti
           Brugia malayi
Snail → *Schistosoma* species
Copepod → *Dracunculus medinensis*
Fly → Sand fly: *Leishmania* species
      Tsetse: *Trypanosoma* species
      Chrysops: Loa loa
      Simulium: Onchocerca
        volvulus

Two Intermediate Hosts

Paragonimus westermani — OVUM — MIRACIDIUM — SNAIL — CERCARIA — CRUSTACEAN — FISH

Diphyllobothrium — OVUM — MIRACIDIUM — CYCLOPS — FISH

Clonorchis sinensis — OVUM — MIRACIDIUM — SNAIL — CERCARIA — FISH

Fasciola species — OVUM — MIRACIDIUM — SNAIL — CERCARIA — PLANT

Fig. 14-1. An overview of the life cycles of parasites of importance to man.

if microbiologists are to be effective diagnostic parasitologists, they must at least understand enough of the various life cycles so that they can recognize the infective and diagnostic forms of those parasites that commonly cause disease in man, and know something about the intermediate hosts that may play a significant role. It is probably not possible, or necessary, to retain in memory every facet of all the life cycles, with the possible exception of those individuals who are actively teaching the subject on a continuous basis.

Figure 14-1 has been designed to assist those readers whose exposure to parasitic diseases is infrequent. Parasites can be divided into three major groups on the basis of their life cycles: (1) those having no intermediate host; (2) those having one intermediate host; and (3) those in which two intermediate hosts are necessary.

Parasites having no intermediate hosts are transmitted directly from man to man (or animal to animal), usually through fecally contaminated food or water supplies. With an organism such as *Enterobius vermicularis,* transmission is virtually directly anal to oral. With the amebas and flagellates and many of the intestinal helminths, transmission is via the formation of a cyst stage or the production of ova which can survive environmental conditions outside of the host until ingested by a second individual. Although the hookworms do not require an intermediate host, their life cycle is somewhat more complex in that the ova that are passed into the environment by an infected host must first undergo two larval molts (rhabditiform and filariform larvae) before a second host can become infected.

Parasites requiring one intermediate host commonly select either a large mammal, a crustacean, or an insect vector within which to complete their life cycle. This may be either a simple or a complex process. For example, man serves as the primary host for *Taenia solium* and *Taenia saginata* since the adult forms of these tapeworms reside in the intestine, while the pig or cow, respectively, serve as secondary hosts (the larvae reside in

skeletal muscle). However, man may also serve as the secondary host for *T. solium* in the form of cysticercosis, which is the larval form of this tapeworm. Similarly, man serves as an inadvertent secondary host for echinococcal disease, within whom the encysted larval forms develop in the liver, brain, and other organs. The dog is the primary host harboring the adult echinococcal tapeworm in the intestine.

When crustaceans or insects are used as the intermediate host, the parasite usually goes through a complex series of developmental stages before the infective form is released. In malaria, for example, the plasmodia undergo sexual gametogenesis within the mosquito before infective sporozoites are injected back into a new host. With the schistosomes, the snail serves as the intermediate host within which a series of maturation stages (redia) occur by which a single miracidium that is ingested is multiplied into numerous infective cercaria.

Parasites that require two intermediate hosts follow essentially the same life cycles. Ova that are passed from the primary host hatch in a suitable aqueous environment and release miracidia. These are ingested usually by a snail (a cyclops with *Diphyllobothrium latum*) which in turn produces numerous cercaria. The cercaria, in turn, are ingested by crustaceans or fish, where they become encysted as metacercaria in the somatic musculature. A second host becomes infected by ingesting the metacercaria in raw or inadequately cooked crab or fish. With *Fasciola* or *Fasciolopsis,* the metacercaria become attached to water plants, and man becomes infected by ingesting uncooked, contaminated water chestnuts.

For these reasons, when taken from the perspective as presented in Chart 14-1, the need to memorize individual facts becomes less and the broad concepts are better retained and serve as working models for practical applications. For an in-depth review of the life cycles of the common parasites of man, refer to the publications of Brooke and Melvin[1]; Melvin, Brooke, and Healey[12]; and Melvin, Brooke, and Sadun.[13]

COLLECTION, TRANSPORT, AND PROCESSING OF SPECIMENS

As with the study of other microorganisms in the laboratory, appropriate specimens must be collected from the patient and transported to the laboratory in a condition sufficiently preserved to allow the detection and identification of any parasitic forms that may be present. The diagnosis of parasitic infections relies in large part on the macroscopic and/or microscopic examination of feces, urine, blood, sputum, and tissues. The implementation of reliable laboratory processing techniques is an integral step in this process. In this chapter it is not possible to review more than a few of the commonly used laboratory procedures that can aid in the recovery and identification of parasitic forms in clinical specimens. For a succinct and practical overview of these procedures, the reader is referred to the manual of Garcia and Ash[5] and the publication of Melvin and Brooke.[11] The texts by Brown,[2] Faust and associates,[3] and Markell and Voge[10] are recommended for a more in-depth coverage of parasitic diseases.

Fecal Specimens

Since medications containing mineral oil, bismuth, antibiotics, antimalarials, or other chemical substances may compromise the detection of intestinal protozoa, examination of specimens must be delayed for one or more weeks after therapy is stopped. Patients who have received a barium enema may not reveal organisms in their stools for at least one week following the enema.

Stool specimens should be collected directly during bowel movement in a clean, wide-mouthed container with a tight-fitting lid. Specimens that are admixed with water (for example, from the toilet bowl) or with urine are unsuitable because the motility of intestinal protozoa may be retarded. The lid should be tightly fitted to the container immediately after collection of the sample to maintain adequate moisture. Every specimen container must be properly labeled as outlined in Chaper 1.

Three fecal specimens are sufficient to make the diagnosis of intestinal parasitic disease, two collected during normal bowel movement and a third after a Fleet's Phospho-Soda or magnesium sulfate purge. Cathartics with an oil base should be avoided since oils tend to retard motility of trophozoites and distort the morphology of the parasites. A total of six specimens may be required if intestinal amebiasis is suspected. Collections should be spaced at least one day apart. More than six samples in a 10-day period rarely yield additional information.

Preservation of Specimens. Many stool specimens for examination of ova and parasites are collected either at home, in the physician's office, or in a clinic some distance from the laboratory that performs the examination. Optimally, specimens should be examined within one hour after collection for the detection of motile trophozoites. If this is not possible, a portion of the specimen should be placed in a suitable fixative which will preserve both cyst and trophozoite forms for examination at a later time. Polyvinyl alcohol (PVA) is highly recommended for this purpose and is commercially available in liquid form from Delkote, Inc.,* and is used in a ratio of 3 parts PVA to 1 part fecal specimen.

Merthiolate-iodine-formalin (MIF) solution can also be used as a fixative. Two solutions must be prepared, and stored separately:

Solution I:

Merthiolate, 1:1000	250 ml.
Formaldehyde, 10% aqueous	25 ml.
Glycerol	5 ml.
Distilled water	250 ml.

Solution II:

Distilled water	100 ml.
Potassium iodide	10 g.
Iodine crystals	5 g.

The shelf life of each solution is many months if stored at room temperature in a brown bottle. Immediately before use, 1 ml. of solution II is added to 13 ml. of solution I

_____

*Delkote, Inc., Penns Grove, New Jersey, 08069

and about one fourth of a teaspoon of fresh feces is added and mixed.

Many parasitologists prefer PVA over MIF and other formalin-containing fixatives because permanent-stained slides can be made from PVA-preserved material while MIF material can be examined as a wet mount only.

Garcia and Ash[5] have proposed the assembly of a fecal collection kit for clinical use along with specific instructions for the collection and preservation of fecal specimens. The kit includes all containers, vials, applicator sticks, preservatives, and labels necessary for proper collection.

Specimens for parasitic examinations shipped in the mail must meet the CDC regulations as outlined in Chapter 1.

Visual Examination of Specimens. Stool specimens should be visually examined for the presence of barium, oils, or other materials that may render them unacceptable for further processing. Patches of blood or mucin should be specifically selected for microscopic study because they may be derived directly from ulcers or purulent abscesses where the concentration of amebae may be highest.

Liquid stools should be microscopically examined immediately for the presence of trophozoites. If the stool is formed or semisolid, there is little chance that motile trophozoites are present and immediate microscopic examination may not be helpful. It is acceptable to hold formed stools in the refrigerator overnight for further processing because the cyst forms of protozoa and the ova of intestinal helminths retain morphologic detail.

Microscopic Examination of Specimens. For optimal detection and identification of parasites in feces, wet mounts for microscopic examination should be prepared directly from fresh unfixed fecal material and from sediment prepared from a concentration procedure. Garcia and Ash[5] also strongly recommend the examination of permanent-stained preparations as well as a wet mount, especially in the identification of intestinal protozoa.

Direct wet mounts should be prepared both in saline and in a small amount of Lugol's iodine. The saline mount is made by emulsifying a small portion of fecal material in a drop of physiologic saline on a microscope slide and overlaying the mixture with a coverslip. Saline mounts have the advantage of retaining the motility of trophozoites; however, definitive identification of either trophozoites or protozoan cysts in these preparations is difficult because the internal structures are often poorly defined.

Iodine is used to highlight the internal structures of the parasites present. Iodine mounts alone, however, may not be satisfactory because the motility of trophozoites is often destroyed. Both types of mounts can be prepared and studied on a single microscope slide.

The Velat-Weinstein-Otto (VWO)[16] supravital stain can prove useful in demonstrating the internal structures of motile trophozoites if the identification cannot be made on the basis of the saline or iodine mounts. The active ingredients of this stain are crystal violet and hematoxylin. The stain is particularly useful in demonstrating nuclear chromatin which appears as a deep purple-black against the light purple background of the cytoplasm. The preparation and use of this stain have been described by Koneman, Richie, and Tiemann.[8]

Concentration Methods. Since parasitic eggs, cysts, and trophozoites in fecal material are often in such low numbers that they are difficult to detect in direct smears or mounts, procedures to concentrate these structures should always be performed. Two types of concentration procedures are commonly used: (1) flotation and (2) sedimentation. These procedures are designed to separate intestinal protozoa and helminth eggs from excess fecal debris. It must be remembered that both of these procedures inactivate the motile forms of protozoa.

When fecal material is concentrated by the flotation technique, ova and cysts float to the top of the zinc sulfate solution having a specific gravity of 1.180. The specific gravity of the protozoa and many of the helminth ova is lower than 1.180. For example, the specific gravity of hookworm ova is 1.055; of ascaris ova, 1.110; of trichuris ova, 1.150; and of giardia cysts, 1.060. Therefore, these and other forms of parasites will float to the top of the fecal suspension and can be collected by placing a coverslip on the meniscal surface of the fecal suspensions. Many of the trematode and cestode ova as well as the infertile eggs of *Ascaris lumbricoides* have a specific gravity greater than 1.180 and do not float. Although it is possible to use solutions of greater specific gravity, the cysts or ova recovered may be distorted.

Concentration of intestinal parasites by sedimentation techniques, using either gravity or light centrifugation, will lead to the recovery of essentially all protozoa, eggs, and larvae present in the specimen. Preparations concentrated by sedimentation contain more debris than those obtained by flotation. If only one technique is to be used, the sedimentation procedure is generally recommended. Both of these concentration procedures are described in Chart 14-1.

Permanent-Stained Smears of Fecal Specimens. Although temporary wet mounts of fecal material for direct microscopic examination facilitate the rapid detection of intestinal parasites in stool specimens, the detection of *Entamoeba histolytica* or other protozoan infections can be greatly enhanced by preparing permanently stained slides. Permanently stained smears are superior to wet mounts for demonstrating detailed morphology of the parasitic forms and are well suited to keep as slide sets for future study, teaching, or reference work. In addition, a fixed smear can be sent either stained or unstained to a reference laboratory for examination and consultation.

Two permanent stains for demonstration of intestinal protozoa are commonly used: (1) the Gomori's trichrome stain and (2) the iron hematoxylin stain. The iron hematoxylin stain is the time-honored technique used for the most exacting definition of the morphology of intestinal parasites. The staining procedure is complicated and must be performed by an experienced person in order to achieve the best results. The trichrome stain is generally recommended for use in most diagnostic laboratories because it is simple and easy to perform and produces uniformly good results with both fresh and PVA-preserved fecal material. The trichrome staining procedure is reviewed in detail in Chart 14-2.

Permanent-stained smears are prepared by spreading a thin film of fecal material on the surface of a glass slide. Smears should be prepared from specimens that are as fresh as possible and the smears must not be too thick. An old, thick smear that has been inadequately fixed may result in failure of the organisms to stain. The fixatives recommended are either PVA or Schaudinn's solution as described in Chart 14-2. When properly stained, the organisms have a blue-green to purple cytoplasm and red to purple-red chromatin against a green staining background material. Helminth eggs and larvae have a red to purple appearance.

Garcia and Ash[5] have described other techniques for the examination of fecal specimens, particularly useful for detection of infections with hookworms, Strongyloides, and Trichostrongylus. These include the Harada-Mori filter paper strip culture technique, the filter paper/slant culture method, charcoal culture technique, Baermann procedure for culture of Strongyloides larvae, methods for performing egg counts, and the hatching of schistosome eggs. Although these techniques are beyond the needs of most clinical laboratories, there may be occasions when one or more may be useful.

Examination of Intestinal Specimens Other than Stools

Parasites such as *Giardia lamblia* and *Strongyloides stercoralis* commonly inhabit the duodenum and jejunum and in some infections may not appear in stool specimens. Duodenal drainage may be required to demonstrate these organisms. The aspirated material should be examined microscopically by preparing a saline wet mount. If motile organisms are seen, a second preparation in a drop of iodine may be helpful in highlighting the characteristic internal structures and facilitating a definitive identification (Tables 14-2 and 14-3).

Dudoenal contents can be most easily obtained through the use of a weighted gelatin capsule within which is coiled a length of nylon string. This string protrudes from one end of the capsule and the free end is taped to the face of the subject. The capsule is swallowed and peristalsis carries the weighted string into the duodenum. After leaving in place for 4 to 6 hours, the string is removed and any bile-stained mucus adhering to the distal end is sampled and examined microscopically by a direct saline mount and a stained smear.

Enterobius vermicularis infection of the rectal-anal canal is best detected through the use of the cellulose tape technique. The adhesive surface of a 3- or 4-inch strip of clear cellulose tape is applied to the perianal folds of the patient suspected of having pinworm infection. A tongue blade can be used to provide a firm backing for the tape. The tape is then placed adhesive side down on a glass microscope slide and examined for the characteristic ova of *E. vermicularis* (Table 14-3). Optimal recovery of eggs is achieved if specimens are collected in the early morning soon after the patient arises.

In some cases of *Entamoeba histolytica* infection where repeated stool examinations have failed to reveal the organisms, examination of sigmoid biopsy material may be helpful. The biopsy should be scheduled so that parasitic examination can be made immediately. If a delay in the examination cannot be prevented, the specimen should be placed in PVA fixative, not in formalin. Several smears of a portion of the material should be prepared and stained with the trichrome stain as described above. The remaining portion of the biopsy should be submitted to a histology laboratory for the preparation of tissue sections.

Examination of Specimens Other than from the Intestinal Tract

Sputum. On rare occasions, sputum samples may be submitted to detect the larval stages of hookworm, Ascaris, or Strongy- loides, or the ova of *Paragonimus westermani.* Usually a direct saline mount is sufficient. If the sputum is unusually thick or mucoid, an equal quantity of 3 per cent n-acetyl cystine can be added, mixed for two or three minutes, and the specimen centrifuged. After centrifugation the sediment is examined microscopically in a wet mount.

Urine and Body Fluids. Fluid specimens should be centrifuged and the sediment examined in wet mount preparations. If suspicious parasites are seen, a second iodine preparation may be helpful in highlighting the diagnostic internal structures.

Tissue Biopsy. It is important that biopsy tissue be submitted to the laboratory without the addition of formalin, or, if necessary, be placed in PVA fixative if a delay in processing cannot be prevented. If the specimen is soft, a small portion should be scraped free and placed in a drop of saline for a wet mount examination. Impression smears should also be prepared by pressing a freshly cut surface of the tissue against a glass slide, which should be placed immediately in Schaudinn's fixative prior to preparation of the trichrome stain. The remaining portion of the biopsy material should be submitted for the preparation of histologic tissue slides.

Muscle. The characteristics spiral larval forms of *Trichinella spiralis* are best demonstrated in a tease mount made from a skeletal muscle biopsy (Fig. 14-2). Garcia and Ash[5] suggest that the biopsy material should be treated with an artificial digestive fluid prepared by placing 5 g. of pepsin in 1000 ml. of distilled water to which has been added 7 ml. of concentrated HCl prior to examination. The tissue is added to this digestion mixture in the ratio of 1 part tissue to 20 parts fluid in an Erlenmeyer flask and incubated at 37° C. in an incubator for 12 to 24 hours. After digestion, examine a few drops under the microsocpe for the presence of larvae. If none are seen, centrifuge a 15-ml. aliquot of the mixture and examine the sediment for larvae.

Blood. Microfilariae and trypanosomes may be detected and identified by their char-

(Text continues on p. 440)

Chart 14-1. *Fecal Concentration Techniques for the Recovery of Intestinal Parasites*

Introduction	The number of parasitic forms in fecal specimens is often too low to be observed microscopically in direct wet mounts or stained smear preparations. Concentration procedures must therefore be employed in order to detect them. The two most commonly used techniques are *flotation* and *sedimentation*.
Principle	*Flotation:* Protozoan cysts and helminth ova of low specific gravity can be made to float to the surface of a solution with a high specific gravity. Zinc sulfate solution, specific gravity 1.180, is most commonly used. The parasites that float to the surface can be collected either by skimming the top of the solution with a wire loop or capillary pipette, or from the undersurface of a coverslip that is placed on the meniscus. *Sedimentation:* The specific gravity of protozoan cysts and helminth ova is greater than water or saline-fecal suspensions and they tend to settle out. This process can be accelerated by light centrifugation. A formalin-ether mixture is often added to the fecal suspension to clear out the fecal debris and fix the parasitic forms that may be present.
Media and Reagents	*Flotation Technique:* The zinc sulfate solution ($ZnSO_4$, specific gravity 1.180) is prepared by adding 331 g. of $ZnSO_4$ to 1000 ml. of warm tap water. Check the specific gravity with a hygrometer and adjust to 1.180 by adding zinc sulfate or water as needed. *Sedimentation Technique:* 1. Formalin, 10% solution 2. Ether (Do not use around open flames.)
Procedure	*Flotation:* 1. Transfer about ½ teaspoon of stool specimen to a test tube containing 3 to 4 ml. of water. Mix thoroughly using applicator sticks. 2. Fill the tube to within 5 mm. of the top, mix again, and centrifuge at 1500 r.p.m. for 1 minute. Discard supernatant. 3. Add 1 to 2 ml. of the zinc sulfate solution to the sediment and resuspend the material using an applicator stick. 4. Fill the tube to within 2 to 3 mm. of the top with additional zinc sulfate solution. 5. Strain the suspension through gauze into a clean beaker. Discard the gauze and return the filtrate to the tube. Again add zinc sulfate solution to within 2 or 3 mm. of the top of the tube. 6. At this point, do one of the following: A. Raise the fluid level in the tube to the brim with zinc sulfate, not allowing any runover. Place a clean no. 1 coverslip on top of the tube so that the undersurface touches the meniscus. Leave undisturbed for 10 minutes.

Procedure *(Continued)*	B. Centrifuge the suspension for 1 minute at 1500 r.p.m. Without shaking or spilling the contents, carefully remove the tube from the centrifuge and place upright in a rack.

7. Place 1 or 2 drops of iodine or water on a microscope slide. Either deftly remove the coverslip and lower it on the drop of iodine, or, with a wire loop or capillary pipette, remove a drop of the top film to the drop of iodine, depending upon which procedure (A or B) in step 6 was used.
8. Examine microscopically for the presence of cysts or ova.

Sedimentation:

1. Thoroughly mix a portion of stool specimen about the size of a walnut into 10 ml. of saline. This suspension should yield about 1 or 2 ml. of sediment.
2. Filter the emulsion through fine mesh gauze into a conical centrifuge tube.
3. Centrifuge at 2000 r.p.m. for 1 minute.
4. Decant the supernatant and wash the sediment with 10 ml. of saline. Centrifuge again and repeat washing until supernatant is clear.
5. After the last wash, decant the supernatant and add 10 ml. of 10% formalin to the sediment. Mix and let stand for 5 minutes.
6. Add 1 or 2 ml. of ether. Stopper the tube and shake vigorously.
7. Centrifuge at 1500 r.p.m. for one minute. Four layers should result as follows: (1) top layer of ether; (2) plug of debris; (3) layer of formalin; and (4) sediment.
8. Free the plug of debris from the sides of the tube by ringing with an applicator stick. Carefully decant the top three layers.
9. With a pipette, mix the remaining sediment with the small amount of remaining fluid and transfer one drop each to a drop of saline and iodine on a glass slide, and coverslip and examine microscopically for the presence of parasitic forms.

Comments

The sedimentation procedure can also be used to process PVA-fixed specimens. The flotation mount should be prepared within one hour after preparation because cysts and ova will begin to resettle after that time. If the sediment cannot be immediately examined when first prepared in the sedimentation technique, a small amount of 10% formalin can be added and the tube tightly capped to prevent drying for examination at a future time.

Bibliography

Garcia, L. S., and Ash, L. R.: Diagnostic Parasitology: Clinical Laboratory Manual. St. Louis, C. V. Mosby, 1975.

Koneman, E. W., Richie, I. E., and Tiemann, C.: Practical Laboratory Parasitology. Baltimore, Williams & Wilkins, 1974.

Pruneda, R. C., Cartwright, G. W., and Melvin, D. M.: Laboratory diagnosis of parasitic diseases. *In* Koneman, E. W., and Britt, M. S. (eds.): Clinical Microbiology. Lecture 15. Bethesda, Md., Health and Education Resources, Inc., 1977.

Chart 14-2. *Trichrome Staining Technique for Fecal Smears*

Introduction	Since the refractive index of protozoan cysts and some helminth ova is near that of water, staining techniques are required to study the details of their internal structures. Permanent stains also enhance the detection of *Entamoeba histolytica,* the most important disease-producing ameba for humans. The trichrome staining technique is a rapid procedure which gives good results for routine purposes.
Principle	The trichrome stain is a modification of the Gomori's stain which contains chromotrope 2R and light green SF as the primary staining agents. Smears made from fresh fecal material must be fixed; PVA samples are already fixed and need no additional treatment.

Media and Reagents

Schaudinn's Fixative Solution:
1. Saturated mercuric chloride ($HgCl_2$):

Mercuric chloride	110 g.
Distilled water	1000 ml.

In a fume hood, boil until the $HgCl_2$ is dissolved and let stand to cool until crystals form.
2. Stock solution:

Mercuric chloride	600 ml.
Ethyl alcohol, 95%	300 ml.

Immediately prior to use add 5 ml. of glacial acetic acid per 100 ml. of stock solution.

Gomori's Trichrome Stain:

Chromotrope 2R	0.6 g.
Light green SF	0.3 g.
Phosphotungstic acid	0.7 g.
Acetic acid (glacial)	1.0 ml.
Distilled water	100.0 ml.

Add 1.0 ml. of glacial acetic acid to the dry components. Allow mixture to stand for 15 to 30 minutes and then add the 100 ml. of distilled water. A good stain is purple in color.

Procedure

Preparation of Material:
1. With a small portion of the fresh stool specimen, prepare two smears on microscope slides using sticks or a brush. The material should be spread thin enough so that newsprint can be read through the smear.
2. Immerse the smears immediately in Schaudinn's fixative solution and allow to fix for a minimum of 30 minutes. Overnight fixation is preferred.

Chart 14-2. *Trichrome Staining Technique for Fecal Smears (Continued)*

Procedure *(Continued)*	3. If the specimen is liquid, mix several drops of fecal material with 3 or 4 drops of PVA on a slide and let dry for several hours in a 37° C. incubator. 4. If the specimen is in PVA, pour some of the mixture onto a paper towel to absorb out the PVA. Prepare slides of the material from the paper towels as described above. Let dry for several hours in a 37° C. incubator. *Staining Technique:* 1. After smears have properly fixed and dried, place the slides in 70% ethyl alcohol and leave for 5 minutes. 2. Wash with two changes of 70% alcohol, one for 5 minutes and one for 2 to 5 minutes. 3. Wash with trichrome stain solution for 10 minutes. 4. Wash with 90% ethyl alcohol, acidified (1% acetic acid) for up to 5 seconds. 5. Dip once in 100% ethyl alcohol. 6. Wash with two changes of 100% ethyl alcohol for 2 to 5 minutes each. 7. Remove alcohol with two changes of xylene or toluene, 2 to 5 minutes each. 8. Add mounting medium and overlay with a no. 1 thickness coverslip. 9. Examine under oil immersion for parasitic forms.
Interpretation	The cytoplasm of thoroughly fixed and well-stained cysts and trophozoites is blue-green, tinged with purple. The nuclear chromatin, chromatoid bodies, and ingested red blood cells appear red or red-purple. Background material is green.
Controls	It is recommended that a number of smears of material containing parasitic organisms of known staining properties be stained in parallel with each unknown smear. If staining is of poor quality, make sure the smears are not too thick and that the staining procedure was followed exactly. Poor fixation of smears and failure to completely remove residual mercuric chloride with alcohol prior to staining are two possible reasons for poor staining.
Bibliography	Garcia, L. S., and Ash, L. R.: Diagnostic Parasitology: Clinical Laboratory Manual. St. Louis, C. V. Mosby, 1975. Pruneda, R. C., Cartwright, G. W., and Melvin, D. M.: Laboratory diagnosis of parasitic diseases. *In* Koneman, E. W., and Britt, M. S. (eds.): Clinical Microbiology. Lecture 15. Bethesda, Md., Health and Education Resources, Inc., 1977.

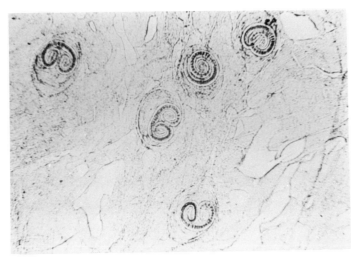

Fig. 14-2. Tease preparation of skeletal muscle showing infestation with *Trichinella spiralis* larvae (high power).

acteristic motility and shape by examining a sample of blood directly under the microscope. Plasmodia can be detected only in properly stained blood smears. Two types of smears must be prepared, a thin smear and a thick smear.

The thin blood smear, used primarily for specific species identification of plasmodia and other intraerythrocytic parasites, is prepared exactly as one used for a differential blood count. The same care must be taken to see that the feather edge is evenly spread and free of holes, streaks, or other such artifacts.

Thick blood smears are especially useful in detecting parasites in light infections because they allow examination of a larger quantity of blood than do thin smear preparations. Blood should be obtained from a finger stick and allowed to flow freely, avoiding "milking" of the finger. Place two or three drops of fresh blood on an alcohol-cleaned slide. Using a circular motion, with the corner of another glass slide or an applicator stick spread drops to cover an area the size of a dime. Continue stirring the drop for about 30 seconds to prevent formation of fibrin clots. Allow film to air dry in a dust-free area. Once the film is dry, the blood should be laked by placing the slide in water

or a buffer solution immediately prior to staining.

Both thin and thick smears should be stained with Giemsa or Wright's stain. Smears should be stained as soon as possible after preparation, and always within 48 hours. Thick smears may require a slightly longer exposure to the stain than the time used for the thin smear preparations.

THE IDENTIFICATION AND DIFFERENTIATION OF PARASITES

Although certain clinical signs and symptoms may suggest the possibility of a parasitic disease, the final diagnosis is made by demonstrating the causative organism in properly collected specimens. Because there are many artifacts which may resemble parasitic forms, the final identification must always rest on well-established morphologic criteria. Microscopic interpretations in particular cannot be left to guesswork, and a laboratory diagnosis of a parasitic disease should not be rendered until adequate identifying features can be clearly and objectively demonstrated.

One problem faced by both the new student and the teacher of clinical parasitology is the lack of a unified approach to the taxonomy of the parasites. The traditional approach of separating the parasites into various morphologic groups (protozoa, nematodes, cestodes, trematodes, *etc.*) will also be followed in this text. This is with the full realization that a certain degree of clinical and laboratory correlation is lost using this approach. For example, even though hookworms and pinworms are taxonomically included with the nematodes, there is considerable difference in their life cycles, modes of infection, and seriousness of the disease processes they cause. In fact, each species of parasite is unique unto itself and any attempts to group them by whatever criteria will meet with some degree of failure.

Before briefly discussing some of the more commonly encountered parasites, one more essential must be mentioned. It is virtually mandatory that each laboratory have at least one microscope equipped with an accurately calibrated ocular micrometer. The procedure for calibrating an ocular micrometer is shown in Chart 14-3. The ability to measure the exact size of parasitic forms encountered in clinical specimen is sometimes essential for the identification of the parasites. In the discussion that follows, emphasis will be placed on the size ranges of the various diagnostic parasite forms that are reviewed.

Intestinal Protozoa

There are three broad groups of intestinal protozoa: (1) the amebae; (2) the flagellates; and (3) the ciliates. The task of learning the features of these protozoa is somewhat lessened in that there are only five species of amebae, three species of flagellates, and one ciliate that are of practical significance. Of these, only *Entamoeba histolytica* and *Giardia lamblia* are pathogenic for man, although in rare instances *Balantidium coli* may cause primary enteritis in debilitated hosts. The other species are of interest only in that they must not be confused with the two species known to be pathogenic.

The Amebae

The five species of amebae of laboratory importance are: *Entamoeba histolytica, Entamoeba coli, Iodamoeba bütschlii, Endolimax nana,* and *Dientamoeba fragilis.* Table 14-1 summarizes the key morphologic features by which each of these species can be identified in the laboratory.

It is essential that all laboratory personnel involved in the study and identification of parasites in clinical specimens be able to identify *Entamoeba histolytica.* All of the amebae listed above, with the exception of *Dientamoeba fragilis* which does not encyst, have two stages in their life cycle, a motile trophozoite form and a cyst. There are only two absolute criteria by which a definitive identification of *Entabmoeba histolytica* can be made, that is, (1) by the demonstration of unidirectional, purposeful motility by the trophozoites, and (2) by the presence of ingested erythrocytes within the trophozoite cytoplasm. The secondary characteristics listed in Table 14-1, namely, the size and position of the karyosome, the distribution of the nuclear chromatin, and the consistency of the cytoplasm, are variable in both the trophozoite and cyst forms, so that they can be used only in making a presumptive identification. It is for this reason that it is essential that liquid or semiliquid stool specimens be examined while they are still warm when any trophozoites that may be present are still active. Trophozoites are not found in formed stools.

There are some morphologic features by which *Entamoeba histolytica* can be excluded if seen in any given parasitic form. The presence of more than four nuclei in a cyst is exclusive, and is one of the more helpful characteristics in the recognition of *Entamoeba coli* cysts. An ameba in which the nuclear membrane is invisible excludes members of the Entamoeba genus. The "entamoeba type" nucleus is characterized by the peripheral margination of the nuclear chromatin on the inner surface of a distinct nuclear membrane. Other genera of amebae

(Text continues on p. 444)

Chart 14-3. *Calibration of the Ocular Micrometer*

Introduction

An ability to accurately measure the size of trophozoites, ova, or other parasitic forms through microscopic examination is often necessary in making a species identification. This measurement can be made using a calibrated scale called a micrometer. The ocular micrometer, a small round glass disk that is etched with a fixed scale, is inexpensive and easy to use and is strongly recommended for routine laboratory use.

Principle

Ocular micrometers are etched with a fixed scale, usually consisting of 50 parallel lines. Depending upon the magnifying power of the set of objectives used in a compound microscope, each division in the ocular micrometer will represent different measurements. Therefore, for each set of oculars and objectives used, the ocular scale must be compared with a known calibrated scale. A stage micrometer etched with a scale of 0.1 and 0.01-mm. divisions is commonly used. It is important to remember that once a calibration has been made for a given set of oculars and objectives, they cannot be interchanged with corresponding components from another microscope.

Materials

Ocular micrometer with fixed scale*
Stage micrometer scaled with 0.1- and 0.01-mm. divisions*
Standard compound microscope

Procedure

1. Remove the ocular from the microscope to be used. If a binocular microscope is used, it is customary to remove the right $10\times$ ocular.
2. Unscrew the eye lens (top lens) of the ocular and insert the micrometer wafer so that it rests on the diaphragm ring inside the ocular. Place the micrometer with the engraved side down. The micrometer should be handled with lens paper and every effort made to prevent lint from adhering to the surface.
3. Replace the ocular in the housing. When viewed through the ocular, the micrometer scale will appear as a series of lined divisions, illustrated as follows:

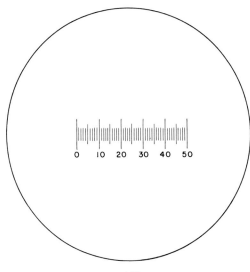

442

Chart 14-3.   *Calibration of the Ocular Micrometer* (Continued)

Procedure
(Continued)

4. Place the stage micrometer under the objective of the microscope that is to be calibrated. Bring into view the stage micrometer scale, which will appear as a series of lines divided into 0.1 and 0.01 millimeter (mm.) divisions, as shown in the following simulated view through the microscope:

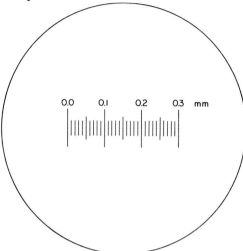

5. Adjust the stage micrometer so that the "0" line on the ocular micrometer scale is exactly superimposed with the "0" line on the stage micrometer scale. When viewed under high magnification (× 450), the superimposition of the two scales will appear as shown in this simulated view through the microscope.

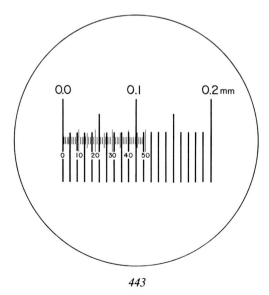

(Continued)

Chart 14-3.   *Calibration of the Ocular Micrometer (Continued)*

6. Without further manipulation, look across the two scales and find the next pair of lines that exactly coincide. In the drawing below (a simulated high-power ×450 view), the coinciding lines are the 40 mark on the ocular scale and the 0.09 mm. mark on the stage micrometer scale (arrow).

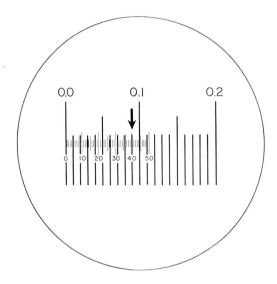

Calculation   The object of the calibration is to determine the width in micrometers (μ) of each ocular scale division, when calibrated against the stage micrometer scale. Thus, as illustrated in the above drawing, 40 units on the ocular scale are equal to 0.09 mm. on the stage micrometer scale.

Therefore, each ocular division is equal to 0.09 mm./40, or 0.00225 millimeters (mm.). Since there are 1000 micrometers (μ) in each millimeter (mm.), each ocular micrometer division in the calibration illustrated here is equal to 0.00225 × 1000, or 2.25 micrometers (μ).

Thus, if an object that is viewed under the microscope occupies 10 ocular scale divisions, it would measure 2.25 × 10, or 22.5 micrometers (μ).

This same calculation can be used for the calibration of any set of oculars and objectives, substituting the appropriate numbers.

Bibliography   Garcia, G. S., and Ash, L. R.: Diagnostic Parasitology: Clinical Laboratory Manual. Pp. 42 and 43. St. Louis, C. V. Mosby, 1975.
Koneman, E. W., Richie, L. E., and Tiemann, C.: Practical Laboratory Parasitology. Pp. 12–14. New York, Medcom Press, 1974.

*Available from American Optical Company, Scientific Instrument Division, Buffalo, N.Y. 14215.

do not have this nuclear chromatin margination and the nuclear membrane remains indistinct or invisible. Many of these species have a large central or eccentric karyosome having the appearance of a "ball in socket" within an empty space.

Iodamoeba bütschlii can usually be easily identified by demonstrating the large glycogen vacuole in the cyst (it is from this structure that the genus name is derived); however, early precysts of *Entamoeba coli* and, less frequently, *E. histolytica* may also have a similar-appearing cytoplasmic inclusion. As mentioned above, demonstrating an "entamoeba type" nucleus would favor one of the Entamoeba species whereas the nuclei of *I. bütschlii* would have the "ball-in-socket" appearance.

Endolimax nana is usually not difficult to identify because of its extremely small size (as small as 5 μ.). Only the small race of *E. histolytica,* known as *Entamoeba hartmanni,* has cysts approaching this small size. Again, the determination of whether or not the nucleus is an "endamoeba type" or not will be helpful.

Dientamoeba fragilis is difficult to identify in wet mounts and a permanently stained preparation is virtually mandatory if morphologic details are to be studied. Nevertheless, this ameba can be identified if a trophozoite is seen that has two nuclei. *D. fragilis* is the only trophozoite that has two nuclei, and the possibility of an early cyst form can be excluded because this organism never develops a cyst stage.

In summary, the features by which *E. histolytica* can be identified in clinical materials include the demonstration of trophozoites with unidirectional motility that have ingested erythrocytes. Secondary features include a smooth, finely granular cytoplasm free of debris, and nuclei both in the trophozoites and cysts that have tiny centrally placed karyosomes and nuclear chromatin evenly distributed in beadlike fashion on the nuclear membrane. The number of nuclei in the cysts will never exceed four.

Intestinal Flagellates

As the name would imply, all flagellates possess flagella, which serve as the means for locomotion. Other structures also serve as an integral part of the locomotor organ, namely, the kinetoplast to which the flagella are attached and the axostyle and parabasal bodies which have a neuromuscular function. Therefore, when any of these structures are identified in a parasite form, the parasite can be tentatively grouped with the flagellates. Unlike the amebae which assume variable shapes, the flagellates are more rigid and tend to retain distinctive shapes, a feature often helpful in their identification.

Giardia lamblia, Chilomastix mesnili, and *Trichomonas hominis* are the only species of flagellates that are commonly seen in stool specimens. *Giardia lamblia* is the only one of the three known to cause disease and also has the distinction of residing in the small rather than the large intestine. For this reason, duodenal drainage specimens may be required to make a diagnosis if stool specimens are negative.

The differential features by which these three flagellates can be identified are listed in Table 14-2. *Giardia lamblia* is bilaterally symmetrical, and the typical trophozoite with two nuclei, one on either side of a central axostyle giving the appearance of a "monkey face," is usually easy to recongize. In wet mounts, demonstrating the graceful "falling-leaf" motility can be a helpful identifying feature, distinguishing it from *C. mesnili,* which has a slower, stiff motion, and from *T. hominis,* which is quick, jerky, and darting.

The most helpful feature in identifying *C. mesnili* is the large anterior nucleus and the presence of the prominent cytostome. *T. hominis* may be somewhat more difficult to definitively identify because it is fragile and does not stain well. The demonstration of an undulating membrane that extends the full length of the organism is a helpful finding. *T. vaginalis,* which can contaminate fecal specimens, can be differentiated because its undulating membrane only extends one half the distance of the body.

Ciliates: *Balantidium coli*

Balantidium coli is the only member of the ciliates known to infect man. It is quite easy to recognize because of its large size (100 μ. or greater in diameter), an outer membrane covered with short cilia, and its large kidney-shaped macronucleus. When observed in wet mounts, the trophozoite has a rotary, boring motility.

(Text continues on p. 451)

Table 14-1. Intestinal Amebae: Key Features for Laboratory Identification

Species	Trophozoites	Illustration	Cysts	Illustration
Entamoeba histolytica	*Size:* 12–60 μ., asymmetrical *Motility:* purposeful, directional *Nucleus:* single and spherical. Karyosome tiny and centrally placed. Chromatin is delicate and evenly distributed in beadlike arrangement on nuclear membrane. *Cytoplasm:* finely granular and contractile vacuoles are inconspicuous. Ingested bacteria and yeasts are absent. Presence of ingested RBC's is diagnostic.	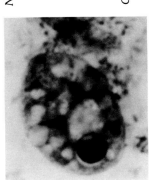	*Size:* 10–20 μ., spherical *Nucleus:* four are present in mature cyst. May be less than four in immature cysts, but never more than four. Karyosome tiny and usually centrally located but may vary in position. Chromatin is delicate and evenly distributed in beadlike fashion along the nuclear membrane. *Cytoplasm:* 10% of cysts may have chromatoidal bars with smooth, rounded ends. Glycogen vacuole may be seen in early precyst.	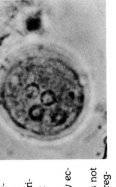
Entamoeba coli	*Size:* 15–50 μ., asymmetrical *Motility:* sluggish and nonpurposeful. Short pseudopodia extend in many directions. *Nucleus:* single, spherical. Karyosome relatively large and eccentrically located. Chromatin is irregularly distributed in uneven clumps along nuclear membrane. *Cytoplasm:* tends to be "junky" with many contractile vacuoles, undigested bacteria, yeasts, and other debris. RBC's are never ingested.	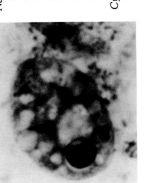	*Size:* 10–35 μ., usually spherical, rarely oval or triangular *Nucleus:* mature cyst may contain 8 or rarely 16 nuclei. Immature cysts have 1 to 8 nuclei. Peripheral chromatin is coarse and granular, unevenly distributed in clumps; somewhat more even than in trophozoites. Karyosome is usually eccentric, but may be central. *Cytoplasm:* chromatoidal bars not frequently seen but have irregular, splintered ends. Glycogen vacuole may be seen in precyst.	

Iodamoeba bütschlii

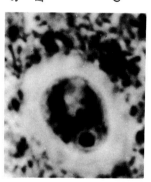

Size: 6–25 μ., asymmetrical
Motility: somewhat sluggish but directional
Nucleus: very large, densely staining karyosome occupying about ½ intranuclear space. No peripheral chromatin on nuclear membrane, giving a "ball-in-socket" appearance. Delicate strands may be seen radiating from karyosome to nuclear membrane.
Cytoplasm: undigested bacteria and food vacuoles may be seen. Glycogen vacuoles are rarely seen.

Size: 6–15 μ., asymmetrical, ovoidal, or elliptical
Nucleus: single. No peripheral chromatin on nuclear membrane. Large karyosome may rest on edge of nuclear membrane, appearing as a ball in socket.
Cytoplasm: characteristic single large glycogen vacuole that stains deep yellow brown with iodine. There are no other inclusions.

(Continued)

[447]

Table 14-1. Intestinal Amebae: Key Features for Laboratory Identification *(Continued)*

Species	Trophozoites	Illustration	Cysts	Illustration
Endolimax nana	*Size:* 6–15 μ., asymmetrical *Motility:* sluggish, forming many short, blunt pseudopodia *Nucleus:* single, with no peripheral chromatin on nuclear membrane. There is a large central karyosome giving a ball-in-socket appearance. *Cytoplasm:* finely vacuolated and may contain undigested bacteria		*Size:* 5–14 μ., oval *Nucleus:* one to four may be present. Immature cysts with less than four are rarely seen. No peripheral chromatin on nuclear membrane. There are large, deeply staining central karyosomes giving distinct ball-in-socket effect. *Cytoplasm:* not distinctive. It is fairly uniform with small granules or oval masses rarely seen.	
Dientamoeba fragilis	*Size:* 5–12 μ., asymmetrical. There may be considerable variation in size and shape in same specimen. *Motility:* active and purposeful *Nucleus:* one or two. Very delicate nuclear membrane and may be difficult to see. Karyosome is composed of 4 to 8 chromatin granules that tend to become separated, giving a shattered appearance. *Cytoplasm:* very vacuolated and may contain numerous undigested bacteria *Note:* permanent stain is necessary for identification		No cyst stage known	

Table 14-2. Intestinal Flagellates and *Balantidium coli*: Key Features for Laboratory Identification

Species	Trophozoites	Illustration	Cysts	Illustration
Giardia lamblia	*Size:* 9–21 μ. long, 5–15 μ. wide; pear-shaped with tapering end. *Motility:* active, "falling leaf" *Nucleus:* two, laterally placed. No peripheral chromatin and difficult to see in unstained mounts. There are small, central karyosomes. *Cytoplasm:* uniform and finely granular. Two median bodies appear as a "mustache" on the axostyle. Sucking disks occupy ½ of ventral surface. *Flagella:* four lateral, two ventral. They are often difficult to see.		*Size:* 8–12 μ. long, 7–10 μ. wide; oval in shape *Nucleus:* 4 in number. Karyosomes are smaller than in trophozoites and tend to be eccentrically placed. There is no peripheral chromatin on nuclear membrane. *Cytoplasm:* clear space between cyst wall and cytoplasm gives an easy to recongize "halo" effect. Ill-defined longitudinal fibrils may be seen. Four median bodies are present.	
Chilomastix mesnili	*Size:* 6–20 μ. long, 5–7 μ. wide; round on one end, tapering to other *Motility:* stiff, rotary *Nucleus:* single and large, placed anteriorly at rounded end. Small central or eccentric karyosome with radiating filaments. It is difficult to see in unstained mounts. *Cytoplasm:* may be vacuolated. Prominent cytostome extending ½ of body length. Spiral groove across ventral surface is hard to see. *Flagella:* three anterior and one in cytostome		*Size:* 6–10 μ. long, 4–6 μ. wide; lemon-shaped with anterior knob *Nuclei:* one, difficult to see in unstained mounts. There is an indistinct central karyosome. *Cytoplasm:* curved cytostome with fibrils, appearing as a "shepherd's crook," often difficult to see	

(Continued)

Table 14-2. Intestinal Flagellates and *Balantidium coli*: Key Features for Laboratory Identification (Continued)

	Trophozoites	Illustration	Cysts	Illustration
Trichomonas hominis	*Size*: 7–15 μ. long, 4–7 μ. wide; teardrop in shape *Motility*: active, nervous, jerky. *Nucleus*: single, anterior. Small central karyosome. There is uneven distribution of chromatin on nuclear membrane. *Cytoplasm*: central, longitudinal axostyle. There is a longitudinal impression (costa) at attachment of undulating membrane that runs the full length of the body (*T. vaginalis* extends only half the body distance). *Flagella*: 3 to 5 anterior, 1 posterior. They are usually difficult to see.		No cyst form known	
Balantidium coli	*Size*: 50–100 μ. long, 40–70 μ. wide; oval in shape *Motility*: rotary, boring *Nucleus*: one kidney-bean-shaped macronucleus; one tiny round micronucleus immediately adjacent to the macronucleus *Cytoplasm*: many food vacuoles and contractile vacuoles; distinct anterior cytostome *Cilia*: body surface is covered with spiral, longitudinal rows of cilia.		*Size*: 50–75 μ. in diameter; spherical to ellipsoid *Nucleus*: one kidney-shaped macronucleus; one tiny spherical micronucleus lying within "hoff" of macronucleus (may be difficult to see) *Cytoplasm*: small vacuoles persist. Cilia retracted within. There is a thick, tough cyst wall.	

Color Plate
14-1

Plate 14-1
Identification of Blood Parasites

The photomicrographs in this plate illustrate the intracellular and extracellular parasitic forms that characterize human infections with malaria, leishmaniasis, and trypanosomiasis.

A.
Plasmodium vivax infection. In the left upper portion of the field, there is an enlarged erythrocyte infected with a single, relatively large-ring trophozoite. The stippled effect seen in the erythrocyte cytoplasm is due to Schüffner's dots, a characteristic finding in *P. vivax* infection.

B.
Plasmodium vivax infection. At left center is an enlarged erythrocyte containing Schüffner's dots and a mature trophozoite with abundant, flowing, ameboid cytoplasm. Note the adjacent erythrocyte with the tiny, single-ring trophozoite.

C.
Plasmodium vivax infection, illustrating a schizont form. The schizonts of *Plasmodium vivax* routinely have more than 12 segments and may have as many as 24. The schizont in the photomicrograph has about 15 segments, each of which represents the precursor of an infective merozoite which is released into the circulation upon rupture of the infected erythrocyte.

D.
Plasmodium malariae infection, illustrating an infected erythrocyte in the central portion of the photograph. Note the bandlike configuration of the trophozoite, which reaches across the diameter of the erythrocyte from one membrane to another. The infected erythrocyte is not enlarged and is free of Schüffner's dots.

E.
Plasmodium malariae, illustrating an erythrocyte infected with the schizont stage. Note that there are only five segments. *P. malariae* rarely produces schizonts with more than 12 segments, a helpful differential characteristic from *P. vivax*.

F.
Plasmodium falciparum infection. Note the heavy infection, with many erythrocytes containing multiple-ring forms. Also note the tiny size of the trophozoite ring forms, many of which tend to attach to the inner surface of the erythrocyte membrane in the so-called "applique" fashion.

G and H.
Plasmodium falciparum infection, illustrating the characteristic banana-shaped gametocytes. *P. falciparum* is further characterized by the lack of intermediate trophozoite forms or schizonts in the peripheral blood; rather, only early ring forms and gametocytes are commonly seen.

I.
Photomicrograph of a lymph node, illustrating the heavy infection of enlarged histocytes with intracellular parasitic forms of *Leishmania donovani*. These appear in this photograph as minute red dots within the cytoplasm, as best illustrated by the large histiocyte in the center of the field.

J.
Peripheral blood, illustrating multiple extracellular trypanosome forms. The organisms are long and spindle-shaped, possessing a single nucleus and a delicate flagellum projecting anteriorly.

K.
Peripheral blood in a case of *Trypanosoma cruzi* infection, illustrating a single "C"-form trypanosome.

L.
Photomicrograph of heart muscle, illustrating infection with the leishmanial forms of *T. cruzi*. Note the tiny leishmanial forms within the swollen cytoplasm of the cardiac muscle fiber. Cardiac involvement in *C. cruzi* infection may lead to heart failure and death.

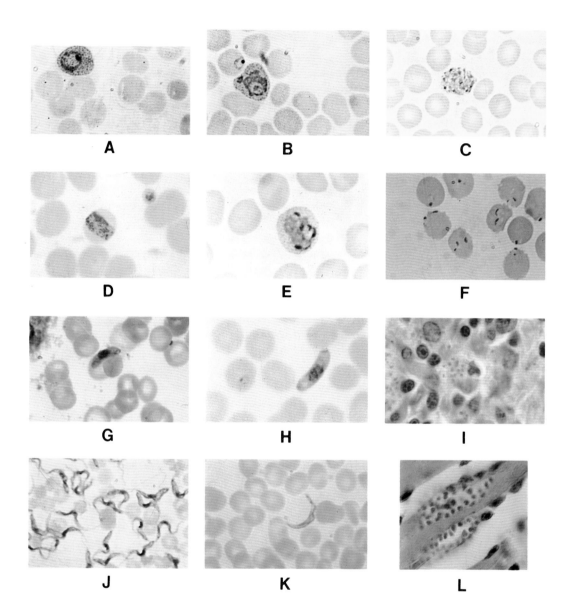

Nematodes

The species of nematodes that infect man include *Ascaris lumbricoides, Trichuris trichiura, Enterobius vermicularis,* the hookworms, and *Strongyloides stercoralis.* The key features for laboratory identification are listed in Table 14-3.

The life cycles of this group of helminths vary considerably in complexity. These nematodes do not have an intermediate host in their life cycle as shown in Figure 14-1; however, most require a stage outside of the human host to develop into an infective form. *Enterobius vermicularis* is an exception in that embryonated eggs are shed which develop into an infective stage within about 6 hours, permitting an anus-to-mouth infective cycle. In contrast, the ova of Trichuris species require 3 weeks or longer to mature into an infective stage, depending upon the temperature, moisture, and soil conditions into which ova are passed. The ova of *Ascaris lumbricoides* require an intermediate time of development in the external environment, averaging 14 days.

The hookworms and *Strongyloides stercoralis* have somewhat more complex life cycles even though an intermediate host is not required. *Ancylostoma duodenale* is the Old World hookworm and *Necator americanus* is the New World species based on their respective areas of endemic disease; however, since their life histories are essentially the same and the two species cannot be differentiated based on the morphology of their ova, the general term "hookworm" is commonly used for both species. Hookworm ova that are passed in the feces are already in an early cleavage stage and soon hatch into the rhabdoid larva stage. This term (*rhabdo,* "rod") is derived from the presence of an active striated muscular esophagus which allows the larvae to feed and live freely in nature. In about 5 to 7 days, the muscular esophagus is lost and the filariform larval stage is formed. Well-aerated, moist soil in a temperature range of 23° to 33°C. is optimal for development of the filariform larvae. Since these larvae cannot feed for themselves,

they soon die out unless they can find a human host. Filariform larvae represent the infective stage for man, and are able to penetrate the skin of the bare foot when contact is made with infected soil. These larvae enter the circulation and pass through the right side of the heart and lungs. They are coughed up and swallowed and then take up final residence in the small intestine where they develop into adult worms.

The life cycle of *Strongyloides stercoralis* is similar to that of the hookworms except that the ova have generally hatched by the time they are passed in the feces, so that the rhabditiform larvae are the common diagnostic form in the stool. The rhabditiform larvae of strongyloides can be distinguished from those of the hookworms by the criteria shown in Table 14-3. In most instances, the cycle continues as with the hookworms; however, in some instances filariform larvae may develop in the lower intestine and direct self-infection through the bowel wall is possible.

Except for *S. stercoralis,* the laboratory identification of these nematodes is dependent upon identifying their characteristic ova in stool specimens. These are illustrated in Table 14-3. In some cases the characteristic thick albuminous coat of *A. lumbricoides* is digested away by pancreatic enzymes in the upper small intestine, and these decorticate forms may be more difficult to identify. Their identification is particularly difficult if the ova are also unfertilized. However, there generally are a few ova which have retained their classic features and their identification is not difficult.

It is important that the ova of the hookworms be identified because of the potentially severe disease they may cause. The adult worms which measure up to 1.5 cm. in length reside in the upper intestine where they are firmly attached to the mucosa by the biting action of cutting mouth parts. The nutrition of the hookworm adults is derived from the blood of the host, which is leached. It is estimated that each adult hookworm

(Text continues on p. 456)

Table 14-3. Intestinal Nematodes: Key Features for Laboratory Identification

Species	Habitat of Adult	Infective Form	Diagnostic Forms	Illustration
Ascaris lumbricoides	Small and large intestine of man; may migrate into bile duct or pancreatic duct	Fertile ova	*Fertile ova:* 60 μ. × 45 μ., round or ovoid, with thick shell covered by a thick albuminous coat; inner cell in various stages of cleavage. They have a brown color. Digestive enzymes may dissolve the albuminous coat, leaving the ovum with a smooth decorticate surface. *Infertile ova:* 90 μ. × 40 μ., elongated. Shell often thin with loss of mamillated albuminous covering. Internal material is a mass of nondescript globules. *Adult worms:* 25 to 35 cm. in length. Males are smaller than females and have a curved tail. White longitudinal streaks on either side of body and lack of muscular segments are helpful identifying features.	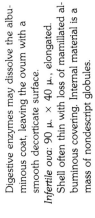 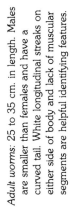 

Trichuris trichiura		Fertile ova	Large intestine	Ova: 54 μ. × 22 μ., elongate, barrel-shaped with polar hyaline "plug" at either end. Shell is yellow to brown; plugs are colorless.
Enterobius vermicularis		Ova	Large intestine, appendix, perianal area	Ova: 55 μ. × 26 μ., elongate, asymmetrical with one side flattened, other side convex. Shell is thin and smooth, and fully developed larvae are usually contained, particularly in cellulose tape preparations.
Hookworms: *Ancylostoma duodenale*		Filariform larvae that penetrate skin	Small intestine. Scolex of adult is firmly attached to mucosa by two pairs of chitinous teeth.	Ova: 60 μ. × 40 μ., oval or ellipsoid. Shells are thin-walled, smooth, and colorless. Internal cleavage is usually well developed at 4 to 8-cells stage, which characteristically pulls away from the shell leaving an empty space.

(Continued)

Table 14-3. *Intestinal Nematodes: Key Features for Laboratory Identification (Continued)*

Species	Habitat of Adult	Infective Form	Diagnostic Forms	Illustration
Hookworms; Ancylostoma duodenale (Continued)			*Rhabditiform larvae:* occasionally seen in stool specimens that have sat at room temperature for many days before being examined. They can be distinguished from rhabditiform larvae of *Strongyloides stercoralis* by the hookworm's long buccal cavity. *Scolex of adult:* presence of two pairs of chitinous teeth.	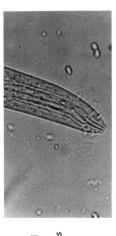 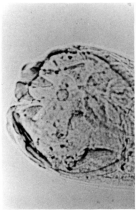
Necator americanus	Small intestine. Scolex of adult is firmly attached to mucosa by cutting plates.	Filariform larvae that penetrate skin	*Ova:* 65 μ × 40 μ. They are morphologically similar to *A. duodenale.* *Rhabditiform larvae:* similar to *A. duodenale* *Scolex of adult:* mouth part fitted with sharp cutting plates.	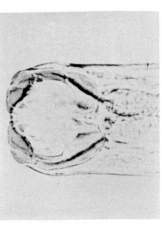

Strongyloides stercoralis Small intestine

Filariform larvae
Self infections may
occur if passage of
stool is delayed
and filariform lar-
vae develop.

Ova: usually not seen in stool specimens,
but similar to those of hookworm
Rhabditiform larvae: this is the form most
commonly seen in stool specimens. By
the time the ova reach the large intestine,
most have hatched. The larvae measure
0.75 to 1.0 mm. in length and 75 μ. in
diameter.

Strongyloides rhabditiform larvae can be
distinguished from those of hookworm
by their short buccal cavity.
Filariform larve: rarely found. Long slender
form with a notched tail. They are highly
infectious.

[455]

leaches about 0.15 ml. of blood each day; therefore, in a heavy infection of 500 worms, the host could be bled the equivalent of 1 pint of blood per week. This is sufficient to cause severe anemia and marked erythroid hyperplasia of the bone marrow.

Hookworm ova cannot be differentiated morphologically from those of Strongyloides species; however, Strongyloides ova are rarely seen in stool specimens because they have hatched into their rhabditiform larva stage by the time they are passed. It may be necessary to estimate the number of hookworm ova present in the stool since there is a direct relationship between numbers and severity of infection. Clinically significant infection is indicated by 2500 to 5000 ova per gram of feces. In contrast, it takes up to 30,000 ova of Trichuris species before the host may have symptoms, whereas even an occasional ascaris ovum is significant because the adult worms have a propensity to migrate into intestinal orifices such as the common bile duct or pancreatic duct and it may require only a few of these to produce major complications. Egg-counting techniques have been described by Garcia and Ash.[5]

Since the ova of the two species of hookworms appear similar, differentiation must be made by demonstrating the mouth parts, as illustrated in Table 14-3. *A. duodenale* has two pairs of chitinous teeth while *N. americanus* has sharp cutting plates; however, this differentiation is of academic interest only in most instances.

The diagnosis of pinworm infection is most commonly made through the use of the cellulose acetate test technique described above. In rare instances the thin-walled ova, flattened on one side, may be found in stool specimens and the identification can be made directly by this means. The adult worms, measuring up to 5 mm. in length, may be seen on the perianal skin, particularly in the late evening.

Cestodes

There are six cestodes (tapeworms) that are of importance to man: the pork tapeworm *Taenia solium;* the beef tapeworm *Taenia saginata;* the fish tapeworm *Diphyllobothrium latum;* the dwarf tapeworm *Hymenolepis nana;* the rat tapeworm *Hymenolepis diminuta;* and the dog tapeworm *Dipylidium caninum.*

With the exception of *Hymenolepis nana,* where infection in man occurs through ingestion of infective ova, the life cycles of the cestodes involve one or more intermediate hosts in which a stage of larval development is necessary. For example, ova that are passed in the feces of the human host harboring the adult tapeworm contaminate vegetative matter, soil, or fresh water, where they are ingested by cows, pigs, fish, *etc.* The ova hatch in the intestines of these intermediate hosts, liberating larvae that in the course of about 2 months find their way into the skeletal muscle or flesh of these hosts. In the muscle, the larvae become encysted in a form called cysticerci, or "bladder worms," and man becomes infected by eating raw or poorly cooked beef, pork, or fish infested with the cysticerci.

It is particularly important that infections with *Taenia solium,* the pork tapeworm, be identified because man may also be the inadvertent host of the larval form of the disease should he ingest infected ova. A disease called cysticercosis may develop, in which cysticercoids, the name given to these larval cystic forms, may develop in virtually any viscera; however, there is a particular predilection for the central nervous system where brain cysts of potentially grave severity may develop. Cysticercosis does not develop with any of the other cestodes that cause tapeworm infection in man.

Unfortunately the ova of *T. solium* cannot be differentiated from those of *T. saginata* on the basis of their morphology. Both have characteristic thick striated shells and a hexacanth (six-hooked) embryo within (Table 14-4). Identification of these two species can only be made by either recovering the proglottids or by retrieving the minute scoleces after the tapeworm is forced to leave its host. The differential features of these two tapeworms are illustrated in Table 14-4. The proglottids of *T. saginata* have more than 13 lateral uterine branches and the scolex is devoid of an armed rostellum, in contrast to *T. solium* which has proglottids with less than 13 uterine segments and a rostellum with a double row of hooklets.

Diphyllobothrium latum, the giant fish tapeworm of man, utilizes two intermediate hosts in the development of its larval forms outside the human host. Ova are passed with fecal material into bodies of fresh water. After several days these hatch and release free-swimming ciliated forms known as miracidia. These in turn are ingested by invertebrate copepods which serve as the first intermediate host. Copepods serve as one of the major food sources for a variety of freshwater fish, pike, turbot, and carp in North America. These fish serve as the second intermediate hosts. The Plerocercoid larvae of the parasite develop within the flesh of the fish. Man becomes infected with the adult tapeworm by ingesting these plerocercoid larvae in raw or poorly cooked fish.

Diagnosis of *D. latum* infection in man is most commonly made by identifying the characteristic opercular ova of the parasite in stool specimens. The morphology of this ovum is illustrated in Table 14-4. The differentiation of *D. latum* ova from those of *Paragonimus westermani*, to be discussed below, may be difficult although the distinctly shouldered operculum of the latter is usually sufficiently visible to enable a differential diagnosis. Proglottids of *D. latum* are rarely passed in the stools. However, they are distinctive when found since their individual segments are broader than they are long (*latum*, "broad"). The scolex, with its shallow longitudinal groove and lateral lip-like folds, is also distinctive when seen.

The dwarf tapeworm *Hymenolepis nana*, which measures no longer than 1.5 inches in length, has a direct cycle in which man is infected by ingesting infected ova in contaminated food or water. The diagnostic ova are illustrated in Table 14-4. These can be distinguished from the ova of *H. diminuta*, the rat tapeworm, by their distinctive polar filaments emanating from the membrane of the internal oncosphere embryo.

Man is an accidental host of *H. diminuta*, the rat tapeworm, and *Dipylidium caninum*,

the dog tapeworm. Man becomes infected by ingestion of the arthropods (meal beetles and dog fleas or lice, respectively) which serve as the intermediate larval hosts. *H. diminuta* may cause cachexia and diarrhea in infected humans, and the diagnosis is made by detecting the characteristic ova in the feces. *D. caninum* usually infects children, and direct symptoms are generally mild, although secondary toxic reactions such as loss of appetite, nervousness, and low-grade fever may be experienced in reaction to the metabolic wastes of the worm. The diagnosis is made by the identification of the characteristic egg packets in the stool as illustrated in Table 14-4.

Trematodes

Seven species of trematodes (flukes) are important parasites of man. These include: the three schistosomes, the adult male and female residing in peripheral sites of the portal vein system; two liver flukes, *Fasciola hepatica* and *Clonorchis sinensis;* the giant intestinal fluke *Fasciolopsis buski;* and the lung fluke *Paragonimus westermani*. The key features for the laboratory identification of these flukes are listed in Table 14-5.

The life cycles of the flukes are similar and, with the exception of the schistosomes, require two intermediate hosts before becoming infective for man (Fig. 14-1). The initial stages are virtually identical; that is, ova are passed into water (with feces passed through open privies on bridges over canals or lakes, for example), hatch either immediately or after a short period of embryonation and release a free-swimming miracidium which is either ingested or penetrates into the flesh of the first intermediate host, a snail. Within the snail the miracidia transform into a developmental larval stage known as a sporocyst for the schistosomes and a rediae for the other trematodes. This developmental period may be as short as 3 weeks for *C. sinensis* and as long as 12 weeks for *F. hepatica*. The purpose of this larval stage in the snail is a process of replication through which many hundreds of cercariae are produced.

(Text continues on p. 465)

Table 14-4. Intestinal Cestodes: Key Features for Laboratory Identification

Species	Habitat of Adult	Infective Form	Diagnostic Forms	Illustration
Taenia saginata	Small intestine	Cysticercoid larvae in beef muscle	*Ova:* 31 μ. to 43 μ., spherical or subspherical, with thick shell with prominent radial striations. Embryonated oncosphere possessing three pairs of hooklets within the shell is diagnostic of the genus (*Tinea* species identification cannot be made on the basis of ova morphology). *Proglottids:* longer than wide. Gravid segments have a central uterine stem with 15 to 20 lateral branches on each side. They are motile when first passed. *Scolex:* 4 suckers characteristic of the genus with a bare crown devoid of a hooked rostellum. Scolex is 2 mm. in diameter.	
Taenia solium	Small intestine (Extraintestinal cysticercosis may develop in man with this species, with encysted larvae infecting the heart, eye, central nervous system, and other viscera.)	Cysticercoid larvae in pork muscle Ova or gravid proglottids leading to cysticercus infection	*Ova:* identical to those of *T. saginata.* Species identification is not possible based on ova morphology. *Proglottids:* longer than wide. Gravid segments have a central uterine stem with only 8 to 13 lateral stems, a major differential feature from those of *T. saginata,* which have more.	

Taenia solium
(Continued)

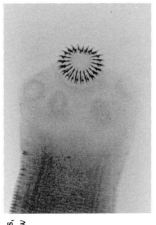

Scolex: 4 suckers characteristic of the genus, with a rostellum armed with a double row of 25 to 30 large and small, brown chitinous hooks

Hymenolepis nana

Small intestine

Ova

Ova: 40 to 60 μ., oval or subspherical. Shell consists of two distinct membranes: the outer membrane is relatively thin and has a smooth surface; the inner membrane has two opposite poles from which 4 to 8 filaments arise and spread between the two membranes (differential feature from *H. diminuta* which is devoid of filaments). Inside the inner membrane is an oncosphere with three pairs of hooklets (hexacanth ovum).

Adult worm: may rarely be found in stool specimens where it can be confused with thin strands of mucin. It measures less than 40 mm. (1.5 inches) in length, and has indistinct, broader than long proglottids, a tiny scolex with 4 suckers, and a protruding rostellum armed with a ring of 20 to 30 spines.

(Continued)

Table 14-4. Intestinal Cestodes: Key Features for Laboratory Identification (Continued)

Species	Habitat of Adult	Infective Form	Diagnostic Forms	Illustration
Hymenolepis diminuta	Small intestine	Cysticercoid larvae that develop in the body cavity of cat or dog fleas and lice which are inadvertently ingested by man	*Ova:* 70 to 85 μ. long, 60 to 80 μ. wide, round to oval. They have similar appearance to the ova of *H. nana* except inner membrane has no protruding filaments. *Adult worm:* Rarely found in stool specimens. Proglottid segments are not distinctive. The scolex is tiny and spherical and has 4 deep, spherical suckers and a rounded rostellum devoid of hooklets.	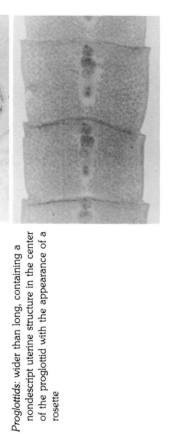
Diphyllobothrium latum	Small intestine	Plerocercoid larvae in flesh of fresh-water fish	*Ova:* 60 to 75 μ. long, 40 to 50 μ. wide. The ova are oval or elliptical. An inconspicuous nonshouldered operculum is seen at one of the lateral ends with a small terminal knob at the other. Shell is thin and smooth. Internal cleavage is not organized and extends to the shell completely filling the inner area. *Proglottids:* wider than long, containing a nondescript uterine structure in the center of the proglottid with the appearance of a rosette	

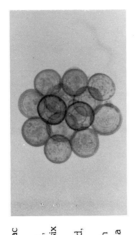

Diphyllobothrium latum
(Continued)

Scolex: shaped like a rounded spatula, with a longitudinal shallow groove bordered on either side by liplike folds

Dipylidium caninum

Small intestine

Cysticercoid larvae in various species of "meal beetles" which are inadvertently ingested by man

Ova: egg packet usually seen, which is a sac enclosing 5 to 15 spherical ova. Each ovum measures 35 to 40 μ., is spherical, and encloses an inner oncosphere with six delicate hook ets.

Proglottids: longer than wide, barrel-shaped, and have a double set of reproductive structures with two genital pores for each segment, one on each side (Other Taenia species have only one genital pore.)

Scolex: conical or ovoid rostellum with 30 to 150 small thornlike hooks arranged in several rows. Rostellum may retract into a depression in upper portion of scolex.

Table 14-5. Trematodes: Key Features for Laboratory Identification

Species	Habitat of Adult	Infective Form	Diagnostic Forms	Illustration
Schistosoma mansoni	Male and female flukes reside together in the portal system venules, primarily those of the large intestine. (See first illustration.)	Forked tailed cercariae that penetrate skin of man wading in fresh-water canals (See second illustration.)	*Adult flukes:* The female is 1.6 cm. long and resides in the gynecophoral canal of the male, which is 1 cm. long. Rarely seen as a diagnostic form. *Cercaria:* The infective cercaria of the Shistosomes has a forked tail. Not used as a diagnostic form. *Ova:* 115 to 180 μ. long, 45 to 70 μ. wide. Elongated with a prominent lateral spine near the more rounded posterior end; anterior end tends to be somewhat pointed and slightly curved. When embryonated the ovum may contain a mature miracidium. (See third illustration.)	
Schistosoma haematobium	Male and female flukes reside together in the portal system venules, primarily those of the urinary bladder.	Same as *S. mansoni*	*Ova:* 110 to 170 μ. long, 40 to 70 μ. wide. Elongated with a rounded anterior end and a prominent spine from the terminus of the more tapered posterior end. Embryonated ovum may contain a mature miracidium.	

Schistosoma japonicum	Male and Female flukes reside together in the portal system venules, primarily those of the small intestine.	Same as *S. mansoni*	*Ova:* 70 to 100 μ. long, 55 to 65 μ. wide. Oval or subspherical. A small rudimentary knob or delicate spine may be seen on the lateral wall. Because it is often located in a depression in the shell, this small spine is often difficult to see.
Fasciola hepatica	Bile ducts	Metacercaria encysted on water plants that are ingested by man	*Ova:* 150 μ. × 80 μ. Elliptical to oval in shape. The shell is thin with a smooth surface and an indistinct operculum is located at one end. The internal cleavage material is poorly organized and extends to the shell margins without leaving a clear space. Ova cannot be distinguished from those of *F. buski*. *Adult fluke:* It measures 3.0 × 1.2 cm. and can be distinguished from that of *F. buski* because of the elongated cone-shaped anterior end.
Fasciolopsis buski	Small intestine	Same as *F. hepatica*	*Ova:* indistinguishable from *F. hepatica* *Adult fluke:* It measures 2.0–7.5 cm. long by 0.8–2.0 cm. wide and in addition to its generally larger size, it can be distinguished from *F. hepatica* by its broadly rounded anterior end.

(Continued)

Table 14-5. *Trematodes: Key Features for Laboratory Identification* (Continued)

Species	Habitat of Adult	Injective Form	Diagnostic Forms	Illustration
Clonorchis sinensis	Bile ducts	Metacercaria encysted in flesh of fresh-water fish	*Ova:* 27 to 35 μ. long, 14 to 17 μ. wide. Broad, round posterior end and concave constriction at anterior end give this relatively small ovum an urnlike appearance. A convex operculum is seen resting on shoulders. A small knob often is present at the posterior end.	
Paragonimus westermani	Cystic cavities in the lung	Metacercaria encysted in flesh of various crustaceans (crayfish, crabs)	*Ova:* 65 to 120 μ. long, 40 to 70 μ. wide. Ovoid or elongated, somewhat asymmetrical, with one edge flattened. A flattened operculum rests into shouldered area, a distinguishing feature from *D. latum.* Shell is thick but smooth and internal cleavage fills entire area. *Adult fluke:* It measures about 12 × 7 mm. and up to 5 mm. in thickness, making it rounder than other flukes. Rarely seen as a diagnostic form.	

Mature schistosome cercariae, upon release from the snail, infect man by directly penetrating the skin without requiring a second intermediate host. A condition known as "swimmers itch" occurs at the sites of penetration. The free-swimming cercariae live only 1 to 3 days, after which they die if a host is not found.

The cercariae of other trematodes are not capable of infecting man directly, but rather seek out a second, intermediate host. For example, *C. sinensis* cercariae infect freshwater fish, those of *P. westermani* use various species of crustaceans (crabs or crayfish), while those of *F. hepatica* and *F. buski* attach to a variety of water plants such as water chestnut where they form encysted structures known as metacercariae. These encysted metacercariae are more resistant than free-swimming cercariae and can survive for several weeks or months. Man becomes infected by ingesting the metacercariae, whether in the form of raw fish or crabmeat or in contaminated leaves or stems of water plants.

The prevalence of a particular trematode disease in a given part of the world is directly dependent upon the presence of the parasites' intermediate hosts in the local inland bodies of water.

The name schistosome is derived from the appearance of the adult male whose "split body," a genital groove or canal, serves as a receptacle for the female during copulation. Depending upon the species, schistosome adults are found in different sites in the portal vein system; that is, in the portal veins of the large intestine for *S. mansoni,* the small intestine for *S. japonicum,* and in the veins of the urinary bladder for *S. haematobium.* Although the adult worms, which measure about 2.5 to 3 cm. in length when mature, may cause portal vein obstruction at the sites where they reside, disease is caused by the extensive tissue damage that results from the deposition of the myriads of ova that are produced daily by the females. These ova have a propensity to invade the walls of the veins and penetrate into the adjacent viscera, eliciting a severe suppura-

tive and granulomatous inflammation and ultimately fibrosis and scarring. This results in marked thickening of the walls of the intestine or urinary bladder, with loss of function of these organs. Bloody diarrhea, bowel obstruction, urinary bladder contraction, and hematuria are the major complications. Ova that are swept upstream in the portal veins and lodge in the liver, particularly in the case of *S. mansoni* infections, can produce severe scarring of the liver and cirrhosis of the liver.

The laboratory diagnosis of schistosomiasis is made by detecting the distinctive large ova in stool specimens or in the urine of patients with *S. haematobium* infection. *S. mansoni* ova measure about 150 × 60 μ. and have a large, lateral, subterminal spine; *S. haematobium* ova are about the same size but have a more delicate terminal spine, while the ova of *S. japonicum* are more spherical and smaller (about 80 to 100 μ.) and have an indistinct knoblike lateral spine (Table 14-5).

The condition known as swimmers itch that develops in the skin at the site of penetration of the infective cercariae may be produced by one of the schistosomes causing the human infections described above or by other species of schistosomes that have animals other than man as their definitive hosts, in which case only the skin manifestations occur in man without further development of the adult forms.

Fasciola hepatica is a large fluke measuring about 2.5 cm × 1.5 cm. that primarily infects the liver of sheep. Man becomes an accidental definitive host when he ingests uncooked water plants such as water chestnuts that are infected with fasciola metacercariae. Laryngopharyngitis, called "halzoun," may result if man ingests raw liver infected with adolescent worms which can attach to the pharyngeal mucosa. Human fascioliasis is manifested by headache, chills, fever, and right upper quadrant pain with hepatomegaly. Jaundice, diarrhea, and anemia may occur in severe infections, and hepatic cirrhosis of the biliary type is a late complication.

The adult flukes are seen only if removed at surgery. The *F. hepatica* adults have a cone-shaped anterior protrusion, while those of *F. buski* are rounded anteriorly. The laboratory diagnosis is most commonly made by detecting the large, 150 μ. × 80 μ., thin-walled operculated ova in feces (Table 14-5). *Fasciolopsis buski* is a fluke similar to *F. hepatica* and resides in the small intestine. It is attached to the mucosa of the small intestine where it may produce mucosal ulcers, resulting in varying degrees of epigastric pain, nausea, and diarrhea. Ascites and intestinal obstruction may be seen in heavy infections. The laboratory diagnosis is made by identifying the large ova in the feces, which are identical in appearance to those of *F. hepatica.*

Clonorchis sinensis, the Chinese liver fluke, is a relatively small fluke that measures about 15 mm. × 5 mm. when mature and resides within the biliary ducts or in the gallbladder. Although the bile ducts may become thickened and dilated, particularly at the points where the flukes are attached to the inner lining, biliary obstruction and jaundice are rare except in extremely heavy infections. The disease tends to remain low-grade and chronic, lasting for years with only minor symptoms of abdominal distress, intermittent diarrhea, and liver pain or tenderness. The disease is most common in a wide area of the Far East, particularly in Indochina, Japan, Korea, Formosa, and Southern China.

Man becomes infected with *C. sinensis* by ingesting raw or poorly cooked fresh-water fish. The laboratory diagnosis is made by identifying the characteristic, small, urn-shaped ova in stool specimens. The ovum measures about 25 μ. × 35 μ. and has a prominent, convex operculum that rests on "shoulders" (Table 14-5). This ovum is similar in morphology to the ova of *Heterophyes heterophyes* and *Metagonimus yokogawai,* two small intestinal flukes that also are prevalent in the Far East and contracted by man through the ingestion of raw or pickled fish.

Paragonimus westermani, the lung fluke of man, may be confused with *Diphyllobothrium latum* in the examination of stool specimens because the ova are similar. *P. westermani* ova, however, have a shouldered operculum, in contrast to the opercula of *D. latum* which are devoid of a shoulder (Table 14-5). The adult fluke, measuring about 12 mm. × 7 mm., resides within the lung, where it forms small pseudocysts that characteristically communicate with branches of the bronchial tree. When one of these pseudocysts ruptures into a bronchus, coughing with hemoptysis is common. This allows the ova to be discharged into the bronchi and ultimately swallowed when they are coughed up into the pharynx. The disease tends to become chronic with varying degrees of fibrosis and scarring of the lungs, leading to intermittent chest pain, fever, chills. Man contracts the disease by eating raw or poorly cooked crayfish or crabmeat.

Blood and Tissue Parasites

Parasites found in the blood or in different organs are usually discussed separately from those inhabiting the gastrointestinal tract. However, it should be understood that blood and tissue parasites, including nematodes, cestodes, and flagellates, are morphologically similar to their intestinal counterparts. The plasmodia responsible for malaria are sporozoa that also have an intestinal counter-part, *Isospora hominis,* which may be found in feces but is not thought to cause primary disease in humans.

In general, the life cycles of the blood and tissue parasites are more complex than those of their intestinal counterparts. Most of the blood parasites involve an arthropod vector as well as a human host, while many of the parasites causing tissue infections utilize various insects for the development of intermediate stages of their life cycles.

Parasites that infect the blood may be found intracellularly in erythrocytes or extracellularly in the plasma. The large extracellular organisms such as microfilariae and trypanosomes can be seen in direct microscopic examinations of unstained mounts of whole blood. The smaller intracellular parasitic forms require a stained preparation in order to visualize their internal structures.

Tissue parasites may be either intracellular or extracellular depending upon the species and the phase of the parasitic cycle or form assumed in any given case. An outline of the blood and tissue parasites that infect man is as follows:

I. Blood Parasites
  A. Intracellular
    1. Plasmodium species
    2. Babesia species (rare)
    3. Theileria species (rare)
  B. Extracellular
    1. Microfilaria
      a. Pathogenic (sheathed)
        (1) *Wuchereria bancrofti*
        (2) *Brugia malayi*
        (3) *Loa loa*
      b. Nonpathogenic (nonsheathed)
        (1) *Dipetalonema perstans*
        (2) *Dipetalonema streptocerca*
        (3) *Mansonella ozzardi*
    2. Trypanosomes
      a. *Trypanosoma gambiense*
      b. *Trypansoma rhodesiense*
      c. *Trypanosoma cruzi*
II. Tissue Parasites
  A. Intracellular
    1. Cutaneous
      a. *Leishmania tropica*
      b. *Leishmania braziliensis*
      c. *Trypansoma cruzi* (leishmanial form)
    2. Visceral
      a. *Leishmania donovani*
      b. *Toxoplasma gondii*
      c. *Trypansoma cruzi* (leishmanial form)
  B. Extracellular
    1. Cutaneous
      a. *Onchocerca volvulus*
      b. *Dracunculus medinensis*
    2. Visceral
      a. *Trichinella spiralis*
      b. *Echinococcus granulosus*
      c. *Cysticerus cellulosae*
      d. *Toxocara* species
      e. *Pneumocystis carinii*

Malaria

In the United States most cases of malaria are found in travelers or foreign-born students who have been recently exposed to mosquitoes in endemic areas. Occasional cases result from blood transfusions or from syringes and needles shared by drug addicts. The diagnosis of malaria often is clinically overlooked in the United States because of the rarity of the disease in most locales. Another reason for misdiagnosis is that the cardinal symptoms, chills and fever, are associated with so many other diseases that malaria is rarely considered. The higher mortality rate in American civilian hospitals (8.5%) for falciparum malaria as compared to that in military hospitals (0.7%) can be attributed to delays in identifying the disease and making the diagnosis.

The life cycle of the plasmodium parasite has two phases: a sexual cycle known as sporogony, which takes place within the intestinal tract of the mosquito, and an asexual cycle known as schizogony occurring in the human host. Sporozoites, the infective form for man, are found in the salivary glands of female anopheline mosquitos. Saliva containing infective sporozoites is injected into the bloodstream of man through the mosquito proboscis. After circulating for about 20 to 30 minutes, the sporozoites enter the parenchymal cells of the liver where they begin to multiply in what is known as the exoerythrocytic cycle. In about 10 days, multiple small forms called merozoites break out of the liver cells and are released into the circulation, where they seek out and penetrate the erythrocytes.

Within the erythrocytes (intraerythrocytic cycle) a series of developmental stages takes place, as illustrated in Plate 14-1. The organisms develop into a ring form known as the trophozoite, which, depending on the species, enlarges and divides into a segmented stage known as the schizont. The individual segments of the schizonts are called merozoites, which, when mature, rupture the erythrocytes and are released into the circu-

Table 14-6. *Plasmodia: Key Features for Laboratory Identification*

Species	Appearance of Erythrocytes	Trophozoites	Schizonts	Gametocytes	Special Features
Plasmodium vivax	Enlarged and pale; Schüffner's dots usually prominent	*Early:* ring relatively; large (⅓ the size of RBC). Rings with two nuclei or cells with two or three rings may be seen. *Mature:* ameboid, with delicate pseudopodia that flow to fill the entire RBC.	12 to 24 segments (merozoites); pigment is fine grained and inconspicuous.	Round to oval and almost completely fill RBC when mature. Large chromatin mass. Pigment is coarse and evenly distributed.	Length of asexual cycle (fever cycle) is 48 hours (benign tertian). Be alert for possibility of mixed species infection when *P. vivax* forms are identified.
Plasmodium falciparum	Normal size; Maurer's dots or clefts are rarely seen.	Rings forms extremely small, occupying no more than ⅕ of the RBC. Double nuclei are common and multiple rings per RBC are usual. Applique forms plastered on the RBC membrane are virually diagnostic.	Not normally seen except in fulminant disease; 24 or more segments are the characteristic.	Characteristic crescent or banana-shaped forms are virtually diagnostic. Microgametocytes stain lighter blue than macrogametocytes.	The ratio of infected to normal red blood cells is high. Intermediate ring forms or schizonts are not commonly seen in the peripheral blood. Length of asexual cycle is 48 hours (malignant tertian).
Plasmodium malariae	Normal size; no dots or clefts form.	*Early:* similar to *P. vivax* except staining is deeper blue and the cytoplasm of the ring is broader. Double rings are rare. *Mature:* less tendency to become ameboid; rather it forms a ribbon or band.	More than 12 segments are rarely seen. Merozoites arrange in rosettes. Pigment is abundant and coarse, often in aggregates within "hof" of rosettes.	Not distinctive and resemble those of *P. vivax*. Red cells are not enlarged. Pigment usually more abundant in *P. vivax* and tends to be coarse and unevenly distributed.	Asexual cycle lasts 72 hours (quartan). Identification of *P. malariae* is often made by exclusion after either *P. vivax* or *P. falciparum* have been excluded by their more distinctive morphologic features.

lation. These merozoites then seek out uninfected erythrocytes and the cycle continues.

The periodicity of the release of merozoites into the circulation accounts for the fever cycle experienced by the infected host: every two days (tertian) for *P. vivax* and *P. falciparum* malaria, and every three days (quartan) for *P. malariae* malaria.

After a few of these erythrocytic cycles have occurred, some of the merozoites change into sexual forms called macro- (female) and micro- (male) gametocytes. When a mosquito that is not infected with plasmodium parasites bites an infected host, these gametocytes are ingested. In the stomach of the mosquito the male microgametocytes develop six to eight flagella which break free to penetrate into the female macrogametocytes. These fertilized zygotes enter the stomach of the mosquito wall where they eventually break out into the body cavity and migrate to the salivary glands.

The laboratory identification of the malarial parasites in humans is made by studying stained thin and thick peripheral blood smears. Smears should be obtained at different times of the day from patients with suspected infections because parasitemia may be intermittent and the relative numbers of circulating parasites may vary.

There are three species of Plasmodium that most commonly cause human malaria: *P. vivax, P. falciparum,* and *P. malariae.* A fourth species, *P. ovale,* is extremely rare and will not be discussed here. The differential features of the three common species are reviewed in Table 14-6. The microscopic morphology of the intraerythrocytic forms are illustrated in Plate 14-1. It is particularly important that infections with *P. falciparum* are recognized because the disease can be particularly severe, with a fatal outcome. Some of the microscopic features of *P. falciparum* are:

1. The ratio of infected to normal red cells is high.
2. Ring forms with double dots or red cells with two or more ring forms are commonplace.
3. Intermediate forms (schizonts) are not seen or are rarely present; only early ring forms and gametocytes are commonly seen in peripheral blood smears.
4. The erythrocytes are not enlarged and Schüffner's dots are not present.
5. Characteristic large banana-shaped gametocytes are diagnostic of the species, and may be particularly prevalent in thick preparations.

The early ring forms of *P. falciparum* are smaller and more delicate than other Plasmodium species, occupying only about one fifth of the cell volume. The so-called applique forms, in which the rings appear plastered against the inner surface of the red blood cell membrane, may be seen.

The identification of *P. vivax* is usually not difficult. The infected erythrocytes are enlarged and appear pale, and stippling with Schüffner's dots is present even in the early ring stages of infection. Schizonts with as many as 24 segments may be observed. Large oval to round gametocytes may be seen but are not easily distinguished from those of *P. malariae* except when observed by experienced parasitologists.

The identification of *P. malariae* is usually made after *P. vivax* and *P. falciparum* have been excluded by the features described above which are more definitive and easier to observe. Specific features of *P. malariae* include the lack of erythrocyte enlargement or presence of Schüffner's dots, the development of trophozoites that tend to form narrow bands across the erythrocytes, and the formation of schizonts with less than 12 segments and a coarse clumping of malarial pigment within the center of the schizont, which takes on a rosette appearance.

Two other intraerythrocytic parasites that resemble Plasmodium species have been rarely reported in man; namely, Babesia species and Theileria species. These organisms belong to the class Piroplasma and are known to cause Texas fever in cattle. The disease is transmitted to cattle and deer by the bites of arachnid hard ticks, the intermediate hosts within which the sexual cycle takes place. These piroplast organisms are considerably smaller than the plasmodium parasites and do not form malarial pigment within the infected erythrocytes. The Babesia piroplasts tend to form packets of two's or three's, simulating rabbit ears or Maltese crosses; the organisms of Theileria are extremely small, and may show rodlike or commalike forms.

Hemoflagellates: Leishmania and Trypanosomes

There are two types of hemoflagellates that cause disease in man: the leishmanial organisms and the trypanosomes. *Leishmania donovani,* the cause of visceral kala-azar in man, and *L. tropica,* the agent of tropical sores, are the common leishmanial diseases. *Trypanosoma gambiense* and *T.*

Table 14-7. *Hemoflagellates: Key Features for Laboratory Identification*

Species	Arthropod vector	Sites of Infection	Diagnostic Forms	Plate Reference
Leishmania donovani *Leishmania tropica*	Phlebotomus fly Leptomonad forms in fly proboscis are infective forms for man.	Intracellular parasites of reticuloendothelial cells: bone marrow, liver, spleen, and lymph nodes. Organisms do not circulate in peripheral blood. *L. tropica* causes cutaneous and mucocutaneous disease called "tropical sore."	Ovoid 2- to 4-μ. forms seen intracellularly in tissue sections. Morphology best seen in Giemsa-stained touch preparations. Presence of rod-shaped kinetoplast adjacent to nucleus is helpful distinguishing feature from the fungus *H. capsulatum*.	Plate 14-1, *I*
Trypanosoma gambiense *Trypanosoma rhodesiense*	Tsetse fly Metacyclic trypanosomal form in salivary gland of fly is infective form for man.	In early infection, the parasites may be found in lymph nodes and circulating in the bloodstream. In later infections, the organism may invade the central nervous system, producing "sleeping sickness."	The organisms are long, slender, spindle-shaped forms, measuring 15 to 30 μ. in length and 1.5 to 4.0 μ. in width. A single flagellum takes origin from a dotlike kinetoplast located posterior to the central nucleus. The flagellum runs along an undulating membrane that projects beyond the anterior point of the organism.	Plate 14-1, *J*
Trypanosoma cruzi	Triatomid bug (known also as reduvid or "kissing" bug) Metacyclic trypanosome in bug feces is infective form for man.	Early in infection, the trypanosome form is found circulating in the blood. A chronic disease form is characterized by leishmanial forms in reticuloendothelial cells or in heart muscle. The leishmanial forms do not circulate in the blood but can be seen only in stained tissue sections.	The circulating trypanosomal organisms are similar to those of *T. gambiense* and *T. rhodesiense*. Characteristic "C" forms are not commonly seen and are rarely diagnostic. Leishmanial forms in tissues are morphologically similar to those of *L. donovani*.	Plate 14-1, *K and L*.

rhodesiense cause African sleeping sickness, while *T. cruzi* is responsible for South American trypanosomiasis. *T. cruzi* can also occur in a leishmanial form in which the fibers of the myocardium or the cells of other visceral organs may be invaded in addition to the circulating, extraerythrocytic form of the infection.

The key features for the laboratory identification of these organisms are reviewed in Table 14-7. In the human host, Leishmania species exist only as intracellular parasites in the leishmanial form, involving the reticuloendothelial cells of the bone marrow, spleen, liver, and lymph nodes. A diagnosis can be made by demonstrating the 2- to 4-μ. ovoid forms with a characteristic rod-shaped kinetoplast in stained tissue sections. Cutaneous leishmaniasis, or tropical sore, is most commonly caused by *L. tropica*, although *L. braziliensis, L. peruana, L. guyanensis,* and *L. mexicana* also cause this condition in Peru, Panama, Mexico, and other countries in Latin America.

The arthropod vectors include certain species of the Phlebotomus (sand flies) group. When the leishmanial forms are taken up from an infected human host by the fly, they transform into leptomonad forms within the midgut of the insect. Within 3 to 5 days this single flagellated leptomonad form migrates into the proboscis of the fly and becomes the infective form for a second human host when the fly again bites.

The life cycles of the two African trypanosomes are similar. Species of the tsetse fly serve as the insect vector. The infective stage in the fly consists of the metacyclic trypansomal forms that are present in the salivary glands. These infective-stage trypanosomes are introduced into the human host when the fly bites again. In man, only the trypanosomal forms are seen microscopically, either in direct mounts of blood or in stained smears. They are long, spindle-shaped, nucleated forms measuring up to 30 μ. in length, and have a single flagellum that runs along an undulating membrane (Table 14-7). In addition to circulating in the

peripheral blood, these trypanosomes have a propensity to invade the tissue of the central nervous system, particularly the brain, which results in the sleeping sickness syndrome.

The life cycle of *T. cruzi* differs from the other trypanosomes in that a species of the triatomid or reduviid bug serves as the arthropod vector. Man becomes infected when the fecal matter that is discharged when the bug feeds, containing the infective trypanosomal forms, is rubbed into the bite wound. In the human host, the trypanosome form occurs in the bloodstream during the early acute phase of the disease and during the intermittent febrile periods. In the more chronic forms the leishmanial stage is found in the tissues, usually either in the reticuloendothelial cells or in heart muscle cells. This form is known as Chagas' disease, and in endemic areas cardiomyopathy is the leading cause of death.

Both domestic animals including dogs, cats, pigs, and rodents serve as reservoir hosts for *T. cruzi*. In the United States, oppossums and raccoons are commonly infected. Occasional infections with *T. cruzi* occur in the southern United States. The prevalence rate is low, however, because the insect vectors are rare and housing conditions are better than in some of the rural areas in South and Latin America. Houses built of adobe, mud, or vegetative material, where there are numerous cracks in the walls, provide the optimal breeding places for the reduviid bugs. The bugs are nocturnal and attack their sleeping victims at night. Prevention of the disease is therefore aimed at improving housing conditions.

Microfilariae

Filariasis, or elephantiasis, is a disease of man that is caused by species of roundworms that inhabit the lymphatic channels and may cause obstruction, inflammation, and swelling of the surrounding tissues. The thread-like adult nematodes, male and female, lie tightly intertwined in the lymphatic chan-

Table 14-8. Filaria: Key Features to Laboratory Identification

Species	Arthropod Vector	Sites of Infection	Diagnostic Forms	Illustration
Wuchereria bancrofti	Mosquitoes: third stage larvae in mosquito proboscis is infective form for man.	The long, slender adult male and female nematodes reside in the lymphatic channels throughout the body, primarily in the legs and pelvis. The lymphatics become blocked, leading, in the chronic form of the disease, to marked swelling and edema of the legs, arms, and scrotum, a condition known as elephantiasis	*Microfilariae:* measuring 245 to 295 μ. in length by 7.5 to 10 μ. in width. These forms can be easily seen in direct microscopic examinations of blood, especially when collected at night. *W. bancrofti* microfilariae have a sheath, and the column of nuclei terminate 15 to 20 μ. proximal to the tail, leaving a clear space.	
Brugia malayi	Mosquitoes: the infective form, is the same as *W. bancrofti*.	Similar disease as with *W. bancrofti*	*Microfilariae:* appear similar to those of *W. bancrofti* and are also released into the bloodstream with nocturnal periodicity. They differ from *W. bancrofti* in that two nuclei, spaced about 10 μ. apart from the main column, extend into the tip of the tail.	
Loa loa	Tabanid flies (*Chrysops dimidiata*): the infective form is the same as for *W. bancrofti*.	The adult worms migrate through the subcutaneous tissue, and particularly may be visualized beneath the thin conjunctival epithelium of the eye (for this reason, *Loa loa* is known as the eye worm). Calabar swellings may occur in other parts of the skin, a helpful diagnostic clue.	*Microfilariae:* appear similar to those of *W. bancrofti* and *B. malayi* except that the column of nuclei extends completely to the tip of the tail section. A sheath is present. Microfilariae are released by the adult worms on a diurnal schedule.	

nels. The diagnosis, however, is usually made by demonstrating the prelarval forms, called microfilariae, in the peripheral blood rather than the adult forms in the tissues.

There are three species of filariae that commonly cause disease in man: *Wuchereria bancrofti, Brugia malayi,* and *Loa loa.* The key features for laboratory identification of these filariae are listed in Table 14-8. Two other species, *Acanthocheilonema perstans* and *Mansonella ozzardi,* also may be found in man but do not cause disease. One of the major differences between the pathogenic and nonpathogenic species is that the former have sheathed microfilariae.

The arthropod hosts include mosquitoes for *W. bancrofti* and *B. malayi* and tabanid flies for *Loa loa.* Biting midges serve as the intermediate hosts for *A. perstans* and *M. ozzardi.* When the vector bites an infected human host, the microfilariae are ingested and penetrate the stomach wall of the insect. An infective third stage develops in the thoracic muscles of the insect and ultimately migrates to the proboscis. When the insects bite again, these infective larvae move down the proboscis to the skin and enter the human host through the wound.

Within the human host the worms mature slowly, requiring several months before the diagnostic microfilariae are formed. This represents a subclinical or symptomless stage of the disease. The diagnosis can be made when the microfilariae are discovered either swimming in the blood when observed microscopically in mounts of whole blood or in stained peripheral blood smears. At the time microfilariae are present, the patient may complain of relapsing fever, headache, malaise, and lymphadenopathy. Elephantiasis is the chronic stage of the disease, when the lymphatic channels become completely obstructed by the host's reaction to the filariae, resulting in marked swelling of the extremities and scrotum.

Microfilariae measure up to 200 μ. in length and about 7 μ. in width. They are ribbonlike in form and can be seen swimming in the blood with an undulating motion, displacing the red blood cells from side to side as they move. The pathogenic species have a prominent sheath that extends beyond the tail section, representing the remnants of the ovum's membrane from which it was derived. Species identification of the three pathogenic species can be made by observing the morphology of the tail sections as illustrated in Table 14-8.

The adult worms produce microfilariae with a regular periodicity. Those of *W. bancrofti* and *B. malayi* are produced in the middle of the night; *Loa loa* have diurnal periodicity. Therefore, to diagnose bancroftian filariasis, it is best to obtain blood smears for examination between midnight and 2:00 A.M.

Loa loa causes a disease in which the adult worm migrates through the subcutaneous tissue and may be observed as a small serpiginous elevation of the thin parts of the skin or beneath the conjunctival lining of the eye. The skin reaction at the site of worm migration produced what are known as Calabar swellings. The propensity to infiltrate beneath the conjunctival epithelium has lead to the term "eye worm" for this organism.

Onchocerca volvulus is another tissue nematode that is related to the filarial worms. However, circulating microfilariae of this species are never observed. The disease is transmitted to man through the bite of simulium black flies. The adult worms develop at the cutaneous site of bite where the infective larval forms are deposited within a dense fibrous nodule (Fig. 14-3). In time they produce microfilariae which remain localized to the infective site (Fig. 14-4). The adults form tangled masses that can be observed in skin nodules, and the microfilariae cause an itchy dermatitis. If the eye is involved, the reaction produced can only be made by demonstrating the tangled mass of adult worms, measuring up to 400 mm. × 0.3 mm., within surgically removed skin nodules.

Dracunculus medinensis is also a tissue roundworm that is often grouped with the filariae. *D. medinensis* is the guinea worm that probably represents the "fiery serpent" of biblical lore. Man acquires the infection

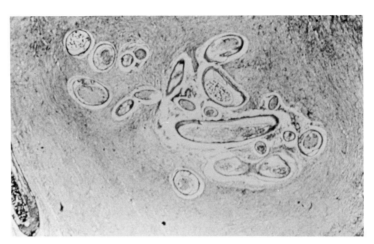

Fig. 14-3. Histologic section of subcutaneous nodule of *Onchocerca volvulus* infestation. The irregular circular structures in the center of the photograph are many nematode adults cut in cross-section (high power).

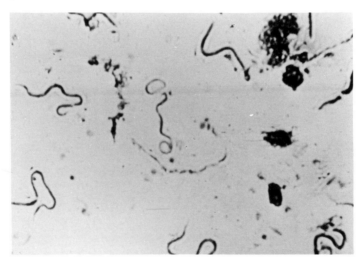

Fig. 14-4. Microfilariae of *Onchocerca volvulus*. These larval forms do not circulate in the peripheral blood, but rather remain localized to the sites of infection (high power).

through ingestion of infected copepods. The larvae develop into adult worms in the serous cavities and the gravid females migrate to the subcutaneous tissue where they produce a burning sensation and ulceration of the skin. These female worms can measure as long as 100 cm. on length (40 inches) and can be removed from the subcutaneous tissue by surgical intervention and winding them slowly on a stick until they are completely removed. The life cycle is completed when the larvae produced by the female worm escapes from the skin blister and are discharged into water in which the copepods live.

Other Tissue Parasites

Three other species of helminths which inhabit human body tissues in the larval forms will be discussed here. The adults of these species normally reside in the intesti-

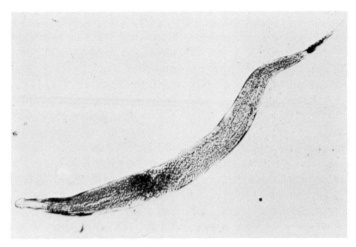

Fig. 14-5. Photograph of adult *Trichinella spiralis* worm.

nal tract of the definitive hosts. Many may be either an intermediate host (Echinococcus), both an intermediate and a definitive host (Trichinella), or an accidental host (Toxocara).

Trichinella spiralis. Trichinella is a disease caused by infection with the nematode *Trichinella spiralis* as a result of ingesting raw or poorly cooked meat of any carnivore, notably pork or pork products containing the parasite. Infections have also been reported following ingestion of poorly cooked bear meat. Smoking, salting, or drying the meat does not destroy the infective larval forms, although prolonged freezing (20 days in the average home freezer) will decontaminate the meat. The disease has worldwide distribution, and in the United States 4 per cent of human cadavers were found to be infected in 1968.[6] Only approximately 100 new cases are currently reported each year in the United States, a tribute to the meat inspection program and the stringent laws against feeding uncooked garbage to pigs.[6]

The cycle in man is initiated by the ingestion of the infective larval form in poorly cooked meat. The larvae are released in the intestine where they burrow into the villi. After molting, the trichinellae develop into adult male and female worms, measuring up to 2 to 4 mm. in length (Fig. 14-5). The average life span of the adults in the intestine is about 4 months; however, during that time,

each female may have released as many as 3000 larval offspring. These larvae enter the circulation and are deposited throughout the tissues of the body; however, those reaching the skeletal muscle become encysted and survive. These larval forms develop into a spiral, coiled 2½ times on themselves, within the muscle fibers (Fig. 14-2). The larvae produce a tissue reaction and the adjacent muscle fibers undergo degeneration so that a cyst measuring about 0.25 to 0.50 mm. develops. In time these cystic lesions may undergo calcification.

The majority of infections are subclinical. The minimal number of larvae required to produce symptoms is about 100 and a fatal dose is estimated at 300,000.[6] Fever, muscle pain and aching, periorbital edema, and peripheral blood eosinophilia are the cardinal features by which a clinical diagnosis can be made.

In the laboratory, trichinosis is diagnosed by detecting the spiral larvae in muscle tissue. The deltoid muscle of the upper arm or the gastrocnemius muscle of the calf are usually selected as muscle biopsy sites. The specimen may be examined by first digesting the muscle fibers with trypsin and mounting some of the digested tissue on a microscope slide, or by preparing a tease preparation of the muscle tissue in a drop of saline and squeezing it between two microscope slides. The presence of linear or spiral larval forms when examined may also be observed in stained tissue sections, although their morphology is not as well delineated.

Visceral Larva Migrans. Larva migrans is a condition in which the larvae of nematode parasites of lower animals migrate into the tissues of man without further development. Larval migrans may be caused by many different species of parasites and may affect either cutaneous or visceral tissues, depending on the body areas affected and the parasites involved. Cutaneous larval migrans is commonly caused by filariform larvae of dog or cat hookworms which in man are unable to proceed beyond the subcutaneous tissue at the sites of penetration. This condition is known as "creeping eruption."

Visceral larval migrans is most commonly caused by *Toxocara canis,* the dog intestinal roundworm which has a life cycle not unlike that of human *A. lumbricoides.* Man becomes an accidental and abnormal host through ingestion of embryonated eggs in the soil. The disease is most common in children because of their close association with dogs and their tendency to consume soil. The embryonated ova hatch in the intestine of the human host, liberating larvae which penetrate into the bowel wall and enter the circulation. However, because man is an abnormal host, the lung cycle is not completed; rather, the larvae are filtered out in various organs, chiefly the liver. They may cause local tissue reaction or granulomas, but the larvae eventually die out with no sequelae. The infection can be suspected in a child with hepatomegaly, nonspecific pulmonary disease, and a high peripheral blood eosinophilia.

Echinococcal Disease. Echinococcosis, or hydatid disease in man, is possibly one of the more difficult parasitic diseases to understand because of the peculiar cystic larval forms that form in the viscera. Man actually serves as an accidental host, since the normal life cycle of this parasite involves dogs or foxes as the definitive hosts, and sheep, cattle, or swine as the intermediate hosts. If man is infected, he also is an intermediate host in whom the larval form of the parasite is harbored.

Echinococcus granulosus and *E. multilocularis* are tapeworms that are found in the intestines of dogs and related carnivores, including wolves, foxes, and jackels. They measure about 3 to 6 mm. in length and possess three proglottids and a scolex armed with a double row of hooklets (Fig. 14-6).

Hexacanth ova, closely resembling those of the Tinea species of human cestodes, are passed in the dog feces and become embryonated in the soil. Under normal circumstances, these ova are ingested by the natural intermediate hosts—sheep, cattle, or swine. The larvae are released from the ova in the intestines of the intermediate hosts, and by means of their hooklets bore through the bowel wall and enter in the circulation.

The circulating embryos are filtered out in the capillaries of various organs, usually the liver since it is the first organ to drain the mesenteric blood. Within the organ, the larvae develop into small cysts called "bladder worms." These cysts may reach as large as 5 cm. within three months and visually have the appearance of "hailstones," as observed by Aristotle (Fig. 14-7).

The wall of the cyst is in fact a germinal membrane from which numerous daughter embryos develop. These form as tiny polypoid structures that line the inner membrane (Fig. 14-8). When they break free from the membrane and float in the fluid within the cyst, they are known as hydatid sand. If examined under the microscope, each grain of sand is in fact a tiny embryonic beginning of a new tapeworm, complete with an inverted scolex with a rostellum armed with hooklets (Fig. 14-9).

The life cycle is complete when infected viscera of the definitive host is eaten by a dog, fox, or other related carnivore.

The infection in man is similar to that found in the herbivorous animals, and the disease is acquired through ingestion of vegetative material or soil infested with ova-bearing dog feces. The laboratory diagnosis is made by demonstrating the daughter cysts that are surgically removed. In man *E. granulosis* infection most commonly results in the formation of a solitary, unilocular cyst in the liver. Brain cysts also may occasionally

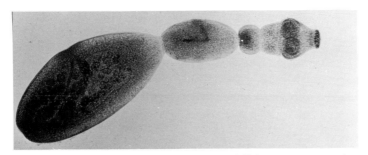

Fig. 14-6. Photograph of adult tapeworm of *Echinococcus granulosus.*

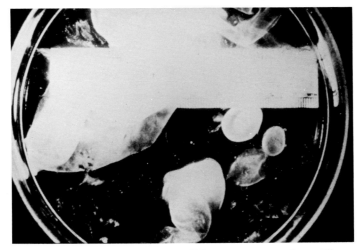

Fig. 14-7. Larval, cystic forms of *Echinococcus granulosus.* These structures were termed "hailstones" by Aristotle.

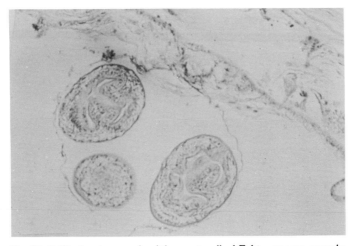

Fig. 14-8. Photomicrograph of the cyst wall of *Echinococcus granulosus* showing development of embryonic "daughter cysts."

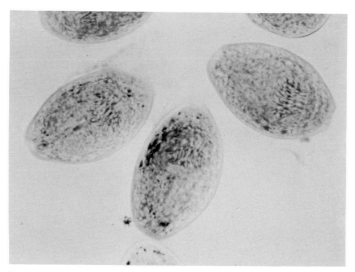

Fig. 14-9. Photomicrograph of "hydatid sand," composed of embryonic larval forms of *Echinococcus granulosus.* Note inverted hooklets within each sand granule.

be found. If the cyst should rupture, either spontaneously in the body or during surgery, there is great danger of death from anaphylactic shock. Metastatic cystic lesions can also develop in virtually any of the visceral organs if the primary cyst ruptures.

E. multilocularis infection manifest more commonly as multioculated cysts, called alveolar cysts because they closely simulate the air sacs of the lung. They may be confused with mucin-secreting carcinoma by the pathologist because they are often free of brood capsules and the characteristic hooked scoleces within the hydatid sand granules may not be visible.

Man may also become the intermediate host of the larval stages of *Taenia solium* as discussed above, in which "bladder worms" or cysticercoid lesions can develop in virtually any of the visceral organs.

For comprehensive yet relatively brief reviews of the laboratory diagnosis of parasitic diseases, the reader is referred to the recent publications by Koneman and associates[8] McQuay,[9] and Pruneda and coauthors.[14] The three-volume *Atlases of Diagnostic Medical Parasitology,* by Smith and coworkers,[15] recently published by the American Society of Clinical Pathologists, provide over 300 color transparancies of diagnostic parasitic forms that are quite valuable for self study or for lecture purposes.

REFERENCES

1. Brooke, M. M., and Melvin, D. M.: Common Intestinal Protozoa of Man: Life Cycle Charts. Public Health Service, Publication No. 1140, Washington, D.C., U.S. Government Printing Office, 1964 and 1969.
2. Brown, H. W.: Basic Clinical Parasitiology. ed. 3. New York, Appleton-Century-Crofts, 1969.
3. Faust, E. C., Russell, P. F., and Jung, R. C.: Craig and Faust's Clinical Parasitology. ed. 8. Philadelphia, Lea & Febiger, 1970.
4. Foster, W. D.: A History of Parasitology. Edinburgh, Livingston, 1965.
5. Garcia, L. S., and Ash, L. R.: Diagnostic Parasitology: Clinical Laboratory Manual. St. Louis, C. V. Mosby, 1975.
6. Gould, S. E.: The story of trichinosis. Am. J. Clin. Pathol., *55:*2–11, 1970.
7. Hoeppli, R.: Parasites and Parasitic Infections in Early Medicine and Science. Singapore, University of Malaya Press, 1959.

8. Koneman, E. W., Richie, I. E., and Tiemann, C.: Practical Laboratory Parasitology. New York, Medcom Press, 1974.

9. McQuay, R. M.: Medical parasitology. *In* Davidsohn, I., and Henry, J. B. (eds.): Clinical Diagnosis by Laboratory Methods. ed. 15, Chap. 19. Philadelphia, W. B. Saunders, 1974.

10. Markell, E. K., and Voge, M.: Medical Parasitology. ed. 3. Philadelphia, W. B. Saunders, 1971.

11. Melvin, D. M., and Brooke, M. M.: Laboratory Procedures for the Diagnosis of Intestinal Parasites. DHEW Publication No. (CDC) 75-8282, Washington, D.C., U.S. Government Printing Office, 1974.

12. Melvin, D. M., Brooke, M. M., and Healy, G. R.: Common Blood and Tissue Parasites of Man: Life Cycle Charts. DHEW Publication No. 1234, Washington, D.C., U.S. Government Printing Office, 1969.

13. Melvin, D. M., Brooke, M. M., and Sadun, E. H.: Common Intestinal Helminths of Man: Life Cycle Charts. DHEW Publication No. 1234, Washington, D.C., U.S. Government Printing Office, 1974.

14. Pruneda, R. C., *et al.*: Laboratory diagnosis of parasitic diseases. *In* Koneman, E. W., and Britt, M. S. (eds.): Clinical Microbiology. Lecture 15. Bethesda, Md., Health and Education Resources, Inc., 1977.

15. Smith, J. W., *et al.*: Atlases of Diagnostic Medical Parasitology. Chicago, American Society of Clinical Pathologists, 1976.

16. Velat, C. A., Weinstein, P. P., and Otto, G. F.: A stain for the rapid differentiation of the trophozoites of intestinal amoeba in fresh, wet preparations. Am. J. Trop. Med. Hyg., *30:*43–51, 1950.

17. Warren, K. S., and Mahmoud, A. A. F.: Algorithms in the diagnosis and management of exotic diseases. XII: Prevention of exotic diseases: advice to travelers. J. Infect. Dis., *133:*596–601, 1976.

Index

Numbers in *italics* indicate a figure; "t" in *italics* following a page number indicates a table.